Intracellular Mechanisms for Neuritogenesis

Intracellular Mechanisms for Neuritogenesis

Edited by

Ivan de Curtis
Laboratory of Cell Adhesion - Dibit
San Raffaele Scientific Institute
Milano, Italy

Springer

Ivan de Curtis (Ed.)
Unit of Cell Adhesion
San Raffaele Scientific Institute
Department of Molecular Biology and Functional Genomics
Via Olgettina, 58
20132 Milano
Italy

ISBN-1-4899-9022-4
ISBN-978-1-4899-9022-8

Softcover re-print of the Hardcover 1st edition 2007

9 8 7 6 5 4 3 2 1

springer.com

Foreword

This book explores the intracellular mechanisms that allow a neuron to differentiate to the highly complex morphology required for its function as fundamental component of the nervous system. Prominent experts in each of the areas considered have generously provided the most up-to-date understanding of the intracellular processes regulating neuronal differentiation during the formation of neural circuits necessary for the function of the brain and the nervous system in general. This book is aimed at graduate and postgraduate students, and to principal investigators, but it could also be of use to complement the preparation of science and medical undergraduate students, and to those who need to understand basic neuronal processes useful for further analysis in any of the many aspects of neuroscience.

Ivan de Curtis
January 2006

Preface

A major issue of modern neurobiology is to understand how neurons extend their neurites to form the highly complex functional networks that shape the nervous system. During development, each neuron has to establish the correct connections with its targets by sending out axons and dendrites to form synapses to set up a functional nervous system. While a great deal of information is available on the extracellular cues driving neuritogenesis, the study of the intricate molecular machinery underlying the intracellular mechanisms has only recently been addressed, and large gaps still exist in our understanding of the corresponding molecular events. The purpose of Intracellular Mechanisms for Neuritogenesis is to present an up-to-date review of the molecular events responsible for several aspects of axonal and dendritic development, to help fill in some of the gaps, and to propose working hypotheses helpful to researchers interested in neuronal development and differentiation. Another goal is to present these topics to the general readership of students, neurobiologists, and molecular and cellular biologists to further stimulate the interest in this exciting area of neuroscience. Each chapter represents the generous effort of experts on distinct but related aspects of the processes required for the development of neurites, including adhesion, cytoskeletal organization, membrane traffic, and signal transduction, all of which are required for the development of synaptic sites between axonal and dendritic terminals.

The growth cone is a sensory and protruding structure of the neurite and is equipped with a large number of different plasma membrane receptors. This allows the growth cone to respond to the variety of extracellular cues encountered while traveling to its final destination. The combination of attractive and repulsive signals delivers the information to the intracellular environment, inducing either protrusion or retraction via actin polymerization and depolymerization, respectively. Therefore, actin dynamics represent an important target of the signaling downstream of the different receptors during growth cone navigation. In this respect, Chapters 1, 9, and 10 focus on characterization of the signaling events downstream of adhesive, neurotrophin, and Wnt receptors, respectively. The growth cone has to integrate the different signals received in parallel from the complex extracellular environment. Chapter 2 describes the mechanisms regulating actin dynamics that follow such integration. Actin dynamics during growth cone navigation evolve into stabilization and neurite elongation, where microtubules and associated proteins play a major role, as discussed in Chapter 3. Similar mechanisms must also regulate the formation and extension of branches along neurites as described in Chapter 12. A number of molecules that play a crucial role in several aspects of the development of neuronal morphology are illustrated in Chapters 7, 8, and 11.

Next to these chapters dealing with well-established intracellular mechanisms driving neurite extension, Chapters 4, 5, and 6 deal with more recently investigated issues, such as the connections between cytoskeleton and membrane traffic, by analyzing the contribution of both exocytic and endocytic pathways to growth cone function, neuronal morphogenesis, and the establishment of neuronal polarity. Moreover, Chapter 13 compares the migration of neural crest cells with that of neuronal growth cones, which both face the problem of migrating over intricate territories to reach their appropriate targets, with the aim of establishing similarities and differences between the two systems. Finally, Chapter 14 analyzes the mechanisms responsible for the inability of injured central neurons to regenerate their axons. This issue is fundamental for research aimed at investigating the possibilities of rebuilding functional neuronal connections to help prevent the permanent loss of function following injury of the central nervous system.

The analysis presented in this volume indicates that multiple signals from the environment converge on common intracellular networks to modulate adhesion, cytoskeleton, and membrane traffic. Some of these networks are very well characterized (e.g., those linked to the Rho family of GTPases, mentioned in several chapters), while others have been identified more recently and will require the enthusiasm of researchers to be thoroughly explored.

Acknowledgments

I want to thank all the authors of this book who have made a generous effort to generate perceptive chapters in their respective fields of competence. Their work will certainly be useful to the large number of researchers interested in understanding neuronal differentiation and to those who would like to approach the comprehension of the molecular basis of the cellular processes at the origin of the formation of the astonishingly intricate nervous system. I also gratefully acknowledge support from Telethon-Italy.

Ivan de Curtis
January 2006

Table of Contents

List of Contributors

José Abad-Rodríguez Fondazione Cavalieri Ottolenghi, A.O. San Luigi Gonzaga, Regione Gonzole 10, 10043 Orbassano, Italy

Philipp Alberts Department of Cell Biology, Yale University School of Medicine, P.O. Box 208002, New Haven, CT 06520, USA

Nariko Arimura Department of Cell Pharmacology, Graduate School of Medicine, Nagoya University, 65 Tsurumai, Showa-ku, Nagoya, Aichi, 466-8550, Japan

James R. Bamburg Department of Biochemistry and Molecular Biology, 1870 Campus Delivery, Colorado State University, Fort Collins, CO 80523-1870, USA

Dario Bonanomi Unit of Experimental Neuropharmacology, San Raffaele Scientific Institute, Via Olgettina, 58, 20132 Milano, Italy

Alfredo Cáceres Instituto Investigacion Medica Mercedes y Martín Ferreyra, INIMEC-CONICET, Av. Friuli 2434, 5000 Cordoba, Argentina

John K. Chilton Institute of Biomedical and Clinical Sciences, Peninsula Medical School, Tamar Science Park, Research Way, Plymouth PL6 8BU, UK

Sara Corbetta San Raffaele Scientific Institute, Dibit, Via Olgettina, 58, 20132 Milano, Italy

Ivan de Curtis San Raffaele Scientific Institute, Dibit, Via Olgettina, 58, 20132 Milano, Italy

Erik W. Dent Department of Biology, Massachusetts Institute of Technology, Cambridge, MA 02139, USA

Sarah Escuin Institut de Génétique et de Biologie Moléculaire et Cellulaire, CNRS/INSERM/ULP, BP10142, 67404 Illkirch, C.U. de Strasbourg, France

Adriana Ferreira Institute for Neuroscience and Department of Cell and Molecular Biology, Northwestern University, Chicago, IL 60611, USA

Kevin Flynn Department of Biochemistry and Molecular Biology, and Molecular, Cellular and Integrative Neuroscience Program, Colorado State University, Fort Collins, CO 80523, USA

Thierry Galli Team "Avenir" INSERM Membrane Traffic in Neuronal & Epithelial Morphogenesis, Institut Jacques Monod, UMR7592, CNRS, Universités Paris 6 & 7, 2, place Jussieu, F-75251 Paris Cedex 05, France

Elisabeth Georges-Labouesse Institut de Génétique et de Biologie Moléculaire et Cellulaire, CNRS/INSERM/ULP, BP10142, 67404 Illkirch, C.U. de Strasbourg, France

Phillip R. Gordon-Weeks King's College London, MRC Centre for Developmental Neurobiology, New Hunts House, Guy's Campus, London SE1 1UL, UK

Zhigang He Division of Neuroscience, Children's Hospital, 320 Longwood Avenue, Harvard Medical School, 320 Longwood Avenue, Boston, MA 02115, USA

Kozo Kaibuchi Department of Cell Pharmacology, Graduate School of Medicine, Nagoya University, 65 Tsurumai, Showa-ku, Nagoya, Aichi, 466-8550, Japan

Katherine Kalil Department of Anatomy and Neuroscience Training Program, University of Wisconsin-Madison, 1300 University Ave, Madison, WI 53706, USA

Anthony J. Koleske Department of Molecular Biophysics and Biochemistry, Department of Neurobiology, Interdepartmental Neuroscience Program,Yale University, SHMC-E31, 333 Cedar Street, New Haven, CT 06520-8024, USA

Paul M. Kulesa Stowers Institute for Medical Research, Kansas City, MO 64110, USA

Frances Lefcort Department of Cell Biology and Neuroscience, Montana State University, Bozeman, MT 59717, USA

Gabriela Paglini Instituto Investigacion Medica Mercedes y Martín Ferreyra, INIMEC-CONICET, Córdoba, Argentina

Chi Pak Department of Biochemistry and Molecular Biology, and Molecular, Cellular and Integrative Neuroscience Program, Colorado State University, Fort Collins, CO 80523, USA

Tim O'Connor Department of Cellular and Physiological Sciences, Member of ICORD, University of British Columbia, Vancouver, BC V6T 1Z3, Canada

Santiago Quiroga Departamento Quimica Biologica, Facultad Ciencias Quimicas (UNC), CIQUIBIC-CONICET, Cordoba, Argentina

Silvana B. Rosso Department of Anatomy and Developmental Biology, University College London, Rockefeller Building, University Street, London, WC1E 6BT, UK

Patricia C. Salinas Department of Anatomy and Developmental Biology, University College London, Rockefeller Building, University Street, London, WC1E 6BT, UK

Jan M. Schwab Center for Experimental Therapeutics and Reperfusion Injury, Department of Anesthesiology, Perioperative and Pain Medicine, Brigham and Women's Hospital and Harvard Medical School, 75 Francis Street, MA 02115, USA

Hameeda Sultana Department of Molecular Biophysics and Biochemistry, Yale University SHMC-E30, 333 Cedar Street, New Haven, CT 06520, USA

Fangjun Tang Department of Molecular, Cellular, and Developmental Biology, Yale University, New Haven, CT 06520-8103, USA

Flavia Valtorta Unit of Experimental Neuropharmacology, San Raffaele Scientific Institute, Via Olgettina, 58, 20132 Milano, Italy

Takeshi Yoshimura Department of Cell Pharmacology, Graduate School of Medicine, Nagoya University, 65 Tsurumai, Showa-ku, Nagoya, Aichi, 466-8550, Japan

1
Adhesion-Induced Intracellular Mechanisms of Neurite Elongation

Sarah Escuin and Elisabeth Georges-Labouesse

1.1. General Presentation of Adhesion Molecules Important in Neurite Elongation

During nervous system development, neurons extend processes that navigate long distances to reach their tissue targets. This navigation occurs through a very complex tissue environment, and axon progression depends on permissive or instructive cues either attractive or repellent. Cell adhesion molecules have long been known to be active in neuritogenesis at all steps including sprouting from the neuronal membrane, neurite extension, and directional navigation as reviewed about 15 years ago (Hynes and Lander, 1992; Rathjen, 1991; Reichardt and Tomaselli, 1991; Reichardt et al., 1990). Cell interactions with the substrate or with neighboring cells are mediated by membrane-associated receptors. Extracellular signals will then be conveyed through these receptors to highly regulated intracellular signaling pathways converging on cytoskeletal networks. Adhesion receptors fall into several categories, integrins, which mediate cell adhesion with the extracellular matrix (ECM), and cadherins and the immunoglobulin superfamily of cell adhesion molecules (IgCAMs), which mediate principally cell–cell adhesion but can also recognize and bind some ECM molecules. In this chapter, we will discuss the current state of knowledge on ECM and their receptors and cite studies that have addressed the question of intracellular mechanisms, in particular *in vivo*.

1.1.1. Extracellular Matrix and Their Receptors in the Nervous System

1.1.1.1. ECM Molecules

Extracellular matrix is a complex molecular network mainly composed of glycoproteins and proteoglycans from different families. ECM in the central nervous system (CNS) differs from other ECMs in that it does not contain fibrous elements or much organized basal lamina (BL) with few exceptions (i.e., the BL underlying the pia mater). In the peripheral nervous system, well-organized BLs encircle nerve bundles. A nonexhaustive list of ECM molecules that are present in the nervous system comprise laminins, tenascins, fibronectin, thrombospondins, and chondroitin sulfate proteoglycans (CSPGs) or heparan sulfate proteoglycans (HSPGs). It should be noted that each of these ECM components in fact comes in several forms. Diversity of fibronectin in tissues is produced by alternative splicing (Vogelezang et al., 1999). As for laminins, which are heterotrimers of α, β, and γ chains, 15 forms are produced, arising from different combinations of the 5 α, 4 β, and 3 γ that have been identified (Miner and Yurchenco, 2004). Diversity is also encountered in other families, such as tenascins, where several forms (tenascin-C, tenascin-R, tenascin-W) are encoded by independent genes, each product being itself subject to alternative splicing and posttranslational modifications (Gotz et al., 1997; Joester and Faissner, 2001).

The distribution of ECM molecules in the nervous system has been studied in several organisms, in the developing nervous system, in adult tissues, and during tissue repair (Gotz et al., 1997; Joester and Faissner, 2001; O'Shea et al., 1990; Sheppard et al., 1991, 1995; Yin et al., 2003). For example, a dynamic localization of fibronectin and CSPG has been observed, first in the proliferative zone of the telencephalon and later in the so-called preplate, suggestive of their roles during migration in the developing cerebral cortex (Sheppard et al., 1991). Liesi (1985a) found that a punctate laminin signal is associated with radial fibers in the developing cortex, where it may serve also as substrate for migrating neurons. This punctate distribution disappears at more advanced stages. In fact, in many cases, laminin expression appears to be transient, occurring at high levels during periods of axonal growth and diminishing later in development. Two notable exceptions are the continued expression of laminin isoforms in the BL of peripheral nerve and the optic nerves of goldfish (Hopkins et al., 1985; Liesi, 1985b), two favorable sites in the adult for either continued axonal growth or nerve regeneration.

After the discovery of new members of the laminin family, very detailed analyses of the expression of laminin chains in the developing and adult nervous system have been performed (Lentz et al., 1997; Patton et al., 1997; Yin et al., 2003).

1.1.1.2. Integrins

As mentioned, the main receptors that bind ECM ligands belong to the integrin family. Integrins are transmembrane molecules with a large extracellular ligand-binding domain, a single transmembrane domain, and a generally short cytoplasmic domain. They are heterodimers of α and β chains. To date 18 α chains and 8 β chains have been identified, leading to around 25 integrin heterodimers. As other cell types, neuronal cells express a repertoire of integrins. It is believed that functionality of these integrins requires an activation step which involves a conformational change correlated with a high binding affinity to the ligand and clustering into cell junctions (Hynes, 2002). Among the 25 integrin heterodimers present in mammals, many belonging to the β1 subgroup (α1β1, α3β1, α4β1, α5β1, α6β1, α7β1) and from the αv subgroup have been identified in neuronal cells and may be functioning in neurite elongation (Pinkstaff et al., 1999).

Integrin α5β1 and α4β1 are primarily fibronectin receptors. Integrin α6β1 and α7β1 are laminin receptors, while integrin α1β1 is a collagen/laminin receptor, and α3β1 can bind mainly laminins and also other ligands.

1.1.2. Intercellular Adhesion Molecules: Cadherins and IgCAMs

Adhesion molecules of the cadherin and Ig superfamily also play major roles in neuritogenesis. These molecules mediate cell–cell interactions through homophilic or heterophilic binding to other cell surface molecules. The function of cell–cell adhesion molecules has been reviewed extensively recently, and readers are referred to these excellent reviews (Huber et al., 2003; Kiryushko et al., 2004; Ranscht, 2000; Rougon and Hobert, 2003; Skaper et al., 2001). Only a very short description of intercellular adhesion molecules is provided here, as a few studies will cite these molecules in the context of ECM and receptors.

Cadherins are Ca^{2+}-dependent transmembrane molecules. Classical cadherins, a group of single-pass transmembrane proteins, are characterized by an extracellular domain of 5 cadherin repeats and a highly conserved cytoplasmic region. The cadherin cytoplasmic domain associates with a group of intracellular proteins, the catenins, which link cadherin molecules to the actin-based cytoskeleton. A recently discovered group of atypical cadherins, which may play important roles in axon growth, are the protocadherins. They are proteins larger than 3000 amino acids with N-terminal cadherin repeats, a G-protein–coupled receptor proteolysis site, 7 transmembrane segments, and a cytoplasmic domain divergent from that of classical cadherins. In the vertebrate nervous system, about 80 different cadherin types have been identified. These include the classical cadherins, such as N-, E- and R-cadherin (Ranscht, 2000). Among those, the main cadherin studied in neurite outgrowth is N-cadherin.

IgCAMs are membrane-associated molecules, either GPI anchored or transmembrane, with a large extracellular domain characterized by repeated segments homologous to the Ig repeat found in immunoglobulins. The IgCAM family is divided in subfamilies, which differ by the number of Ig repeats and other structural features like the fibronectin type III repeat (Kiryushko et al., 2004). Besides NCAM, the principal other IgCAMs that are implicated in neurite elongation are L1/NgCAM, NrCAM, neurofascin, and TAG-1/axonin. While IgCAMs bind essentially other CAMs, a number of studies have illustrated that they can also bind ECM molecules. For example, L1 and NCAM can bind proteoglycans, ECM, and integrins (for L1) (Kiryushko et al., 2004).

1.2. ECM and Integrins in Neuritogenesis: Cellular Studies

The diversity of adhesion molecules suggests that specific behaviors should be brought about by specific molecules. *In vivo*, neurites extend in a very complex and dynamic environment and contact a network of molecules which modulate each other's activity. Much of our knowledge on the importance of adhesion in neuritogenesis comes thus from cellular studies, where answers to the question of specificity are obtained by experimental assays using defined extracellular environments.

1.2.1. Specific Activities of ECM and Integrins

Many ECM molecules stimulate the outgrowth of neuronal processes *in vitro*, while a few have restrictive effects. Molecules with promoting effects include laminins, fibronectin, several collagens, and thrombospondin (Manthorpe et al., 1983; Rogers et al., 1986). As for tenascins, it has been shown that they can have dual activities (Wehrle-Haller and Chiquet, 1993; Wehrle and Chiquet, 1990). For example, tenascin-C carries distinct repulsive or neurite outgrowth promoting sites for neurons (Gotz et al., 1996). CSPGs are very well known to act as inhibitory molecules and provide barriers to cell migration and axon growth both *in vivo*, during development and regeneration, and *in vitro* (Carulli et al., 2005).

Results from de Miguel and Vargas (2000) illustrate that during development or regeneration of the nervous system, particular sets of ECM proteins have specific effects at all steps in regulating the number, direction, extension, or retraction of neurites. Attachment of cultured anterior pagoda neurons to the ECM inside the ganglionic capsules or to CNS homogenates induces a characteristic T-shaped outgrowth pattern, different from patterns of anterior pagoda neurons plated outside the ganglionic capsule or on leech laminin extracts.

Is there a relationship between the identity of neurons and their substrate preference? In a study, Guan et al. (2003) addressed the question of substrate preference and integrin expression by different classes of sensory neurons before target innervation. They showed that presumptive cutaneous neurons show a preference for laminin, while presumptive proprioceptive neurons extend neurites similarly on fibronectin and laminin. Accordingly, proprioceptive neurons express higher levels of fibronectin receptors ($\alpha3\beta1$, $\alpha4\beta1$, $\alpha5\beta1$), and presumptive cutaneous afferents higher levels of the laminin receptor $\alpha7\beta1$.

By plating cells on fibronectin, laminin, or endogenous matrix, Hynds and Snow (2001) showed that fibronectin-supported neurite outgrowth was comparable to that of endogenous matrix, whereas laminin significantly increased outgrowth in comparison to fibronectin. CSPGs inhibited growth cone advance from cells on fibronectin or laminin, but growth cones on laminin preferentially stop/stall upon CSPG contact, whereas those on fibronectin predominantly turn. Differential CSPG-induced behaviors were correlated with increased growth cone spreading and decreased migration rates in cells on laminin but not fibronectin. In these experiments, it seems that growth cone turning in response to CSPGs must be regulated in part by nonintegrin-mediated mechanisms, since contact with CSPG neither upregulated integrin protein expression nor promoted cellular redistribution of integrin subunits. Furthermore, a function-blocking anti-integrin $\beta1$ antibody did not affect growth cone behavior on contact with CSPGs (Hynds and Snow, 2001).

1.2.2. New Ligands and Receptors

After the discovery of new members of ECM or receptor families, progress has been made in the definition of their roles in the nervous system or in the relation between ligands and receptors. A detailed analysis of motor axon growth on different laminins has shown that while axons grow freely on laminin-1 and 2, the classical laminins used in cellular studies, they fail to cross from laminin-4 to laminin-1, and they stop when in contact with laminin-11 (Patton et al., 1997). It was shown that laminin-5, mostly known for its expression at the dermal–epidermal junction, is also able to promote neurite outgrowth. In fact, detailed analysis of laminin-5 expression has revealed its presence also in the embryonic nervous system, in the retina. Laminin-5 promotes neurite outgrowth through its binding to integrin $\alpha 3\beta 1$. This has been shown in several types of neurons: embryonic chick neurons, PC12 neuronal cells, and neuronal cells derived from the embryonal carcinoma cells NT2 (Culley et al., 2001; Mechai et al., 2005; Stipp and Hemler, 2000). In PC12 cells, the cytoplasmic domain of integrin $\alpha 3$ fused to the extracellular domain of the interleukin-2 receptor was able to inhibit neurite outgrowth on laminin-5 (Mechai et al., 2005). In NT2N cells, $\alpha 3\beta 1$ is associated with tetraspanins (TM4SF) for its function in neurite outgrowth, and antibodies against the TM4SF CD151 and CD81 inhibit neurite outgrowth (Stipp and Hemler, 2000).

As an example of a new ligand for an integrin, a report has shown that in neuroblastoma cells B104, integrin $\alpha 1\beta 1$, known to bind collagen and laminin, can recognize a new member of the thrombospondin family, SCO-spondin, which does not contain a RGD peptide. SCO-spondin incorporated into the matrix provides a permissive environment for the pathfinding of commissural axons (Bamdad et al., 2004). Another example is the binding of $\alpha 7\beta 1$ integrin to the alternatively spliced region of human tenascin-C during neurite outgrowth (Mercado et al., 2004).

An important finding is that integrins can also serve as receptors for a member of the semaphorin family (Nakamoto et al., 2004). Semaphorins are secreted or transmembrane axon guidance molecules that act as repulsive cues. Main receptors for the semaphorin family are plexins whose activity is regulated by neuropilins. Now, one of the semaphorins, semaphorin-7A has been shown to bind a $\beta 1$-containing integrin in olfactory bulb neurons in mouse (Pasterkamp et al., 2003).

1.3. Adhesion and Cellular Mechanisms of Neurite Extension

As proposed by da Silva and Dotti (2002), the initial sprouting of a neurite can be seen as a three-step event. First, the original round shape is broken down to make a bud. Second, the bud is transformed into a neurite and third, the neurite is transformed into an axon or a dendrite. Following these initials steps, elongation or retraction, turning and branching, fasciculation, and ultimately target innervation take place. While much needs to be learned, in particular in the precise cascade relationship and molecular mechanisms that link membrane receptors to cytoskeleton, progress has been made to define how adhesion to the substrate influences these steps.

1.3.1. Neurite Sprouting and Elongation

The process of neuritogenesis begins immediately after neuronal commitment. The activation of membrane receptors by extracellular cues could be at the origin of the initial breakdown of symmetry. It has been proposed that adhesion induces reduction of membrane tension locally and that these sites will correspond to the extension of filopodia or lamellipodia (da Silva and Dotti, 2002). Once formation of the neurite is initiated, membrane attachment to the substrate via integrins is required, as in other cells, for movement to occur. The mobile part of the axon that is responsible for axonal elongation is the growth cone. One model that has been discussed for growth cone motility is the motor and clutch model (Goldberg, 2003; Suter and Forscher, 2000). The motor for motility is mainly the actin cytoskeleton, and a balance between actin polymerization and myosin-mediated retrograde flow determines the rate of advance. When the growth cone contacts an adhesive substrate, an increasingly strong interaction takes place between cell surface receptors and the actin cytoskeleton, thus engaging the clutch (Goldberg, 2003).

In general, in stationary cells, ligand binding by integrins leads to their accumulation in stable focal adhesions. Integrin association with adhesive zones is stable comparatively to that in motile cells, where integrins have been shown to diffuse rapidly in the membrane, thereby indicating that associations with the ECM and cytoskeleton are transient (Duband et al., 1988). In contrast to other cell types, neurons extend neurites over a wide range of ECM ligand concentrations, suggesting that they regulate their degree of adhesion to the ECM. Neurons could alter their interactions with the matrix to preserve growth cone motility either by changing receptor affinity (Hynes and Lander, 1992) or regulating the amount or the subcellular distribution of receptors. Condic and Letourneau (1997) by looking at integrin $\alpha 6\beta 1$ levels in chick dorsal root ganglia neurons on laminin observed that on low concentrations of laminin, the level of cell surface-associated integrin was higher, while total integrin mRNA and protein was lower. It was proposed that the availability of ligand determines the cell surface level of receptors by altering the rate of receptor internalization or degradation (Condic and Letourneau, 1997).

Regulation of integrin activity is one of the important mechanisms that controls neurite outgrowth. For example, embryonic retinal ganglion neurons lose the ability to extend neurites on laminin-1 as they develop while still being able to do so on other laminins. Ivins et al. (1998) showed a few years ago that using antibodies against the short arm of laminin-1, or removing the short arm by proteolytic cleavage, can restore neurite extension. Responses to this activated laminin are mediated by the classical laminin receptors $\alpha 3\beta 1$ and $\alpha 6\beta 1$. Following up with the same system, the same group showed that activation of integrin $\alpha 6\beta 1$ either by treatment with Mn^{2+} or an activated version of Ras results in the promotion of neurite extension (Ivins et al., 2000). Thus, while integrins can be present on cells, their activation state is regulated so that response on laminin is obtained only after activation.

1.3.2. Choice Between Axon and Dendrite

One fundamental event in neuritogenesis is the decision of whether a neurite will become an axon or a dendrite. Many factors influence this choice including extracellular signals and intrinsic properties of the neurite (Goldberg, 2004). Is this choice regulated by adhesion? Esch et al. (1999) showed that when a neuron is plated on alternating stripes of laminin and NgCAM, the axon develops on the opposite material as the cell body, which suggests that a change from one axon-promoting substrate to another can be sufficient to initiate axon specification. This group has also shown that N-cadherin has axon-promoting properties similar to laminin and NgCAM, while N-cadherin and NgCAM have opposite effects on dendritic growth. N-cadherin promotes whereas NgCAM reduces dendritic growth (Esch et al., 2000). Using a microcontact printing technique to stamp alternating stripes of polylysine and laminin on glass coverslips, it has been shown that substrates containing laminin favor axon development while those composed of polylysine promote adhesion and development of the somata and dendrites (Corey et al., 1997; Wheeler et al., 1999). This is consistent with results of Lochter et al. (1995), which suggest that axon specification may occur in the neurite that simply grows the fastest, and is not restrained by local inhibiting substrate cues.

Another factor that could provide a mechanism for polarization is the local stability of the actin cytoskeleton. The role of localized instability of the actin network in specifying axonal fate was examined in rat hippocampal neurons in culture (Bradke and Dotti, 1999). During normal neuronal development, actin dynamics and instability polarized to a single growth cone before axon formation. Consistently, global application of actin-depolymerizing drugs and of the Rho-signaling inactivator toxin B to nonpolarized cells produced neurons with multiple axons. Moreover, disruption of the actin network in one individual growth cone induced its neurite to become the axon. Thus, local instability of the actin network restricted to a single growth cone is a physiological signal specifying neuronal polarization (Bradke and Dotti, 1999). While no direct experimental evidence is available, engagement of adhesion receptors is likely to influence locally actin dynamics.

1.3.3. Neurite Extension, Turning of Growth Cones, Pathfinding

What favors the continuous extension of neurites and which factors will induce branching or changes in direction? One early analysis was performed by Letourneau (1975), who studied the preferential attachment and growth of dorsal root ganglion neurons on patterned substrates. Letourneau (1975) hypothesized that if the growth cone encounters more than one type of surface, it will turn toward and follow the more adhesive surface. The same laboratory showed that growth cones preferentially turn up at surface-bound gradient of a peptide derived from the laminin-1 $\alpha 1$ chain (Adams et al., 2005). The results support the hypothesis that growth cone migration *in vivo* can be directed for long distances by surface bound gradients of ligands.

Another laboratory (Lemmon et al., 1992) investigated adhesiveness as a factor influencing growth cone pathfinding using patterns of various adhesion molecules. They examined the choices made by retinal ganglion cells on stripes of L1, N-cadherin, and laminin patterned by a microfluidic method. In contrast to the results of Letourneau (1975), their results illustrated that adhesiveness is not always a good predictor of the rate of axonal growth or of the degree of fasciculation. Since neurites show little selectivity between the three substrates, these authors concluded that some adhesion molecules may serve as permissive substrates but do not provide information about which path to take at a choice point or about which direction to go (Lemmon et al., 1992). Whether adhesion molecules provide permissive or instructive cues is still a debated issue, since they also cooperate with several families of axonal guidance molecules in some circumstances.

Experimental evidence that integrins and ligands could be involved not only during continuous elongation in the same direction but also in branching or turning, for example, when a different substrate is encountered, or if an obstacle is present, comes from our own studies on *Caenorhabditis elegans* (*C. elegans*) integrins. We have observed that a partial loss of function of the INA-1/PAT-3 integrin, a laminin receptor in *C. elegans*, leads to premature arrest and ectopic branching of commissural axons most probably when the axon reaches the dorsal muscle and has to extend beyond it (Figure 1.1) (Poinat et al., 2002).

1.4. Intracellular Pathways Triggered by Adhesion

1.4.1. Adhesion Molecules and Growth Factors in Neurite Outgrowth

The extension of neurites is induced by a variety of extracellular factors, in particular by trophic factors such as nerve growth factor (NGF), brain-derived neurotrophic factor, neurotrophins, and others. As has been illustrated in many studies, growth factor-induced and adhesion-induced signaling pathways converge for various cellular responses (ffrench-Constant and Colognato, 2004). Trophic factors cooperate with adhesion molecules also in the induction of neurite growth (Goldberg, 2003; Guan and Rao, 2003; Huber et al., 2003). A study by Tucker et al. (2005) has illustrated such cooperation between ECM/integrin and NGF in primary adult neurons. Adult dorsal root ganglia neurons plated on surfaces coated with a thin film of laminin exhibited integrin-dependent neurite

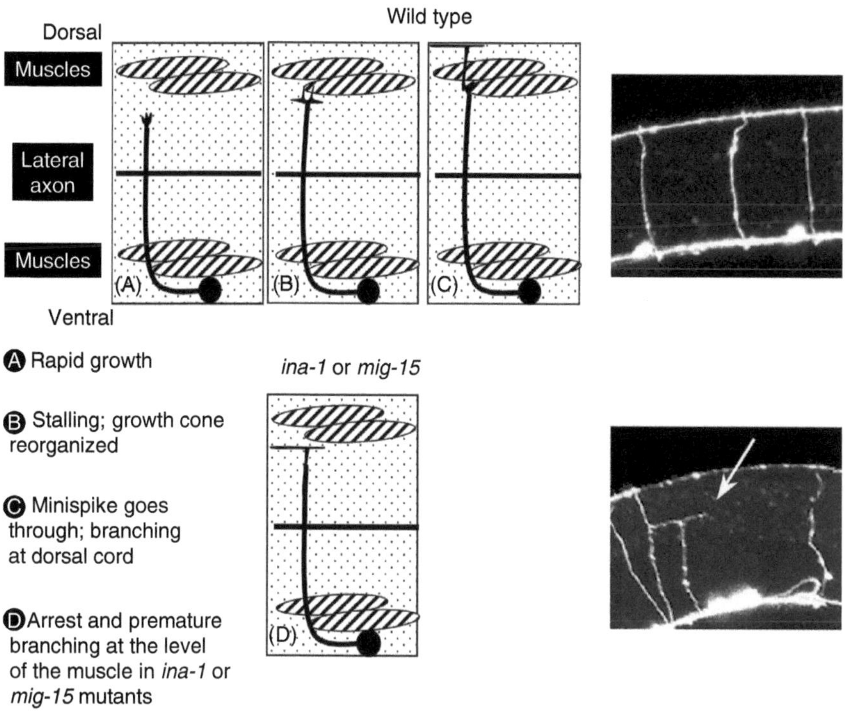

FIGURE 1.1. Top: GABAergic neuron cell bodies are located in the ventral nerve cord of the *C. elegans* larva. In the normal situation, they send a short projection anteriorly (A), which branches to form a commissure. When the commissure reaches the dorsal nerve cord, it further branches to make a short anterior–posterior extension (C). During their navigation to the dorsal cord, commissures meet obstacles; in particular, they have to navigate between muscles (cross-hatched ellipses) and the epidermis (A, B). As illustrated in the right picture, GABAergic axons and cell bodies can be visualized with a GFP expressed under the control of a GABAergic-specific promoter (Knobel et al., 1999). Bottom: In viable *αina-1*, or *mig-15* mutants, commissures often fail to reach the dorsal cord and stop when they meet an obstacle (such as a muscle), in which case they extend a short anterior–posterior branch as they normally do in the dorsal cord (D). A typical premature arrest in a viable *αina-1* mutant, as visualized with the GFP expressed in GABAergic neurons, is illustrated in the right image (arrow). [Reproduced from Poinat et al. (2002) with permission from Elsevier for pictures.]

outgrowth. The addition of NGF resulted in a significant increase in the integrin-dependent outgrowth, correlated with increased expression of integrin subunits and activation of known downstream signaling intermediates such as focal adhesion kinase (FAK), Src, and Akt.

A different situation is encountered in postnatal retinal ganglion cells, which, as CNS neurons, do not extend axons after nerve injury and die when severed from their targets *in vivo*. Most cellular studies thus use growth factors to support survival in culture, which makes it difficult to dissociate trophic factor effect on survival and axon growth. Goldberg et al. (2002) have examined the effect of ECM or CAMs alone on the promotion of axon growth using clonal cultures from purified postnatal day 8 rat retinal ganglion cells. In their hands, ECM molecules or CAMs used as culture substrates or provided as soluble forms were each insufficient to induce significant axon growth but instead potentiated the effects of peptide growth factors.

1.4.2. Intracellular Signaling

Our current knowledge on the mechanisms of integrin signaling pathways comes from studies on the molecules that are known to be associated with cell-ECM junctions in various cell types (Renaudin et al., 1999). Molecules that are present in focal adhesions include talin, vinculin, α-actinin that can be found in most cases. However, junctions can show variations in their composition depending on their adhesion status (Geiger et al., 2001; Li et al., 2005; Renaudin et al., 1999). Expectedly, major integrin partners, such as the FAK, the integrin linked kinase (ILK), the adaptor protein paxillin, and others play a role in neurite outgrowth.

1.4.2.1. FAK and Neurite Outgrowth

FAK is a nonreceptor protein tyrosine kinase which plays a major role in integrin signaling pathways. Its activation by phosphorylation is one of the first events triggered by integrin engagement (Parsons, 2003). FAK is also known as a substrate for the kinase Src. In addition to its function as a protein tyrosine kinase, FAK is a large adaptor protein with binding sites for many proteins involved in cell signaling and motility, including several growth factor receptors, integrins, phosphoinositide 3-kinase (PI3K), Src, p130CAS, Rho GTPases regulators (Graf, Trio, 190RhoGEF), ArfGAP, and cytoskeletal proteins such as paxillin (Hildebrand et al., 1995; Mitra et al., 2005). Previous studies have shown that activation of FAK after costimulation of growth factor receptors and integrins promotes the outgrowth of neurites from PC12 and SH-SY5Y neuronal cells (Ivankovic-Dikic et al., 2000). Indeed, this has received confirmation from the study of Rico et al. (2004), who have examined the effects of deleting FAK *in vivo* in the mouse nervous system and found that FAK is necessary to promote normal growth cone motility. The absence of FAK results in an increase of axon terminals in Purkinje cells and an excess of axonal arborization by hippocampal neurons in cellular assays (Rico et al., 2004). FAK would thus be acting as a negative regulator of axonal branching. It is interesting that the FAK knockout did not inhibit initial axon outgrowth on laminin and polyD lysine, at least at early time points. This finding indicates that the role of FAK is more specific than previously thought or that other kinases may be compensating the absence of FAK in basal neurite outgrowth. One such kinase may be proline-rich tyrosine kinase-2 (PYK2), a closely related molecule.

While $FAK^{-/-}$ mutant phenotypes show clear similarities to integrin mutant phenotypes, in particular in the developing cerebral cortex, FAK activity is also regulated by other extracellular signals. A recent finding is that FAK associates directly with one of the netrin receptors, Deleted in Colorectal Carcinoma (DCC) (Li et al., 2004; Ren et al., 2004). FAK is activated by netrin, and in *Xenopus laevis* its inhibition blocks netrin-induced neurite outgrowth (Li et al., 2004). Attraction by netrin seems to involve FAK and Src family kinases in murine or chicken axons. It has been shown that netrin can stimulate axon outgrowth on polylysine or laminin. FAK-null neurons, however, do not respond to this stimulation (Li et al., 2004; Liu et al., 2004).

Many of the proteins that can bind FAK regulate activities of the Rho GTPase family, which in turn control axon growth guidance and branching through regulation of the cytoskeleton (Billuart et al., 2001). Inhibition of RhoA has been shown to reduce neurite retraction and increase branch extension, similar to phenotypes observed after loss of FAK (Thies and Davenport, 2003). Interactions between FAK and Rho are complex as they have been shown to regulate each other's activity. On one hand, Rho activation induces focal adhesion formation, thereby promoting FAK activation (Barry et al., 1997). On the other hand, activation of FAK induces Rho downregulation, which reduces focal adhesions stability (Ren et al., 2000). Loss of this regulatory loop is thought to explain the perturbations observed in $FAK^{-/-}$ mutant neurons. Rico et al., (2004) suggest that FAK may control axonal branching dynamics by coordinating activation of a RhoGEF (p190RhoGEF), recruitment of a Rho effector (Graf or another protein), and inhibition of a RhoGAP (possibly p190RhoGAP).

1.4.2.2. Paxillin

Paxillin was originally identified as a substrate for the kinase Src and is a member of a larger superfamily that includes the paralogues Hic-5/Ara55 and leupaxin (Turner, 2000). It localizes primarily to specialized sites of adhesion between cells and the ECM and functions as an adaptor molecule that recruits signaling and structural proteins to these sites through its multiple domains. Paxillin directly interacts with multiple structural and signaling proteins such as tubulin, p120RasGAP, PKL, PTP-PEST, FAK, Src, Crk, and Csk (Turner, 2000).

In PC12 and SH-SY5Y cells, association of paxillin with FAK seems to be necessary for normal neurite extension, since overexpression of a paxillin mutant that cannot associate with FAK inhibits neurite outgrowth (Ivankovic-Dikic et al., 2000). Tyrosine phosphorylation of paxillin does not seem to be essential. In contrast, p38MAPK-mediated serine phosphorylation of paxillin seems to be required in NGF-induced neurite outgrowth in PC12 cells (Huang et al., 2004). Experiments by Rico et al. (2004) bring the possibility that paxillin has a role in regulating axon growth or branch formation by hippocampal neurons.

An interesting finding has been that paxillin interaction with integrin $\alpha 4$ has a role in the local activation of Rac GTPase in the lamellipodia of non-neuronal cells (Goldfinger et al., 2003; Nishiya et al., 2005). It has been shown that integrin $\alpha 4$ phosphorylation is restricted to the leading edge. Since paxillin can associate with the nonphosphorylated form of integrin $\alpha 4$, paxillin-$\alpha 4$ association occurs only at the sides and rear of the migrating cells where it inhibits Rac activation. As a result, Rac activation will be limited to the lamellipodia (Nishiya et al., 2005). This illustrates one important and general mechanism whereby very fine and precise local regulation of activity might be obtained. While these studies were not performed in neurons, integrin α4 and paxillin are also expressed in several neuronal cell lines, and for α4 during nerve repair, so that this mechanism may apply (Huang et al., 2004; Vogelezang et al., 2001).

1.4.2.3. Integrin-Linked Kinase

ILK is a serine–threonine kinase first identified by virtue of its interaction with the cytoplasmic domain of β1 integrin (Hannigan et al., 1996). It is composed of three structurally distinct domains: three ankyrin repeats near the N-terminus, a short linker sequence, and a kinase domain at the C-terminus. More recent studies have shown that ILK binds partners, such as PINCH and Parvin, and participates to several intracellular pathways (Akt/PKB, PI3K) (Grashoff et al., 2004). While there is a general agreement that ILK plays a central role in the reorganization of the actin cytoskeleton and attachment to focal adhesions (Grashoff et al., 2004), the function of the kinase domain is still unclear.

A role for ILK in neurite outgrowth has been suggested by several cellular studies. In NIE-115 cells, ILK requirement was demonstrated in β1 integrin-dependent neurite outgrowth on laminin (Ishii et al., 2003), while in PC12 or rat dorsal root ganglia neurons, a similar role was revealed in NGF-induced neurite growth (Mills et al., 2003). Both studies point to a role of ILK in the inhibition of GSK-3β, with the implication that ILK could protect against aberrant phosphorylation of tau by GSK-3β. A role for ILK in promoting increased stabilization of axonal structures during neurite outgrowth has thus been proposed (Mills et al., 2003).

1.4.2.4. Other Signaling Molecules

Several well-known signaling molecules mediate the effect of ECM and integrins on neurite growth. Recent results in this field are the implication of Abelson (Abl) family kinases into the maintenance of dendrite branch in mouse. In mutant mice for Abl and Abl-related gene (Arg) kinases, Moresco et al. (2005) observed deficient dendrite arbor maintenance. In *in vitro* assays, the authors found that Arg is required for laminin-induced neurite branching in cortical neurons (Moresco et al., 2005).

A major element of integrin signaling pathway is the Src kinase. In a report by Robles et al. (2005), the Src kinase was shown to be responsible of the high levels of tyrosine phosphorylation (PY) present at growth cone periphery and in filopodial tips of *X. laevis* spinal neurons. In addition, locally reduced Src activity on one side of the growth cone generates an asymmetry in filopodial motility and PY signaling that induces repulsive turning. This suggests that local changes in filopodial PY levels may underlie growth cone pathfinding decisions. An enrichment of the p21-activated kinase PAK, a known effector of the Cdc42 GTPases, was observed at the tips of PY-positive filopodia. The authors propose that PAK acts as a downstream target of Cdc42 and Src in filopodia. In other systems, an enrichment of integrin β1 has also been reported at filopodial tips (Grabham and Goldberg, 1997; Wu et al., 1996).

Our own studies have implicated another serine-threonine kinase, MAP4K4, which belongs to the STE20/GCK family (also known as Nck interacting kinase [NIK]), in axonal navigation (Poinat et al., 2002). We first found that MAP4K4 interacts molecularly with the cytoplasmic domain of the β1 integrin chain and localizes at tips of cellular processes in fibroblasts (Poinat et al., 2002) or in NGF-treated PC12 cells (our unpublished observations). By genetic analysis in *C. elegans*, we have shown that the orthologue of MAP4K4, MIG-15, and the integrin INA-1/PAT-3 are required during commissural axon navigation. Loss of function of MIG-15 or integrin leads to premature branching or arrest of the growth cone. In Figure 1.1, a schematic drawing illustrates how commissural axons project from the ventral nerve fiber to the dorsal nerve fiber. These axons encounter several types of obstacles during their navigation. It has been observed by time-lapse confocal microscopy that the growth cone stops and stalls when meeting such obstacles, for a short period before resuming its navigation (Knobel et al., 1999). During this short period, filopodia appear and retract until one extends and crosses the obstacle (Knobel et al., 1999). When integrin or MIG-15 is reduced or absent, the axon cannot resume its proper navigation after pausing. While the precise cellular mechanisms are still to be defined, this is in favor of a role for MIG-15 and integrins in axon pathfinding.

1.5. Adhesion and the Cytoskeleton

A wealth of information has come in the recent years on the regulation of the two major actin and microtubule cytoskeletal networks in neuritogenesis. As will be described in other chapters, progress has been made, in particular in the elucidation of the pathways that lead to the Rho family of small G-proteins, Cdc42, and Rac, via various intermediates such as protein tyrosine kinases and phosphatidylinositol 4,5-bisphosphate (Dent and Gertler, 2003; Goldberg, 2003; Hall, 1998; Li et al., 2005; Nobes and Hall, 1995) which regulate cytoskeletal dynamics. As far as cell adhesion is concerned, our understanding of the articulation between adhesion and the cytoskeleton has progressed, particularly in non-neuronal cell types, and models of cell motility have been proposed (Li et al., 2005; Ridley et al., 2003). Only a few elements will be described here, since the topics of cytoskeleton and small G-proteins will be covered in other chapters.

While the actin cytoskeleton plays the leading role in neurite elongation, there is some evidence for interdependence of actin and microtubules (Dent and Kalil, 2001; Gibney and Zheng, 2003). Grabham et al. (2003) in examining chick sympathetic neurons have observed that reorganization of microtubules in growth cones occurs very rapidly after exposure to laminin and this effect is mediated by Rac. These results suggest that part of the laminin-induced neurite elongation may be linked to microtubules.

WASP and WAVE family proteins have received attention (Pollard et al., 2000; Takenawa and Miki, 2001). Nozumi et al. (2003) found that WAVE1 distributed continuously along the leading edge, and WAVE2 and WAVE3 showed a discrete and dynamic localization at the initiation sites of microspikes on the leading edge and remained at the tips of filopodia during elongation. Similar subcellular localizations at the tips of filopodia or at the leading edge of lamellipodia have been reported for N-WASP, Ena/VASP, mammalian Ena (Mena) (Lanier et al., 1999; Nakagawa et al., 2001; Rottner et al., 1999). This family most likely plays important roles in the formation or elongation of filopodia and in cell motility. However, filopodia can be formed even in the absence of N-WASP in fibroblasts or of Ena/VASP in the growth cone (Bear et al., 2002; Snapper et al., 2001). An interaction between FAK and WASP, as well as a phosphorylation of N-WASP by FAK, has been reported in fibroblasts (Wu et al., 2004). This observation may provide us a direct link between cell adhesion and remodeling of the actin cytoskeleton (Wu et al., 2004). However, this type of interaction has not been established yet in neurons.

It has been shown that laminin can guide axonal outgrowth by affecting the localization of myosin IIB in growth cones. The authors propose that depending on the level and location of activation by integrins, local myosin II contractile activity may cause growth cones to turn, branch, retract, stall, or advance with precision along a laminin border (Turney and Bridgman, 2005).

1.6. *In Vivo* Functions of Adhesion Molecules in Neuronal Cells

What are the evidences for a function of integrins and ligands in neurite elongation *in vivo*? In invertebrate model organisms, *Drosophila* and *C. elegans*, loss-of-function mutations in integrin subunits or ECM ligands result in abnormal axonal navigation. For example, in *Drosophila*, mutations into genes encoding the main integrin β chain, βPS, or the laminin α chain lead to defects in targeting of some classes of axons (Garcia-Alonso et al., 1996; Hoang and Chiba, 1998). Similarly, mutations into genes encoding the integrin α chain INA-1 or neuronal-specific knockdown of the β chain PAT-3 by RNA interference in *C. elegans* lead to abnormal navigation of commissural axons, which display premature arrest and branching (Baum and Garriga, 1997; Poinat et al., 2002). Furthermore, mutations into laminin genes result into marked axonal guidance defects (Huang et al., 2003). In vertebrates, Lilienbaum et al. (1995) examined integrin function in retinal ganglion cell development *in vivo* by transfecting genes encoding various dominant forms of the chicken β1 integrin subunit into intact eye primordia of *Xenopus* embryos. They could show that β1 integrins play an important role in regulating the outgrowth of axons and dendrites from retinal ganglion cells in the retina but that chimeric integrins do not impair growth cone steering in general. Genetic screens in the zebrafish have led to the identification of mutations in genes for laminin or tenascin that result in axon guidance defects (Paulus and Halloran, 2006).

In the mouse, while a number of mutations have been produced by gene targeting, phenotypes have essentially revealed more early functions in the developing nervous system, for example, in neuronal migration. Abolishing the expression of β1 in the mouse brain by a conditional knockout approach has revealed a role in the formation of the radial glia network and attachment of the glial endfeet to the BL underlying the pia mater (Graus-Porta et al., 2001). Absence of integrin β1 leads to cortical layer disorganization as is the case also for brain-specific knockout of FAK or ILK (Beggs et al., 2003; Niewmierzycka et al., 2005). No strong evidence was found for axonal migration defects linked to the mutation. In contrast, when β1 integrin was inactivated in neural crest cells and derivatives, alterations were observed in the PNS (transient abnormal trajectories of cranial nerves, defects in branching, and fasciculation of nerves innervating the limbs) (Pietri et al., 2004). Derivation of neural cells from $\beta1^{-/-}$ embryonic stem cells has allowed some studies *in vitro* (Andressen et al., 1998, 2005; Rohwedel et al., 1998). Indeed, $\beta1^{-/-}$ neurons show a decreased capacity to extend neurites on different substrates (Andressen et al., 2005). In another study, Rohwedel et al. (1998) found an acceleration of neuronal differentiation in $\beta1^{-/-}$ neurons. These cell lines could constitute good models to examine intracellular pathways and their alterations in the absence of integrin β1.

Knockout of the α6 integrin and of the α6 and α3 in the mouse leads to neuronal ectopias, with clusters of cells extruding through breaches of the BL (De Arcangelis et al., 1999; Georges-Labouesse et al., 1998), with some similarities with the phenotypes associated with brain-specific inactivation of integrin β1 as described previously (Graus-Porta et al., 2001). Here again, defects may originate from BL disorganization or alterations at some sites of the attachment or radial glia endfeet. In addition, the role of integrin α3β1 may also be linked to its ability to bind reelin, as proposed by Dulabon et al. (2000). In all these examples, a role of these integrin in neurite elongation in the knockout animals awaits further studies. This illustrates that while global integrin inactivation may not reveal a role in neurite elongation, more precise experiments in specific tissues may be informative. A combination of *ex vivo* (explant cultures) and *in vitro* studies using mutant cells may be required to access to dynamic aspects of neurite elongation.

The mutation of integrin α7 in mouse has revealed its role in axonal regeneration. Indeed, it had been shown that integrin α7 is strongly upregulated in axotomized neurons in various injury models during peripheral nerve regeneration but not after CNS injury in the adult nervous system. Consistent with these studies, the deletion of the α7 subunit leads to an impairment in axonal outgrowth and a delayed target reinnervation of regenerating facial motoneurons (Werner et al., 2000). However, the absence of the α7 integrin subunit causes only a partial reduction in the speed of nerve fiber regeneration, which suggests a partial functional compensation by other molecules. The strong increase of the β1 integrin chain level after axotomy in the $\alpha7^{-/-}$ mice clearly supports such a compensatory mechanism by other associated α subunits.

Concerning the main ECM ligands, mutations in mice have not been very informative to date, in several cases because total inactivation results in early embryonic lethality. In few cases, defects in the nervous system have been described. First, alterations similar to the one described previously (cortical and retinal layer disorganization) are observed in the brain of mice carrying a mutation of the nidogen-binding site present on the γ1 chain arm of laminin (Halfter et al., 2002). Laminin-α5 knockout leads to exencephaly due to a failure of neural tube associated with congenital muscular dystrophy in human. In mice carrying laminin-2 mutations, peripheral neuropathies have been observed, but at the moment, no precise study on neurite elongation has been reported. While tenascin-C and tenascin-R knockout mice do not present developmental phenotypes, facial nerve repair seems to be affected differently by the two mutations (Guntinas-Lichius et al., 2005). As discussed already, more precise experiments than total inactivation, such as knockout in specific tissues, may be required.

1.7. Cross Talk with Other Families

In the recent years, important families that act as attractive or repulsive cues in axonal guidance have been discovered including the netrins, semaphorins, ephrins, and robo and Slit. While a detailed description of their properties would be beyond the scope of this chapter, several studies have highlighted cross talks and possible intersections or cross regulations between these families and adhesion molecules (Nakamoto et al., 2004). As already described, semaphorin-7A has been shown to bind a β1-containing integrin in olfactory bulb neurons in mouse (Pasterkamp et al., 2003). In other tissues, it has been shown that semaphorins can regulate integrin functions (Nakamoto et al., 2004).

Slits are a family of secreted ligands that bind the transmembrane molecules roundabout (robo), which mediate axonal guidance at the midline. A genetic analysis in *Drosophila* showed that integrins and their ligands tiggrin and laminin regulate axonal responses to Slit. When Slit and integrin functions are reduced, growth cones respond more to attractive signals (Stevens and Jacobs, 2002).

Several connections with the netrin family have also been reported. A first study concerned retinal axons. These are attracted by netrin-1, but it has been shown that in the presence of laminin, this attractive response is converted into a repulsive response (Hopker et al., 1999). It has been proposed that this effect is related to an effect of laminin in lowering the level of cAMP. One of the netrin receptors, DCC, which mediates growth cone attraction, has been shown to activate FAK, as already mentioned. Thus, netrin and adhesion receptors share common downstream signaling pathways in neurons.

1.8. Conclusions

In the recent years, progress in our understanding of adhesion molecules has been brought about by the development of cellular and genetic approaches in several organisms. Given the complexity of the regulation and the convergence of many different types of molecules on the process of neuritogenesis, it is still a challenge to approach the question of specificity and cooperation and understand the molecular and cellular mechanisms of this process. We are currently witnessing the development of very powerful and sophisticated imaging technologies that allow to access to dynamic parameters in living cells or tissues. A very recent example comes from the study of neuronal migration in the mouse developing cortex, where time-lapse microscopy has allowed to visualize and compare cortical neuron migration in mutant versus control animals (Sanada et al., 2004). The near future should certainly see important progress by the combination of precisely designed manipulation of genes and live imaging.

Acknowledgments. A large number of important articles could not be cited here owing to space restrictions, and the authors apologize. Research in E.G-L laboratory is funded by institutional funds from CNRS, INSERM and Université Louis Pasteur, and by funds from the Association pour la Recherche sur le Cancer, the Association Française contre les Myopathies and La Ligue Régionale contre le Cancer (Haut-Rhin). S.E. is supported by a fellowship from the French Ministry of Research.

References

Adams, D.N., Kao, E.Y., Hypolite, C.L., Distefano, M.D., Hu, W.S., and Letourneau, P.C., 2005, Growth cones turn and migrate up an immobilized gradient of the laminin IKVAV peptide, *J. Neurobiol.* **62:** 134–147.

Andressen, C., Arnhold, S., Puschmann, M., Bloch, W., Hescheler, J., Fassler, R., et al., 1998, Beta1 integrin deficiency impairs migration and differentiation of mouse embryonic stem cell derived neurons, *Neurosci. Lett.* **251:** 165–168.

Andressen, C., Adrian, S., Fassler, R., Arnhold, S., and Addicks, K., 2005, The contribution of beta1 integrins to neuronal migration and differentiation depends on extracellular matrix molecules, *Eur. J. Cell Biol.* **84:** 973–982.

Bamdad, M., Volle, D., Dastugue, B., and Meiniel, A., 2004, Alpha1beta1-integrin is an essential signal for neurite outgrowth induced by thrombospondin type 1 repeats of SCO-spondin, *Cell Tissue Res.* **315:** 15–25.

Barry, S.T., Flinn, H.M., Humphries, M.J., Critchley, D.R., and Ridley, A.J., 1997, Requirement for Rho in integrin signalling, *Cell Commun. Adhes.* **4:** 387–398.

Baum, P.D., and Garriga, G., 1997, Neuronal migrations and axon fasciculation are disrupted in ina-1 integrin mutants, *Neuron* **19:** 51–62.

Bear, J.E., Svitkina, T.M., Krause, M., Schafer, D.A., Loureiro, J.J., Strasser, G.A., et al., 2002, Antagonism between Ena/VASP proteins and actin filament capping regulates fibroblast motility, *Cell* **109:** 509–521.

Beggs, H.E., Schahin-Reed, D., Zang, K., Goebbels, S., Nave, K.A., Gorski, J., et al., 2003, FAK deficiency in cells contributing to the basal lamina results in cortical abnormalities resembling congenital muscular dystrophies, *Neuron* **40:** 501–514.

Billuart, P., Winter, C.G., Maresh, A., Zhao, X., and Luo, L., 2001, Regulating axon branch stability: The role of p190 RhoGAP in repressing a retraction signaling pathway, *Cell* **107:** 195–207.

Bradke, F., and Dotti, C.G., 1999, The role of local actin instability in axon formation, *Science* **283:** 1931–1934.
Carulli, D., Laabs, T., Geller, H.M., and Fawcett, J.W., 2005, Chondroitin sulfate proteoglycans in neural development and regeneration, *Curr. Opin. Neurobiol.* **15:** 116–120.
Condic, M.L., and Letourneau, P.C., 1997, Ligand-induced changes in integrin expression regulate neuronal adhesion and neurite outgrowth, *Nature* **389:** 852–856.
Corey, J.M., Brunette, A.L., Chen, M.S., Weyhenmeyer, J.A., Brewer, G.J., and Wheeler, B.C., 1997, Differentiated B104 neuroblastoma cells are a high-resolution assay for micropatterned substrates, *J. Neurosci. Methods* **75:** 91–97.
Culley, B., Murphy, J., Babaie, J., Nguyen, D., Pagel, A., Rousselle, P., et al., 2001, Laminin-5 promotes neurite outgrowth from central and peripheral chick embryonic neurons, *Neurosci. Lett.* **301:** 83–86.
da Silva, J.S., and Dotti, C.G., 2002, Breaking the neuronal sphere: Regulation of the actin cytoskeleton in neuritogenesis, *Nat. Rev. Neurosci.* **3:** 694–704.
De Arcangelis, A., Mark, M., Kreidberg, J., Sorokin, L., and Georges-Labouesse, E., 1999, Synergistic activities of alpha3 and alpha6 integrins are required during apical ectodermal ridge formation and organogenesis in the mouse, *Development* **126:** 3957–3968.
de Miguel, F.F., and Vargas, J., 2000, Native extracellular matrix induces a well-organized bipolar outgrowth pattern with neurite extension and retraction in cultured neurons, *J. Comp. Neurol.* **417:** 387–398.
Dent, E.W., and Gertler, F.B., 2003, Cytoskeletal dynamics and transport in growth cone motility and axon guidance, *Neuron* **40:** 209–227.
Dent, E.W., and Kalil, K., 2001, Axon branching requires interactions between dynamic microtubules and actin filaments, *J. Neurosci.* **21:** 9757–9769.
Duband, J.L., Nuckolls, G.H., Ishihara, A., Hasegawa, T., Yamada, K.M., Thiery, J.P., et al., 1988, Fibronectin receptor exhibits high lateral mobility in embryonic locomoting cells but is immobile in focal contacts and fibrillar streaks in stationary cells, *J. Cell Biol.* **107:** 1385–1396.
Dulabon, L., Olson, E.C., Taglienti, M.G., Eisenhuth, S., McGrath, B., Walsh, C.A., et al., 2000, Reelin binds alpha3beta1 integrin and inhibits neuronal migration, *Neuron* **27:** 33–44.
Esch, T., Lemmon, V., and Banker, G., 1999, Local presentation of substrate molecules directs axon specification by cultured hippocampal neurons, *J. Neurosci.* **19:** 6417–6426.
Esch, T., Lemmon, V., and Banker, G., 2000, Differential effects of NgCAM and N-cadherin on the development of axons and dendrites by cultured hippocampal neurons, *J. Neurocytol.* **29:** 215–223.
ffrench-Constant, C., and Colognato, H., 2004, Integrins: Versatile integrators of extracellular signals, *Trends Cell Biol.* **14:** 678–686.
Garcia-Alonso, L., Fetter, R.D., and Goodman, C.S., 1996, Genetic analysis of Laminin A in Drosophila: Extracellular matrix containing laminin A is required for ocellar axon pathfinding, *Development* **122:** 2611–2621.
Geiger, B., Bershadsky, A., Pankov, R., and Yamada, K.M., 2001, Transmembrane crosstalk between the extracellular matrix—cytoskeleton crosstalk, *Nat. Rev. Mol. Cell Biol.* **2:** 793–805.
Georges-Labouesse, E., Mark, M., Messaddeq, N., and Gansmuller, A., 1998, Essential role of alpha 6 integrins in cortical and retinal lamination, *Curr. Biol.* **8:** 983–986.
Gibney, J., and Zheng, J.Q., 2003, Cytoskeletal dynamics underlying collateral membrane protrusions induced by neurotrophins in cultured Xenopus embryonic neurons, *J. Neurobiol.* **54:** 393–405.
Goldberg, J.L., 2003, How does an axon grow? *Genes Dev.* **17:** 941–958.

Goldberg, J.L., 2004, Intrinsic neuronal regulation of axon and dendrite growth, *Curr. Opin. Neurobiol.* **14:** 551–557.

Goldberg, J.L., Espinosa, J.S., Xu, Y., Davidson, N., Kovacs, G.T., and Barres, B.A., 2002, Retinal ganglion cells do not extend axons by default: Promotion by neurotrophic signaling and electrical activity, *Neuron* **33:** 689–702.

Goldfinger, L.E., Han, J., Kiosses, W.B., Howe, A.K., and Ginsberg, M.H., 2003, Spatial restriction of alpha4 integrin phosphorylation regulates lamellipodial stability and alpha4beta1-dependent cell migration, *J. Cell Biol.* **162:** 731–741.

Gotz, B., Scholze, A., Clement, A., Joester, A., Schutte, K., Wigger, F., et al., 1996, Tenascin-C contains distinct adhesive, anti-adhesive, and neurite outgrowth promoting sites for neurons, *J. Cell Biol.* **132:** 681–699.

Gotz, M., Bolz, J., Joester, A., and Faissner, A., 1997, Tenascin-C synthesis and influence on axonal growth during rat cortical development, *Eur. J. Neurosci.* **9:** 496–506.

Grabham, P.W., and Goldberg, D.J., 1997, Nerve growth factor stimulates the accumulation of beta1 integrin at the tips of filopodia in the growth cones of sympathetic neurons, *J. Neurosci.* **17:** 5455–5465.

Grabham, P.W., Reznik, B., and Goldberg, D.J., 2003, Microtubule and Rac 1-dependent F-actin in growth cones, *J. Cell Sci.* **116:** 3739–3748.

Grashoff, C., Thievessen, I., Lorenz, K., Ussar, S., and Fassler, R., 2004, Integrin-linked kinase: Integrin's mysterious partner, *Curr. Opin. Cell Biol.* **16:** 565–571.

Graus-Porta, D., Blaess, S., Senften, M., Littlewood-Evans, A., Damsky, C., Huang, Z., et al., 2001, Beta1-class integrins regulate the development of laminae and folia in the cerebral and cerebellar cortex, *Neuron* **31:** 367–379.

Guan, K.L., and Rao, Y., 2003, Signalling mechanisms mediating neuronal responses to guidance cues, *Nat. Rev. Neurosci.* **4:** 941–956.

Guan, W., Puthenveedu, M.A., and Condic, M.L., 2003, Sensory neuron subtypes have unique substratum preference and receptor expression before target innervation, *J. Neurosci.* **23:** 1781–1791.

Guntinas-Lichius, O., Angelov, D.N., Morellini, F., Lenzen, M., Skouras, E., Schachner, M., et al., 2005, Opposite impacts of tenascin-C and tenascin-R deficiency in mice on the functional outcome of facial nerve repair, *Eur. J. Neurosci.* **22:** 2171–2179.

Halfter, W., Dong, S., Yip, Y.P., Willem, M., and Mayer, U., 2002, A critical function of the pial basement membrane in cortical histogenesis, *J. Neurosci.* **22:** 6029–6040.

Hall, A., 1998, Rho GTPases and the actin cytoskeleton, *Science* **279:** 509–514.

Hannigan, G.E., Leung-Hagesteijn, C., Fitz-Gibbon, L., Coppolino, M.G., Radeva, G., Filmus, J., et al., 1996, Regulation of cell adhesion and anchorage-dependent growth by a new beta 1-integrin-linked protein kinase, *Nature* **379:** 91–96.

Hildebrand, J.D., Schaller, M.D., and Parsons, J.T., 1995, Paxillin, a tyrosine phosphorylated focal adhesion-associated protein binds to the carboxyl terminal domain of focal adhesion kinase, *Mol. Biol. Cell* **6:** 637–647.

Hoang, B., and Chiba, A., 1998, Genetic analysis on the role of integrin during axon guidance in Drosophila, *J. Neurosci.* **18:** 7847–7855.

Hopker, V.H., Shewan, D., Tessier-Lavigne, M., Poo, M., and Holt, C., 1999, Growth-cone attraction to netrin-1 is converted to repulsion by laminin-1, *Nature* **401:** 69–73.

Hopkins, J.M., Ford-Holevinski, T.S., McCoy, J.P., and Agranoff, B.W., 1985, Laminin and optic nerve regeneration in the goldfish, *J. Neurosci.* **5:** 3030–3038.

Huang, C., Borchers, C.H., Schaller, M.D., and Jacobson, K., 2004, Phosphorylation of paxillin by p38MAPK is involved in the neurite extension of PC-12 cells, *J. Cell Biol.* **164:** 593–602.

Huang, C.C., Hall, D.H., Hedgecock, E.M., Kao, G., Karantza, V., Vogel, B.E., et al., 2003, Laminin alpha subunits and their role in C. elegans development, *Development* **130:** 3343–3358.

Huber, A.B., Kolodkin, A.L., Ginty, D.D., and Cloutier, J.F., 2003, Signaling at the growth cone: Ligand-receptor complexes and the control of axon growth and guidance, *Annu. Rev. Neurosci.* **26:** 509–563.

Hynds, D.L., and Snow, D.M., 2001, Fibronectin and laminin elicit differential behaviors from SH-SY5Y growth cones contacting inhibitory chondroitin sulfate proteoglycans, *J. Neurosci. Res.* **66:** 630–642.

Hynes, R.O., 2002, Integrins: Bidirectional, allosteric signaling machines, *Cell* **110:** 673–687.

Hynes, R.O., and Lander, A.D., 1992, Contact and adhesive specificities in the associations, migrations, and targeting of cells and axons, *Cell* **68:** 303–322.

Ishii, T., Furuoka, H., Muroi, Y., and Nishimura, M., 2003, Inactivation of integrin-linked kinase induces aberrant tau phosphorylation via sustained activation of glycogen synthase kinase 3beta in N1E-115 neuroblastoma cells, *J. Biol. Chem.* **278:** 26970–26975.

Ivankovic-Dikic, I., Gronroos, E., Blaukat, A., Barth, B.U., and Dikic, I., 2000, Pyk2 and FAK regulate neurite outgrowth induced by growth factors and integrins, *Nat. Cell Biol.* **2:** 574–581.

Ivins, J.K., Colognato, H., Kreidberg, J.A., Yurchenco, P.D., and Lander, A.D., 1998, Neuronal receptors mediating responses to antibodyactivated laminin-1, *J. Neurosci.* **18:** 9703–9715.

Ivins, J.K., Yurchenco, P.D., and Lander, A.D., 2000, Regulation of neurite outgrowth by integrin activation, *J. Neurosci.* **20:** 6551–6560.

Joester, A., and Faissner, A., 2001, The structure and function of tenascins in the nervous system, *Matrix Biol.* **20:** 13–22.

Kiryushko, D., Berezin, V., and Bock, E., 2004, Regulators of neurite outgrowth: Role of cell adhesion molecules, *Ann. N Y Acad. Sci.* **1014:** 140–154.

Knobel, K.M., Jorgensen, E.M., and Bastiani, M.J., 1999, Growth cones stall and collapse during axon outgrowth in Caenorhabditis elegans, *Development* **126:** 4489–4498.

Lanier, L.M., Gates, M.A., Witke, W., Menzies, A.S., Wehman, A.M., Macklis, J.D., et al., 1999, Mena is required for neurulation and commissure formation, *Neuron* **22:** 313–325.

Lemmon, V., Burden, S.M., Payne, H.R., Elmslie, G.J., and Hlavin, M.L., 1992, Neurite growth on different substrates: Permissive versus instructive influences and the role of adhesive strength, *J. Neurosci.* **12:** 818–826.

Lentz, S.I., Miner, J.H., Sanes, J.R., and Snider, W.D., 1997, Distribution of the ten known laminin chains in the pathways and targets of developing sensory axons, *J. Comp. Neurol.* **378:** 547–561.

Letourneau, P.C., 1975, Possible roles for cell-to-substratum adhesion in neuronal morphogenesis, *Dev. Biol.* **44:** 77–91.

Li, S., Guan, J.L., and Chien, S., 2005, Biochemistry and biomechanics of cell motility, *Annu. Rev. Biomed. Eng.* **7:** 105–150.

Li, W., Lee, J., Vikis, H.G., Lee, S.H., Liu, G., Aurandt, J., et al., 2004, Activation of FAK and Src are receptor-proximal events required for netrin signaling, *Nat. Neurosci.* **7:** 1213–1221.

Liesi, P., 1985a, Do neurons in the vertebrate CNS migrate on laminin? *EMBO J.* **4:** 1163–1170.

Liesi, P., 1985b, Laminin-immunoreactive glia distinguish regenerative adult CNS systems from non-regenerative ones, *EMBO J.* **4:** 2505–2511.

Lilienbaum, A., Reszka, A.A., Horwitz, A.F., and Holt, C.E., 1995, Chimeric integrins expressed in retinal ganglion cells impair process outgrowth in vivo, *Mol. Cell. Neurosci.* **6:** 139–152.

Liu, G., Beggs, H., Jurgensen, C., Park, H.T., Tang, H., Gorski, J., et al., 2004, Netrin requires focal adhesion kinase and Src family kinases for axon outgrowth and attraction, *Nat. Neurosci.* **7:** 1222–1232.

Lochter, A., Taylor, J., Braunewell, K.H., Holm, J., and Schachner, M., 1995, Control of neuronal morphology in vitro: Interplay between adhesive substrate forces and molecular instruction, *J. Neurosci. Res.* **42:** 145–158.

Manthorpe, M., Engvall, E., Ruoslahti, E., Longo, F.M., Davis, G.E., and Varon, S., 1983, Laminin promotes neuritic regeneration from cultured peripheral and central neurons, *J. Cell Biol.* **97:** 1882–1890.

Mechai, N., Wenzel, M., Koch, M., Lucka, L., Horstkorte, R., Reutter, W., et al., 2005, The cytoplasmic tail of the alpha(3) integrin subunit promotes neurite outgrowth in PC12 cells, *J. Neurosci. Res.* **82:** 753–761.

Mercado, M.L., Nur-e-Kamal, A., Liu, H.Y., Gross, S.R., Movahed, R., and Meiners, S., 2004, Neurite outgrowth by the alternatively spliced region of human tenascin-C is mediated by neuronal alpha7beta1 integrin, *J. Neurosci.* **24:** 238–247.

Mills, J., Digicaylioglu, M., Legg, A.T., Young, C.E., Young, S.S., Barr, A.M., et al., 2003, Role of integrin-linked kinase in nerve growth factor-stimulated neurite outgrowth, *J. Neurosci.* **23:** 1638–1648.

Miner, J.H., and Li, C., 2000, Defective glomerulogenesis in the absence of laminin alpha5 demonstrates a developmental role for the kidney glomerular basement membrane, *Dev. Biol.* **217:** 278–289.

Miner, J.H., and Yurchenco, P.D., 2004, Laminin functions in tissue morphogenesis, *Annu. Rev. Cell Dev. Biol.* **20:** 255–284.

Mitra, S.K., Hanson, D.A., and Schlaepfer, D.D., 2005, Focal adhesion kinase: In command and control of cell motility, *Nat. Rev. Mol. Cell Biol.* **6:** 56–68.

Moresco, E.M., Donaldson, S., Williamson, A., and Koleske, A.J., 2005, Integrin-mediated dendrite branch maintenance requires Abelson (Abl) family kinases, *J. Neurosci.* **25:** 6105–6118.

Nakagawa, H., Miki, H., Ito, M., Ohashi, K., Takenawa, T., and Miyamoto, S.,2001, N-WASP, WAVE and Mena play different roles in the organization of actin cytoskeleton in lamellipodia, *J. Cell Sci.* **114:** 1555–1565.

Nakamoto, T., Kain, K.H., and Ginsberg, M.H., 2004, Neurobiology: New connections between integrins and axon guidance, *Curr. Biol.* **14:** R121–R123.

Niewmierzycka, A., Mills, J., St-Arnaud, R., Dedhar, S., and Reichardt, L.F., 2005, Integrin-linked kinase deletion from mouse cortex results in cortical lamination defects resembling cobblestone lissencephaly, *J. Neurosci.* **25:** 7022–7031.

Nishiya, N., Kiosses, W.B., Han, J., and Ginsberg, M.H., 2005, An alpha4 integrin-paxillin-Arf-GAP complex restricts Rac activation to the leading edge of migrating cells, *Nat. Cell Biol.* **7:** 343–352.

Nobes, C.D., and Hall, A., 1995, Rho, rac, and cdc42 GTPases regulate the assembly of multimolecular focal complexes associated with actin stress fibers, lamellipodia, and filopodia, *Cell* **81:** 53–62.

Nozumi, M., Nakagawa, H., Miki, H., Takenawa, T., and Miyamoto, S., 2003, Differential localization of WAVE isoforms in filopodia and lamellipodia of the neuronal growth cone, *J. Cell Sci.* **116:** 239–246.

O'Shea, K.S., Rheinheimer, J.S., and Dixit, V.M., 1990, Deposition and role of thrombospondin in the histogenesis of the cerebellar cortex, *J. Cell Biol.* **110:** 1275–1283.

Parsons, J.T., 2003, Focal adhesion kinase: The first ten years, *J. Cell Sci.* **116:** 1409–1416.

Pasterkamp, R.J., Peschon, J.J., Spriggs, M.K., and Kolodkin, A.L., 2003, Semaphorin 7A promotes axon outgrowth through integrins and MAPKs, *Nature* **424:** 398–405.

Patton, B.L., Miner, J.H., Chiu, A.Y., and Sanes, J.R., 1997, Distribution and function of laminins in the neuromuscular system of developing, adult, and mutant mice, *J. Cell Biol.* **139:** 1507–1521.

Paulus, J.D., and Halloran, M.C., 2006, Zebrafish bashful/laminin-alpha1 mutants exhibit multiple axon guidance defects, *Dev. Dyn.* **235:** 213–224.

Pietri, T., Eder, O., Breau, M.A., Topilko, P., Blanche, M., Brakebusch, C., et al., 2004, Conditional beta1-integrin gene deletion in neural crest cells causes severe developmental alterations of the peripheral nervous system, *Development* **131:** 3871–3883.

Pinkstaff, J.K., Detterich, J., Lynch, G., and Gall, C., 1999, Integrin subunit gene expression is regionally differentiated in adult brain, *J. Neurosci.* **19:** 1541–1556.

Poinat, P., De Arcangelis, A., Sookhareea, S., Zhu, X., Hedgecock, E.M., Labouesse, M., et al., 2002, A conserved interaction between beta1 integrin/PAT-3 and Nck-interacting kinase/MIG-15 that mediates commissural axon navigation in C. elegans, *Curr. Biol.* **12:** 622–631.

Pollard, T.D., Blanchoin, L., and Mullins, R.D., 2000, Molecular mechanisms controlling actin filament dynamics in nonmuscle cells, *Annu. Rev. Biophys. Biomol. Struct.* **29:** 545–576.

Ranscht, B., 2000, Cadherins: Molecular codes for axon guidance and synapse formation, *Int. J. Dev. Neurosci.* **18:** 643–651.

Rathjen, F.G., 1991, Neural cell contact and axonal growth, *Curr. Opin. Cell Biol.* **3:** 992–1000.

Reichardt, L.F., and Tomaselli, K.J., 1991, Extracellular matrix molecules and their receptors: Functions in neural development, *Annu. Rev. Neurosci.* **14:** 531–570.

Reichardt, L.F., Bossy, B., Carbonetto, S., de Curtis, I., Emmett, C., Hall, D.E., et al., 1990, Neuronal receptors that regulate axon growth, *Cold Spring Harb. Symp. Quant. Biol.* **55:** 341–350.

Ren, X.D., Kiosses, W.B., Sieg, D.J., Otey, C.A., Schlaepfer, D.D., and Schwartz, M.A., 2000, Focal adhesion kinase suppresses Rho activity to promote focal adhesion turnover, *J. Cell Sci.* **113:** 3673–3678.

Ren, X.R., Ming, G.L., Xie, Y., Hong, Y., Sun, D.M., Zhao, Z.Q., et al., 2004, Focal adhesion kinase in netrin-1 signaling, *Nat. Neurosci.* **7:** 1204–1212.

Renaudin, A., Lehmann, M., Girault, J., and McKerracher, L., 1999, Organization of point contacts in neuronal growth cones, *J. Neurosci. Res.* **55:** 458–471.

Rico, B., Beggs, H.E., Schahin-Reed, D., Kimes, N., Schmidt, A., and Reichardt, L.F., 2004, Control of axonal branching and synapse formation by focal adhesion kinase, *Nat. Neurosci.* **7:** 1059–1069.

Ridley, A.J., Schwartz, M.A., Burridge, K., Firtel, R.A., Ginsberg, M.H., Borisy, G., et al., 2003, Cell migration: Integrating signals from front to back, *Science* **302:** 1704–1709.

Robles, E., Woo, S., and Gomez, T.M., 2005, Src-dependent tyrosine phosphorylation at the tips of growth cone filopodia promotes extension, *J. Neurosci.* **25:** 7669–7681.

Rogers, S.L., Edson, K.J., Letourneau, P.C., and McLoon, S.C., 1986, Distribution of laminin in the developing peripheral nervous system of the chick, *Dev. Biol.* **113:** 429–435.

Rohwedel, J., Guan, K., Zuschratter, W., Jin, S., Ahnert-Hilger, G., Furst, D., et al., 1998, Loss of beta1 integrin function results in a retardation of myogenic, but an acceleration of neuronal, differentiation of embryonic stem cells in vitro, *Dev. Biol.* **201:** 167–184.

Rottner, K., Behrendt, B., Small, J.V., and Wehland, J., 1999, VASP dynamics during lamellipodia protrusion, *Nat. Cell Biol.* **1:** 321–322.

Rougon, G., and Hobert, O., 2003, New insights into the diversity and function of neuronal immunoglobulin superfamily molecules, *Annu. Rev. Neurosci.* **26:** 207–238.

Sanada, K., Gupta, A., and Tsai, L.H., 2004, Disabled-1-regulated adhesion of migrating neurons to radial glial fiber contributes to neuronal positioning during early corticogenesis, *Neuron* **42:** 197–211.

Sheppard, A.M., Hamilton, S.K., and Pearlman, A.L., 1991, Changes in the distribution of extracellular matrix components accompany early morphogenetic events of mammalian cortical development, *J. Neurosci.* **11:** 3928–3942.

Sheppard, A.M., Brunstrom, J.E., Thornton, T.N., Gerfen, R.W., Broekelmann, T.J., McDonald, J.A., et al., 1995, Neuronal production of fibronectin in the cerebral cortex during migration and layer formation is unique to specific cortical domains, *Dev. Biol.* **172:** 504–518.

Skaper, S.D., Moore, S.E., and Walsh, F.S., 2001, Cell signalling cascades regulating neuronal growth-promoting and inhibitory cues, *Prog. Neurobiol.* **65:** 593–608.

Snapper, S.B., Takeshima, F., Anton, I., Liu, C.H., Thomas, S.M., Nguyen, D., et al., 2001, N-WASP deficiency reveals distinct pathways for cell surface projections and microbial actin-based motility, *Nat. Cell Biol.* **3:** 897–904.

Stevens, A., and Jacobs, J.R., 2002, Integrins regulate responsiveness to slit repellent signals, *J. Neurosci.* **22:** 4448–4455.

Stipp, C.S., and Hemler, M.E., 2000, Transmembrane-4-superfamily proteins CD151 and CD81 associate with alpha 3 beta 1 integrin, and selectively contribute to alpha 3 beta 1-dependent neurite outgrowth, *J. Cell Sci.* **113:** 1871–1882.

Suter, D.M., and Forscher, P., 2000, Substrate-cytoskeletal coupling as a mechanism for the regulation of growth cone motility and guidance, *J. Neurobiol.* **44:** 97–113.

Takenawa, T., and Miki, H., 2001, WASP and WAVE family proteins: Key molecules for rapid rearrangement of cortical actin filaments and cell movement, *J. Cell Sci.* **114:** 1801–1809.

Thies, E., and Davenport, R.W., 2003, Independent roles of Rho-GTPases in growth cone and axonal behavior, *J. Neurobiol.* **54:** 358–369.

Tucker, B.A., Rahimtula, M., and Mearow, K.M., 2005, Integrin activation and neurotrophin signaling cooperate to enhance neurite outgrowth in sensory neurons, *J. Comp. Neurol.* **486:** 267–280.

Turner, C.E., 2000, Paxillin interactions, *J. Cell Sci.* **113:** 4139–4140.

Turney, S.G., and Bridgman, P.C., 2005, Laminin stimulates and guides axonal outgrowth via growth cone myosin II activity, *Nat. Neurosci.* **8:** 717–719.

Vogelezang, M.G., Scherer, S.S., Fawcett, J.W., and ffrench-Constant, C., 1999, Regulation of fibronectin alternative splicing during peripheral nerve repair, *J. Neurosci. Res.* **56:** 323–333.

Vogelezang, M.G., Liu, Z., Relvas, J.B., Raivich, G., Scherer, S.S., and ffrench-Constant, C., 2001, Alpha4 integrin is expressed during peripheral nerve regeneration and enhances neurite outgrowth, *J. Neurosci.* **21:** 6732–6744.

Wehrle, B., and Chiquet, M., 1990, Tenascin is accumulated along developing peripheral nerves and allows neurite outgrowth in vitro, *Development* **110:** 401–415.

Wehrle-Haller, B., and Chiquet, M., 1993, Dual function of tenascin: Simultaneous promotion of neurite growth and inhibition of glial migration, *J. Cell Sci.* **106:** 597–610.

Werner, A., Willem, M., Jones, L.L., Kreutzberg, G.W., Mayer, U., and Raivich, G., 2000, Impaired axonal regeneration in alpha7 integrin-deficient mice, *J. Neurosci.* **20:** 1822–1830.

Wheeler, B.C., Corey, J.M., Brewer, G.J., and Branch, D.W., 1999, Microcontact printing for precise control of nerve cell growth in culture, *J. Biomech. Eng.* **121:** 73–78.

Wu, D.Y., Wang, L.C., Mason, C.A., and Goldberg, D.J., 1996, Association of beta 1 integrin with phosphotyrosine in growth cone filopodia, *J. Neurosci.* **16:** 1470–1478.

Wu, X., Suetsugu, S., Cooper, L.A., Takenawa, T., and Guan, J.L., 2004, Focal adhesion kinase regulation of N-WASP subcellular localization and function, *J. Biol. Chem.* **279:** 9565–9576.

Yin, Y., Kikkawa, Y., Mudd, J.L., Skarnes, W.C., Sanes, J.R., and Miner, J.H., 2003, Expression of laminin chains by central neurons: Analysis with gene and protein trapping techniques, *Genesis* **36:** 114–127.

2 Regulation of Growth Cone Initiation and Actin Dynamics by ADF/Cofilin

KEVIN FLYNN, CHI PAK, AND JAMES R. BAMBURG

2.1. Summary

During the process of neural development, growth cones are formed from reorganization of a lamella surrounding the soma and are consigned to find their apposite synaptic targets by decoding an assortment of guidance cues. Be they diffusible or substrate-bound, attractive or repulsive, small molecule or peptide, these guidance cues share the ability to regulate the growth cone (Huber et al., 2003). To successfully reach its target, a growth cone must appropriately integrate signals from the multitude of cues that impinge upon it, translating external signals into internal responses. Important targets of guidance signaling are the cytoskeleton systems, which underlie both growth cone structure and behavior (Lin et al., 1994). In particular, the actin cytoskeleton that constitutes a major portion of the growth cone infrastructure plays a central and fundamental role. As a result, actin-binding proteins, which largely determine the constitution and dynamics of actin-dependent systems, are important targets of growth cone regulation (Letourneau, 1996). The actin depolymerizing factor (ADF)/cofilin family of proteins are among these targets (Fass et al., 2004; Meberg, 2000; Meyer and Feldman, 2002; Sarmiere and Bamburg, 2004). This chapter reviews over two decades of studies that have defined a niche for the ADF/cofilin (AC) family proteins in regulating actin dynamics that are important to growth cone formation, borrowing heavily from work performed in nonneuronal systems. Because more mechanistic studies have been performed on the role of AC proteins in growth cone structure and motility than on their role in neuritogenesis, these studies will be discussed to provide a framework for understanding how modulating AC activity could lead to the establishment of a growth cone and neurite outgrowth.

2.2. Growth Cones: Structure and Function

Growth cones are labile, pleiomorphic protrusions at the distal tips of axon and dendrites, which on a flat substratum often have a "hand-like" morphology (Goldberg and Burmeister, 1989). The growth cone is classically subdivided into two distinct domains based on the segregation of filament networks and cytoplasmic organelles (Bridgman and Dailey, 1989; Forscher and Smith, 1988; Lin et al., 1994; Smith, 1988). The most proximal portion or "palm" of the growth cone is referred to as the central domain or C-domain (Figure 2.1). The C-domain is contiguous with the neurite proper ("wrist") and therefore contains a sheath of densely bundled microtubules, most of which terminate within this domain (Dailey and Bridgman, 1991; Forscher and Smith, 1988; Letourneau, 1983; Tanaka and Kirschner, 1991). The C-domain also contains and corrals cytoplasmic organelles (Dailey and Bridgman, 1991; Tsui et al., 1983). Surrounding the C-domain is the peripheral domain or P-domain. The P-domain is highly structured, containing both actin-based filopodia and lamella (Forscher and Smith, 1988). Protrusive, spike-like bundles of actin, sometimes referred to as ribs, radiate outward from the distal circumference of the C-domain into the P-domain where they often extend like "fingers" radiating away from the growth cone as filopodia (Figure 2.1) (Bray and Chapman, 1985; Forscher and Smith, 1988). Broad, sheet-like lamella enjoin the regions between filopodia as a thin "veil" much like the webbing on duck feet. The distinct structures of filopodia and lamella are attributed to the cytoskeleton and the extracellular matrix to which the growth cone is adhered. Filopodia are composed of parallel, filamentous actin (F-actin) bundles that have most if not all of the filament barbed (plus) ends at their distal tips (Svitkina et al., 2003; Vignjevic et al., 2003). The lamella, in contrast, contains an entangled meshwork of F-actin with the majority of filaments angled to grow outward (Lewis and Bridgman, 1992; Svitkina et al., 1997). Microtubules are mostly excluded from the P-domain, although occasionally and importantly microtubules may penetrate into this region guided by F-actin cables (Figure 2.1) (Letourneau, 1983; Zhou et al., 2002). Microtubule penetration into the P-domain promotes leading edge extension of the growth cone (Buck and Zheng, 2002; Lee et al., 2004; Zhou and Cohan, 2003). A third domain has also been described that exists at the interface between the C-domain and the P-domain, referred to as the transition zone (T-zone) (Schaefer et al., 2002). F-actin, assembled into ribs and meshwork of the lamellipodium, undergoes retrograde transport into the T-zone and undergoes compression, creating a population of transversely oriented actin filaments called actin arcs (Lin and Forscher, 1995; Lin et al., 1996; Schaefer et al., 2002). Actin disassembly also occurs within this region. The retrograde translocation of actin arcs into the C-domain contributes to the packing of C-domain microtubules and inhibits microtubule penetration.

Growth cones are extremely dynamic. For instance, growth cone lamella and filopodia extend and retract at a rate up to ten times faster than axon extension. These dynamic changes require a highly dynamic actin cytoskeleton system, which is maintained by constant recycling (Okabe and Hirokawa, 1991). Turning over the actin network in growth cones requires both new actin assembly and disassembly, which are important for maintaining the extant growth cone structure. Inhibiting new actin assembly with cytochalasin B resulted in collapse of the growth cone due to the loss of lamella and filopodia (Forscher and Smith, 1988; Marsh and Letourneau, 1984). Likewise, inhibiting actin disassembly with

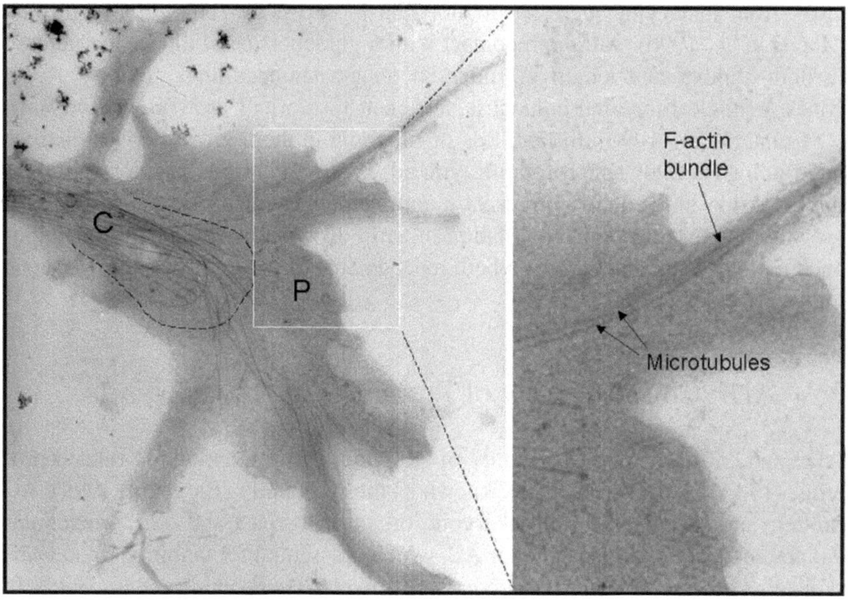

FIGURE 2.1. Transmission electron micrograph of a thin section from a glutaraldehyde fixed hippocampal growth cone. Panel on left shows the microtubules mostly terminating in or near the C-domain (dashed outline) with some microtubules penetrating the P-domain where they often align with actin filament bundles that extend into filopodia (right panel from boxed area). Special treatments to preserve F-actin were not utilized in these images in which microtubules were well preserved. (Image courtesy of Judy Boyle.)

jasplakinolide also resulted in collapse of the growth cone due to a myosin-dependent retraction of the stabilized actin network (Gallo et al., 2002). In addition to its role in the maintenance of growth cone structure, actin disassembly is also important for neurite growth and growth cone guidance. Although neurite elongation per se does not require the actin filament system as it can be driven entirely by a microtubule-based process (Marsh and Letourneau, 1984), the rate and direction of neurite growth are regulated by the focal disassembly of an organized actin cytoskeleton (Bentley and Toroian-Raymond, 1986; Zhou et al., 2002). For example, focal disassembly of actin bundles can induce growth cone turning by reorienting microtubule growth away from the affected region. In fact, the coincident global loss of actin bundles and leading edge actin actually leads to growth cone collapse. Therefore, the global or localized disassembly of actin bundles may represent a general mechanism for growth cone collapse or repulsive growth cone turning, respectively. However, actin disassembly may additionally contribute to growth cone guidance by promoting filopodia dynamics, which may require recycling of assembled actin.

Filopodia likely operate as environmental sensors that when stabilized can direct directional growth. For example, more filopodia are observed on the growth cone edge that faces toward the direction of a glutamate gradient than away from it and the increase in filopodia is necessary for attractive turning (Zheng et al., 1996). Although contact with a guidepost cell induces an increase in both number and length of filopodia on grasshopper limb pioneer growth cones, a single filopodial contact is sufficient to reorient growth cone advance (O'Conner et al., 1990). In fact, loss of filopodia in these growth cones disrupts pathfinding (Bentley and Toroian-Raymond, 1986). Therefore, promoting filopodia extension may equally represent a complementary mechanism for attractive growth cone turning. Thus, regulation of actin disassembly by the various guidance molecules may act through both mechanisms to ensure proper growth cone guidance and neurite growth.

2.3. ADF/Cofilin Family of Proteins

ADF and cofilin belong to a family of structurally and functionally related proteins ~19 kD in size, collectively known as the AC family (Bamburg, 1999). AC proteins are expressed in all eukaryotic organisms. Yeast and other unicellular eukaryotes carry only one gene for AC, which is essential for viability. Metazoans typically have multiple genes encoding AC proteins (Lappalainen et al., 1998). In mammals, there are three separate genes encoding AC proteins: ADF, the ubiquitously expressed nonmuscle cofilin (cof1), and the muscle-specific cofilin (cof2b). Cof2 mRNA can undergo alternative splicing to yield a minor isoform (cof2a), which is expressed in some nonmuscle cells (Thirion et al., 2001).

ADF was the first member of this family identified. Isolated from embryonic chick brain, ADF was characterized as an actin-binding protein that depolymerizes actin filaments (Bamburg et al., 1980). Cofilin was later identified and named from its ability to form cofilamentous structures with actin (Nishida et al., 1984). Shortly thereafter, ADF and cofilin were discovered to share sequence homology with other related proteins across phylogeny and have been referred to as the AC protein family. The AC proteins bind to actin in a 1:1 stoichiometry (Giuliano et al., 1988) but exhibit a higher affinity for ADP-actin than for ATP- or ADP-Pi-actin subunits either in F-actin or as monomers (Carlier et al., 1997; Maciver and Weeds, 1994; Rosenblatt et al., 1997). The AC proteins enhance actin dynamics by two mechanisms; they facilitate monomer loss from the actin filament pointed end and induce severing of the filaments into shorter oligomers (Carlier et al., 1997; Maciver et al., 1998). Although ADF and cofilin are very similar from a functional perspective and each can rescue basic actin-dependent events when expression of the other has been silenced (Hotulainen et al., 2005), they exhibit some nuanced differences that may prove physiologically relevant (Chen et al., 2004; Ono and Benian, 1998; Vartiainen et al., 2002; Yamashiro and Ono, 2005; Yeoh et al., 2002). For example, *in vitro*, ADF enhances pointed end monomer loss more efficiently and maintains a larger unassembled actin pool, whereas cofilin severs F-actin more efficiently and thus may provide nucleation sites for additional filament growth. Thus, activation of AC within a cell may lead to depolymerization of actin filaments in one region and a net assembly of F-actin in another region, depending on the localized pools of actin monomer, actin sequestering proteins, and barbed-end capping proteins available in each locale. As a result, the AC proteins should be considered actin dynamizing proteins (Carlier et al., 1997) that can disassemble the existing actin infrastructure to allow for the resurrection of a newer one.

2.4. Mechanisms of AC Regulation

There are multiple mechanisms for regulating the AC proteins, which include phosphorylation/dephosphorylation (phosphoregulation), binding to phosphatidyl inositol phosphates, sequestering and delivery by 14-3-3 scaffolding proteins, and pH. In addition, the effects of AC on actin dynamics can either be inhibited or synergized by the activities of other actin-binding proteins. Homologues of yeast actin interacting protein-1 (Aip1), also called WD repeat-1 (WDR1) in chicken and mammals and unc78 in *Caenorhabditis elegans*, enhance the severing activity of cofilin (Ono et al., 2004), whereas the Srv2/CAP protein enhances nucleotide exchange on AC-bound ADP-actin (Balcer et al., 2003; Moriyama and Yahara, 2002). However, considering that phosphoregulation has been the dominant point of focus within growth cones, this chapter devotes itself exclusively to this mechanism. By doing so, we do not imply that the other regulatory mechanisms are irrelevant or uninteresting. On the contrary, we hope to encourage future research in these areas, which have been largely overlooked.

The phosphoregulation of AC proteins was first identified by recognizing that dephosphorylated ADF retained its ability to depolymerize F-actin, whereas the phosphorylated species did not (Morgan et al., 1993). The single regulatory site of phosphorylation, serine-3 on the immature mammalian protein, was later identified for both ADF (Agnew et al., 1995) and cofilin (Moriyama et al., 1996). Phosphorylation at this penultimate residue of the demethionated and *N*-acetylated protein (Agnew et al., 1995; Kanamuri et al., 1998) compromises an important AC-actin interaction without significantly altering AC structure resulting in the severely reduced ability of the AC proteins to bind to F-actin (Blanchoin et al., 2000). Given the importance of AC phosphoregulation, mutants that harbor a disruption of the phosphoregulatory site were designed, which include a serine-3 to alanine (S3A) mutant and a serine-3 to glutamate or aspartate (S3E or S3D) mutant. The unphosphorylatable S3A mutant is considered to be "constitutively active" as it precludes inactivation by phosphorylation, although regulation by other mechanisms (pH, PIP_2-binding, etc) can still occur. The phosphomimetic S3E or S3D mutant is nonactivatable and may exhibit inhibitory effects possibly through its sequestering of regulatory proteins that have a high affinity for phosphorylated AC species, such as 14-3-3 proteins (Gohla and Bokoch, 2002) or the AC phosphatases. From a naïve perspective, the S3A and S3E mutants are often regarded diametrically and are utilized and interpreted in such a context. However, in particular situations, the S3A and S3E mutants may actually induce comparable effects, or the S3E mutant may have no effect at all. Therefore, the most reliable means of probing the effects of prolonged AC inactivation is accomplished by the experimental downregulation or genetic loss of function of endogenous AC proteins. The S3A, S3E, and S3D mutants will be collectively referred to as the phospho-mutants.

Phosphorylation (inactivation) of AC proteins is subserved by two AC specific protein kinase families: the LIM kinase (LIMK) and the TES kinase (TESK). However, of these kinases, only the two isoforms of the LIM kinases, LIM kinase-1 (LIMK1) and LIM kinase-2 (LIMK2), have been demonstrated to be appreciably expressed in neurons (Foletta et al., 2004). In fact, the highest level of LIMK mRNAs is found in the brain (Mizuno et al., 1994). The LIMK contain two repeated LIM domains and a PDZ domain. The LIM (lin-11, isl-1, and mec-3) domain is a stretch of 50–60 amino acids rich in cysteines and histidines, which organize into double zinc fingers (Dawid et al., 1998). The PDZ (PSD-95, Discs-large, and ZO-1) domain contains two important submotifs, a guanylate kinase-like motif and SH3 motif (McGee and Bredt, 1999). Both the LIM and PDZ domains mediate protein–protein interactions. For the LIM kinases, the LIM domain proves especially critical for its self-association into a dimer complex (Hiraoka et al., 1996). Many signaling mechanisms exist to regulate the LIM kinases. In growth cones of cultured primary hippocampal neurons, LIMK1 is regulated in part by ubiquitination through the ubiquitin ligase, Rnf6 (Tursun et al., 2005). In addition, the LIMK are activated by two upstream kinase families, the Rho kinases (ROCK) and the p21-activated kinases (PAK) (Edwards et al., 1999; Maekawa et al., 1999). The ROCK and PAK family of kinases are stimulated by the Rho family of small GTPases (Leung et al., 1995; Manser et al., 1994). These participants comprise a complete signaling mechanism that eventually leads to AC protein inactivation by phosphorylation.

Dephosphorylation (activation) of the AC proteins is subserved by two AC specific protein phosphatases: the slingshot phosphatases (SSH) (Niwa et al., 2002) and the chronophin phosphatases (Gohla et al., 2005). Although AC specific protein phosphatases have been identified, nonspecific protein phosphatases are sometimes capable of dephosphorylating the AC proteins (Samstag and Nebl, 2003). Considerably less is known regarding dephosphorylation signaling mechanisms. For example, virtually nothing is known regarding the signaling mechanisms that regulate chronophin. Even for SSH, regulatory mechanisms have been elucidated that affect only the SSH-1L isoform. Although a comprehensive map of regulatory mechanisms that activate or inactivate the entire family of SSH phosphatases is lacking, it still appears that regulation of SSH-1L operates through divergent but parallel pathways. SSH-1L is activated in a Ca^{2+}-dependent manner by calcineurin, also known as protein phosphatase-2B (PP2B), and inhibited by 14-3-3 (Nagata-Ohashi et al., 2004; Wang et al., 2005). Both of these mechanisms possibly affect the ability of SSH-1L to bind to F-actin, which is required for its activity (Nagata-Ohashi et al., 2004; Soosairajah et al., 2005). SSH-1L is negatively regulated *in vitro* through phosphorylation by PAK4 (Soosairajah et al., 2005). Active SSH-1L also dephosphorylates (inactivates)

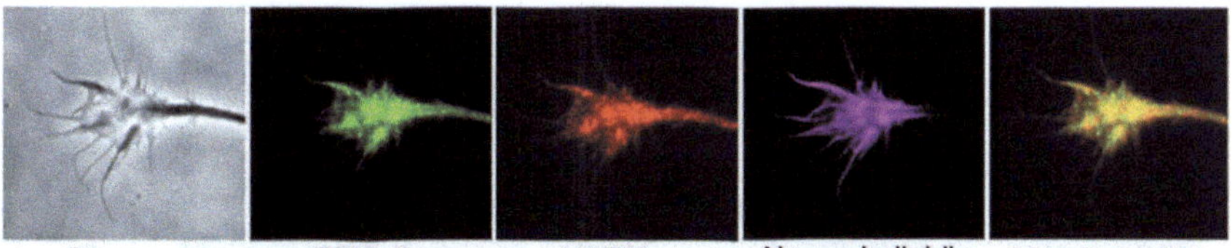

FIGURE 2.2. Localization of slingshot (SSH-1L), LIMK1, and F-actin in hippocampal growth cone. Phase image of the growth cone shows extensive filopodia, which do stain with Alexa phalloidin but are not very visible at the intensity used to print that image. SSH-1L (fluorescein) and LIMK1 (Texas Red) show enrichment in similar regions as evidenced by the overlay, but there are some regions where one or the other stains more intensely. (Modified from Soosairajah et al., 2005.)

LIMK1 *in vitro*. Therefore, in some cells, PAK4 may function as a master switch downstream of the Rho family of small GTPases, leading to phosphorylation (inactivation) of AC proteins by the concomitant activation of the LIMK1 and inactivation of SSH-1L phosphatase. Both SSH-1L and LIMK1 are enriched in similar regions of growth cones (Figure 2.2).

2.5. Putative Role of AC Proteins in Neuritogenesis

Neuritogenesis for most neurons occurs shortly after the final mitotic division and results in the initiation of nascent neurite processes, which later differentiate into mature axons and dendrites. Although the process of neuritogenesis has been well characterized for only selected neuronal populations within the brain, it has been studied in cultured hippocampal neurons and in cell culture models using neuroblastoma cells. When these cells are plated onto a growth permissive substrate, they extend a lamella surrounding most or all of the soma (Figure 2.3) (see also Dehmelt et al., 2003). Neurites form from the localized collapse of the lamella around bundled microtubules and the remaining noncollapsing lamella forms the growth cones at the end of these neurites. The formation of growth cones is accompanied by microtubule growth into these regions, which begin their advance away from the soma. This process requires the coordinated regulation of actin-based and microtubule-based dynamics. *In vivo* there may be multiple mechanisms for neuritogenesis. For example, in the cerebellum the granule cell soma migrates along glial extensions while extending behind it an initial neurite process, which eventually becomes the axon (Sanes et al., 2000). Nevertheless, neuritogenesis inevitably requires regulation of the peripheral actin cytoskeleton system, which implicates the AC proteins (DaSilva and Dotti, 2002). Studies from the *Drosophila* twinstar mutants (Ng and Luo, 2004) and the cofilin null mouse embryo (Gurniak et al., 2005) suggest that AC proteins are required for normal neuritogenesis. However, due to the lack of published information

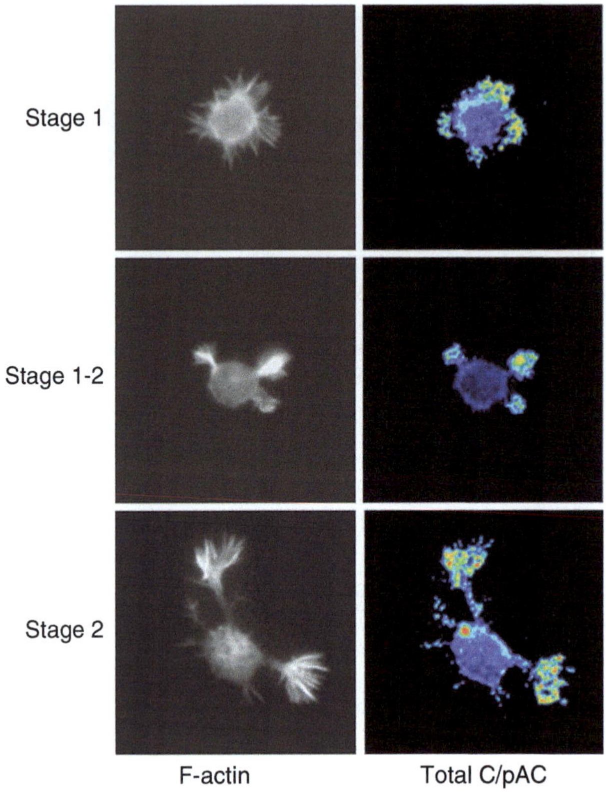

FIGURE 2.3. Localization of F-actin and active AC in hippocampal neurons undergoing neuritogenesis. Stage 1 neuron has extended a surrounding lamellipodium in which active AC (brighter colors) is enriched. As growth cones form from the remaining lamellipodium after collapse around microtubule bundles, the growth cones remain enriched with active AC as neurites elongate during stage 2. The AC activity is obtained by ratio imaging of cells immunostained for total cofilin with a mouse monoclonal antibody and phospho cofilin with a rabbit antibody.

showing a direct AC involvement within neuronal cell types, we will draw heavily from studies done on nonneuronal cells for studying cell polarization.

A general role for the AC proteins in the establishment and maintenance of cell polarization has been characterized within nonneuronal cells, which also extend cortical lamella similarly to plated neurons. This was demonstrated by overexpressing LIM kinase in fibroblasts, which resulted in the inactivation of AC at the leading edge and the loss of polarity (Dawe et al., 2003). Furthermore, LIMK1 and slingshot are required for directional migration in fibroblasts (Nishita et al., 2005). Polarity was restored by coexpressing the AC S3A mutant. Studies using nonneuronal cells also suggest that for neuronal polarization the AC proteins are involved not only in the initial extension of somatic lamella but also the localized collapse of lamellar regions into growth cone-like structures. AC activation is required for lamellipod formation triggered by epidermal growth factor (EGF) treatment of metastatic MTLn3 cells (Chan et al., 2000). The ability of AC to promote lamellipod formation required its severing activity to increase the number of free barbed ends at the leading edge. In addition, the localized collapse of lamella necessary to form the nascent growth cone-like structures probably requires the regionalized inhibition of AC. In monocytes, the RhoA-ROCK signaling pathway inactivates AC to limit the protrusions of membrane to the leading edge (Worthylake and Burridge, 2003). Thus, the engenderment and maintenance of neuronal cell polarization likely requires the modulation of AC at nearly every stage.

In general, it appears that globalized RhoA activation inhibits neuritogenesis, whereas globalized Rac1 and cdc42 activation stimulate neuritogenesis. In cultured hippocampal neurons, the exogenous expression of RhoA reduced the number of neurites and the inhibition of RhoA by C3 exoenzyme had the opposite effect (DaSilva et al., 2003). Nerve growth factor (NGF) treatment of PC12 cells, which resulted in the formation of neurite-like processes, involved a decrease in RhoA activity (Yamaguchi et al., 2001). Rac1 and cdc42 exhibited increased activity and become localized to the cell periphery when activated by NGF treatment (Aoki et al., 2004). Although Rac1 and cdc42 activation is necessary, they actually cycle between an active and inactive state at the periphery. It is possible that this cycling plays an important part in regulating the dynamics of AC for proper activity.

NGF treatment of PC12 cells induced the translocation of AC to ruffling membranes (Meberg et al., 1998). Furthermore, in freshly cultured hippocampal neurons that still exhibit a symmetrical lamella, activated AC colocalizes with F-actin to the cortical edge, whereas the cell body is largely devoid of activated AC (Figure 2.3). In neurons that exhibit localized dissolution of the lamella, AC is again colocalized to nascent growth cone-like structures with F-actin. In elongating neurites, activated AC continues to be localized to the growth cones, which migrate away from the cell body. A study published only in abstract form has shown a direct role of AC proteins and their upstream regulators in neuritogenesis in PC12 cells (Endo et al., 2005). Silencing either cofilin or ADF expression in PC12 cells had only slight effects on the numbers of neurite bearing cells following NGF treatment, but silencing expression of both AC proteins abolished neurite outgrowth. Similarly, silencing of SSH-1L and SSH-2L alone only modestly decreased neurite formation whereas silencing both together inhibited neuritogenesis. Silencing LIMK1 and LIMK2 together decreased neuritogenesis by 50%. These results suggest that AC phosphocycling is essential for neurite initiation. However, no mechanistic studies are available to show what activities of AC proteins are required for neurite formation. On the other hand, there have been detailed experimental studies on the role of AC proteins and their upstream regulators in growth cone behavior, including filopodial elongation, that may help shed light upon their role in neuritogenesis. These are discussed below.

2.6. AC Proteins: Effects on Growth Cone Structure and Physiology

ADF was first localized to growth cones in chick dorsal root ganglion (DRG) neurons (Bamburg and Bray, 1987) but AC proteins have since been identified in growth cones of many and diverse neuronal types. Endogenous AC is enriched in the peripheral domain in filopodia and lamella, and its association to these regions likely reflects its presumed role in actin disassembly (Figure 2.4). A less appreciated, although by no means ignored, aspect of actin dynamics in the peripheral growth cone is its indirect consequence on microtubule assembly. The actin filament meshwork can act as a physical barrier for advancing microtubules (Rodriguez et al., 2003; Forscher and Smith, 1988). AC proteins are often localized along microtubules invading the growth cone periphery (Figure 2.4) and the overexpression of AC proteins in cortical neurons increases the incidence of microtubules in growth cones (Meberg and Bamburg, 2000) presumably by dissolution of the peripheral F-actin barrier. Differential staining for active and inactive AC demonstrates that active (dephosphorylated) AC is localized to the P-domain including the periphery of lamella and proximal filopodia, whereas

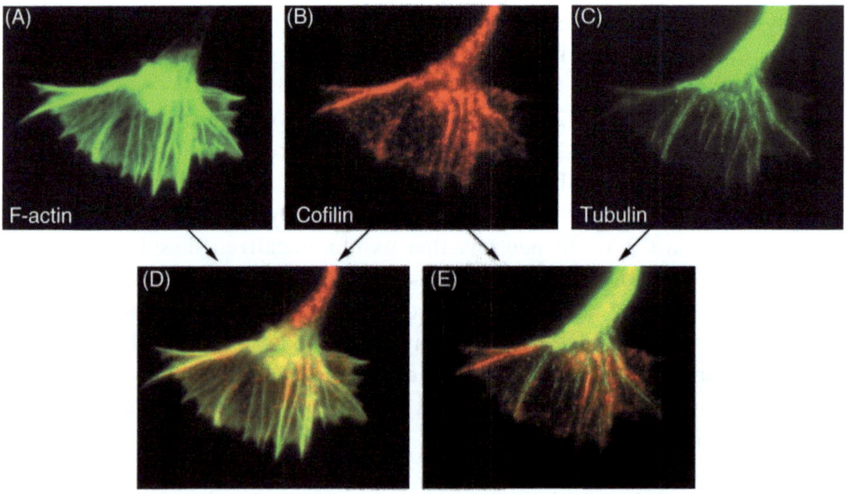

FIGURE 2.4. Localization of F-actin, AC, and microtubules in rat hippocampal neuronal growth cones by fluorescence microscopy. (A) Fluorescein-conjugated phalloidin staining reveals F-actin is enriched in the peripheral region of the growth cone (P-domain). (B) Indirect immunofluorescence of both ADF and cofilin reveals their distribution throughout the growth cone but accumulation along filamentous structures. (C) Microtubules, detected with a β-tubulin antibody, are mainly localized in the neurite shaft and the central domain of the growth cone, although some penetrate into the P-domain. (D) Overlay of panels A and B showing AC proteins localize with several, but not all, F-actin filament bundles. (E) Overlay of panels B and D showing AC is notably localized in areas adjacent to microtubules that penetrate into the P-domain.

inactive (phosphorylated) pAC is diffused throughout the growth cone (Figure 2.5). The pAC pool likely represents recently inactivated AC that is dissociated from peripheral F-actin and can be rapidly mobilized to active regions of actin disassembly. Although a role for AC in regulating growth cones was suggested very early on, its involvement was only recently demonstrated.

The AC proteins regulate various aspects of growth cone structure and physiology; however, the generalized effects of AC activity upon stimulation by guidance cues have been ambivalent. Therefore, the effects of AC activity within growth cones cannot be categorically defined and are probably regulated in subtly different ways. This has proven especially true for its effects on growth cone structure. For instance, global AC activation within mouse DRG growth cones induced by semaphorin-3A (Sema3A) resulted in the collapse of the growth cone and dissolution of peripheral F-actin (Aizawa et al., 2001) (see later), whereas local application causes a turning response (Fan and Raper, 1995). However, in another instance, AC activation within chick retinal ganglion cell (RGC) growth cones, this time induced by global application of brain derived neurotrophic factor (BDNF), resulted in filopodia extension and the stimulation of actin polymerization at filopodia tips (Gehler et al., 2004b) (see later). As a result, it appears that AC activation can lead to both actin polymerization and depolymerization depending on the guidance cue. However, the differences in the responses may also depend on the engagement of other non-AC–dependent signaling mechanisms, not on AC activation per se. We will first review the direct effects

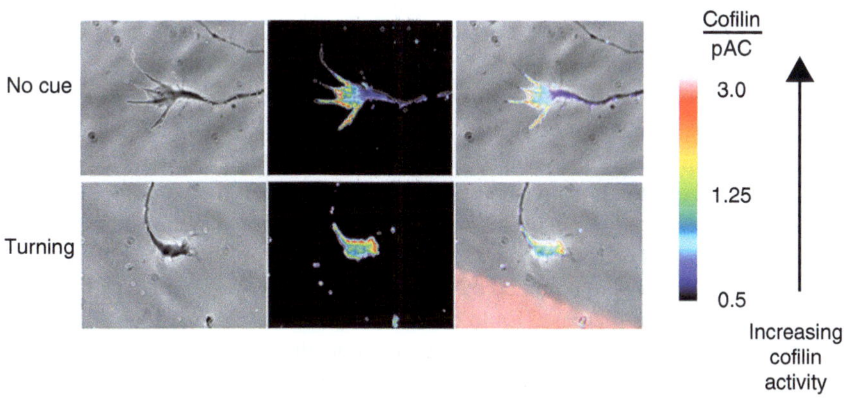

FIGURE 2.5. AC activity is localized to the peripheral leading edge of advancing and turning growth cones. Ratio imaging of relative AC activity (determined by ratio imaging as described in legend of Figure 3) in a hippocampal neuronal growth cone that does not experience contact with a guidance cue (top series) and a growth cone that has contacted and is turning away from an aggrecan stripe (red fluorescence barrier in lower right panel). Left panels are phase contrast images and center panels are the ratio images with overlays on the right. Note the highest level of AC activity is along the periphery of the leading edge in the direction of migration.

of AC activity on growth cone structure and physiology that has utilized the AC mutants.

The effects of AC activity on growth cone structure studied by expressing or direct loading of the various AC mutants have been largely consistent in suggesting that AC activity regulates growth cone structure, in part, by promoting filopodia dynamics. This has been observed by overexpressing human cofilin (h-cofilin) S3A in chick DRG growth cones and direct loading of *Xenopus* ADF/cofilin (XAC) S3A protein in chick RGC growth cones, which resulted in increases in filopodia length. Although not quantified in chick DRG growth cones, filopodia length and number increased by 30% and 15%, respectively, in chick RGC growth cones loaded with XAC S3A (Gehler et al., 2004b). In addition, in chick DRG growth cones expressing h-cofilin S3A, filopodia branched more frequently and appeared more tortuous (Endo et al., 2003). Because filopodial elongation requires new actin assembly, one effect of AC activity is to stimulate regionalized actin polymerization. However, growth cones under both conditions also appeared spindly and were unable to maintain a normal "hand" like morphology (Endo et al., 2003; Gehler et al., 2004b). This morphological disturbance was accompanied by an indefinite loss of the C-domain and reductions in lamellar protrusions. Thus, a concurrent effect of AC activity is to depolymerize existing actin networks. Taken together, the combined effect of actin assembly and disassembly actually suggests that the role of AC in growth cones is to stimulate rapid actin turnover. This interpretation may be consistent with the structural phenotype observed. The amplified disassembly of retrogradely transported actin filaments within the T-zone and proximal boundary of the P-domain may disrupt the maintenance of structured actomyosin filament networks, leading to reductions of these regions. Growth cones of myosin IIB$^{-/-}$ superior cervical ganglion neurons exhibit a similar phenotype with reduced lamellar protrusions and shrinkage of the C-domain (Brown and Bridgman, 2003). Retrograde flow of F-actin is also enhanced in myosin IIB$^{-/-}$ growth cones, which may additionally contribute to the packing of microtubules in the C-domain. Faster disassembly may stoke polymerization by increasing the pool of polymerization-competent actin (monomer or oligomer). This is supported by a 38% increase in protrusion rates of chick DRG growth cones overexpressing h-cofilin S3A (Endo et al., 2003). This effect was maximized by sustained AC activity as wild-type h-cofilin resulted in a smaller increase (Endo et al., 2003).

Although these observations offer insights into the molecular mechanisms of AC activity within growth cones, they do not address its significance to growth cone migration and neurite outgrowth. Short-range growth cone migration is spurred by AC activity. Chick DRG growth cones expressing h-cofilin S3A migrated 38% faster than control growth cones (Endo et al., 2003), which was the maximum observed. This effect is most likely explained by an increase in actin-based protrusions alone. The ability of the S3A mutant to optimally promote migration is not observed for longer term neurite growth, which probably relies more heavily on the dynamics of the microtubule cytoskeleton. This result was demonstrated in both cultured rat cortical neurons and *Drosophila* mushroom body (MB) neurons homozygous null for twinstar (tsr^{null} MB neurons), the *Drosophila* AC protein orthologue. Overexpression of the XAC S3A mutant in rat cortical neurons resulted in an increase in median neurite length by 17%; however, this increase was significantly less than the 60% increase observed for neurons expressing wild-type XAC (Meberg and Bamburg, 2000). Likewise, overexpression of the twinstar S3A mutant in tsr^{null} MB neurons rescued normal axon growth to a lesser degree than wild-type twinstar (Ng and Luo, 2004). The inability of the S3A mutant to rescue outgrowth as effectively as the wild type could potentially be explained by the need for a coupled phosphocycling of AC activity to maximally enhance the F-actin turnover and allow for microtubule penetration into the P-domain to promote microtubule-dependent neurite growth. Alternatively, optimal F-actin turnover may follow a bell-shaped dose–response curve for AC activity, and the S3A mutant lowers the response because of its higher activity. Certainly some degree of AC activity is clearly necessary for neurite growth. The downregulation of endogenous AC with antisense oligonucleotides resulted in the potent reduction of total neurite length by at least 70% in rat cortical neurons (Kuhn et al., 2000). Similarly, tsr^{null} MB neurons exhibited severe axon growth defects with only 10% of neurons within the normal axon growth range (Ng and Luo, 2004). Thus, suppressing actin turnover also appears to inhibit microtubule-dependent neurite growth. Collectively, these observations suggest that the AC proteins are indeed necessary for neurite growth but require phosphoregulation for optimal growth. Optimal growth might simply require an intermediate level of AC activity or a finer control of AC activity spatially and temporally.

Investigations into the role of AC proteins in growth cones have focused primarily on their effects during processive motility and advancement. However, compelling evidence exists to suggest that the AC proteins may additionally play a role in growth cone turning. Considering the presumed role for the AC proteins in growth cone migration and neurite growth, the asymmetric and differential activation of the AC proteins might well lead to a directional bias in advancement. This hypothesis is indirectly supported by the observation that the asymmetric activation of AC proteins at the leading edge in nonneuronal cells can lead to directional migration (Ghosh et al., 2005). Some preliminary evidence in growth cones maintains this view as well. Aggrecan, which is a chemorepulsive signal, produced the asymmetric activation of AC at the growth cone peripheral edge, which faces away from the aggrecan source (Fass et al., 2004) (Figure 2.5). Asymmetric activation of AC was also associated only with turning growth cones; in the absence of guidance cues AC activity is present uniformly in the leading edge of the P-domain. Thus, it may be that the asymmetric and localized activation of AC in response to guidance cues represents a generalized mechanism for remodeling and redirecting the growth cone. Unfortunately, only snapshots of localized AC activity can be obtained through fixation and staining methods and these do not tell the full story about the rapidity of AC phosphocycling in different growth cone domains.

The AC proteins could potentially exert their effects through two distinct but overlapping mechanisms, by increasing monomer disassembly or by severing filaments, or both. To discern the relative contributions of each of these activities different activity mutants were designed: a tyrosine-82 to phenylalanine (Y82F) mutation and a serine-94 to aspartate (S94D) mutation. The Y82F mutant retains the ability to sever actin filaments but only minimally disassembles filaments into monomers (Moriyama and Yahara, 1999). Conversely, the S94D mutant retains much of its ability to disassemble filaments into monomers but only weakly severs (Moriyama and Yahara, 2002). The Y82F and S94D mutants have been used in an S3A background to determine their ability to reverse the growth inhibition of LIMK1 overexpression in chick DRG growth cones. Results of these studies demonstrated that filament severing plays the dominant if not exclusive role in exerting AC's effects. Overexpressing the Y82F S3A double mutant reversed LIMK1's inhibitory effects on protrusion rates and growth cone migration to a degree matched only by the S3A mutant alone (Endo et al., 2003). Monomer disassembly does not appear to contribute to this rescue as the S94D S3A double mutant did not rescue either process significantly above that of a S120A S3A double mutant, which carries disruptions of both actin dynamizing activities (Endo et al., 2003). Although it remains unclear where exactly in the growth cone filament severing by AC occurs, severed filaments could account for the enhanced actin turnover observed in more rapidly growing growth cones (Meberg and Bamburg, 2000). It should not be forgotten that severed filaments may also provide new nucleation sites for enhanced filopodial or lamellipodial growth.

2.7. BDNF Signaling Pathway to AC for Regulation of Growth Cone Filopodial Length

BDNF is a member of the neurotrophins, which also include NGF, neurotrophin-3 (NT3), and neurotrophin-4/5 (NT4/5). BDNF functions as a chemoattractant guidance molecule for chick RGCs and stimulates filopodial elongation and filopodial numbers (Gehler et al., 2004a). The mechanism of filopodia extension required increased polymerization of F-actin at distal tips; therefore, a role for the AC proteins was again suggested to facilitate actin turnover. The effect on filopodia growth can be completely explained by the dephosphorylation of the AC proteins (Gehler et al., 2004b). The direct loading into growth cones of the XAC S3A mutant resulted in an increase in filopodia length by 30%, which was not enhanced with subsequent BDNF treatment. In addition, BDNF treatment of control growth cones resulted in the same degree of filopodia extension as the XAC S3A mutant alone. Conversely, the XAC S3E mutant completely blocked the stimulatory effect of BDNF on filopodia elongation, demonstrating that the pathway is AC dependent. As further evidence, BDNF treatment resulted in the dephosphorylation (activation) of AC with a time-course that paralleled filopodial elongation (Figure 2.6).

Dephosphorylation of the AC proteins required signaling through the p75 pan-neurotrophin ($p75^{NTR}$) receptor and the inactivation of Rho-activated kinase (ROCK). Inhibition of ROCK alone would give the full BDNF response. Because

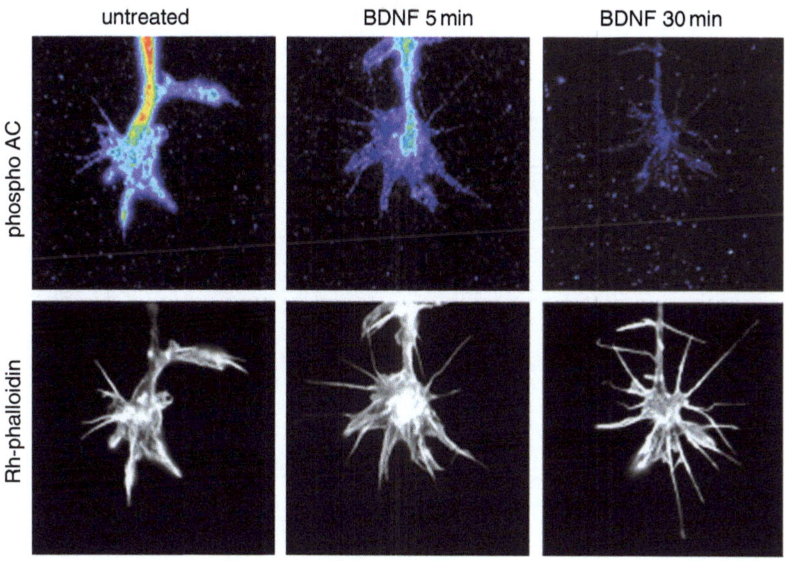

FIGURE 2.6. BDNF stimulation causes an increase in the number and length of filopodia and a concomitant decrease in phospho-AC in retinal ganglion growth cones. Images of chick retinal ganglion growth cones fixed and stained before treatment or 5 min and 30 min after BDNF treatment. The upper panels are images of growth cones immunostained for the inactive (phospho) AC. Brightest colors represent the highest amounts and all images were obtained using identical settings. The lower panels are the same growth cones as in the upper panels but stained for F-actin with rhodamine phalloidin. (Adapted from Gehler et al., 2004b.)

BDNF, NGF, and NT3 resulted in comparable increases in filopodia length, this supported the notion that $P75^{NTR}$ was involved and not the Trk receptors (Gehler et al., 2004a), and this interpretation was further supported by specifically activating the $p75^{NTR}$ with a chick antibody, which mimicked the BDNF effects. The effect of BDNF on filopodial length was not blocked by a TrkB-specific antibody. AC activation downstream of $p75^{NTR}$ required a progressive inhibition of RhoA. This is believed to occur through the physical binding of BDNF to $p75^{NTR}$, which inhibits its basal activation of RhoA. BDNF treatment of RGC growth cones induced a progressive reduction in RhoA activity for over 30 min, and these effects could be mimicked with the *Clostridium botulinum* C3 exoenzyme, which specifically inhibits RhoA (Chardin et al., 1989). Similar to what was observed for Sema3A signaling, BDNF also engages another GTPase besides RhoA. For BDNF, cdc42 activation was also demonstrated, which facilitates the inhibition of RhoA as demonstrated by loading growth cones with an active form of cdc42 (Chen et al., 2006). These effects were shown to be independent of myosin activity (Gehler et al., 2004b), which has an additive effect on filopodial elongation. A summary of some of these signaling pathways that impact AC regulation is shown in Figure 2.7.

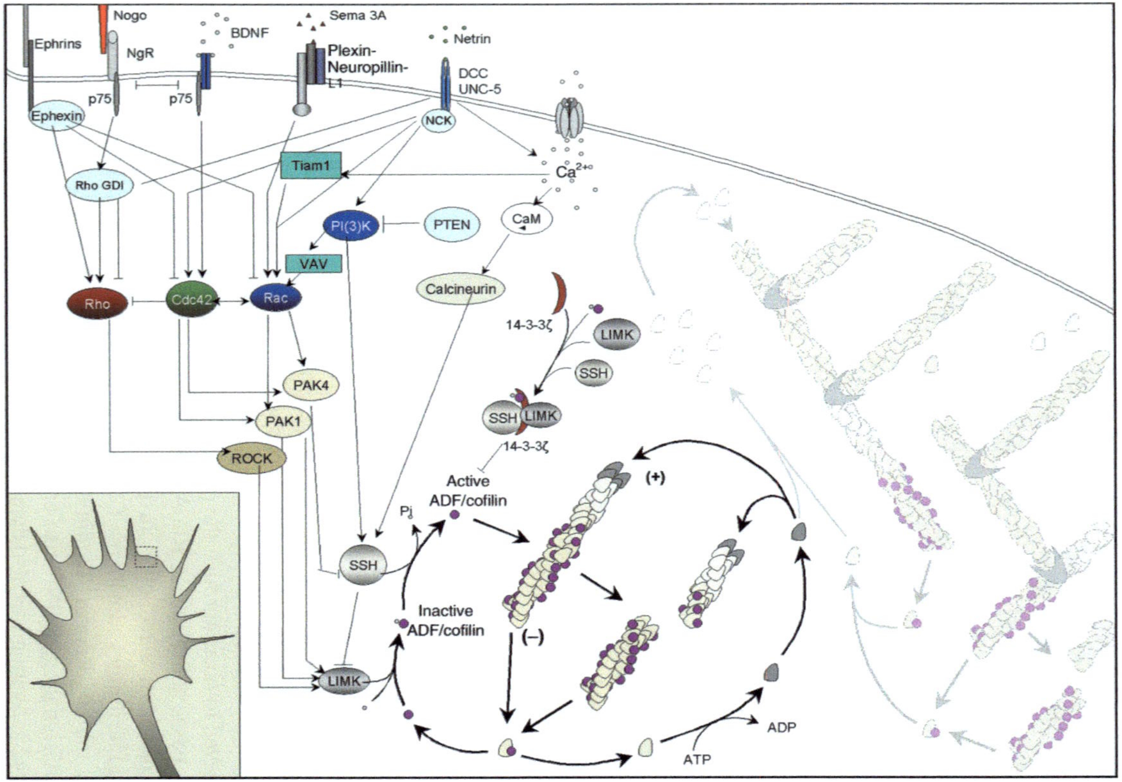

FIGURE 2.7. Signaling pathways affecting AC activity in neuronal growth cones. Schematic diagram of the neuronal growth cone with receptors for several known guidance cues and their downstream signaling pathways regulating AC phosphorylation. The AC phosphocycle is also shown as well as how it is tied in with the turnover of the F-actin network in the leading edge. In nonneuronal cells, the branched actin filament network is formed from the Arp2/3 complex, which in growth cones is mostly confined to the central domain. The major protein for forming actin meshworks in growth cones is probably filamin.

To determine if AC severing activity was necessary for filopodial elongation, a lysine 95,96 to glutamine 95,96 (KK95,96QQ) mutation in AC was made. This mutation was severing deficient similar to the S94D mutation described above (Moriyama and Yahara, 2002; Pope et al., 2000). Loading this protein into growth cones did not increase filopodial elongation but did block the BDNF-induced elongation as efficiently as the S3E mutant, suggesting that AC severing activity is required (Gehler et al., 2004b).

2.8. Role of Microtubules in Regulation of AC Activity

In the nonneuronal chick cardiac fibroblasts (CCFs) that undergo spontaneous polarization (Dawe et al., 2003), microtubules penetrate the leading lamellipodium, but multiple lamellipodia on nonpolarized cells are almost devoid of microtubules. Thus, microtubules appear to play a major role in the establishment of cell polarity, a function ascribed to them many years ago (Vasiliev et al., 1970). Thus, it was of no surprise to observe that nocodazole-treated CCFs (no microtubules) fail to polarize (Cramer et al., submitted). Because these nonpolarized cells had an actin organizational structure similar to nonpolarized CCFs that were overexpressing LIMK1 or AC S3E, experiments were undertaken to see if polarity could be rescued by expressing the active AC S3A, which rescues polarity in CCFs expressing LIMK1 (Dawe et al., 2003). Surprisingly, a persistent polarized migration was observed in nocodazole treated CCFs expressing the AC S3A. Furthermore, in polarized migrating CCFs, zones of active AC appear around dynamic microtubule plus ends in the central regions of the lamellipodium whereas distal to the microtubule plus-ends phospho-AC was plentiful (Cramer et al., submitted). These findings strongly suggest that one role of dynamic microtubules in polarized migration is to deliver molecules that regulate the spatial and temporal activity of AC proteins. Once polarity has been established by the coordinated interaction of microtubules and actin filaments, the presence of active AC can sustain the pool of actin needed to maintain protrusive growth. Although these same observations have not as yet been made in growth cones, it would be surprising if a similar microtubule-dependent regulation of AC activity is not found. Such microtubule-based regulation of local AC activity is likely to be of extreme importance in the establishment of the initial growth cone during neuritogenesis when the MAP2c-dependent bundles of microtubules are established (Dehmelt et al., 2003) and their plus ends penetrate the lamellipodial regions that are destined to become the nascent growth cones.

2.9. Effects of LIMK1 and SSH-1L Overexpression on Growth Cone Structure and Physiology

The use of the AC phospho-mutants underscores the singular importance of the AC phosphorylation site, serine-3. Because this single residue represents an important target of regulation, elucidating the full complement of regulatory partners and pathways that modulate its state of phosphorylation will represent a significant step in determining its general physiological significance to growth cone behavior and the guidance molecules that regulate it. Fortunately, identification of some of the immediate regulatory partners in growth cones has been accomplished. These include LIMK1 and SSH-1L (Aizawa et al., 2001; Endo et al., 2003; Nagata-Ohashi et al., 2004; Rosso et al., 2004). However, the presence of these two proteins does not categorically exclude the other known AC protein kinases and phosphatases, which include alternative isoforms of the LIM kinases, the SSH phosphatases and chronophin.

Similar to the localization observed for endogenous AC and F-actin, endogenous LIMK1 is enriched in both growth cone filopodia and lamella (Endo et al., 2003) (Figure 2.2). In both normal and engorged growth cones of rat hippocampal neurons, phosphorylated (activated) LIMK1 is enriched at the peripheral rim (Rosso et al., 2004). However, the effects of LIMK1 on growth cone structure cannot always be explained by its ability to inactivate the AC proteins. Overexpressing the constitutively active h-LIMK1 GL177,178EA mutant in mouse DRG neurons actually augmented protrusions of thick filopodia and reduced lamellar protrusions (Aizawa et al., 2001). To further confound the matter, inhibition of endogenous LIMK1 within these same neurons with 20 μg/ml of a pseudosubstrate LIMK1 inhibitor peptide (S3-peptide) likewise caused the elongation of filopodia and reductions in lamella (Aizawa et al., 2001). Additionally, overexpressing wild-type LIMK1 in rat hippocampal neurons resulted in larger growth cones and stimulated the accumulation of the polarity complex Par3/Par6 and receptor tyrosine kinases, which are factors known to promote axon specification and growth (Rosso et al., 2004; Shi et al., 2003). Overexpressing both the h-LIMK1 GL177,178EA mutant and wild-type LIMK1 resulted in the increased phosphorylation of AC; therefore, the inactivation of AC cannot explain these phenotypic differences (Aizawa et al., 2001; Rosso et al., 2004). Still, in some cases, the effects of LIMK1 can indeed be explained by the inactivation of AC. In chick DRG growth cones, overexpressing wild-type chick LIMK1 (chLIMK1) resulted in the severe inhibition of filopodia and lamella (Endo et al., 2003). In fact, the growth cone appeared compacted with severe condensations of the central and peripheral domains. The opposite effects were observed by overexpressing the kinase dead D467A mutant, which resulted in longer filopodia and an irregular growth cone shape, although central and peripheral domains were still evident. Moreover, lamellar and filopodial protrusions persisted on the neurite shaft even after the growth cone migrated, possibly suggesting defects in de-adhesion (Endo et al., 2003).

Despite the confounding effects of LIMK1 on growth cone structure, these studies nevertheless suggested that LIMK1 would have equally dramatic effects on growth cone dynamics. Indeed, this was the case. Overexpressing wild-type chLIMK1 in chick DRG growth cones severely inhibited protrusion rates by 73% compared to control growth cones (Endo et al., 2003). Growth cone migration rates were also severely inhibited by 82% by overexpressing wild-type chLIMK1, which was not observed for the D467A mutant. These results are consistent with the inactivation of AC activity. In several transgenic clones of Drosophila MB neurons overexpressing LIMK1, severe defects in axon growth were observed (Ng and Luo, 2004). Only nominal defects in axon growth were observed with the kinase-inactive LIMK1. Thus, the severe inhibition of neurite growth due to LIMK1 results from its kinase activity, presumably through suppressing rapid actin turnover, which would inhibit both actin-based protrusions and microtubule-based growth. However, the conflicting nature of LIMK1's effects argues that LIMK1 may engage divergent mechanisms that do not depend on AC alone. Empirical support for this possibility comes from the observation that the robust overexpression of wild-type LIMK1 in MB neurons can actually result in more severe axon growth defects than occur by knocking out twinstar alone (Ng and Luo, 2004).

In contrast to LIMK1, the reported effects of SSH-1L within growth cones can be fully explained by its capacity to dephosphorylate AC. The effects of overexpressing wild-type SSH-1L are similar to that of overexpressing cofilin S3A. Overexpressing wild-type SSH-1L in chick DRG growth cones resulted in a spindly appearance with ambiguously defined C- and P-domains (Endo et al., 2003). Likewise, filopodia were longer and exhibited more frequent branching. An increase in the protrusion rate and growth cone migration rate was observed with wild-type SSH-1L that was comparable to the cofilin S3A mutant. These effects required phosphatase activity, as expression of the phosphatase-inactive SSH-1L (CS) mutant did not result in any of these effects. Therefore, the effects of SSH-1L within growth cones can be attributed predominantly to its ability to dephosphorylate AC, as these effects are completely consistent with the activation of AC. Furthermore, because SSH-1L dephosphorylates LIMK1 on its activation loop and should thus inactivate it, its expression should reverse the effects of LIMK1 overexpression on growth cone morphology and dynamics. As mentioned before, wild-type LIMK1 induced chick DRG growth cones to become undersized and compacted (Endo et al., 2003). Co-expression of SSH-1L rescued filopodia protrusions, although the normal "hand"-like morphology was not restored. The lack of complete structural rescue, however, may merely reflect differences in expression levels. Wild-type SSH-1L expression also reversed the inhibition of protrusion rates and growth cone migration rates by LIMK1 (Endo et al., 2003). SSH-1L antagonism of LIMK1 was also observed for neurite growth in *Drosophila* MB neurons. The simultaneous expression of wild-type SSH-1L and LIMK1 in transgenic MB clones entirely rescued normal axon growth, which was gravely disrupted when only wild-type LIMK1 was expressed (Ng and Luo, 2004). Within these systems, the effects of overexpressing SSH-1L have been broadly consistent with AC overactivation alone suggesting that involvement of other target proteins is unlikely.

These results are consistent with studies cited above on neuritogenesis in PC12 cells in which LIMK1/2 silencing had less dramatic effects on neurite initiation than did combined silencing of ADF and cofilin or the silencing of SSH-1L/2L (Endo et al., 2005) and suggest that slingshot functions exclusively through AC proteins whereas LIMKs may have other effects, some of which may be unrelated to its activity as a kinase. However, SSH signaling mechanisms may indeed prove to be more complicated than currently perceived as its effects become more extensively characterized. The future characterization of SSH within growth cones will also need to address the outstanding issue of its regulation. It will be of particular interest to determine if the calcium-regulated activation of slingshot via calcineurin, which was identified as the neuregulin-dependent mechanism of its activation (Nagata-Ohashi et al., 2004), is a general mechanism for growth cones in which calcium signaling is so important (Henley and Poo, 2004).

2.10. Rho Proteins: Effects on Growth Cone Structure and Physiology

The Rho family of small GTPases (Rho GTPases) functions as important molecular intermediaries that signal between receptor–ligand interactions and target effector proteins that regulate alterations of the cytoskeleton (Govek et al., 2005; Song and Poo, 2001). A review of the Rho GTPases signaling is covered elsewhere in this volume. Here we will only review aspects of the Rho GTPases that relate to their effects on the regulation of the AC proteins within growth cones. In addition, only three members of the Rho GTPases, RhoA, Rac1, and cdc42, will be of primary focus.

Although complete signaling pathways that link the Rho GTPases to the AC proteins have been biochemically characterized (e.g., Arber et al., 1998; Yang et al., 1998), only the RhoA-ROCK signaling pathway has been reasonably linked to AC regulation within growth cones. Modulation of the RhoA-ROCK signaling pathway is required for the regulation of AC downstream of two distinctive guidance molecules, Sema3A and BDNF. From these studies, it was demonstrated that the bidirectional modulation of RhoA was necessary for either the phosphorylation or dephosphorylation of AC. Activation of RhoA-ROCK was required for AC phosphorylation, and conversely, RhoA-ROCK inhibition was required for AC dephosphorylation. This generally agrees with the notion that ROCK acts as a negative regulator of growth cone motility. The involvement of RhoA-ROCK in both the phosphorylation and dephosphorylation of AC suggests that this signaling pathway may represent the major mechanism for modulating AC activity within some types of growth cones, even under basal conditions. This idea is mildly supported by the fact that in addition to Sema3A, other growth cone collapsing factors, such as lysophosphatidic acid (LPA) and ephrin-A, also engage the RhoA-ROCK signaling pathway in neurons (Dash et al., 2004; Sahin et al., 2005). Various other Rho GTPase-dependent signaling pathways have also been identified within growth cones; however, their association with AC regulation has only been suggestive (Kuhn et al., 1999). For example, Rac-PAK signaling has only been linked circumstantially to AC regulation, as overexpressing wild-type Rac1 exacerbated LIMK1's inhibitory effects and required Rac's ability to bind to PAK (Ng and Luo, 2004). Therefore, a Rac-PAK-LIMK1 signaling pathway may also play a complementary role in inactivating AC but may be specific to growth cone type.

Although it was initially thought that the Rho GTPases would actually induce similar effects in different types of growth cones, this has not been the case, probably because of divergence in Rho GTPase signaling, which in some cases may regulate SSH activity as well as LIMK1 (Tanaka et al., 2005). As an example, Rac activation can facilitate neurite growth as well inhibit it. Whereas the inhibitory effect of Rac on neurite growth required PAK-dependent signaling to LIMK1, the stimulatory effect of Rac required PAK-independent signaling (Ng and Luo, 2004). In support of this, Rac1 Y40F, which only weakly binds to PAK, actually rescued normal axon growth from the inhibitory effects of LIMK1. Even within PAK-dependent signaling, Rac can engage divergent signaling mechanisms that do not impinge on AC. This includes the PAK-dependent phosphorylation and inhibition of myosin light chain kinase (MLCK), which is normally required to activate myosin II (Sanders et al., 1999). In fact, the potential of RhoA-ROCK signaling and Rac-PAK signaling to oppositely regulate myosin II represents a major point of divergence. Thus, whereas PAK can inhibit myosin II activity through the inhibition of MLCK, ROCK actually stimulates myosin II activity through the phosphorylation and inactivation of myosin light chain phosphatase (MLCP) (Kimura et al., 1996). Therefore, even assuming that the Rho GTPases actually regulate the AC proteins equally, the ability of each Rho GTPase to engage multiple and distinct signaling mechanisms is sufficient to explain their divergent and sometimes opposing effects.

An additional intricacy of Rho GTPase signaling is apparent because of cross talk between the individual Rho GTPases. Activated Rho GTPases are capable of regulating the other Rho GTPases either synergistically or antagonistically. For example, RhoA is believed to inhibit Rac under basal conditions, which is possibly relieved through p75 receptor signaling (Yamashita and Tohyama, 2003). As described above, RhoA inhibition induced by BDNF is actually mediated through cdc42 activation (Chen et al., 2005). In certain instances, Rho GTPase signaling inevitably involves the coordinate regulation of all three Rho GTPase members through cross talk, although this may not be categorically true (Govek et al., 2005). The intricacies of Rho GTPase signaling allow it to participate in a wide range of responses; therefore, it should not be surprising that all the known families of guidance molecules engage at least one member of the Rho GTPases (Govek et al., 2005). Modulation of Rho GTPase signaling is accomplished either through a direct interaction with the Rho GTPase itself or through the Rho GTPase regulatory proteins, Rho guanine nucleotide exchange factors (RhoGEFs) or Rho GTPase activating proteins (RhoGAPs).

2.11. Sema3A-Induced Regulation of AC Proteins in Growth Cones

The semaphorin superfamily of proteins plays a central role in patterning the developing nervous system (Nakamura et al., 2000). In particular, the effects and mechanisms of Sema3A as a chemorepulsive guidance cue have been characterized *in vitro* and *in vivo*. The targeted knockout of Sema3A in mice resulted in the abnormal projection of peripheral nerve axons into regions normally avoided (Tanaguchi et al., 1997). In cultured chick DRG neurons, Sema3A treatment led to neurite retraction by inducing growth cone collapse, which also involved the progressive loss of peripheral F-actin (Aizawa et al., 2001). The dramatic nature of this dissolution suggested that the AC proteins might be involved in the rapid disassembling of the peripheral actin cytoskeleton system. Sema3A treatment first resulted in the rapid phosphorylation and then gradual dephosphorylation of the AC proteins. After 1 min of Sema3A treatment, relative phosphorylated AC levels were increased by 3.5-fold. After 5 min, however, the relative level of phosphorylated AC dropped to 0.16-fold. By 10 min of Sema3A treatment, growth cone collapse was complete.

The phosphorylation state of AC affected not only its localization within the growth cone but also the distribution of F-actin. In untreated mouse DRG neurons, dephosphorylated AC is localized to peripheral filopodia and lamella along with F-actin. However, the rapid phosphorylation of AC induced its localization to the central domain, concomitant with the transient polymerization of F-actin in the P-domain 3 min after treatment. Ten minutes following Sema3A treatment, the AC proteins were gradually dephosphorylated and peripheral F-actin levels decreased. Strangely, after 10 min of treatment, total AC levels within growth cones were also dramatically reduced by at least 60% as measured by immunostaining (Aizawa et al., 2001). The mechanism for this precipitous downregulation is not obvious but could arise either from epitope blockage, transport out of the growth cone (secretion or retrograde transport), or degradation of the AC protein. Additionally, a mechanism for the subsequent dephosphorylation of AC was not addressed, although the initial phosphorylation likely involves RhoA activation and ROCK, as both dominant-negative RhoA and ROCK inhibition protected growth cones from collapse induced by the related semaphorin, Sema4D (Gherardi et al., 2004; Oinuma et al., 2003). In addition, the translation of RhoA is locally upregulated in response to Sema3A treatment (Wu et al., 2005). Active Rac1 is also required for growth cone collapse. In fact, the activation of RhoA probably depends on the binding of active Rac1 to plexin, which functions as a coreceptor for the semaphorins (Hu et al., 2001; Rohm et al., 2000; Vikis et al., 2000). Downstream of RhoA and ROCK, the LIMK are recruited, which was demonstrated with a LIM kinase pseudosubstrate peptide (S3 peptide) that was designed as a potent cell-permeable (penetratin linked) inhibitor of the LIMK (Aizawa et al., 2001). Pretreatment of mouse DRG neurons with the S3 peptide for 30 min before treating with Sema3A blocked growth cone collapse in a dose-dependent manner.

2.12. Nongrowth Cone Effects of AC That can Impact Neuritogenesis, Growth Cone Behavior, and Neurite Outgrowth

Textbook models of Golgi function in membrane processing are generally confined to delivery of coatamer protein (COP) II-decorated vesicles from the endoplasmic reticulum (ER) to the Golgi, retrograde flow of COP I vesicles from the

Golgi to ER, and budding of clathrin-coated vesicles from the *trans*-Golgi network (e.g., Alberts et al., 2002). Often excluded is the highly dynamic Golgi membrane tubule system, first observed following expression of fluorescent membrane protein chimeras (Lippincott-Schwartz et al., 2000). Elongated Golgi membrane tubules with high surface/volume ratio sort membrane proteins. These tubules can enhance the rates at which materials transcend the Golgi and are sorted and transferred between compartments (Lippincott-Schwartz et al., 2000). Rosso et al. (2004) first demonstrated that LIMK1 overexpression in neurons suppresses formation of Golgi derived tubules, whereas a dominant negative kinase dead LIMK1 had the opposite effect. In hippocampal neurons LIMK1 is targeted to the Golgi through the LIM domains, whereas the PDZ domain targets LIMK1 to the growth cone. Inhibiting cofilin at the Golgi by targeted overexpression of ΔPDZ-LIMK1 accelerates axon formation and increases delivery of Par3/Par6, IGF1 receptors, and NCAM at the growth cone, while inhibiting the export of synaptophysin-containing membrane. Overexpressing the cofilin S3A mutant rescues LIMK1-suppressed Golgi tubule extensions, demonstrating that cofilin is the target of LIMK1. Even in the presence of cofilin S3A, Golgi tubules are repressed by the actin-stabilizing drug jasplakinolide, suggesting actin filaments are the downstream target of cofilin. These studies suggest that Golgi membrane tubules compete with other delivery systems used for sorting membrane proteins. Enhancing membrane vesicle targeting can influence development and neurite outgrowth rates.

Within the region of the Golgi, associated with its vesicles, are short actin filaments, some of which contain the tropomyosin TM5NM2 isoform (Percival et al., 2004). Following treatment with Brefelden A (an inhibitor of ER to Golgi transport) or nocodazole (an inhibitor of microtubule assembly), localized Golgi disappeared and TM5NM2 dispersed. After cytochalsin D treatment to block assembly of actin, TM5NM2 remains perinuclear (Percival et al., 2004) even though the Golgi fragments (Rosso et al., 2004). One possible interpretation of these results is that there may be more than one actin filament network associated with the Golgi, one containing TM5NM2 that is stable to cytochalasin D (probably because of a very slow turnover), and another that regulates Golgi assembly and association with the microtubule system, which is cytochalasin D sensitive. It is this latter system that is likely to be regulated by cofilin.

2.13. Conclusions and Perspectives

Additional functions of AC proteins in regulating aspects of cell behavior other than plasma membrane mediated actin dynamics have been recently discovered. The targeting of AC proteins along with actin to the outer mitochondrial membrane has been shown to be both necessary and sufficient for cytochrome *C* leakage in activating mitochondrial-dependent apoptosis pathways (Chua et al., 2003). The AC proteins in some single-cell organisms and in all metazoans contain a nuclear localization signal, which is cryptic under most conditions because AC proteins are normally cytoplasmic. Although the function of nuclear AC has yet to be established, cofilin and actin are targeted to the nucleus of rat peritoneal mast cells under conditions that disrupt F-actin (latrunculin-B) and during ATP depletion (Pendleton et al., 2003). Nuclear translocation of cofilin following its dephosphorylation has been reported during T-cell proliferation and production of interleukin-2 (Samstag et al., 1994). Within the nucleus, several functions for actin have been recently identified. Nuclear actin has been shown to play a role in export of unspliced HIV genomic RNA and the rev responsive element HIV mRNA (Hofmann et al., 2001; Kimura et al., 2000). In human cells, chromatin remodeling is regulated in part by the human Brahma-related gene-1 (Brg1), a catalytic subunit of the Swi/Snf chromatin remodeling complex, which has been implicated in gene regulation and cell proliferation as well as a candidate tumor suppressor (Hendricks et al., 2004). Brg1 regulates genes important for T-lymphocyte differentiation (Gebuhr et al., 2003) and thus could be the functional linkage with cofilin nuclear targeting in T-cell proliferation referred to earlier. Brg1 regulates the expression of the cdk inhibitor p21 and is necessary for formation of flat cells, growth arrest, and cell senescence (Kang et al., 2004). Furthermore, the C-terminus of nuclear DNA helicase II (NDH II) also binds to F-actin implicating the actin nucleoskeleton in RNA processing, transport, or other actin-related processes (Zhang et al., 2002). Substantial evidence has also accumulated that actin is required for the transcriptional activity of all three eukaryotic RNA polymerases (reviewed in Visa, 2005). Along with its ability to regulate delivery of Golgi derived membrane, AC functioning through these other pathways can have major effects on cell behavior that are unrelated to its ability to locally dynamize actin filaments. Future studies on its effects within growth cones may have to take these other targets into account.

Acknowledgements. The authors would like to thank Drs. Barbara Bernstein, O'Neil Wiggan, Ms. Laurie Minamide, Alisa Shaw, and Janel Funk, and Mr. Michael Maloney for valuable discussions. Support from the National Institutes of Health grants NS40371, HL58064, DK69408 (JRB), and NS43115 (JRB and KF), and the National Science Foundation grant DGE0234615 (CP) is gratefully acknowledged.

References

Agnew, B.J., Minamide, L.S., and Bamburg, J.R., 1995, Reactivation of phosphorylated actin depolymerizing factor and identification of the regulatory site, *J. Biol. Chem.* **270:** 17582–17587.

Alberts, B., Bray, D., Lewis, J., Raff, M., Roberts, K., and Watson, J.D., 2002, *Molecular Biology of the Cell*, 4th ed., Garland Publishing, New York.

Aizawa, H., Wakatsuki, S., Ishii, A., Moriyama, K., Sasaki, Y., Ohashi, K., et al., 2001, Phosphorylation of cofilin by LIM-kinase is necessary for semaphorin 3A-induced growth cone collapse, *Nature Neurosci.* **4:** 367–373.

Aoki, K., Nakamura, T., and Matsuda, M., 2004, Spatio-temporal regulation of Rac1 and Cdc42 activity during nerve growth factor-induced neurite outgrowth in PC12 cells, *J. Biol. Chem.* **279:** 713–719.

Arber, S., Barbayannis, F.A., Hanser, H., Schneider, C., Stanyon, C.A., Bernard, O., et al., 1998, Regulation of actin dynamics through phosphorylation of cofilin by LIM-kinase, *Nature* **398:** 805–809.
Balcer, H.I., Goodman, A.L., Rodal, A.A., Smith, E., Kugler, J., Heuser, J.E., et al., 2003, Coordinated regulation of actin filament turnover by a high molecular weight Srv2/CAP complex, cofilin, profilin, and Aip1, *Curr. Biol.* **13:** 2159–2169.
Bamburg, J.R., 1999, Proteins of the ADF/cofilin family: Essential regulators of actin dynamics, *Annu Rev. Cell Dev. Biol.* **15:** 185–230.
Bamburg, J.R., and Bray, D., 1987, Distribution and cellular localization of actin depolymerizing factor, *J. Cell Biol.* **105:** 2817–2825.
Bamburg, J.R., Harris, H.E., and Weeds, A.G., 1980, Partial purification and characterization of an actin depolymerizing factor from brain, *FEBS Lett.* **121:** 178–181.
Bentley, D., and Toroian-Raymond, A., 1986, Disoriented pathfinding by pioneer neuron growth cones deprived of filopodia by cytochalasin treatment, *Nature* **323:** 712–715.
Blanchoin, L., Pollard, T.D., and Mullins, R.D., 2000, Interactions of ADF/cofilin, Arp2/3 complex, capping protein, and profilin in remodeling of branched actin filament networks, *Curr. Biol.* **10:** 1273–1282.
Bray, D., and Chapman, K., 1985, Analysis of microspike movements on the neuronal growth cone, *J. Neurosci.* **5:** 3204–3213.
Bridgman, P.C., and Dailey, M.E., 1989, The organization of myosin and actin in rapid rozen nerve growth cones, *J. Cell Biol.* **108:** 95–109.
Brown, M.E., and Bridgman, P.C., 2003, Retrograde flow rate is increased in growth cones from myosin IIB knockout mice, *J. Cell Sci.* **116:** 1087–1094.
Buck, K., and Zheng, J.Q., 2002, Growth cone turning induced by local modification of microtubule dynamics, *J. Neurosci.* **22:** 9358–9367.
Carlier, M., Laurent, V., Santolini, J., Melki, R., Didry, D., Xia, G.X., et al., 1997, Actin depolymerizing factor (ADF/Cofilin) enhances the rate of filament turnover: Implication in actin-based motility, *J. Cell Biol.* **136:** 1307–1322.
Chan, A.Y., Bailly, M., Zebda, N., Segall, J.E., and Condeelis, J.S., 2000, Role of cofilin in epidermal growth factor stimulted actin polymerization and lamellipod protrusion, *J. Cell Biol.* **148:** 531–542.
Chardin, P., Boquet, P., Madaule, P., Popoff, M.R., Rubin, E.J., and Gill, D.M., 1989, The mammalian G protein rhoC is ADP-ribosylated by *Clostridium botulinum* exoenzyme C3 and affects actin microfilaments in Vero cells, *EMBO J.* **8:** 1087–1092.
Chua, B.T., Volbracht, C., Tan, K.O., Li, R., Yu, V.C., and Li, P., 2003, Mitochondrial translocation of cofilin is an early step in apoptosis induction, *Nature Cell Biol.* **5:** 1083–1089.
Chen, H., Bernstein, B., Sneider, J.D., Boyle, J.A., Minamide, L.S., and Bamburg, J.R., 2004, *In vitro* activity differences between proteins of the ADF/cofilin family define two distinct subgroups, *Biochemistry* **43:** 7127–7142.
Chen, T., Gehler, S., Shaw, A.E., Bamburg, J.R., and Letourneau, P., 2006, Cdc42 participates in the regulation of ADF/cofilin and retinal growth cone filopodia by brain derived neurotrophic factor, *J. Neurobiol.* **66:** 103–114.
Cramer, L.P., Bamburg, J.R., and Mseka, T. (submitted), Microtubules spatially regulate ADF/cofilin activity to control cell polarity and directed migration.
Dailey, M.E., and Bridgman, P.C., 1991, Structure and organization of membrane organelles along distal microtubule segments in growth cones, *J. Neurosci. Res.* **30:** 242–258.
Dash, P.K., Orsi, S.A., Moody, M., and Moore, A.N., 2004, A role for hippocampal Rho-ROCK pathway in long-term spatial memory, *Biochem. Biophys. Res. Commun.* **322:** 893–898.

DaSilva, J.S., and Dotti, C., 2002, Breaking the neuronal sphere: Regulation of the actin cytoskeleton in neuritogenesis, *Nature Rev. Neurosci.* **3:** 694–704.

DaSilva, J.S, Medina, M., Zuliani, C., Di Nardo, A., Witke, W., and Dotti, C., 2003, RhoA/ROCK regulation of neuritogenesis via profilin IIa-mediated control of actin stability, *J. Cell Biol.* **162:** 1267–1279.

Dawe, H.R., Minamide, L.S., Bamburg, J.R., and Cramer, L.P., 2003, ADF/cofilin controls cell polarity during fibroblast migration, *Curr. Biol.* **13:** 252–257.

Dawid, I.B., Breen, J.J., Toyama, R., 1998, LIM domains: Multiple roles as adapters and functional modifiers in protein interactions, *Trends Genet.* **14:** 156–162.

Dehmelt, L., Smart, F.M., Ozer, R.S., and Halpain, S., 2003, The role of microtubule associated protein 2c in the reorganization of microtubules and lamellipodia during neurite initiation, *J. Neurosci.* **23:** 9479–9490.

Edwards, D.C., Sanders, L.C. Bokoch, G.M., and Gill, G.N., 1999, Activation of LIM-kinase by pak1 couples rac/cdc42 GTPase signalling to the cytoskeletal dynamics, *Nature Cell Biol.* **1:** 253–259.

Endo, M., Ohshi, K., Sasaki, Y., Goshima, Y., Niwa, R., Uemura, T., et al., 2003, Control of growth cone motility and morphology by LIM kinase and slingshot via phosphorylation and dephosphorylation of cofilin, *J. Neurosci.* **23:** 2527–2537.

Endo, M., Ohashi, K., and Mizuno, K., 2005, Role of cofilin phosphocycle by LIM-kinase and slingshot in NGF-induced neurite outgrowth, *Mol. Biol. Cell* **16:** 676a.

Fan, J., and Raper, J.A., 1995, Localized collapsing cues can steer growth cones without inducing their full collapse, *Neuron* **14:** 263–274.

Fass, J., Gehler, S., Sarmiere, P., Letourneau, P., and Bamburg, J.R., 2004, Regulating filopodial dynamics through actin-depolymerizing factor/cofilin, *Anat. Sci. Internat.* **79:** 173–183.

Foletta, V.C., Moussi, N., Sarmiere, P.D., Bamburg, J.R., and Bernard, O., 2004, LIM kinase, a key regulator of actin dynamics, is widely expressed in embryonic and adult tissues, *Exp. Cell Res.* **294:** 392–405.

Forscher, P., and Smith, S.J., 1988, Actions of cytochalasins on the organization of actin filaments and microtubules in a neuronal growth cone, *J. Cell Biol.* **107:** 1505–1516.

Gallo, G., Yee, H.F., and Letourneau, P., 2002, Actin turnover is required to prevent axon retraction driven by endogenous actomyosin contractility, *J. Cell Biol.* **158:** 1219–1228.

Gebuhr, T.C., Kovaley, G.I. Bultman, S., Godfrey, V., Su, L., and Magnuson, T., 2003, The role of Brg1, a catalytic subunit of mammalian chromatin remodeling complexes, in T cell development, *J. Exp. Med.* **198:** 1937–1949.

Gehler, S., Gallo, G., Veien, E., and Letourneau, P.C., 2004a, p75NTR signaling regulates growth cone filopodial dynamics through modulating RhoA activity, *J Neurosci.* **24:** 4363–4372.

Gehler, S., Shaw, A.E., Sarmiere, P.D., Bamburg, J.R., and Letourneau, P.C., 2004b, Brain-derived neurotrophic factor regulation of retinal growth cone filopodial dynamics is mediated through actin depolymerizing factor/cofilin, *J. Neurosci.* **24:** 10741–10749.

Gherardi, E., Love, C.A., Esnouf, R.M., and Jones, E.Y., 2004, The sema domain, *Curr. Opin. Struct. Biol.* **14:** 669–678.

Ghosh, M., Song, X., Mouneimne, G., Sidani, M., Lawrence, D.S., and Condeelis, J.S., 2005, Cofilin promotes actin polymerization and defines the direction of cell motility, *Science* **304:** 743–746.

Giuliano, K.A., Khatib, F.A., Hayden, S.M., Daoud, E.W., Adams, M.E., Amorese, D.A., et al., 1988, Properties of purified actin depolymerizing factor from chick brain, *Biochemistry* **27:** 8931–8937.

Gohla, A., and Bokoch, G.M., 2002, 14-3-3 regulates actin dynamics by stabilizing phosphorylated cofilin, *Curr. Biol.* **12:** 1704–1710.

Gohla, A., Birkenfeld, J., and Bokoch, G.M., 2005, Chronophin, a novel HAD-type serine protein phosphatase, regulates cofilin-dependent actin dynamics, *Nat. Cell Biol.* **7:** 21–29.

Goldberg, D.J., and Burmeister, D.W., 1989, Looking into growth cones, *Trends Neurosci.* **12:** 503–506.

Govek, E., Newey, S.E., and Van Aelst, L., 2005, The role of the rho GTPases in neuronal development, *Genes Dev.* **19:** 1–49.

Gurniak, C.B., Perlas, E., and Witke, W., 2005, The actin depolymerizing factor n-cofilin is essential for neural tube morphogenesis and neural crest cell migration, *Dev. Biol.* **278:** 231–241.

Hendricks, K.B., Shanahan, F., and Lees, E., 2004, Role for BRG1 in cell cycle control and tumor suppression, *Mol. Cell Biol.* **24:** 362–376.

Henley, J., and Poo, M., 2004, Guiding neuronal growth cones using Ca2+ signals, *Trends Cell Biol.* **14:** 320–330.

Hiraoka, J., Okano, I., Higuchi, O., Yang, N., and Mizuno, K., 1996, Self-association of LIM kinase 1 mediated by the interaction between an N-terminal LIM domain and a C-terminal kinase domain, *FEBS Lett.* **399:** 117–121.

Hofmann, W., Reichart, B., Ewald, A., Muller, E., Schmitt, I., Stauber, R.H., et al., 2001, Cofactor requirements for nuclear export of Rev response element (RRE)- and constitutive transport element (CTE)-containing retroviral RNAs. An unexpected role for actin, *J.Cell Biol.* **152:** 895–910.

Hotulainen, P., Paunola, E., Vartianen, M.K., and Lappalainen, P., 2005, Actin-depolymerizing factor and cofilin-1 play overlapping roles in promoting rapid F-actin depolymerization in mammalian nonmuscle cells, *Mol. Biol. Cell* **16:** 649–664.

Hu, H., Marton, T.F., and Goodman, C.S., 2001, Plexin B mediates axon guidance in *Drosophila* by simultaneously inhibiting active Rac and enhancing RhoA signaling, *Neuron* **32:** 39–51.

Huber, A.B., Kolodkin, A.L., Ginty, D.D., and Cloutier, J.F., 2003, Signaling at the growth cone: Ligand-receptor complexes and the control of axon growth and guidance, *Annu. Rev. Neurosci.* **26:** 509–563.

Kanamuri, T., Suzuki, M., and Titani, K., 1998, Complete amino acid sequences and phosphorylation sites, determined by Edman degradation and mass spectrometry, of rat parotid destrin- and cofilin-like proteins, *Arch. Oral. Biol.* **43:** 955–967.

Kang, H., Cui, K., and Zhao, K., 2004, BRG1 controls the activity of the retinoblastoma protein via the regulation of p21CIP1/WAF1/SDI, *Mol. Cell Biol.* **24:** 1188–1199.

Kimura, K., Ito, M., Amano, M., Chihara, K., Fukata, Y., Nakafuku, M., et al., 1996, Regulation of myosin phosphatase by Rho and Rho-associated kinase (Rho-kinase), *Science* **273:** 245–248.

Kimura, T., Hashimoto, I., Yamamoto, A., Nishikawa, M., and Fujisawa, J.I., 2000, Rev-dependent association of the intron-containing HIV-1 gag mRNA with the nuclear actin bundles and the inhibition of its nucleocytoplasmic transport by latrunculin-B, *Genes Cells.* **5:** 289–307.

Kuhn, T.B., Brown, M.D., Wilcox, C.L., Raper, J.A., and Bamburg, J.R., 1999, Myelin and collapsin-1 induce motor neuron growth cone collapse through different pathways: Inhibition of collapse by opposing mutants of rac-1, *J. Neurosci.* **19:** 1965–1975.

Kuhn, T.B., Meberg, P.J., Brown, M.D., Bernstein, B.W., Minamide, L.S., Jensen, J.R., et al., 2000, Regulating actin dynamics in neuronal growth cones by ADF/cofilin and rho family GTPases, *J. Neurobiol.* **44:** 126–144.

Lappalainen, P., Kessels, M.M., Cope, M.J., and Drubin, D.G., 1998, The ADF homology (ADF-H) domain: A highly exploited actin-binding module, *Mol. Biol. Cell* **9:** 1951–1959.
Lee, H., Engel, U., Rusch, J., Scherrer, S., Sheard, K., and Van Vactor, D., 2004, The microtubule plus end tracking protein Orbit/MAST/CLASP acts downstream of the tyrosine kinase Abl in mediating axon guidance, *Neuron* **42:** 913–926.
Letourneau, P.C., 1983, Differences in the organization of actin in the growth cones compared with the neurites of cultured neurons from chick embryos, *J. Cell Biol.* **97:** 963–973.
Letourneau, P.C., 1996, The cytoskeleton in nerve growth cone motility and axonal pathfinding, *Perspect. Dev. Neurobiol.* **4:** 111–123.
Leung, T., Manser, E., Tan, L., and Lim, L., 1995, A novel serine/threonine kinase binding the ras-related RhoA GTPase which translocates the kinase to peripheral membranes, *J. Biol. Chem.* **270:** 29051–29054.
Lewis, A.K., and Bridgman, P.C., 1992, Nerve growth cone lamellipodia contain two populations of actin filaments that differ in organization and polarity, *J. Cell Biol.* **119:** 1219–1243.
Lin, C.H., and Forscher, P., 1995, Growth cone advance is inversely proportional to retrograde F-actin flow, *Neuron* **14:** 763–771.
Lin, C.H., Thompson, C.A., and Forscher, P., 1994, Cytoskeletal reorganization underlying growth cone motility, *Curr. Opin. Neurobiol.* **4:** 640–647.
Lin, C.H., Espreafico, E.M., Mooseker, M.S., and Forscher, P., 1996, Myosin drives retrograde F-actin flow in neuronal growth cones, *Neuron* **16:** 769–782.
Lippincott-Schwartz, J., Roberts, T.H., and Hirschberg, K., 2000, Secretory protein trafficking and organelle dynamics in living cells, *Annu. Rev. Cell Dev. Biol.* **16:** 557–589.
Maciver, S.K., and Weeds, A.G., 1994, Actophorin preferentially binds momomeric ADP-actin over ATP-bound actin: Consequences for cell locomotion, *FEBS Lett.* **347:** 251–256.
Maciver, S.K., Pope, B.J., Whytock, S., and Weeds, A.G., 1998, The effect of two actin depolymerizing factors (ADF/cofilins) on actin filament turnover: pH sensitivity of F-actin binding by human ADF, but not of *Acanthamoeba* actophorin, *Eur. J. Biochem.* **256:** 388–397.
Maekawa, M., Ishizaki, T., Boku, S., Watanabe, N., Fujita, A., Iwamatsu, A., et al., 1999, Signaling from Rho to the actin cytoskeleton through protein kinases ROCK and LIM kinase, *Science* **285:** 895–898.
Manser, E., Leung, T., Salihuddin, H., and Lim, L., 1994, A brain serine/threonine protein kinase activated by Cdc42 and Rac1, *Nature* **367:** 40–46.
Marsh, L., and Letourneau, P.C., 1984, Growth of neurites without filopodial or lamellipodial activity in the presence of cytochalasin B, *J. Cell Biol.* **99:** 2041–2047.
Mcgee, A.W., and Bredt, D.S., 1999, Identification of an intramolecular interaction beween the SH3 and guanylate kinase domains of PSD-95, *J. Biol. Chem.* **274:** 17431–17436.
Meberg, P.J., 2000, Signal-regulated ADF/cofilin activity and growth cone motility, *Mol. Neurobiol.* **21:** 97–107.
Meberg, P., and Bamburg, J.R., 2000, Increase in neurite outgrowth mediated by overexpression of actin depolymerizing factor, *J. Neurosci.* **20:** 2459–2469.
Meberg, P., Ono, S., Minamide, L., Takahashi, M., and Bamburg, J.R., 1998, Actin depolymerizing factor and cofilin phosphorylation dynamics: Response to signals that regulate neurite extension, *Cell Motil. Cytoskel.* **39:** 172–190.

Meyer, G., and Feldman, E.L., 2002, Signaling mechanisms that regulate actin-based motility processes in the nervous system, *J. Neurochem.* **83:** 490–503.

Mizuno, K., Okano, I., Ohashi, K., Nunoue, K., Kuma, K., Miyata, T., et al., 1994, Identification of a human cDNA encoding a novel protein kinase with two repeats of LIM/double zinc finger motif, *Oncogene* **9:** 1605–1612.

Morgan, T.E., Lockerbie, R.O., Minamide, L.S., Browning, M.D., and Bamburg, J.R., 1993, Isolation and characterization of a regulated form of actin depolymerizing factor, *J. Cell Biol.* **122:** 623–633.

Moriyama, K., and Yahara, I., 1999, Two activities of cofilin, severing and accelerating directional depolymerization of actin filaments, are affected differentially by mutations around the actin-binding helix, *EMBO J.* **18:** 6752–6761.

Moriyama, K., and Yahara, I., 2002, The actin severing activity of cofilin is exerted by the interplay of three distinct sites on cofilin and essential for cell viability, *Biochem. J.* **365:** 147–155.

Moriyama, K., Iida, K., and Yahara, I., 1996, Phosphorylation of Ser-3 of cofilin regulates its essential function on actin, *Genes Cells* **1:** 73–86.

Nagata-Ohashi, K., Ohta, Y., Goto, K., Chiba, S., Mori, R., Nishita, M., et al., 2004, A pathway of neuregulin-induced activation of cofilin phosphatase slingshot and cofilin in lamellipodia, *J. Cell Biol.* **165:** 465–471.

Nakamura, F., Kalb, R.G., and Strittmatter, S.M., 2000, Molecular basis of semaphorin-mediated axon guidance, *J. Neurobiol.* **44:** 219–229.

Ng, J., and Luo, L., 2004, Rho GTPases regulate axon growth through convergent and divergent signaling pathways, *Neuron* **44:** 779–793.

Nishida, E., Maekawa, S., and Sakai, H., 1984, Cofilin, a protein in porcine brain that binds to actin filaments and inhibits their interactions with myosin and tropomyosin, *Biochemistry* **23:** 5307–5313.

Nishita, M., Tomizawa, T., Yamamoto, M., Horita, Y., Ohashi, K., and Mizuno, K., 2005, Spatial and temporal regulation of cofilin activity by LIM kinase and slingshot is critical for directional cell migration, *J. Cell Biol.* **171:** 349–359.

Niwa, R., Nagata-Ohashi, K., Takeichi, M., Mizuno, K., and Uemura, T., 2002, Control of actin reorganization by slingshot, a family of phosphatases that dephosphorylate ADF/cofilin, *Cell* **108:** 233–246.

O'Connor, T.P., Duerr, J.S., and Bentley, D., 1990, Pioneer growth cone steering decisions mediated by single filopodial contacts in situ, *J. Neurosci.* **10:** 3935–3946.

Oinuma, I., Katoh, H., Harada, A., and Negishi, M., 2003, Direct interaction of Rnd1 with Plexin-B1 regulates PDZ-mediated Rho activation by plexin-B1 and induces cell contraction in Cos-7 cells, *J. Biol. Chem.* **278:** 25671–25677.

Okabe, S., and Hirokawa, N., 1991, Actin dynamics in growth cones, *J. Neurosci.* **11:** 1918–1929.

Ono, S., and Benian, G.M., 1998, Two *Caenorhabditis elegans* actin depolymerizing factor/cofilin proteins, encoded by the unc-60 gene, differentially regulate actin filament dynamics, *J. Biol. Chem.* **273:** 3778–3783.

Ono, S., Mohri, K., and Ono, K., 2004, Microscopic evidence that actin-interacting protein 1 actively disassembles actin-depolymerizing factor/cofilin-bound filaments, *J. Biol. Chem.* **279:** 14207–14212.

Pendleton, A., Pope, B., Weeds, A., and Koffer, A., 2003, Latruncluin B or ATP depletion induces cofilin-dependent translocation of actin into the nuclei of mast cells, *J. Biol. Chem.* **278:** 14394–14400.

Percival, J.M., Hughes, J.A., Brown, D.L., Schevzov, G., Heimann, K., Vrhovski, B., et al., 2004, Targeting of tropomyosin isoform to short microfilaments with the Golgi complex, *Mol. Biol. Cell* **15:** 268–280.

Pope, B.J., Gonsior, S.M., Yeoh, S., McGough, A., and Weeds, A.G., 2000, Uncoupling actin filament fragmentation by cofilin from increased subunit turnover, *J. Mol. Biol.* **298:** 649–661.

Rodriguez, O.C., Schaefer, A.W., Mandato, C.A., Forscher, P., Bement, W.M., and Waterman-Storer, C.M., 2003, Conserved microtubule-actin interactions in cell movement and morphogenesis, *Nature Cell Biol.* **5:** 599–609.

Rohm, B., Ottemeyer, A., Lohrum, M., and Puschel, A.W., 2000, Plexin/Neuropilin complexes mediate repulsion by the axonal guidance signal semaphorin 3A, *Mech. Dev.* **93:** 95–104.

Rosenblatt, J., Agnew, B.J., Abe, H., Bamburg, J.R., and Mitchison, T.J., 1997, *Xenopus* actin depolymerizing factor/cofilin (XAC) is responsible for the turnover of actin filaments in *Listeria monocytogenes* tails, *J. Cell Biol.* **136:** 1323–1332.

Rosso, S., Bollati, F., Bisbal, M., Peretti, D., Sumi, T., Nakamura, T., et al., 2004, LIMK1 regulates Golgi dynamics, traffic of Golgi-derived vesicles, and process extension in primary cultured neurons, *Mol. Biol. Cell* **15:** 3433–3449.

Sahin, M., Greer, P.L., Lin, M.Z., Poucher, H., Eberhart, J., Schmidt, S., et al., 2005, Eph-dependent tyrosine phosphorylation of ephexin-1 modulates growth cone collapse, *Neuron* **46:** 191–204.

Samstag, Y., and Nebl, G., 2003, Interaction of cofilin with the serine phosphatases PP1A and PP2A in normal and neoplastic human T-lymphocytes, *Adv. Enzyme Regul.* **43:** 197–211.

Samstag, Y., Eckerskom, C., Wesselborg, S., Henning, S., Wallich, R., and Meuer, S.C., 1994, Costimulatory signals for human T-cell activation induce nuclear translocation of pp19/cofilin, *Proc. Natl. Acad. Sci. USA* **91:** 4494–4498.

Sanders, L.C., Matsumura, F., Bokoch, G.M., and de Lanerolle, P., 1999, Inhibition of myosin light chain kinase by p21-activated kinase, *Science* **283:** 2083–2085.

Sanes, D.H, Reh, T.A., and Harris, W.A., 2000, *Development of the Nervous System*, 1st edn. Academic Press, San Diego, pp. 92–100.

Sarmiere, P.D., and Bamburg, J.R., 2004, Regulation of the neuronal cytoskeleton by ADF/cofilin, *J. Neurobiol.* **58:** 103–117.

Schaefer, A.W., Kabir, N., and Forscher, P., 2002, Filopodia and actin arcs guide the assembly and transport of two populations of microtubules with unique dynamic parameters in neuronal growth cones, *J. Cell Biol.* **158:** 139–152.

Shi, S.H., Jan, L.Y., and Jan, Y.N., 2003, Hippocampal neuronal polarity specified by spatially localized mpar3/mpar6 and PI 3-kinase activity, *Cell* **112:** 63–75.

Smith, S.J., 1988, Neuronal cytomechanics: The actin-based motility of growth cones, *Science* **242:** 708–715.

Song, H., and Poo, M., 2001, The cell biology of neuronal migration, *Nature Cell Biol.* **3:** E81–88.

Soosairajah, J., Maiti, S., Wiggan, O., Sarmeire, P., Moussi, N., Sarcevic, B., et al., 2005, Interplay between components of a novel lim kinase-slingshot phosphatase complex regulates cofilin, *EMBO J.* **24:** 473–486.

Svitkina, T.M., Verkhovsky, A.B., McQuade, K.M., and Borisy, G.G., 1997, Analysis of the actin-myosin II system in fish epidermal keratocytes: Mechanism of cell body translocation, *J. Cell Biol.* **139:** 397–415.

Svitkina, T.M., Bulanova, E.A., Chaga, O.Y., Vignjevic, D.M., Kojima, S., Vasiliev, J. et al., 2003, Mechanism of filopodia initiation by reorganization of a dendritic network, *J. Cell Biol.* **160:** 409–421.

Tanaguchi, M., Yuasa, S., Fujisawa, H., Naruse, I., Saga, S., Mishina, M., et al., 1997, Disruption of semaphorin III/D gene causes severe abnormality in peripheral nerve projection, *Neuron* **19:** 519–530.

Tanaka, E.M., and Kirschner, M.W., 1991, Microtubule behavior in the growth cones of living neurons during axon elongation, *J. Cell Biol.* **115:** 345–363.

Tanaka, K., Okubo, Y., and Abe, H., 2005, Involvement of slingshot in the rho-mediated dephosphorylation of ADF/cofilin during *Xenopus* cleavage, *Zool. Sci.* **22:** 971–984.

Thirion, C., Stucka, R., Mendel, B., Gruhler, A., Jaksch, M., Nowak, K.J., et al., 2001, Characterization of human muscle type cofilin (CFL2) in normal and regenerating muscle, *Eur. J. Biochem.* **268:** 3473–3482.

Tsui, H.C., Ris, H., and Klein, W.L., 1983, Ultrastructural networks in growth cones and neurites of cultured central nervous system neurons, *Proc. Natl. Acad. Sci. USA* **80:** 5779–5783.

Tursun, B., Schluter, A., Peters, M.A., Viehweger, B., and Ostendorff, H.P., Soosairajah, J., et al., 2005, The ubiquitin ligase Rnf6 regulates local LIM kinase 1 levels in axonal growth cones, *Genes Dev.* **19:** 2307–2319.

Vartiainen, M.K., Mustonen, T., Mattila, P.K., Ojala, P.J., Thesleff, I., Partanen, J., et al., 2002, The three mouse actin-depolymerizing factor/cofilins evolved to fulfill cell-type-specific requirements for actin dynamics, *Mol. Biol. Cell* **13:** 183–194.

Vasiliev, J.M., Gelfand, I.M, Domnina, L.V., Ivanova, O.Y., Komm, S.G., Olshevskaja, L.V., 1970, Effect of colcemid on the locomotory behaviour of fibroblasts, *J. Embryol. Exp. Morphol.* **24:** 625–640.

Vignjevic, D., Yarar, D., Welch, M.D., Peloquin, J., Svitkina, T., and Borisy, G.G., 2003, Formation of filopodia-like bundles *in vitro* from a dendritic network, *J. Cell Biol.* **160:** 951–962.

Vikis, H.G., Li, W., He, Z., and Guan, K.L., 2000, The semaphorin receptor plexin-B1 specifically interacts with active Rac in a ligand-dependent manner, *Proc. Natl. Acad. Sci. USA* **97:** 12457–12462.

Visa, N., 2005, Actin in transcription. Actin is required for transcription by all three RNA polymerases in the eukaryotic cell nucleus, *EMBO Rep.* **6:** 218–219.

Wang, Y., Shibiaski, F., and Mizuno, K., 2005, Calcium signal-induced cofilin dephosphorylation is mediated by slingshot via calcineurin, *J. Biol. Chem.* **280:** 12683–12689.

Worthylake, R.A., and Burridge, K., 2003, RhoA and Rock promote migration by limiting membrane protrusions, *J. Biol. Chem.* **278:** 13578–13584.

Wu, K.Y., Hengst, U., Cox, L.J., Macosko, E.Z., Jeromin, A., Urquhart, E.R., et al., 2005, Local translation of RhoA regulates growth cone collapse, *Nature* **436:** 1020–1024.

Yamaguchi, Y., Katoh, H., Yasui, H., Mori, K., and Negishi, M., 2001, RhoA inhibits the nerve growth factor-induced Rac1 activation through Rho-associated kinase-dependent pathway, *J. Biol. Chem.* **276:** 18977–18983.

Yamashiro, S., and Ono, S., 2005, The two *Caenorhabditis elegans* actin-depolymerizing factor/cofilin proteins differently enhance actin filament severing and depolymerization, *Biochemistry* **44:** 14238–14247.

Yamashita, T., and Tohyama, M., 2003, The p75 receptor acts as a displacement factor that releases Rho from Rho-GDI, *Nature Neurosci.* **6:** 461–467.

Yang, N., Higuchi, O., Ohashi, K., Nagata, K., Wada, A., Kagawa, K., et al., 1998, Cofilin phosphorylation by LIM-kinase 1 and its role in Rac-mediated actin organization, *Nature* **393:** 809–812.

Yeoh, S., Pope, B., Mannherz, H.G., and Weeds, A.G., 2002, Determining the differences in actin binding by human ADF and cofilin, *J. Mol. Biol.* **315:** 911–925.

Zhang, S., Buder, K., Burkhardt, C., Schlott, B., Gorlach, M., and Grosse, F., 2002, Nuclear DNA helicase II/RNA helicase A binds to filamentous actin, *J. Biol. Chem.* **277:** 843–853.

Zheng, J.Q., Wang, J., and Poo, M., 1996, Essential role of filopodia in chemotropic turning of nerve growth cone induced by a glutamate gradient, *J. Neurosci.* **16:** 1140–1149.

Zhou, F.Q., and Cohan, C.S., 2003, How actin filaments and microtubules steer growth cones to their targets, *J. Neurobiol.* **58:** 84–91.

Zhou, F.Q., Waterman-Storer, C.M., and Cohan, C.S., 2002, Focal loss of actin bundles causes microtubule redistribution and growth cone turning, *J. Cell Biol.* **157:** 839–849.

3
Role of Microtubules and MAPs During Neuritogenesis

JOHN K. CHILTON AND PHILLIP R. GORDON-WEEKS

3.1. Introduction

Neurons are among the most polarized cells in the animal kingdom commonly possessing two types of cellular processes, usually a single axon and several dendrites, whose structural and functional properties are distinct. Axons in general convey information away from the neuronal cell body whereas dendrites receive information from outside of the cell and convey it toward the cell body. Consequently, axons form presynaptic structures whereas dendrites (and cell bodies) are generally postsynaptic. Our knowledge of the specification of axons and dendrites during the early stages of neuronal differentiation stems mainly from cell culture studies of dissociated, differentiating neurons (Bradke and Dotti, 2000; Craig and Banker, 1994). Such studies have shown that embryonic hippocampal pyramidal neurons (Dotti and Banker, 1987) and embryonic cerebellar macro-neurons (Ferreira et al., 1989) initially extend two or more processes that are bipotential in terms of whether they will become axons or dendrites. These bipotential processes grow slowly until, apparently randomly, one, which is destined to become the axon, extends at a faster rate while the others stall or slow down and then later differentiate into dendrites. One of the earliest distinguishing features of the neurite that becomes the axon is that its growth cone enlarges.

Not all types of neuron in dissociated culture follow this differentiation pattern. For example, sparsely plated cerebellar granule cells first extend an axon and then another and finally dendrites (Powell et al., 1997). Sympathetic neurons, growing on poly-D-lysine in serum-free medium in the absence of nonneuronal cells, extend only a single axon (Bruckenstein and Higgins, 1988; Lein et al., 1995) but can be induced to extend dendrites if, for example, exposed to bone morphogenetic proteins (Lein et al., 1996).

Cell culture experiments suffer from the disadvantage that the embryonic context is lost or disrupted and, therefore, might not exactly reflect events *in vivo*. There are well-documented examples in embryos where neuritogenesis begins with the emergence of a single neurite that becomes the axon and later emerging neurites become dendrites (Jacobson, 1991; Roberts, 1988). Despite this caveat, primary neuronal cell cultures have been used extensively to investigate the molecular mechanisms underlying neuronal polarity—the specification of axons and dendrites—and a number of key polarity molecules have been identified (Wiggin et al., 2005).

3.2. Role of Microtubules in Neuritogenesis

The acquisition of cell polarity in differentiating neurons is discussed in detail in Chapter 11. Here, we limit our discussion to the role of microtubules (MTs) and microtubule-associated proteins (MAPs) in axonal neuritogenesis: neurite extension, branching, and growth cone pathfinding. It has been recognized for a long time that the extension of neurites is critically dependent on the MT cytoskeleton whereas the motility of the growth cone, i.e., the extension and retraction of filopodia and lamellipodia, is dependent on actin filaments (Gordon-Weeks, 2005). The evidence for this includes the long-standing observation that agents which depolymerize MTs or inhibit polymerization impair neurite extension of neurons in culture (Dent and Gertler, 2003), as does downregulation of tubulin with antisense oligodeoxynucleotides to tubulin mRNA (Teichman-Weinberg et al., 1988) and microinjection of colchicine-tubulin, a molecule that is unable to polymerize (Keith, 1990). In contrast, depolymerization of actin filaments interferes with growth cone motility but does not block neurite growth of neurons in culture, provided they are growing on an adherent surface (Gordon-Weeks, 2005) or neurites *in vivo* (Bentley and Toroian-Raymond, 1986; Chien et al., 1993). Although these experiments clearly show that MT dynamics and transport are important for neuritogenesis, they do not provide mechanistic insights into how or where in the neurite MTs are important. Growth of the neurite occurs by incorporation of material at the growth cone (Gordon-Weeks, 2005), and this is probably true of incorporation of MTs into the neurite cytoskeleton. If this is true then the growth cone is predicted to be one site in the neurite where MT depolymerizing agents might act to inhibit neuritogenesis. Indeed, the growth cone is far more sensitive to MT depolymerizing agents than the neurite shaft or neuronal cell body when these agents are applied locally in culture (Bamburg et al., 1986). Contemporary views about the interactions between MTs and actin filaments in growth cones and the role of this interaction in growth cone turning, a fundamental behavior of growth cones during pathfinding, imply that MTs are incorporated into the neurite cytoskeleton at the growth cone. In growth cones, filopodia are the primary entities detecting guidance cues and are dependent on actin cytoskeleton dynamics, whereas growth cone turning depends on MT dynamics (Gordon-Weeks, 2004).

Microtubules are cytoskeletal polymers of tubulin that have diverse roles in cellular functions such as cell division, organelle transport, and cell motility. They are long, hollow filaments composed of a number of protofilament polymers (13 in neurons) of α-tubulin and β-tubulin heterodimers. Microtubules grow and shrink by end-polymerization of heterodimer subunits. They are polar filaments, partly because tubulin heterodimers are intrinsically polar and associate head-to-tail in the protofilaments, which lie parallel to the MT long axis. The kinetics of subunit addition/subtraction is different at the two ends. The fast-growing end of the MT is known as the "plus" end, while the slow-growing end is called the "minus" end. Tubulin is a nucleotide triphosphatase, and both α-tubulin and β-tubulin bind guanosine triphosphate (GTP), however, only the GTP bound to β-tubulin is hydrolyzed to GDP, sometime after assembly (Amos, 2004). Activation of hydrolysis of the β-tubulin-bound GTP occurs when, during assembly, an activation domain in α-tubulin is brought close to the β-tubulin GTP binding site as the protofilaments go from a curved to a straight conformation. Microtubule polymerization behavior is more complex than that seen with simple equilibrium polymers, which are either elongating or shrinking, and this is most likely because of the energy derived from hydrolysis of the GTP bound to β-tubulin; the GTP bound to α-tubulin remains unhydrolyzed and unexchanged because it is trapped between the monomers of the heterodimer (Amos, 2004). Microtubules *in vitro* (Horio and Hotani, 1986; Mitchison and Kirschner, 1984a,b) and in cultured cells (Cassimeris et al., 1988; Hayden et al., 1990; Sammak and Borisy, 1988) undergo cycles of relatively slow, continuous plus-end growth punctuated by random transitions, called catastrophes, to phases of rapid shortening due to curvature of the protofilaments and subsequent heterodimer dissociation (Kirschner and Mitchison, 1986a,b). Rapid shortening may be "rescued" by another transition to sustained growth, or the MT may depolymerize completely. This behavior of switching between phases of growth and shrinkage is known as dynamic instability, and the catastrophes underlying dynamic instability are probably caused by the loss of a GTP-β-tubulin "cap" at the end of the MT. The size of the GTP cap is unknown since it has not been visualized, although it is likely to be as small as one tubulin heterodimer subunit (Amos, 2004). No mechanism for the regulation of the size of the cap has been found. Although MTs in neurons are formed at an MT-organizing center, the centrosome, they are detached by an unknown mechanism and transported into and along the neurites by a process that depends on MT motors (Ahmad et al., 1998; Baas, 1998, 2002; He et al., 2005). Thus, in neurites, most MTs are not attached to the MT-organizing center and are wholly contained within the neurite with their minus ends probably capped. Microtubules in growing axons are all oriented with their plus-ends distal, i.e., facing the growth cone, whereas in growing dendrites, the MT population is mixed in terms of orientation (Baas et al., 1987, 1988; Heidemann et al., 1981). In all growth cones, dendritic or axonal, MT plus-ends are directed toward the growth cone periphery and hence the filopodia and lamellipodia (Baas et al., 1988).

Many of the MTs in growth cones are undergoing dynamic instability, which enables them to explore the peripheral, actin filament-rich domain (Dent and Kalil, 2001; Dent et al., 1999; Stepanova et al., 2003; Tanaka and Kirschner, 1995; Tanaka et al., 1995). This exploratory behavior and the interaction between these dynamic MTs and the bundle of actin filaments within filopodia are essential for growth cone turning (Buck and Zheng, 2002; Challacombe et al., 1997; Gordon-Weeks, 2004; Kalil and Dent, 2005; Schaefer et al., 2002; Williamson et al., 1996; Zhou et al., 2002) and neurite branching (Dent and Kalil, 2001). The exact mechanism underlying this interaction is unknown but probably involves MAPs, especially those generally referred to as +TIPs that locate at the plus ends of MTs (Section 3.4), and actin-binding proteins.

3.2.1. *Microtubule-Associated Proteins*

Proteins that bind directly to MTs were historically referred to as MAPs. During cycles of MT assembly/disassembly, bona fide MAPs ought, by definition, to bind to MTs with a constant stoichiometry, and this, in the past, has formed the basis of a strict test for MAPs. This test is now rarely applied, and the definition has been relaxed to some extent to include proteins that associate with MTs. Microtubule-associated proteins include: the so-called structural MAPs, which alter MT stability and structure, and in neurons include tau, MAP2, and MAP1A/B; MAPs that associate with the plus ends of MTs (+TIPs), including EB1/3, CLIP-170/115, BPAG/MACF7; MT motors (kinesins, cytoplasmic dyneins); kinases and adaptor proteins. A diverse range of MT-binding domains is found within these proteins including basic repeats (tau/MAP2/MAP1A/1B), multichain complexes (MAP1A/B), zinc-finger domains (GEF-H1), helical coiled coils (APC), CAP-Gly domains (CLIP-115/170), and GAS2 domains (BPAG/ACF7).

3.3. Structural MAPs in Neuritogenesis

The structural MAPs include the tau/MAP2/MAP4 family of MT-binding proteins that share homologous, repeat-domain MT-binding sequences and the MAP1A/1B MT-binding proteins that also share repeat-domain MT-binding sequences but ones that are distinct from the tau/MAP2/MAP4 family. With the exception of MAP4, which is found in nonneuronal cells, all of these MAPs are highly enriched in neurons; tau and MAP1B particularly in axons and MAP2 in the somato-dendritic compartment. These MAPs bind uniformly along the MT wall, and their repeat domains allow both an interaction with individual tubulin subunits within the MT lattice and can also stabilize MTs by cross-linking individual tubulin subunits. The molecular details of these interactions will have to await structural studies of the MAP/tubulin complex. The neuronal structural MAPs are multiply phosphorylated, and, in general, phosphorylation influences their binding to MTs.

3.3.1. *MAP1B*

The high molecular weight microtubule-associated protein MAP1B belongs, along with MAP1A, to a family of MAPs that bind to MTs but do not bundle them. Although not uniquely found in neurons, MAP1B expression is highest in these cells and is one of the earliest MAPs to be expressed in them during neuronal differentiation (Gonzalez-Billault et al., 2004). MAP1B expression levels in neurons, at least in the central nervous system (CNS), is highest while neurons are differentiating and is downregulated toward the end of axonogenesis (Gordon-Weeks and Fischer, 2000). However, there are many regions in the adult brain, such as the hippocampus, where neurons continue to express the protein. In the cerebral cortex generally, MAP1B is found in the postsynaptic densities of a subpopulation of asymmetric synapses (Kawakami et al., 2003), where it may function in the organization of neurotransmitter receptors (Hanley et al., 1999). MAP1B expression is also maintained in the adult PNS (Gordon-Weeks and Fischer, 2000). In contrast, MAP1A is essentially absent from the embryonic nervous system, and expression levels are highest in neurons in the adult brain.

MAP1B is a polyprotein and during, or soon after, translation, a proteolytic cleavage event separates an N-terminal heavy chain (HC) from a C-terminal light chain (LC1, Figure 3.1) (Hammarback et al., 1991). MAP1A is also a polyprotein with its own light chain, LC2 (Langkopf et al., 1992). The proteases responsible have not been identified but, since uncleaved protein has never been found, it is probably constitutively activated and unregulated. MAP1B HC and LC1 both possess MT-binding domains and also bind to each other, in regions that have been mapped (Noble et al., 1989; Noiges et al., 2002; Zauner et al., 1992) (Figure 3.1). It is thought that the HC and LC form a MT-binding complex on the MT wall along with LC3, a separate gene product (Mann and Hammarback, 1994). There is no evidence that LC1 functions in isolation, although heterologous transfection experiments reveal that LC1 can bind to actin filaments (Togel et al., 1998). MAP1B HC seems to suppress this activity (Togel et al., 1998). In the C-terminal half of the HC of MAP1B is a highly basic motif, imperfectly repeated 21 times, which binds MTs along their length, probably at the negatively charged external surface (Noble et al., 1989). Regions flanking the repeat-region are also involved in MT binding (Noble et al., 1989; Zauner et al., 1992). This MT-binding domain of MAP1B has

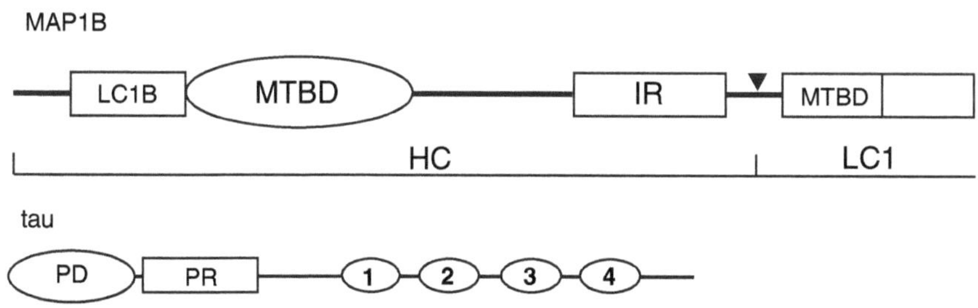

FIGURE 3.1. Domain organization of structural MAPs. Schematic representation of the domain organization of MAP1B and tau. The N-terminus is to the left. The microtubule-binding domains (MTBD) in the HC and LC1 of MAP1B are shown along with the imperfect repeat region (IR) and the region where LC1 binds to the HC (LC1B). The arrowhead indicates the site of proteolytic cleavage that gives rise to the HC and LC1. In tau, the microtubule-binding repeat domains are numbered 1–4, for the tau form with the highest number of repeats. The proline-rich region (PR) and the projection domain (PD) are also shown.

sequence homology with that of MAP1A but is distinct from the MT-binding domain in the tau/MAP2/MAP4 family (see in the following paragraph).

The consequences of MAP1B binding to MTs are not very marked (Vandecandelaere et al., 1996). Although MAP1B can lower the critical concentration of tubulin for plus-end assembly and stabilize MTs, these effects are not as strong as those seen with the tau/MAP2/MAP4 family (see the following text). MAP1B does not induce MT bundles when overexpressed in heterologous cells, unlike the tau/MAP2/MAP4 structural MAP family (Goold et al., 1999; Takemura et al., 1992).

At a cellular level, MAP1B appears not to be necessary for neurite initiation or neuronal polarity but it does play a role in neuritogenesis and possibly in growth cone pathfinding (Gordon-Weeks and Fischer, 2000). Early experiments with primary neuronal cultures showed that antisense oligodeoxynucleotides to MAP1B lower MAP1B expression and reversibly reduce the rate of neurite growth or block neurite outgrowth completely (Brugg et al., 1993; DiTella et al., 1996). These experiments have not yet been confirmed with siRNA technology. There are four mouse lines in which the MAP1B locus has been disrupted by homologous recombination (Edelmann et al., 1996; Gonzalez-Billault et al., 2000; Meixner et al., 2000; Takei et al., 1997), but only one of these lines is a true null (Meixner et al., 2000). A diverse, and only partially overlapping, spectrum of neural abnormalities is seen in these transgenic mice (Gonzalez-Billault and Avila, 2000). These include aberrant formation of major axon tracks, particularly the corpus callosum, in the CNS (Del Rio et al., 2004; Meixner et al., 2000; Takei et al., 2000). The malformation of axon tracks has been interpreted as supporting a role for MAP1B in growth cone pathfinding (Meixner et al., 2000). However, such a phenotype may also result if axons simply grow more slowly than normal and therefore fail to arrive at the appropriate time to respond to growth cone guidance cues at "choice points" where pathfinding decisions are made. Indeed, cultured embryonic dorsal root ganglion neurons from one of the MAP1B knockout mice do extend axons more slowly than normal (Gonzalez-Billault et al., 2002), as do neurons from the tau/MAP1B double knockout mouse (Takei et al., 2000).

More direct evidence for a role for MAP1B in pathfinding comes from two sets of experiments with neurons in culture. In growth cones turning in culture at sharp substrate borders between permissive and nonpermissive molecules, MAP1B, phosphorylated by cyclin-dependent kinase-5 (Cdk5) is restricted to that part of the growth cone that remains on the permissive surface (Hahn et al., 2005; Mack et al., 2000). Micro-CALI inactivation of MAP1B phosphorylated by Cdk5 on one side of the growth cone compromises turning on that side. Similarly, the growth cones of adult dorsal root ganglion cells from MAP1B knockout mice (Meixner et al., 2000) have impaired turning responses in culture (Bouquet et al., 2004).

A considerable body of indirect evidence suggests that MAP1B influences MT stability in growing axons and that this is regulated by phosphorylation, particularly by GSK-3β (Gonzalez-Billault et al., 2004). MAP1B has been shown to be directly phosphorylated by the serine/threonine kinase glycogen synthase kinase-3β (GSK-3β) in embryonic axons (Trivedi et al., 2005) and may also be directly phosphorylated by Cdk5 (Hahn et al., 2005). Inhibition of GSK-3β by Wnts, intercellular signaling proteins that play widespread roles in embryonic cell specification, stem cell differentiation and synaptogenesis (Ciani and Salinas, 2005; see also Chapter 10), or by pharmacological agents, such as lithium chloride or small organic molecule inhibitors, reduces the rate of axon extension in differentiating neurons in culture, induces spread areas along the neurite shaft, enlarges the size of growth cones, and disrupts growth cone filopodia dynamics (Goold et al., 1999; Hall et al., 2000, 2002; Lucas and Salinas, 1997; Lucas et al., 1998; Owen and Gordon-Weeks, 2003; Takahashi et al., 1999). This is associated with an increase in the numbers of stable MTs in growth cones, suggesting that phosphorylation of MAP1B by GSK-3β plays a role in regulating MT dynamics. Furthermore, neurons cultured from MAP1B knockout mice, in which MAP1B expression is absent or greatly reduced, also show striking increases in stable MTs in growth cones, and this is associated with a reduction in axon growth and an increase in growth cone size (Gonzalez-Billault et al., 2001, 2002; Takei et al., 2000; Teng et al., 2001). A role for GSK-3β in regulating the control that MAP1B exerts on MT stability in growth cones is supported by the finding that cotransfection of MAP1B and GSK-3β into nonneuronal cells in culture leads to a loss of stable MTs and this effect is dependent on the phosphorylation of MAP1B by GSK-3β (Goold et al., 1999; Trivedi et al., 2005) and that focal destruction of phosphorylated MAP1B in growth cones in culture using micro-CALI interferes with growth cone turning (Mack et al., 2000). Finally, recent evidence shows that the growth cone guidance molecule netrin-1 regulates MAP1B phosphorylation by Cdk5 and GSK-3β and that, therefore, MAP1B may be a down-stream regulator of growth cone guidance molecules (Del Rio et al., 2004).

3.3.2. Tau Proteins

Tau proteins are low molecular weight MAPs that are highly expressed in the embryonic and adult nervous system (Brandt and Leschik, 2004). A single gene encodes members of the tau protein family, and molecular heterogeneity is produced by splicing of the primary transcript and posttranslational modification, chiefly phosphorylation. There are six isoforms in the mammalian adult nervous system that differ by the presence of three, alternatively spliced exons. In contrast, in the developing CNS, only the shortest isoform is present. In neurons of the peripheral nervous system, additional splice isoforms are expressed, producing high molecular weight taus. Tau proteins are extended molecules that lack secondary structure. However, four regions within tau are recognized: an N-terminal projection domain that extends from the outer surface of the MT wall; a proline-rich region; a repeat-region of three or four (when exon-10 is present) imperfect 18-amino acid repeats separated from each other by stretches of 13 or 14 amino acids; a C-terminal tail domain (Figure 3.1). The repeat-region represents the major MT-binding domain, but a short sequence within the proline-rich region flanking the N-terminal end of the repeat-region strongly influences MT binding and is required for the ability of tau to promote MT assembly (Goode et al., 1997). There is also evidence that the C-terminal flanking region plays a role in MT binding. Structural analysis suggests that the repeat-regions bind to the surface of β-tubulin that faces the MT lumen (Amos, 2004). With the exception of exon-10, which contributes to MT binding, the function of the alternatively spliced exons is unknown.

Tau proteins bind along the length of MTs and promote MT assembly and stability, as revealed by *in vitro* cell injection of tau (Drubin and Kirschner, 1986) and heterologous cell transfection studies (Brandt and Leschik, 2004). In addition, MT bundling is also seen in cells transfected with high levels of tau or the tau isoform containing four imperfect repeats (Chen et al., 1992; Kanai et al., 1989; Lee and Rook, 1992; Takemura et al., 1992). However, there is no direct evidence that tau, or any of the structural MAPs, can cross-link or bundle MTs and no domains that are associated with bundling, as distinct from MT binding, have been identified. Microtubule bundling may occur in transfected cells as an indirect effect of increased MT stability (Lee and Brandt, 1992). More likely, the N-terminal projection domain of tau (and MAP2) probably serves to repel neighboring MTs and thus determine inter-MT distance (Chen et al., 1992). One consequence of this is that tau may facilitate the movement of MT motors, and their cargo, along MTs by creating space (Chen et al., 1992). When MTs bind high levels of tau, for example, when overexpressed in cells, MT motor movement is impeded (Amos and Schlieper, 2005). It is not difficult to see how this function might be very important for neuritogenesis.

The functions of tau proteins are modulated by phosphorylation, and many of the phosphorylated amino acids are located in either the proline-rich region or the C-terminal tail that flank either side of the repeat-region (Billingsley and Kincaid, 1997). Phosphorylation of tau within the MT-binding domains or their flanking regions tends to decrease MT binding and the ability to promote MT assembly (Gustke et al., 1994; Preuss et al., 1997). Tau is hyperphosphorylated in several neurodegenerative diseases and in Alzheimer's disease, tau is the major protein of neurofibrillary tangles, particularly, but not exclusively, at GSK-3β sites (Billingsley and Kincaid, 1997; Brandt and Leschik, 2004).

In differentiating neurons, tau is expressed in both the somato-dendritic and axonal compartments, although more highly in the latter (Binder et al., 1985; Brandt et al., 1995; Gordon-Weeks et al., 1989; Kosik and Finch, 1987). In growing axons, there is a gradient of phosphorylated tau, which is highest near the cell body (Mandell and Banker, 1996). An antibody, Tau-1, specifically labels growing axons because it recognizes tau at a phosphorylation site only when it is unphosphorylated. This antibody has been extensively used as a marker for axons in studies of neuronal polarity.

There is contradictory evidence that tau contributes to neuritogenesis. Experiments in cultured neurons with antisense oligodeoxynucleotides against tau suggest that this MAP is important for neurite extension, particularly axonal extension (Caceres and Kosik, 1990; Caceres et al., 1991; DiTella et al., 1994; Hanemaaijer and Ginzburg, 1991; Shea et al., 1992). Neurites extend more rapidly than normal in PC12 cells overexpressing tau and, conversely, more slowly when tau levels are reduced in PC12 cells using antisense oligodeoxynucleotides (Esmaeli-Azad et al., 1994). When endogenous tau expression is suppressed in cerebellar macro-neurons in culture, although these cells fail to differentiate an axon, neurite formation is unimpaired (Caceres and Kosik, 1990; Caceres et al., 1991). Overexpression of tau in insect Sf9 cells that do not normally express this MAP induces these cells to extend long, neurite-like processes containing parallel arrays of MTs (Baas et al., 1991; Knops et al., 1991; LeClerc et al., 1996). Chromophore-assisted laser inactivation of endogenous tau in chick dorsal root ganglion neurons in culture reduces neurite number and length, when applied to the whole cell, and reduces neurite extension rate when applied to the growth cone (Liu et al., 1999). In contrast, microinjection of tau antibodies into cultured sympathetic neurons has very little effect on axon growth or MT distribution and dynamics (Tint et al., 1998).

Genetic experiments in mice are also contradictory. The first tau knockout mouse to be derived is relatively normal, showing only subtle changes in MT organization in small caliber axons, and cultured neurons from these mice differentiate axons and dendrites normally (Harada et al., 1994). However, embryonic hippocampal neurons cultured from a second, independently derived, tau knockout mouse shows delays in the development of neuronal polarity (Dawson et al., 2001). The emergence of neurites from hippocampal neurons from these knockout mice is delayed compared to neurons from wild-type mice, and the length of both axons and dendrites is shorter as a consequence. It is not clear, however, if the rate of elongation is slower, since it was not measured; although the axons from hippocampal neurons of knockout mice do respond with increased growth rate when grown on laminin. In both knockouts, there was a compensatory increased expression of MAP1A, but not other structural MAPs, including MAP1B and MAP2 (Dawson et al., 2001; Harada et al., 1994). There may, however, be redundancy of function between the structural MAPs, e.g., MAP1B/tau because, although the individual knockouts are viable, the double knockout (MAP1B/tau) is not (Takei et al., 2000, 2001). In the double knockout, the loss of the MAP1B alleles is the more serious in terms of the severity of the phenotype.

3.4. Microtubule Plus-End Proteins (+TIPs) in Neuritogenesis

The MAPs so far described tend to bind uniformly along the MT providing mechanical stabilization, a scaffold for other proteins and regulation of inter-MT distance in MT bundles. During neuritogenesis, the MTs themselves must also extend in order to support the nascent process. This is both physical in terms of mechanical strength and also functional by providing tracks for the rapid directional delivery of molecular cargos to and from the growth cone. The rate of extension of the axonal MTs as a population must be carefully regulated in a vectorial fashion. In terms of magnitude, this is not only to prevent excessive MT accumulation but also to ensure that the net direction of polymerization coordinates with that in which the growth cone is heading. These properties are controlled by a class of proteins known as "+TIPs" which specifically accumulate at the plus end of growing MTs. They can be further divided into families based on their structure and mode of MT binding. Although endogenous +TIPs are localized to MT tips, at increasing levels of overexpression binding occurs along the length of the MT and in most cases eventually causes the MTs to bundle. Binding, in nearly every instance, is sensitive to the addition of the MT-depolymerizing drug nocodazole. The importance of +TIPs as fundamental constituents of the cytoskeleton is reflected in their frequent conservation throughout eukaryotic species (Akhmanova and Hoogenraad, 2005). Despite the ubiquity of many of their members, some appear to have specialized functions in neurons and are essential for specific aspects of neural development. Mutations of +TIPs are associated with a range of neurological disorders of differing severity (Hoogenraad et al., 2004; Kato and Dobyns, 2003). The following section will give a brief overview of the main families of +TIPs and an outline of their structure and localization. Subsequent sections will then address the key issues of how they bind so specifically to the tip of growing MTs and how this is regulated. With this background in place, evidence for the essential role of +TIPs in neuritogenesis will be discussed.

3.4.1. +TIP Families

3.4.1.1. CLIPs

The distinction of being the first +TIP to be discovered belongs to CLIP-170. It was isolated in a search for proteins with ATP-sensitive binding to MTs in the hope of finding factors involved in motor protein function. As the name suggests, a 170 kDa protein was identified, and its novelty lay in the fact that, while it bound MTs, it did so to a subset of them and at discrete sites (Rickard and Kreis, 1990). Closer analysis revealed that this protein linked endocytic vesicles to MTs, thus it was named *c*ytoplasmic *li*nker *p*rotein (CLIP) (Pierre et al., 1992). Human CLIP-170 is an extended molecule, 135 nm long, with three main structural features of a head and a tail linked by a central rod (Pierre et al., 1992, 1994; Scheel et al., 1999). Within the N-terminus head, it contains two "CAP-Gly" motifs (Li et al., 2002), flanked by basic, serine-rich regions which, together, are necessary and sufficient for MT binding (Figure 3.2). However, full-length CLIP-170 binds tubulin with a lower stoichiometry demonstrating the influence of the rod and tail (Scheel et al., 1999). The C-terminus tail contains two "zinc-knuckle" metal-binding motifs which are required for interactions with other proteins and the induction of MT bundles (Goodson et al., 2003). The two ends are linked by a long series of coiled-coil domains, which are sufficiently flexible for the tail domain to interact with the head and inhibit MT binding (Lansbergen et al., 2004). Tissue-specific alternative splicing of vertebrate CLIP-170 allows insertion of an 11 and/or 35 amino acid segment within the coiled-coil stretch (Griparic and Keller, 1998; Griparic and Keller, 1999; Griparic et al., 1998). An independently discovered leucocyte protein, named restin and thought to bind intermediate filaments rather than MTs, is identical to CLIP-170 containing the 35 amino acid insert (Bilbe et al., 1992; Griparic et al., 1998). The functional significance of the inserts is unknown.

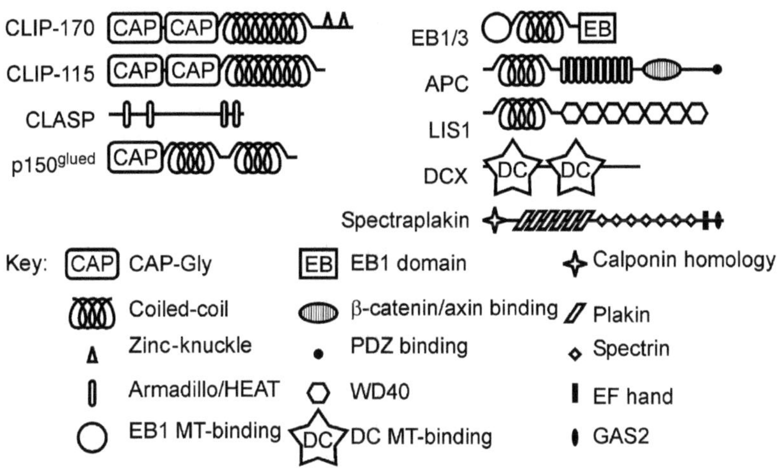

FIGURE 3.2. Structure of +TIP proteins. Schematic representation of the structure of +TIPs. Key regions described in the text are shown. The N-terminus is to the left. Note that the relative sizes of the molecules are not to scale.

CLIP-115 is the other major member of the CLIP family and is nervous system–specific; it has two N-terminal CAP-Gly MT-binding domains with 93% and 97% similarity to those of CLIP-170, also flanked by serine-rich regions. These are followed by a stretch of coiled-coil sequences, but the protein is truncated and lacks the C-terminal metal-binding motifs (De Zeeuw et al., 1997). Although CLIP-115 causes MTs to bundle when it is overexpressed, it does not form aggregates, unlike CLIP-170 (Hoogenraad et al., 2000; Perez et al., 1999).

3.4.1.2. CLASPs

A search for proteins that interact with CLIPs isolated two mammalian proteins termed CLASPs (*CL*IP-*As*sociating *P*roteins), which bind to both MTs and CLIPs (Akhmanova et al., 2001). They are the mammalian homologs of a previously characterized *Drosophila* protein, Orbit/Mast, required for mitosis (Inoue et al., 2000; Lemos et al., 2000). Endogenous CLASPs bind to MT tips while overexpression causes MT bundling and a relocalization of CLIP-170 to the bundles. There are multiple isoforms of both CLASP1 and CLASP2, but they share a common C-terminal domain that interacts with the coiled-coil region of the CLIPs (Akhmanova et al., 2001).

3.4.1.3. End-Binding Proteins

There are three main members of the end-binding 1 (EB1) family, although up to seven members have been postulated (Su and Qi, 2001). The first one to be discovered, EB1, is a 268 amino acid protein isolated in a screen for proteins interacting with the adenomatous polyposis coli (APC) tumor suppressor

(Su et al., 1995). EB1 and its homologs are intimately involved in mitosis, both for positioning the spindle and segregating chromosomes (Tirnauer and Bierer, 2000). Like the CLIPs, EB1 binds to the tips of MTs while they are elongating (Berrueta et al., 1998; Mimori-Kiyosue et al., 2000b; Morrison et al., 1998). Structurally, although much smaller than CLIP-170, EB1 also has a tripartite arrangement of an N-terminal MT-binding head linked by coiled-coils to a distinctive "EB1 family" C-terminal tail, which interacts with other proteins (Figure 3.2). EB1 family members have a single, characteristic MT-binding domain, not of the CAP-Gly variety but one that is surprisingly similar, topographically, to the calponin homology domain, common to many actin-binding proteins (Hayashi and Ikura, 2003). The MT-binding domain is necessary but not sufficient to cause MT bundling (Bu and Su, 2003). Again in a parallel with CLIP-170, there is strong evidence that EB1 can adopt an autoinhibitory conformation in which the carboxy terminal folds back along the flexible intermediate domain to inhibit the MT-binding domain (Hayashi et al., 2005). EB1 is ubiquitous, however EB2 is rapidly upregulated in activated T cells (Renner et al., 1997) and EB3 is enriched in the nervous system (Nakagawa et al., 2000).

3.4.1.4. Adenomatous Polyposis Coli

APC was identified as the gene mutated in familial polyposis and sporadic colorectal cancer thus removing its influence on the degradation of β-catenin and the Wnt signaling pathway (Groden et al., 1991; Nathke, 2004). It is a very large protein of 310 kDa comprising several functional domains, among which is a highly basic MT-binding region with similarity to tau protein (Deka et al., 1998). Fluorescently tagged APC moves along the length of MTs accumulating at the tips rather than directly attaching there (Askham et al., 2000; Mimori-Kiyosue et al., 2000a). There is thus a distinction between those +TIPs that can directly recognize and bind to growing MT ends and those that require some form of assistance. The interactions between +TIPs are numerous and how these multimolecular complexes are regulated is discussed in more detail in the later sections.

3.4.1.5. $p150^{glued}$

Dynein is a motor protein directed toward the minus end of MTs (Vale, 2003). Translocation requires a complex of accessory proteins, and it was during the identification of these that the 150 kDa protein named dynactin, for dynein-interacting protein, was isolated (Gill et al., 1991; Holzbaur et al., 1991). This turned out to share more than 50% sequence identity with a *Drosophila* gene, glued, identified nearly a decade earlier (Harte and Kankel, 1982), hence the title $p150^{glued}$. Dynactin is often used as a generic term for the entire dynein accessory complex, so $p150^{glued}$ will be used here to denote one specific component. $p150^{glued}$ forms a sidearm with a globular head and projects from the core of the dynactin complex, suggesting that it may function as a receptor (Schafer et al., 1994). Structurally, $p150^{glued}$ has similarities with CLIP-170, the most salient being a single N-terminal CAP-Gly domain, followed by a coiled-coil region

(Figure 3.2). Although p150^{glued} binds in a nocodazole-sensitive manner to MT tips and can cause MTs to bundle (Ligon et al., 2003; Vaughan et al., 1999), whether it can do this autonomously is open to debate. There is evidence that CLIPs are required for the binding of p150^{glued} to MTs (Komarova et al., 2002), while conflicting data suggest that it recognizes and binds MT tips independently of CLIPs or indeed EB1 (Vaughan et al., 2002).

3.4.1.6. Lissencephaly Genes: LIS1, DCX

Lissencephaly is one of the most striking examples of the requirement of MT function in neural development. The name literally means "smooth brain" and is a succinct description of the actual phenotype. In affected individuals the ridges of the cerebral cortex, known as gyri, are lost to a variable degree depending on the severity. In mild cases, this leads to clumsiness and seizures but in the worst cases mental retardation and a dramatically shortened life expectancy result. Causative mutations have been identified in a number of genes, each with a subtly different manifestation (Kato and Dobyns, 2003). The first gene isolated as being responsible, LIS1 (Reiner et al., 1993) was initially characterized as a subunit of brain platelet-activating factor acetylhydrolase (Hattori et al., 1994), an enzyme that inactivates platelet-activating factor, a protein involved in neuronal differentiation and signaling. LIS1 is a 47 kDa protein, the bulk of which consists of seven WD40 repeats, a domain associated with the formation of protein scaffolds (Figure 3.2). There is also a short "LIS homology" domain at the N-terminus, which is involved in homodimerization (Cahana et al., 2001). Despite the lack of an identified MT binding motif, it is clear that LIS1 binds to MTs (Faulkner et al., 2000; Sapir et al., 1997; Smith et al., 2000), but whether this occurs in interphase cells has been disputed (Faulkner et al., 2000). The binding to MTs, particularly recruitment to their ends, seems to be dependent on CLIP-170 (Coquelle et al., 2002).

Mutations in doublecortin (DCX) cause lissencephaly phenotypically similar to LIS1 but affecting different cortical areas (desPortes et al., 1998; Gleeson et al., 1998, 2000). DCX is a 40 kDa protein that binds MTs and is enriched in the tips of the processes of migrating neurons. Unlike LIS1, DCX is able to bind MTs independently of other proteins (Francis et al., 1999; Gleeson et al., 1999; Horesh et al., 1999). This ability is conferred by two N-terminal DCX domains (Figure 3.2) (Kim et al., 2003). The C-terminal of DCX is rich in serine and proline residues and may be required for vesicle transport (Friocourt et al., 2001).

3.4.1.7. Spectraplakins

The spectraplakins are a large family of giant proteins. They function as molecular cross-linkers between the cytoskeleton and the plasma membrane, particularly at sites of cellular adhesion, and also between elements within the cytoskeleton itself. Their defining motifs are an N-terminal plakin domain followed by a series of spectrin repeats. The spectraplakin genes are very complex with many potential splice donor and acceptor sites generating large RNAs up to 27 kb in size

(Roper et al., 2002). Interactions with MTs were first demonstrated for a *Drosophila* plakin known as Short Stop (Gregory and Brown, 1998; Prokop et al., 1998). Specific targeting to MT tips has been demonstrated for mammalian ACF-7 (Kodama et al., 2003), while other spectraplakins interact with +TIPs such as EB1 or p150^{glued} (Liu et al., 2003; Slep et al., 2005; Subramanian et al., 2003).

3.4.2. Regulation of +TIP Localization

3.4.2.1. +TIPs That Bind Microtubule Tips Directly

As is clear from the brief introduction to the various +TIPs, they employ different mechanisms to arrive at the end of MTs (Figure 3.3). Some possess an intrinsic affinity for tubulin, however, this is a property shared with the other MAPs described earlier and does not explain their remarkable specificity for the plus ends of growing MTs. Other +TIPs do not interact so readily with tubulin and instead "hitchhike" (Carvalho et al., 2003) a ride to their destination. +TIPs do not simply bind the extreme end of the MT, they extend a short distance along it, in the form of dashes or "comets." These are clearly visible using fluorescently tagged +TIPs, which produce an intracellular fireworks display (Mimori-Kiyosue et al., 2000b; Perez et al., 1999). There are three main mechanisms by which +TIP binding may be regulated in order to create comets. First, +TIPs could be incorporated into the MT structure by copolymerization with tubulin subunits. Second, they may be retrogradely transported a short distance before detaching. Third, they may recognize a biochemical or physical structure specific to MT plus ends. This last hypothesis provides perhaps the best scope for explaining how +TIPs would subsequently dissociate at a short distance from the end. At the plus end of a MT, chains of tubulin

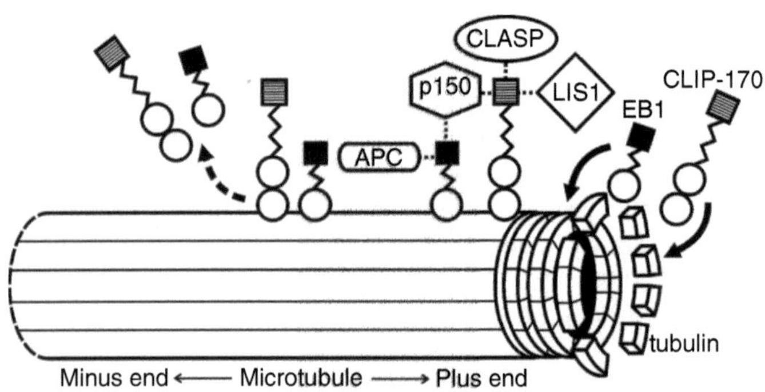

FIGURE 3.3. +TIP MT-binding mechanisms. EB1 and CLIP-170 bind to MT plus ends at the site of tubulin assembly (heavy curved arrows) via their MT-binding domains (circles). These are linked by a coiled-coil stretch (zig-zag) to characteristic C-terminal motifs (black square—EB1; hatched square—CLIP-170), which promote localization of other proteins at the plus end. Examples shown are APC, p150^{glued}, CLASP, and LIS1, interactions are indicated by dotted lines. As MT extension proceeds, EB1 and CLIP-170 dissociate from the older MT (dashed curved arrow).

dimers form protofilaments which align longitudinally into sheets. These sheets then roll up to form the hollow tube that constitutes a MT (Chrétien et al., 1995). The curvature of the sheets is thermodynamically unfavorable, and so MAPs are required to act as molecular staples and maintain the cylindrical structure. For a detailed account of the mechanics of MT ends see Howard and Hyman (2003). As regards a possible biochemical tag, there is thought to be a "cap" of GTP-tubulin at the very tip of the MT, which may be recognized by +TIPs (Chrétien et al., 1995).

Evidence for +TIP association with protofilaments comes from studies of fluorescently labeled EB1 binding to MTs in cell extracts. EB1 most strongly associated with curved structures emanating from the tips of growing MTs. Furthermore, speckles of EB1 remained static with respect to the comet as a whole therefore arguing strongly against its transport (Tirnauer et al., 2002). Thus, a possible localization mechanism is that EB1 binds to the tubulin protofilaments but is dislodged as they roll up into a tube. There is also evidence that EB1 binds to tubulin dimers before their incorporation (Juwana et al., 1999). The procedure by which +TIPs bind to the end of MTs and subsequently dissociate further along after additional polymerization has occurred is termed "treadmilling." Like EB1, CLIP-170 is also believed to treadmill and is not subject to translocation from its absolute point of MT binding (Folker et al., 2005). However, it does not appear to have a preference for a specific physical conformation of the MT end. Instead, it seems to associate with free tubulin oligomers before incorporation into the MT (Diamantopoulos et al., 1999). There is evidence that CLIP-170 exhibits a preference for GTP-tubulin over the GDP-bound form, however, this is of minor significance compared to its affinity for free tubulin dimers (Folker et al., 2005). To address the question of how CLIP-170 subsequently dissociates from older polymer, it has been proposed that the tubulin oligomers to which it binds are initially in a curved configuration favorable to CLIP-170 binding. On incorporation into the growing MT lattice, this curvature is lost along with the CLIP-170 (Arnal et al., 2004). It seems probable that an additional regulatory event occurs to ensure a rapid rate of dissociation (Folker et al., 2005). One candidate for this is the presence of the C-terminal tyrosine residue of tubulin. For reasons still unknown, tubulin normally undergoes a cycle of detyrosination at its C-terminal, which is then reattached by a specific tubulin tyrosine ligase. In the absence of this enzyme, CLIP-170 fails to bind MT tips whereas EB1 is apparently unaffected (Erck et al., 2005).

DCX binds directly to MTs, for which its two DC domains are essential. However, these motifs do not share any sequence similarity with the MT-binding motifs of classical (structural) neuronal MAPs. High-resolution imaging of the structure of DCX using cryoelectron microscopy has revealed the manner in which it binds MTs (Moores et al., 2004). The DC domains are globular (Kim et al., 2003), unlike the coiled structure of the EB1 or CLIP-170 MT-binding domains (Hayashi and Ikura, 2003; Li et al., 2002), and can wedge in between protofilaments. The accessibility of this binding site varies with the number of protofilaments. Thirteen protofilaments naturally constitute a MT, and the optimal binding correlates neatly with this number (Moores et al., 2004). How DCX is regulated to both stabilize MTs but at the same time be located to the distal tips of neuronal processes is unclear (Friocourt et al., 2003; Schaar et al., 2004).

3.4.2.2. +TIPs That Bind Microtubule Tips Indirectly

Other +TIPS require assistance to bind MTs or to find their way to the ends (Figure 3.3). The C-terminus of EB1 acts as a cargo-binding region for recruiting several other proteins to MT ends, indeed spectraplakins and APC share a common motif which recognizes this domain (Slep et al., 2005). APC binds to the C-terminus of EB1 family proteins independently of MTs (Askham et al., 2002; Barth et al., 2002; Bu and Su, 2003; Honnappa et al., 2005). Conversely, APC lacking the EB1-binding region can bind MTs, significantly however, it can no longer accumulate at the tips (Askham et al., 2000). In fact, APC requires further help even to reach the ends of MTs. Fluorescently tagged APC moves along MTs before concentrating at the tips (Mimori-Kiyosue et al., 2000a), and this motion is dependent on a plus-end directed kinesin motor protein (Jimbo et al., 2002). $p150^{glued}$ is also promiscuous in its use of binding partners to reach MT ends. The use of deletion mutants and crystallographic data has revealed a binding site for $p150^{glued}$ at the C-terminus of EB1 family proteins (Askham et al., 2002; Berrueta et al., 1998; Bu and Su, 2003; Hayashi et al., 2005). In addition, $p150^{glued}$ can also hitch a ride on CLIP-170 (Komarova et al., 2002) by binding to the distal zinc knuckle (Goodson et al., 2003). This motif within CLIP-170 is also required for the localization of LIS1 to MT ends (Coquelle et al., 2002). This interaction may mediate the link between LIS1 and dynactin (Faulkner et al., 2000; Smith et al., 2000). In addition, LIS1 can also form complexes with DCX (Caspi et al., 2000). CLASPs were initially identified as proteins that bound to MT tips through the CLIPs, although they were subsequently found to contain a repeated motif that can bind both EB1 and EB3 and is necessary for recruitment to MT ends (Akhmanova et al., 2001; Mimori-Kiyosue et al., 2005). While the CLIP and EB1 families may appear to be discrete master regulators of +TIP complexes, recent data suggests that they actually form an important interaction which edges the EB family ahead in the hierarchy. CLIPs bind to the EB proteins, and the C-terminal tyrosine residue of EB proteins is necessary to slow the dissociation of CLIPs from MTs and allow comets to form (Komarova et al., 2005).

In addition to the features listed previously, through which +TIPs may recognize MT ends, there is extensive regulation of their binding to MTs and to each other, in most cases by phosphorylation (Figure 3.3). EB1, $p150^{glued}$, and DCX are all subject to serine phosphorylation, which reduces their affinity for MTs (Akhmanova et al., 2001; Graham et al., 2004; Hayashi and Ikura, 2003; Tanaka et al., 2004; Vaughan et al., 2002). The C-terminal portion of APC is required for it to bind EB1, but this interaction can be blocked by phosphorylation of APC by kinases including protein kinase A and Cdc2 (Askham et al., 2000; Nakamura et al., 2001). The interaction of APC and CLASPs with MTs is attenuated after phosphorylation by GSK-3β (Akhmanova et al., 2001; Zumbrunn et al., 2001). CLIP-115 is also sensitive to phosphorylation (Hoogenraad et al., 2000) while

CLIP-170 also exists in phosphorylated forms but its regulation is complicated: either phosphorylation or dephosphorylation can increase its binding to MTs (Choi et al., 2002). LIS1 is subject to regulated phosphorylation (Sapir et al., 1999; Schaar et al., 2004), which is required for its recruitment by CLIP-170 (Coquelle et al., 2002).

3.4.3. +TIP Control of MT Dynamics

One further consideration is the importance of the selectivity of +TIPs for growing MT ends and their actual effect once there. As discussed previously, MTs exhibit a property known as dynamic instability, alternating stochastically between periods of polymerization and depolymerization. The switch from the former to the latter is termed a catastrophe while the reverse is named a rescue. Thus, the kinetics of a MT end can be defined by a series of parameters: the rate of growth or shrinkage and the frequency of catastrophes or rescues. +TIPs influence these variables and hence the net behavior of the MT end (Carvalho et al., 2003; Howard and Hyman, 2003). The exact effect of +TIPs is often complicated by whether the analysis is performed *in vitro* with purified MTs and the presence or absence of other +TIPs. EB1 acts to increase MT stability by modulating multiple factors, it decreases the frequency of catastrophes and the rate of depolymerization while also increasing the rescue frequency and the time spent polymerizing (Tirnauer et al., 2002). EB1 does not efficiently induce *de novo* polymerization; it does however increase polymerization of seeded MTs. The reverse is true of $p150^{glued}$; it has little effect on MT seeds but can induce tubulin polymerization in their absence (Ligon et al., 2003). The CLASPs have a stabilizing effect on MTs by decreasing the frequency with which either growth or shortening events occur (Akhmanova et al., 2001; Mimori-Kiyosue et al., 2005). CLIP-170 seems to act more specifically by promoting rescue events. In the absence of the CLIP-170 head, MTs grow out toward the cell periphery but then undergo long periods of depolymerization, suggesting that CLIP-170 enables dynamic plus ends to persist near the cell edge (Komarova et al., 2002). Whereas CLIP-170 promotes rescue, LIS1 reduces the frequency of catastrophes (Sapir et al., 1997), thus the effects of these two proteins might be synergized by their mutual interaction (Coquelle et al., 2002). APC promotes MT polymerization and stabilizes them by slowing the rate of depolymerization (Munemitsu et al., 1994; Zumbrunn et al., 2001).

3.4.4. +TIPs in Neuritogenesis

The generic importance of +TIPs in neural development is underlined by the neurological phenotype of humans and mice harboring mutations or deletions of +TIP genes. In adult mice, CLIP-115 expression is highest in the hippocampus, and in animals lacking CLIP-115, there is a decrease in synaptic plasticity, implying a role in hippocampal memory processes (Hoogenraad et al., 2002). CLIP-115 is also found in the dendrites of a proportion of the Purkinje cells and correspondingly, knockout mice display motor impairments. Additional abnormalities in these mice include decreased corpus callosum and increased ventricle size, however MT dynamics, surprisingly, are normal. There is, though, increased accumulation of CLIP-170 and $p150^{glued}$ at MT ends suggesting that they may compensate for the absence of CLIP-115 (see also downregulation of CLASP) (Hoogenraad et al., 2002). The human CLIP-115 gene, *CYLN2*, lies within the chromosomal region deleted in cases of Williams Syndrome a disorder affecting cardiac, craniofacial, and neural development (Bellugi et al., 1999; Hoogenraad et al., 2004). Many of the neurological deficits exhibited by mice lacking CLIP-115 are also displayed by Williams Syndrome patients (Hoogenraad et al., 2004). Given the central role of CLIP-170 in +TIP localization, it is noteworthy that mice hypomorphic for CLIP-170 display no gross defects, except male sterility as a result of morphologically abnormal sperm (Akhmanova et al., 2005). Mutations in $p150^{glued}$ occur in families with amyotrophic lateral sclerosis (Munch et al., 2004, 2005), which probably arises from disruption of dynein-mediated transport rather than aberrant MT polymerization (LaMonte et al., 2002). The effects of LIS1 and DCX mutations in humans are well-documented and have been described previously (Kato and Dobyns, 2003). Homozygous loss of LIS1 is lethal but, in mice, developing neurons are sensitive to the gene dosage, mirroring the phenotypic variation seen in humans (Smith et al., 2000).

CLIP-115, EB3, LIS1, and DCX are all highly enriched in neurons (De Zeeuw et al., 1997; Francis et al., 1999; Gleeson et al., 1999; Nakagawa et al., 2000; Smith et al., 2000), suggesting that they may play specialized roles in these cells in response to the unique demands that neuritogenesis places on the neuronal cytoskeleton. First, the cell must be specifically polarized in order to generate an axon and dendrites. Second, once their position has been defined, these processes must elongate over long distances many times the size of the soma. Third, neurite extension is directional. The growth cone at its distal tip responds to environmental cues in order to navigate to its target. Therefore, the process of cytoskeletal lengthening must be sufficiently flexible to allow the growth cone to turn and manoeuvre. +TIPs have the capacity to underlie the achievement of all three of these stipulations.

In migrating cells, polarized protrusions at the leading edge are dependent on the action of APC on MTs in response to spatially-regulated kinase activity (Etienne-Manneville and Hall, 2003; Etienne-Manneville et al., 2005). APC is involved in the same process in neurites; it marks the nascent axon tip and is required for its efficient elongation, remaining enriched in those that grow fastest (Shi et al., 2004; Shimomura et al., 2005; Zhou et al., 2004). Overexpression of full-length or truncated mutants of APC interfere with the localization of Par3, a polarity protein in neurons (Chapter 11), at the tip of the neurite destined to become the axon and block neuronal polarity (Shi et al., 2004).

The propensity of +TIPs to promote MT elongation implies that they would also be essential for axon elongation. However, there have been few studies to directly test this link, although EB1 has been claimed to facilitate axonogenesis in MAP1B knockout mice (Jimenez-Mateos et al., 2005b). EB1 and EB3 have both been used to monitor the behavior of MT ends in neurites. These revealed that the majority of comets move toward the growth cone, with only a few moving retrogradely (Ma et al., 2004; Morrison et al., 2002; Stepanova et al., 2003). EB3 is found throughout the neuron on MT ends: in the soma, neurites, and growth cones. While EB3 moves unidirectionally in axons, within dendrites it is bidirectional reflecting the MT organization in these compartments (Stepanova et al., 2003). The average speed at which comets move within a given neuronal type is consistent regardless of the direction and whether EB1, EB3, or CLIP-170 is used as a marker (Stepanova et al., 2003). The differing behavior of +TIPs in axons compared to dendrites has also been studied using N2A neuroblastoma cells. The addition of dibutyric cyclic AMP to these cells induces axon formation whereas retinoic acid causes the production of dendrites. By comparison with the undifferentiated state, production of either type of neurite leads to a marked upregulation of $p150^{glued}$ and α-tubulin. EB1 is only significantly upregulated after axons are induced suggesting a bias toward involvement in axonal formation or transport (Morrison et al., 2002). Whether EB3 is subject to similar changes is unknown. EB1 is required for the recruitment of the motor protein myosin-Va to MT ends (Wu et al., 2005) and so may be involved in the correct distribution of cargo proteins to neuronal compartments.

+TIPs may be particularly relevant to growth cone behavior in a manner described by the "search and capture" hypothesis (Kirschner and Mitchison, 1986a). In this model, the dynamic instability of MTs allows them to constantly probe the intracellular environment. A localized concentration of functional +TIPs would tend to increase net MT extension toward that part of the cell. These MTs are then captured at sites on the cell cortex and stabilized producing a reorientation of the cytoskeleton. Although originally a means to explain the role of MTs in generating varied aspects of cell morphogenesis, it is particularly apt as a foundation for the searching performed by axons. In the case of a neuron, MTs dynamically probe the growth cone P-domain. In response to a positive extracellular signal, through the recruitment of +TIPs, MTs would extend toward the part of the growth cone receiving the cue and be captured there. Sustained signaling would therefore lead to a turning of the growth cone toward the attractant source. Two processes must thus be demonstrated: first, MTs and +TIPs probe the growth cone and its extremities and second, MTs can subsequently be captured and stabilized. The first condition is satisfied by the behavior of labeled EB3 that has been filmed probing the growth cone periphery, even entering filopodia (Stepanova et al., 2003). The second requirement has yet to be demonstrated in neurons, but there is accumulating evidence for such a process in other migratory cells, with the molecular protagonists being found also in neurons. IQGAP is a protein recruited to cell edges by the activity of Rac1 and Cdc42 GTPases. At these cortical foci, IQGAP captures MT ends through an interaction with CLIP-170 (Fukata et al., 2002). IQGAP also recruits APC to the cortex of motile cells, and this association is a prerequisite for migration (Watanabe et al., 2004). EB1 is also necessary for MT capture and stabilization through its interaction with APC. As well as binding to each other, EB1 and APC can both bind to a formin, mDia, itself a target for Rho GTPase. The resulting complex is necessary to stabilize MTs and for cell motility (Wen et al., 2004). CLASPs also play a key role in promoting MT polymerization at the cell edge. They are palmitoylated and can directly bind to a component of the cell cortex, in addition they bind to EB1 independently of the CLIPs (Akhmanova et al., 2001; Mimori-Kiyosue et al., 2005). The binding to both EB1 and the cortex is necessary for MT stabilization suggesting that CLASPs play a pivotal, localized role in maintaining MT ends at the cell edge (Mimori-Kiyosue et al., 2005). In support of this, a study of *Drosophila* CLASP, using genetic tools to circumvent the need for it during oogenesis, pinpointed it at the tips of MTs that were probing the growth cone peripheral domain. Furthermore, *Drosophila* CLASP is also required for correct axon guidance decisions and functions downstream of Abl, a tyrosine kinase which modulates many axonal receptors (Lee et al., 2004).

Therefore, a number of pathways are emerging by which localized signaling at the cell periphery by small GTPases can recruit effector proteins, which in turn bind +TIPs and create stable points of MT attachment with the cortex of motile cells. A similar mechanism is likely to occur in growth cones where GTPases are downstream of many axon guidance receptors (Guan and Rao, 2003). It is probable that in neurons the same process occurs to direct and capture MTs but preferentially uses neuronally enriched +TIPs or isoforms thereof (Nakagawa et al., 2000).

References

Ahmad, F.J., Echeverri, C.J., Vallee, R.B., and Baas, P.W., 1998, Cytoplasmic dynein and dynactin are required for the transport of microtubules into the axon, *J. Cell Biol.* **140:** 391–401.

Akhmanova, A., and Hoogenraad, C.C., 2005, Microtubule plus-end-tracking proteins: Mechanisms and functions, *Curr. Opin. Cell Biol.* **17:** 47–54.

Akhmanova, A., Hoogenraad, C.C., Drabek, K., Stepanova, T., Dortland, B., Verkerk, T., et al., 2001, Clasps are CLIP-115 and -170 associating proteins involved in the regional regulation of microtubule dynamics in motile fibroblasts, *Cell* **104:** 923–935.

Akhmanova, A., Mausset-Bonnefont, A.L., van Cappellen, W., Keijzer, N., Hoogenraad, C.C., Stepanova, T., et al., 2005, The microtubule plus-end-tracking protein CLIP-170 associates with the spermatid manchette and is essential for spermatogenesis, *Genes Dev.* **19:** 2501–2515.

Amos, L.A., 2004, Microtubule structure and its stabilisation,. *Org. Biomol. Chem.* **2:** 2153–2160.

Amos, L.A., and Schlieper, D., 2005, Microtubules and maps, *Adv. Protein Chem.* **71:** 257–298.

Arnal, I., Heichette, C., Diamantopoulos, G.S., and Chretien, D., 2004, CLIP-170/tubulin-curved oligomers coassemble at microtubule ends and promote rescues, *Curr. Biol.* **14:** 2086–2095.

Askham, J.M., Moncur, P., Markham, A.F., and Morrison, E.E., 2000, Regulation and function of the interaction between the APC tumour suppressor protein and EB1, *Oncogene* **19:** 1950–1958.

Askham, J.M., Vaughan, K.T., Goodson, H.V., and Morrison, E.E., 2002, Evidence that an interaction between EB1 and p150(Glued) is required for the formation and maintenance of a radial microtubule array anchored at the centrosome. *Mol. Biol. Cell* **13:** 3627–3645.

Baas, P.W., 1998, The role of motor proteins in establishing the microtubule arrays of axons and dendrites, *J. Chem. Neuroanat.* **14:** 175–180.

Baas, P.W., 2002, Microtubule transport in the axon, *Int. Rev. Cytol.* **212:** 41–62.

Baas, P.W., White, L.A., and Heidemann, S.R., 1987, Microtubule polarity reversal accompanies regrowth of amputated neuritis, *Proc. Natl. Acad. Sci. USA* **84:** 5272–5276.

Baas, P.W., Deitch, J.S., Black, M.M., and Banker, G.A., 1988, Polarity orientation of microtubules in hippocampal neurons: Uniformity in the axon and nonuniformity in the dendrite, *Proc. Natl. Acad. Sci. USA* **85:** 8335–8339.

Baas, P.W., Pienkowski, T.P., and Kosik, K.S., 1991, Processes induced by tau expression in Sf9 cells have an axon-like microtubule organization, *J. Cell Biol.* **115:** 1333–1344.

Bamburg, J.R., Bray, D., and Chapman, K., 1986, Assembly of microtubules at the tip of growing axons, *Nature* **321:** 788–790.

Barth, A.I., Siemers, K.A., and Nelson, W.J., 2002, Dissecting interactions between EB1, microtubules and APC in cortical clusters at the plasma membrane, *J. Cell Sci.* **115:** 1583–1590.

Bellugi, U., Lichtenberger, L., Mills, D., Galaburda, A., and Korenberg, J.R., 1999, Bridging cognition, the brain and molecular genetics: Evidence from Williams syndrome, *Trends Neurosci.* **22:** 197–207.

Bentley, D., and Toroian-Raymond, A., 1986, Disoriented pathfinding by pioneer neurone growth cones deprived of filopodia by cytochalasin treatment, *Nature* **323:** 712–715.

Berrueta, L., Kraeft, S.K., Tirnauer, J.S., Schuyler, S.C., Chen, L.B., Hill, D.E., et al., 1998, The adenomatous polyposis coli-binding protein EB1 is associated with cytoplasmic and spindle microtubules, *Proc. Natl. Acad. Sci. USA* **95:** 10596–10601.

Bilbe, G., Delabie, J., Bruggen, J., Richener, H., Asselbergs, F. A., Cerletti, N., et al., 1992, Restin: A novel intermediate filament-associated protein highly expressed in the Reed-Sternberg cells of Hodgkin's disease, *EMBO J.* **11:** 2103–2113.

Billingsley, M.L., and Kincaid, R.L., 1997, Regulated phosphorylation and dephosphorylation of tau protein: Effects on microtubule interaction, intracellular trafficking and neurodegeneration, *Biochem. J.* **323:** 577–591.

Binder, L.I., Frankfurter, A., and Rebhun, L.I., 1985, The distribution of tau in the mammalian central nervous system, *J. Cell Biol.* **101:** 1371–1378.

Bouquet, C., Soares, S., von Boxberg, Y., Ravaille-Veron, M., Propst, F., and Nothias, F., 2004, Microtubule-associated protein 1B controls directionality of growth cone migration and axonal branching in regeneration of adult dorsal root ganglia neurons, *J. Neurosci.* **24:** 7204–7213.

Bradke, F., and Dotti, C.G., 2000, Establishment of neuronal polarity: Lessons from cultured hippocampal neurons, *Curr. Opin. Neurobiol.* **10:** 574–581.

Brandt, R., and Leschik, J., 2004, Functional interactions of tau and their relevance for Alzheimer's disease, *Curr. Alzheimer Res.* **1:** 255–269.

Brandt, R., Leger, J., and Lee, G., 1995, Interaction of tau with the neural plasma membrane mediated by tau's amino-terminal projection domain, *J. Cell Biol.* **131:** 1327–1340.

Bruckenstein, D.A., and Higgins, D., 1988, Morphological differentiation of embryonic rat sympathetic neurons in tissue culture. I. Conditions under which neurons form axons but not dendrites, *Dev. Biol.* **128:** 324–336.

Brugg, B., Reddy, D., and Matus, A., 1993, Attenuation of microtubule-associated protein 1B expression by antisense oligodeoxynucleotides inhibits initiation of neurite outgrowth, *Neuroscience* **52:** 489–496.

Bu, W., and Su, L.K., 2003, Characterization of functional domains of human EB1 family proteins, *J. Biol. Chem.* **278:** 49721–49731.

Buck, K.B., and Zheng, J.Q., 2002, Growth cone turning induced by direct local modification of microtubule dynamics, *J. Neurosci.* **22:** 9358–9367.

Caceres, A., and Kosik, K.S., 1990, Inhibition of neurite polarity by tau antisense oligonucleotides in primary cerebellar neurons, *Nature* **343:** 461–463.

Caceres, A., Potrebic, S., and Kosik, K.S., 1991, The effect of tau antisense oligonucleotides on neurite formation of cultured cerebellar macroneurons, *J. Neurosci.* **11:** 1515–1523.

Cahana, A., Escamez, T., Nowakowski, R.S., Hayes, N.L., Giacobini, M., von Holst, A., et al., 2001, Targeted mutagenesis of Lis1 disrupts cortical development and LIS1 homodimerization, *Proc. Natl. Acad. Sci. USA* **98:** 6429–6434.

Carvalho, P., Tirnauer, J.S., and Pellman, D., 2003, Surfing on microtubule ends, *Trends Cell Biol.* **13:** 229–237.

Caspi, M., Atlas, R., Kantor, A., Sapir, T., and Reiner, O., 2000, Interaction between LIS1 and doublecortin, two lissencephaly gene products, *Hum. Mol. Genet.* **9:** 2205–2213.

Cassimeris, L., Pryer, N.K., and Salmon, E.D., 1988, Real-time observations of microtubule dynamic instability in living cells, *J. Cell Biol.* **107:** 2223–2231.

Challacombe, J.F., Snow, D.M., and Letourneau, P.C., 1997, Dynamic microtubule ends are required for growth cone turning to avoid an inhibitory guidance cue, *J. Neurosci.* **17:** 3085–3095.

Chen, J., Kanai, Y., Cowan, N.J., and Hirokawa, N., 1992, Projection domains of MAP2 and tau determine spacings between microtubules in dendrites and axons, *Nature* **360:** 674–677.

Chien, C.B., Rosenthal, D.E., Harris, W.A., and Holt, C.E., 1993, Navigational errors made by growth cones without filopodia in the embryonic Xenopus brain, *Neuron* **11:** 237–251.

Choi, J.H., Bertram, P.G., Drenan, R., Carvalho, J., Zhou, H.H., and Zheng, X.F., 2002, The FKBP12-rapamycin-associated protein (FRAP) is a CLIP-170 kinase, *EMBO Rep.* **3:** 988–994.

Chrétien, D., Fuller, S.D., and Karsenti, E., 1995, Structure of growing microtubule ends: Two-dimensional sheets close into tubes at variable rates, *J. Cell Biol.* **129:** 1311–1328.

Ciani, L., and Salinas, P.C., 2005, WNTs in the vertebrate nervous system: From patterning to neuronal connectivity, *Nat. Rev. Neurosci.* **6:** 351–362.

Coquelle, F.M., Caspi, M., Cordelieres, F.P., Dompierre, J.P., Dujardin, D.L., Koifman, C., et al., 2002, LIS1, CLIP-170's key to the dynein/dynactin pathway, *Mol. Cell Biol.* **22:** 3089–3102.

Craig, A.M., and Banker, G., 1994, Neuronal polarity, *Annu. Rev. Neurosci.* **17:** 267–310.

Dawson, H.N., Ferreira, A., Eyster, M.V., Ghoshal, N., Binder, L.I., and Vitek, M.P., 2001, Inhibition of neuronal maturation in primary hippocampal neurons from tau deficient mice, *J. Cell Sci.* **114:** 1179–1187.

De Zeeuw, C.I., Hoogenraad, C.C., Goedknegt, E., Hertzberg, E., Neubauer, A., Grosveld, F., et al., 1997, CLIP-115, a novel brain-specific cytoplasmic linker protein, mediates the localization of dendritic lamellar bodies, *Neuron* **19:** 1187–1199.

Deka, J., Kuhlmann, J., and Muller, O., 1998, A domain within the tumor suppressor protein APC shows very similar biochemical properties as the microtubule-associated protein tau, *Eur. J. Biochem.* **253:** 591–597.

Del Rio, J.A., Gonzalez-Billault, C., Urena, J.M., Jimenez, E.M., Barallobre, M.J., Pascual, M., et al., 2004, MAP1B is required for Netrin 1 signaling in neuronal migration and axonal guidance, *Curr. Biol.* **14:** 840–850.

Dent, E.W., and Gertler, F.B., 2003, Cytoskeletal dynamics and transport in growth cone motility and axon guidance, *Neuron* **40:** 209–227.

Dent, E.W., and Kalil, K., 2001, Axon branching requires interactions between dynamic microtubules and actin filaments, *J. Neurosci.* **21:** 9757–9769.

Dent, E.W., Callaway, J.L., Szebenyi, G., Baas, P.W., and Kalil, K., 1999, Reorganization and movement of microtubules in axonal growth cones and developing interstitial branches, *J. Neurosci.* **19:** 8894–8908.

desPortes, V., Pinard, J.M., Billuart, P., Vinet, M.C., Koulakoff, A., Carrie, A., et al., 1998, A novel CNS gene required for neuronal migration and involved in X-linked subcortical laminar heterotopia and lissencephaly syndrome, *Cell* **92:** 51–61.

Diamantopoulos, G.S., Perez, F., Goodson, H.V., Batelier, G., Melki, R., Kreis, T.E., and Rickard, J.E., 1999, Dynamic localization of CLIP-170 to microtubule plus ends is coupled to microtubule assembly, *J. Cell Biol.* **144:** 99–112.

DiTella, M., Feiguin, F., Morfini, G., and Caceres, A., 1994, Microfilament-associated growth cone component depends upon Tau for its intracellular localization, *Cell Motil. Cytoskeleton* **29:** 117–130.

DiTella, M.C., Feiguin, F., Carri, N., Kosik, K.S., and Caceres, A., 1996, MAP-1B/TAU functional redundancy during laminin-enhanced axonal growth, *J. Cell Sci.* **109:** 467–477.

Dotti, C.G., and Banker, G.A., 1987, Experimentally induced alteration in the polarity of developing neurons, *Nature* **330:** 254–256.

Drubin, D.G., and Kirschner, M.W., 1986, Tau protein function in living cells, *J. Cell Biol.* **103:** 2739–2746.

Edelmann, W., Zervas, M., Costello, P., Roback, L., Fischer, I., Hammarback, J.A., et al., 1996, Neuronal abnormalities in microtubule-associated protein 1B mutant mice, *Proc. Natl. Acad. Sci. USA* **93:** 1270–1275.

Erck, C., Peris, L., Andrieux, A., Meissirel, C., Gruber, A.D., Vernet, M., et al., 2005, A vital role of tubulin-tyrosine-ligase for neuronal organization, *Proc. Natl. Acad. Sci. USA* **102:** 7853–7858.

Esmaeli-Azad, B., McCarty, J.H., and Feinstein, S.C., 1994, Sense and antisense transfection analysis of tau function: Tau influences net microtubule assembly, neurite outgrowth and neuritic stability, *J. Cell Sci.* **107:** 869–879.

Etienne-Manneville, S., and Hall, A., 2003, Cdc42 regulates GSK-3β and adenomatous polyposis coli to control cell polarity, *Nature* **421:** 753–756.

Etienne-Manneville, S., Manneville, J.B., Nicholls, S., Ferenczi, M.A., and Hall, A., 2005, Cdc42 and Par6-PKCzeta regulate the spatially localized association of Dlg1 and APC to control cell polarization, *J. Cell Biol.* **170:** 895–901.

Faulkner, N.E., Dujardin, D.L., Tai, C.Y., Vaughan, K.T., O'Connell, C.B., Wang, Y., et al., 2000, A role for the lissencephaly gene LIS1 in mitosis and cytoplasmic dynein function, *Nat. Cell Biol.* **2:** 784–791.

Ferreira, A., Busciglio, J., and Caceres, A., 1989, Microtubule formation and neurite growth in cerebellar macroneurons which develop in vitro: Evidence for the involvement of the microtubule-associated proteins, MAP-1a, HMW-MAP2 and Tau, *Brain Res. Dev. Brain Res.* **49:** 215–228.

Folker, E.S., Baker, B.M., and Goodson, H.V., 2005, Interactions between CLIP-170, Tubulin, and Microtubules: Implications for the Mechanism of CLIP-170 Plus-End Tracking Behavior, *Mol. Biol. Cell* **16:** 5373–5384.

Francis, F., Koulakoff, A., Boucher, D., Chafey, P., Schaar, B., Vinet, M.C., et al., 1999, Doublecortin is a developmentally regulated, microtubule-associated protein expressed in migrating and differentiating neurons, *Neuron* **23:** 247–256.

Friocourt, G., Chafey, P., Billuart, P., Koulakoff, A., Vinet, M.C., Schaar, B.T., et al., 2001, Doublecortin interacts with mu subunits of clathrin adaptor complexes in the developing nervous system, *Mol. Cell. Neurosci.* **18:** 307–319.

Friocourt, G., Koulakoff, A., Chafey, P., Boucher, D., Fauchereau, F., Chelly, J., et al., 2003, Doublecortin functions at the extremities of growing neuronal processes, *Cereb. Cortex* **13:** 620–626.

Fukata, M., Watanabe, T., Noritake, J., Nakagawa, M., Yamaga, M., Kuroda, S., et al., 2002, Rac1 and Cdc42 capture microtubules through IQGAP1 and CLIP-170, *Cell* **109:** 873–885.

Gill, S.R., Schroer, T.A., Szilak, I., Steuer, E.R., Sheetz, M.P., and Cleveland, D.W., 1991, Dynactin, a conserved, ubiquitously expressed component of an activator of vesicle motility mediated by cytoplasmic dynein, *J. Cell Biol.* **115:** 1639–1650.

Gleeson, J.G., Allen, K.M., Fox, J.W., Lamperti, E.D., Berkovic, S., Scheffer, I., et al., 1998, Doublecortin, a brain-specific gene mutated in human X-linked lissencephaly and double cortex syndrome, encodes a putative signaling protein, *Cell* **92:** 63–72.

Gleeson, J.G., Lin, P.T., Flanagan, L.A., and Walsh, C.A., 1999, Doublecortin is a microtubule-associated protein and is expressed widely by migrating neurons, *Neuron* **23:** 257–271.

Gleeson, J.G., Luo, R.F., Grant, P.E., Guerrini, R., Huttenlocher, P.R., Berg, M.J., et al., 2000, Genetic and neuroradiological heterogeneity of double cortex syndrome, *Ann. Neurol.* **47:** 265–269.

Gonzalez-Billault, C., and Avila, J., 2000, Molecular genetic approaches to microtubule-associated protein function, *Histol. Histopathol.* **15:** 1177–1183.

Gonzalez-Billault, C., Avila, J., Caceres, A., 2001, Evidence for the role of MAPIB in axon formation. *Mol. Biol. Cell* **12:** 2087–2098.

Gonzalez-Billault, C., Demandt, E., Wandosell, F., Torres, M., Bonaldo, P., Stoykova, A., et al., 2000, Perinatal lethality of microtubule-associated protein 1B-deficient mice expressing alternative isoforms of the protein at low levels, *Mol. Cell. Neurosci.* **16:** 408–421.

Gonzalez-Billault, C., Owen, R., Gordon-Weeks, P.R., and Avila, J., 2002, Microtubule-associated protein 1B is involved in the initial stages of axonogenesis in peripheral nervous system cultured neurons, *Brain Res.* **943:** 56–67.

Gonzalez-Billault, C., Jimenez-Mateos, E.M., Caceres, A., Diaz-Nido, J., Wandosell, F., and Avila, J., 2004, Microtubule-associated protein 1B function during normal development, regeneration, and pathological conditions in the nervous system, *J. Neurobiol.* **58:** 48–59.

Goode, B.L., Denis, P.E., Panda, D., Radeke, M.J., Miller, H.P., Wilson, L., and Feinstein, S.C., 1997, Functional interactions between the proline-rich and repeat regions of tau enhance microtubule binding and assembly, *Mol. Biol. Cell* **8:** 353–365.

Goodson, H.V., Skube, S.B., Stalder, R., Valetti, C., Kreis, T.E., Morrison, E.E., et al., 2003, CLIP-170 interacts with dynactin complex and the APC-binding protein EB1 by different mechanisms, *Cell Motil. Cytoskeleton* **55:** 156–173.

Goold, R.G., Owen, R., and Gordon-Weeks, P.R., 1999, Glycogen synthase kinase 3β phosphorylation of microtubule-associated protein 1B regulates the stability of microtubules in growth cones, *J. Cell Sci.* **112:** 3373–3384.

Gordon-Weeks, P.R., 2004, Microtubules and growth cone function, *J. Neurobiol.* **58:** 70–83.

Gordon-Weeks, P.R., 2005, *Neuronal Growth Cones*, Cambridge University Press, Cambridge.

Gordon-Weeks, P.R., and Fischer, I., 2000, MAP1B expression and microtubule stability in growing and regenerating axons, *Microsc. Res. Tech.* **48:** 63–74.

Gordon-Weeks, P.R., Mansfield, S.G., and Curran, I., 1989, Direct visualisation of the soluble pool of tubulin in the neuronal growth cone: Immunofluorescence studies following taxol polymerization, *Brain Res. Dev. Brain Res.* **49:** 305–310.
Graham, M.E., Ruma-Haynes, P., Capes-Davis, A.G., Dunn, J.M., Tan, T.C., Valova, V.A., et al., 2004, Multisite phosphorylation of doublecortin by cyclin-dependent kinase 5, *Biochem. J.* **381:** 471–481.
Gregory, S.L., and Brown, N.H., 1998, kakapo, a gene required for adhesion between and within cell layers in Drosophila, encodes a large cytoskeletal linker protein related to plectin and dystrophin, *J. Cell Biol.* **143:** 1271–1282.
Griparic, L., and Keller, T.C., 1998, Identification and expression of two novel CLIP-170/Restin isoforms expressed predominantly in muscle, *Biochim. Biophys. Acta* **1405:** 35–46.
Griparic, L., and Keller, T.C., III, 1999, Differential usage of two 5' splice sites in a complex exon generates additional protein sequence complexity in chicken CLIP-170 isoforms, *Biochim. Biophys. Acta* **1449:** 119–124.
Griparic, L., Volosky, J.M., and Keller, T.C., III, 1998, Cloning and expression of chicken CLIP-170 and restin isoforms, *Gene* **206:** 195–208.
Groden, J., Thliveris, A., Samowitz, W., Carlson, M., Gelbert, L., Albertsen, H., et al., 1991, Identification and characterization of the familial adenomatous polyposis coli gene, *Cell* **66:** 589–600.
Guan, K.L., and Rao, Y., 2003, Signalling mechanisms mediating neuronal responses to guidance cues, *Nat. Rev. Neurosci.* **4:** 941–956.
Gustke, N., Trinczek, B., Biernat, J., Mandelkow, E.M., and Mandelkow, E., 1994, Domains of tau protein and interactions with microtubules, *Biochemistry* **33:** 9511–9522.
Hahn, C.M., Kleinholz, H., Koester, M.P., Grieser, S., Thelen, K., and Pollerberg, G.E., 2005, Role of cyclin-dependent kinase 5 and its activator P35 in local axon and growth cone stabilization, *Neuroscience* **134:** 449–465.
Hall, A.C., Lucas, F.R., and Salinas, P.C., 2000, Axonal remodeling and synaptic differentiation in the cerebellum is regulated by WNT-7a signaling, *Cell* **100:** 525–535.
Hall, A.C., Brennan, A., Goold, R.G., Cleverley, K., Lucas, F.R., Gordon-Weeks, P.R. et al., 2002, Valproate regulates GSK-3-mediated axonal remodelling and synapsin I clustering in developing neurons, *Mol. Cell. Neurosc.* **20:** 257–270.
Hammarback, J.A., Obar, R.A., Hughes, S.M., and Vallee, R.B., 1991, MAP1B is encoded as a polyprotein that is processed to form a complex N-terminal microtubule-binding domain, *Neuron* **7:** 129–139.
Hanemaaijer, R., and Ginzburg, I., 1991, Involvement of mature tau isoforms in the stabilization of neurites in PC12 cells, *J. Neurosci. Res.* **30:** 163–171.
Hanley, J.G., Koulen, P., Bedford, F., Gordon-Weeks, P.R., and Moss, S.J., 1999, The protein MAP-1B links GABA(C) receptors to the cytoskeleton at retinal synapses, *Nature* **397:** 66–69.
Harada, A., Oguchi, K., Okabe, S., Kuno, J., Terada, S., Ohshima, T., et al., 1994, Altered microtubule organization in small-calibre axons of mice lacking tau protein, *Nature* **369:** 488–491.
Harte, P.J., and Kankel, D.R., 1982, Genetic analysis of mutations at the Glued locus and interacting loci in Drosophila melanogaster, *Genetics* **101:** 477–501.
Hattori, M., Adachi, H., Tsujimoto, M., Arai, H., and Inoue, K., 1994, Miller-Dieker lissencephaly gene encodes a subunit of brain platelet-activating factor acetylhydrolase [corrected], *Nature* **370:** 216–218.

Hayashi, I., and Ikura, M., 2003, Crystal structure of the amino-terminal microtubule-binding domain of end-binding protein 1 (EB1), *J. Biol. Chem.* **278:** 36430–36434.

Hayashi, I., Wilde, A., Mal, T.K., and Ikura, M. 2005, Structural basis for the activation of microtubule assembly by the EB1 and p150Glued complex, *Mol. Cell* **19:** 449–460.

Hayden, J.H., Bowser, S.S., and Rieder, C.L., 1990, Kinetochores capture astral microtubules during chromosome attachment to the mitotic spindle: Direct visualization in live newt lung cells, *J. Cell Biol.* **111:** 1039–1045.

He, Y., Francis, F., Myers, K.A., Yu, W., Black, M.M., and Baas, P.W., 2005, Role of cytoplasmic dynein in the axonal transport of microtubules and neurofilaments, *J. Cell Biol.* **168:** 697–703.

Heidemann, S.R., Landers, J.M., and Hamborg, M.A., 1981, Polarity orientation of axonal microtubules, *J. Cell Biol.* **91:** 61–665.

Holzbaur, E.L., Hammarback, J.A., Paschal, B.M., Kravit, N.G., Pfister, K.K., and Vallee, R.B., 1991, Homology of a 150K cytoplasmic dynein-associated polypeptide with the Drosophila gene Glued, *Nature* **351:** 579–583.

Honnappa, S., John, C.M., Kostrewa, D., Winkler, F.K., and Steinmetz, M.O., 2005, Structural insights into the EB1-APC interaction, *EMBO J.* **24:** 261–269.

Hoogenraad, C.C., Akhmanova, A., Grosveld, F., De Zeeuw, C.I., and Galjart, N., 2000, Functional analysis of CLIP-115 and its binding to microtubules, *J. Cell Sci.* **113:** 2285–2297.

Hoogenraad, C.C., Koekkoek, B., Akhmanova, A., Krugers, H., Dortland, B., Miedema, M., et al., 2002, Targeted mutation of Cyln2 in the Williams syndrome critical region links CLIP-115 haploinsufficiency to neurodevelopmental abnormalities in mice, *Nat. Genet.* **32:** 116–127.

Hoogenraad, C.C., Akhmanova, A., Galjart, N., and De Zeeuw, C.I., 2004, LIMK1 and CLIP-115: Linking cytoskeletal defects to Williams syndrome, *Bioessays* **26:** 141–150.

Horesh, D., Sapir, T., Francis, F., Wolf, S.G., Caspi, M., Elbaum, M., et al., 1999, Doublecortin, a stabilizer of microtubules, *Hum. Mol. Genet.* **8:** 1599–1610.

Horio, T., and Hotani, H., 1986, Visualization of the dynamic instability of individual microtubules by dark-field microscopy, *Nature* **321:** 605–607.

Howard, J., and Hyman, A.A., 2003, Dynamics and mechanics of the microtubule plus end, *Nature* **422:** 753–758.

Inoue, Y.H., do Carmo, A.M., Shiraki, M., Deak, P., Yamaguchi, M., Nishimoto, Y., et al., 2000, Orbit, a novel microtubule-associated protein essential for mitosis in Drosophila melanogaster, *J. Cell Biol.* **149:** 153–166.

Jacobson, M., 1991, *Developmental Neurobiology*, Plenum, New York.

Jimbo, T., Kawasaki, Y., Koyama, R., Sato, R., Takada, S., Haraguchi, K., et al., 2002, Identification of a link between the tumour suppressor APC and the kinesin superfamily, *Nat. Cell Biol.* **4:** 323–327.

Jimenez-Mateos, E.M., Wandosell, F., Reiner, O., Avila, J., and Gonzalez-Billault, C., 2005a, Binding of microtubule-associated protein 1B to LIS1 affects the interaction between dynein and LIS1, *Biochem. J.* **389:** 333–341.

Jimenez-Mateos, E.M., Paglini, G., Gonzalez-Billault, C., Caceres, A., and Avila, J., 2005b, End binding protein-1 (EB1) complements microtubule-associated protein-1B during axonogenesis, *J. Neurosci. Res.* **80:** 350–359.

Juwana, J.P., Henderikx, P., Mischo, A., Wadle, A., Fadle, N., Gerlach, K., et al., 1999, EB/RP gene family encodes tubulin binding proteins, *Int. J. Cancer* **81:** 275–284.

Kalil, K., and Dent, E.W., 2005, Touch and go: Guidance cues signal to the growth cone cytoskeleton, *Curr. Opin. Neurobiol.* **15:** 521–526.

Kanai, Y., Takemura, R., Oshima, T., Mori, H., Ihara, Y., Yanagisawa, M., et al., 1989, Expression of multiple tau isoforms and microtubule bundle formation in fibroblasts transfected with a single tau cDNA, *J. Cell Biol.* **109:** 1173–1184.

Kato, M., and Dobyns, W.B., 2003, Lissencephaly and the molecular basis of neuronal migration, *Hum. Mol. Genet.* **12:** R89–R96.

Kawakami, S., Muramoto, K., Ichikawa, M., and Kuroda, Y., 2003, Localization of microtubule-associated protein (MAP) 1B in the postsynaptic densities of the rat cerebral cortex, *Cell Mol. Neurobiol.* **23:** 887–894.

Keith, C.H., 1990, Neurite elongation is blocked if microtubule polymerization is inhibited in PC12 cells, *Cell Motil. Cytoskeleton* **17:** 95–105.

Kim, M.H., Cierpicki, T., Derewenda, U., Krowarsch, D., Feng, Y., Devedjiev, Y., et al., 2003, The DCX-domain tandems of doublecortin and doublecortin-like kinase, *Nat. Struct. Mol. Biol.* **10:** 324–333.

Kirschner, M.W., and Mitchison, T., 1986a, Beyond self-assembly: From microtubules to morphogenesis, *Cell* **45:** 329–342.

Kirschner, M.W., and Mitchison, T., 1986b, Microtubule dynamics, *Nature* **324:** 621.

Knops, J., Kosik, K.S., Lee, G., Pardee, J.D., Cohen, G.L., and McConlogue, L., 1991, Overexpression of tau in a non-neuronal cell induces long cellular processes, *J. Cell Biol.* **114:** 725–732.

Kodama, A., Karakesisoglou, I., Wong, E., Vaezi, A., and Fuchs, E., 2003, ACF7: An essential integrator of microtubule dynamics, *Cell* **115:** 343–354.

Komarova, Y.A., Akhmanova, A.S., Kojima, S., Galjart, N., and Borisy, G.G., 2002, Cytoplasmic linker proteins promote microtubule rescue in vivo, *J. Cell Biol.* **159:** 589–599.

Komarova, Y., Lansbergen, G., Galjart, N., Grosveld, F., Borisy, G.G., and Akhmanova, A., 2005, EB1 and EB3 Control CLIP Dissociation from the Ends of Growing Microtubules, *Mol. Biol. Cell* **16:** 5334–5345.

Kosik, K.S., and Finch, E.A., 1987, MAP2 and tau segregate into dendritic and axonal domains after the elaboration of morphologically distinct neurites: An immunocytochemical study of cultured rat cerebrum, *J. Neurosci.* **7:** 3142–3153.

LaMonte, B.H., Wallace, K.E., Holloway, B.A., Shelly, S.S., Ascano, J., Tokito, M., et al., 2002, Disruption of dynein/dynactin inhibits axonal transport in motor neurons causing late-onset progressive degeneration, *Neuron* **34:** 715–727.

Langkopf, A., Hammarback, J.A., Muller, R., Vallee, R.B., and Garner, C.C., 1992, Microtubule-associated proteins 1A and LC2. Two proteins encoded in one messenger RNA, *J. Biol. Chem.* **267:** 16561–16566.

Lansbergen, G., Komarova, Y., Modesti, M., Wyman, C., Hoogenraad, C.C., Goodson, H.V., et al., 2004, Conformational changes in CLIP-170 regulate its binding to microtubules and dynactin localization, *J. Cell Biol.* **166:** 1003–1014.

LeClerc, N., Baas, P.W., Garner, C.C., and Kosik, K.S., 1996, Juvenile and mature MAP2 isoforms induce distinct patterns of process outgrowth, *Mol. Biol. Cell* **7:** 443–455.

Lee, G., and Brandt, R., 1992, Microtubule-bundling studies revisited: Is there a role for MAPs? *Trends Cell Biol.* **2:** 286–289.

Lee, G., and Rook, S.L., 1992, Expression of tau protein in non-neuronal cells: Microtubule binding and stabilization, *J. Cell Sci.* **102:** 227–237.

Lee, H., Engel, U., Rusch, J., Scherrer, S., Sheard, K., and Van Vactor, D., 2004, The microtubule plus end tracking protein Orbit/MAST/CLASP acts downstream of the tyrosine kinase Abl in mediating axon guidance, *Neuron* **42:** 913–926.

Lein, P., Johnson, M., Guo, X., Rueger, D., and Higgins, D., 1995, Osteogenic protein-1 induces dendritic growth in rat sympathetic neurons, *Neuron* **15:** 597–605.

Lein, P., Guo, X., Hedges, A.M., Rueger, D., Johnson, M., and Higgins, D., 1996, The effects of extracellular matrix and osteogenic protein-1 on the morphological differentiation of rat sympathetic neurons, *Int. J. Dev. Neurosci.* **14:** 203–215.

Lemos, C.L., Sampaio, P., Maiato, H., Costa, M., Omel'yanchuk, L.V., Liberal, V., and Sunkel, C.E., 2000, Mast, a conserved microtubule-associated protein required for bipolar mitotic spindle organization, *EMBO J.* **19:** 3668–3682.

Li, S., Finley, J., Liu, Z.J., Qiu, S.H., Chen, H., Luan, C.H., et al., 2002, Crystal structure of the cytoskeleton-associated protein glycine-rich (CAP-Gly) domain, *J. Biol. Chem.* **277:** 48596–48601.

Ligon, L.A., Shelly, S.S., Tokito, M., and Holzbaur, E.L., 2003, The microtubule plus-end proteins EB1 and dynactin have differential effects on microtubule polymerization, *Mol. Biol. Cell* **14:** 1405–1417.

Liu, C.W., Lee, G., and Jay, D.G., 1999, Tau is required for neurite outgrowth and growth cone motility of chick sensory neurons, *Cell Motil. Cytoskeleton* **43:** 232–242.

Liu, J.J., Ding, J., Kowal, A.S., Nardine, T., Allen, E., Delcroix, J.D., et al., 2003, BPAG1n4 is essential for retrograde axonal transport in sensory neurons, *J. Cell Biol.* **163:** 223–229.

Lucas, F.R., and Salinas, P.C., 1997, WNT-7a induces axonal remodeling and increases synapsin I levels in cerebellar neurons, *Dev. Biol.* **192:** 31–44.

Lucas, F.R., Goold, R.G., Gordon-Weeks, P.R., and Salinas, P.C., 1998, Inhibition of GSK-3β leading to the loss of phosphorylated MAP-1B is an early event in axonal remodelling induced by WNT-7a or lithium, *J. Cell Sci.* **111:** 1351–1361.

Ma, Y., Shakiryanova, D., Vardya, I., and Popov, S.V., 2004, Quantitative analysis of microtubule transport in growing nerve processes, *Curr. Biol.* **14:** 725–730.

Mack, T.G., Koester, M.P., and Pollerberg, G.E., 2000, The microtubule-associated protein MAP1B is involved in local stabilization of turning growth cones, *Mol. Cell. Neurosci.* **15:** 51–65.

Mandell, J.W., and Banker, G.A., 1996, A spatial gradient of tau protein phosphorylation in nascent axons, *J. Neurosci.* **16:** 5727–5740.

Mann, S.S., and Hammarback, J.A., 1994, Molecular characterization of light chain 3. A microtubule binding subunit of MAP1A and MAP1B, *J. Biol. Chem.* **269:** 11492–11497.

Meixner, A., Haverkamp, S., Wassle, H., Fuhrer, S., Thalhammer, J., Kropf, N., et al., 2000, MAP1B is required for axon guidance and is involved in the development of the central and peripheral nervous system, *J. Cell Biol.* **151:** 1169–1178.

Mimori-Kiyosue, Y., Shiina, N., and Tsukita, S., 2000a, Adenomatous polyposis coli (APC) protein moves along microtubules and concentrates at their growing ends in epithelial cells, *J. Cell Biol.* **148:** 505–518.

Mimori-Kiyosue, Y., Shiina, N., and Tsukita, S., 2000b, The dynamic behavior of the APC-binding protein EB1 on the distal ends of microtubules, *Curr. Biol.* **10:** 865–868.

Mimori-Kiyosue, Y., Grigoriev, I., Lansbergen, G., Sasaki, H., Matsui, C., Severin, F., et al., 2005, CLASP1 and CLASP2 bind to EB1 and regulate microtubule plus-end dynamics at the cell cortex, *J. Cell Biol.* **168:** 141–153.

Mitchison, T., and Kirschner, M.W., 1984a, Microtubule assembly nucleated by isolated centromeres, *Nature* **312:** 232–237.

Mitchison, T., and Kirschner, M.W., 1984b, Dynamic instability of microtubule growth, *Nature* **312:** 237–242.

Moores, C.A., Perderiset, M., Francis, F., Chelly, J., Houdusse, A., and Milligan, R.A., 2004, Mechanism of microtubule stabilization by doublecortin, *Mol. Cell* **14:** 833–839.

Morrison, E.E., Wardleworth, B.N., Askham, J.M., Markham, A.F., and Meredith, D.M., 1998, EB1, a protein which interacts with the APC tumour suppressor, is associated with the microtubule cytoskeleton throughout the cell cycle, *Oncogene* **17:** 3471–3477.

Morrison, E.E., Moncur, P.M., and Askham, J.M., 2002, EB1 identifies sites of microtubule polymerisation during neurite development, *Brain Res. Mol. Brain Res.* **98:** 145–152.

Munch, C., Sedlmeier, R., Meyer, T., Homberg, V., Sperfeld, A.D., Kurt, A., et al., 2004, Point mutations of the p150 subunit of dynactin (DCTN1) gene in ALS, *Neurology* **63:** 724–726.

Munch, C., Rosenbohm, A., Sperfeld, A.D., Uttner, I., Reske, S., Krause, B.J., et al., 2005, Heterozygous R1101K mutation of the DCTN1 gene in a family with ALS and FTD, *Ann. Neurol.* **58:** 777–780.

Munemitsu, S., Souza, B., Muller, O., Albert, I., Rubinfeld, B., and Polakis, P., 1994, The APC gene product associates with microtubules in vivo and promotes their assembly in vitro, *Cancer Res.* **54:** 3676–3681.

Nakagawa, H., Koyama, K., Murata, Y., Morito, M., Akiyama, T., and Nakamura, Y., 2000, EB3, a novel member of the EB1 family preferentially expressed in the central nervous system, binds to a CNS-specific APC homologue, *Oncogene* **19:** 210–216.

Nakamura, M., Zhou, X.Z., and Lu, K.P., 2001, Critical role for the EB1 and APC interaction in the regulation of microtubule polymerization, *Curr. Biol.* **11:** 1062–1067.

Nathke, I.S., 2004, The adenomatous polyposis coli protein: The Achilles heel of the gut epithelium, *Annu. Rev. Cell Dev. Biol.* **20:** 337–366.

Noble, M., Lewis, S.A., and Cowan, J., 1989, The microtubule binding domain of the microtubule-associated protein MAP-1B contains a repeated sequence motif unrelated to that of MAP-2 and tau, *J. Cell Biol.* **109:** 3367–3376.

Noiges, R., Eichinger, R., Kutschera, W., Fischer, I., Nemeth, Z., Wiche, G., et al., 2002, Microtubule-associated protein 1A (MAP1A) and MAP1B: Light chains determine distinct functional properties, *J. Neurosci.* **22:** 2106–2114.

Owen, R., and Gordon-Weeks, P.R., 2003, Inhibition of glycogen synthase kinase 3β in sensory neurons in culture alters actin filament and microtubule dynamics in growth cones, *Mol. Cell. Neurosci.* **23:** 626–637.

Perez, F., Diamantopoulos, G.S., Stalder, R., and Kreis, T.E., 1999, CLIP-170 highlights growing microtubule ends in vivo, *Cell* **96:** 517–527.

Pierre, P., Scheel, J., Rickard, J.E., and Kreis, T.E., 1992, CLIP-170 links endocytic vesicles to microtubules, *Cell* **70:** 887–900.

Pierre, P., Pepperkok, R., and Kreis, T.E., 1994, Molecular characterization of two functional domains of CLIP-170 in vivo, *J. Cell Sci.* **107:** 1909–1920.

Powell, S.K., Rivas, R.J., Rodriguez-Boulan, E., and Hatten, M.E., 1997, Development of polarity in cerebellar granule neurons, *J. Neurobiol.* **32:** 223–236.

Preuss, U., Biernat, J., Mandelkow, E.M., and Mandelkow, E., 1997, The 'jaws' model of tau-microtubule interaction examined in CHO cells, *J. Cell Sci.* **110:** 789–800.

Prokop, A., Uhler, J., Roote, J., and Bate, M., 1998, The kakapo mutation affects terminal arborization and central dendritic sprouting of Drosophila motorneurons, *J. Cell Biol.* **143:** 1283–1294.

Reiner, O., Carrozzo, R., Shen, Y., Wehnert, M., Faustinella, F., Dobyns, W.B., et al., 1993, Isolation of a Miller-Dieker lissencephaly gene containing G protein beta-subunit-like repeats, *Nature* **364:** 717–721.

Renner, C., Pfitzenmeier, J.P., Gerlach, K., Held, G., Ohnesorge, S., Sahin, U., et al., 1997, RP1, a new member of the adenomatous polyposis coli-binding EB1-like gene family, is differentially expressed in activated T cells, *J. Immunol.* **159:** 1276–1283.

Rickard, J.E., and Kreis, T.E., 1990, Identification of a novel nucleotide-sensitive microtubule-binding protein in HeLa cells, *J. Cell Biol.* **110:** 1623–1633.

Roberts, A., 1988, The early development of neurons in Xenopus embryos revelaed by transmitter immunocytochemistry for serotonin, GABA and glycine, in: *Developmental Neurobiology of the Frog*, E.D. Pollack, and H.D. Bibb, eds., Alan R. Liss, New York, pp. 191–205.

Roper, K., Gregory, S.L., and Brown, N.H., 2002, The 'spectraplakins': Cytoskeletal giants with characteristics of both spectrin and plakin families, *J. Cell Sci.* **115:** 4215–4225.

Sammak, P.J., and Borisy, G.G., 1988, Direct observation of microtubule dynamics in living cells, *Nature* **332:** 724–726.

Sapir, T., Elbaum, M., and Reiner, O., 1997, Reduction of microtubule catastrophe events by LIS1, platelet-activating factor acetylhydrolase subunit, *EMBO J.* **16:** 6977–6984.

Sapir, T., Cahana, A., Seger, R., Nekhai, S., and Reiner, O., 1999, LIS1 is a microtubule-associated phosphoprotein, *Eur. J. Biochem.* **265:** 181–188.

Schaar, B.T., Kinoshita, K., and McConnell, S.K., 2004, Doublecortin microtubule affinity is regulated by a balance of kinase and phosphatase activity at the leading edge of migrating neurons, *Neuron* **41:** 203–213.

Schafer, D.A., Gill, S.R., Cooper, J.A., Heuser, J.E., and Schroer, T.A., 1994, Ultrastructural analysis of the dynactin complex: An actin-related protein is a component of a filament that resembles F-actin, *J. Cell Biol.* **126:** 403–412.

Schaefer, A.W., Kabir, N., and Forscher, P., 2002, Filopodia and actin arcs guide the assembly and transport of two populations of microtubules with unique dynamic parameters in neuronal growth cones, *J. Cell Biol.* **158:** 139–152.

Scheel, J., Pierre, P., Rickard, J.E., Diamantopoulos, G.S., Valetti, C., van der Goot, F.G., et al., 1999, Purification and analysis of authentic CLIP-170 and recombinant fragments, *J. Biol. Chem.* **274:** 25883–25891.

Shea, T.B., Beermann, M.L., Nixon, R.A., and Fischer, I. 1992, Microtubule-associated protein tau is required for axonal neurite elaboration by neuroblastoma cells, *J. Neurosci. Res.* **32:** 363–374.

Shi, S.H., Cheng, T., Jan, L.Y., and Jan, Y.N., 2004, APC and GSK-3β are involved in mPar3 targeting to the nascent axon and establishment of neuronal polarity, *Curr. Biol.* **14:** 2025–2032.

Shimomura, A., Kohu, K., Akiyama, T., and Senda, T., 2005, Subcellular localization of the tumor suppressor protein APC in developing cultured neurons, *Neurosci. Lett.* **375:** 81–86.

Slep, K.C., Rogers, S.L., Elliott, S.L., Ohkura, H., Kolodziej, P.A., and Vale, R.D., 2005, Structural determinants for EB1-mediated recruitment of APC and spectraplakins to the microtubule plus end, *J. Cell Biol.* **168:** 587–598.

Smith, D.S., Niethammer, M., Ayala, R., Zhou, Y., Gambello, M.J., Wynshaw-Boris, A., and Tsai, L.H., 2000, Regulation of cytoplasmic dynein behaviour and microtubule organization by mammalian Lis1, *Nat. Cell Biol.* **2:** 767–775.

Stepanova, T., Slemmer, J., Hoogenraad, C.C., Lansbergen, G., Dortland, B., De Zeeuw, C.I., et al., 2003, Visualization of microtubule growth in cultured neurons via the use of EB3-GFP (end-binding protein 3-green fluorescent protein), *J. Neurosci.* **23:** 2655–2664.

Su, L.K., and Qi, Y., 2001, Characterization of human MAPRE genes and their proteins, *Genomics* **71:** 142–149.

Su, L.K., Burrell, M., Hill, D.E., Gyuris, J., Brent, R., Wiltshire, R. et al., 1995, APC binds to the novel protein EB1, *Cancer Res.* **55:** 2972–2977.

Subramanian, A., Prokop, A., Yamamoto, M., Sugimura, K., Uemura, T., Betschinger, J., et al., 2003, Shortstop recruits EB1/APC1 and promotes microtubule assembly at the muscle-tendon junction, *Curr. Biol.* **13:** 1086–1095.

Takahashi, M., Yasutake, K., and Tomizawa, K., 1999, Lithium inhibits neurite growth and tau protein kinase I/glycogen synthase kinase-3β-dependent phosphorylation of juvenile tau in cultured hippocampal neurons, *J. Neurochem.* **73:** 2073–2083.

Takei, Y., Kondo, S., Harada, A., Inomata, S., Noda, T., and Hirokawa, N., 1997, Delayed development of nervous system in mice homozygous for disrupted microtubule-associated protein 1B (MAP1B) gene, *J. Cell Biol.* **137:** 1615–1626.

Takei, Y., Teng, J., Harada, A., and Hirokawa, N., 2000, Defects in axonal elongation and neuronal migration in mice with disrupted tau and map1b genes, *J. Cell Biol.* **150:** 989–1000.

Takemura, R., Okabe, S., Umeyama, T., Kanai, Y., Cowan, N.J., and Hirokawa, N., 1992, Increased microtubule stability and alpha tubulin acetylation in cells transfected with microtubule-associated proteins MAP1B, MAP2 or tau, *J. Cell Sci.* **103:** 953–964.

Tanaka, E., and Kirschner, M.W., 1995, The role of microtubules in growth cone turning at substrate boundaries, *J. Cell Biol.* **128:** 127–137.

Tanaka, E., Ho, T., and Kirschner, M.W., 1995, The role of microtubule dynamics in growth cone motility and axonal growth, *J. Cell Biol.* **128:** 139–155.

Tanaka, T., Serneo, F.F., Tseng, H.C., Kulkarni, A.B., Tsai, L.H., and Gleeson, J.G., 2004, Cdk5 phosphorylation of doublecortin ser297 regulates its effect on neuronal migration, *Neuron* **41:** 215–227.

Teichman-Weinberg, A., Littauer, U.Z., and Ginzburg, I., 1988, The inhibition of neurite outgrowth in PC12 cells by tubulin antisense oligodeoxyribonucleotides, *Gene* **72:** 297–307.

Teng, J., Takei, Y., Harada, A., Nakata, T., Chen, J., and Hirokawa, N., 2001, Synergistic effects of MAP2 and MAP1B knockout in neuronal migration, dendritic outgrowth, and microtubule organization, *J. Cell Biol.* **155:** 65–76.

Tint, I., Slaughter, T., Fischer, I., and Black, M.M., 1998, Acute inactivation of tau has no effect on dynamics of microtubules in growing axons of cultured sympathetic neurons, *J. Neurosci.* **18:** 8660–8673.

Tirnauer, J.S. and Bierer, B.E., 2000, EB1 proteins regulate microtubule dynamics, cell polarity, and chromosome stability, *J. Cell Biol.* **149:** 761–766.

Tirnauer, J.S., Grego, S., Salmon, E.D., and Mitchison, T.J., 2002, EB1-microtubule interactions in Xenopus egg extracts: Role of EB1 in microtubule stabilization and mechanisms of targeting to microtubules, *Mol. Biol. Cell* **13:** 3614–3626.

Togel, M., Wiche, G., and Propst, F., 1998, Novel features of the light chain of microtubule-associated protein MAP1B: Microtubule stabilization, self interaction, actin filament binding, and regulation by the heavy chain, *J. Cell Biol.* **143:** 695–707.

Trivedi, N., Marsh, P., Goold, R.G., Wood-Kaczmar, A., and Gordon-Weeks, P.R., 2005, Glycogen synthase kinase-3β phosphorylation of MAP1B at Ser1260 and Thr1265 is spatially restricted to growing axons, *J. Cell Sci.* **118:** 993–1005.

Vale, R.D., 2003, The molecular motor toolbox for intracellular transport, *Cell* **112:** 467–480.

Vandecandelaere, A., Pedrotti, B., Utton, M.A., Calvert, R.A., and Bayley, P.M., 1996, Differences in the regulation of microtubule dynamics by microtubule-associated proteins MAP1B and MAP2, *Cell Motil. Cytoskeleton* **35:** 134–146.

Vaughan, K.T., Tynan, S.H., Faulkner, N.E., Echeverri, C.J., and Vallee, R.B., 1999, Colocalization of cytoplasmic dynein with dynactin and CLIP-170 at microtubule distal ends, *J. Cell Sci.* **112:** 1437–1447.

Vaughan, P.S., Miura, P., Henderson, M., Byrne, B., and Vaughan, K.T., 2002, A role for regulated binding of p150(Glued) to microtubule plus ends in organelle transport, *J. Cell Biol.* **158:** 305–319.

Watanabe, T., Wang, S., Noritake, J., Sato, K., Fukata, M., Takefuji, M., et al., 2004, Interaction with IQGAP1 links APC to Rac1, Cdc42, and actin filaments during cell polarization and migration, *Dev. Cell* **7:** 871–883.

Wen, Y., Eng, C.H., Schmoranzer, J., Cabrera-Poch, N., Morris, E.J., Chen, M., et al., 2004, EB1 and APC bind to mDia to stabilize microtubules downstream of Rho and promote cell migration, *Nat. Cell Biol.* **6:** 820–830.

Wiggin, G.R., Fawcett, J.P., and Pawson, T., 2005, Polarity proteins in axon specification and synaptogenesis, *Dev. Cell* **8:** 803–816.

Williamson, T.W., Gordon-Weeks, P.R., Schachner, M., and Taylor, J., 1996, Microtubule reorganization is obligatory for growth cone turning, *Proc. Natl. Acad. Sci. USA* **93:** 15221–15226.

Wu, X.S., Tsan, G.L., and Hammer, J.A., III, 2005, Melanophilin and myosin Va track the microtubule plus end on EB1, *J. Cell Biol.* **171:** 201–207.

Zauner, W., Kratz, J., Staunton, J., Feick, P., and Wiche, G., 1992, Identification of two distinct microtubule binding domains on recombinant rat MAP 1B, *Eur. J. Cell Biol.* **57:** 66–74.

Zhou, F.Q., Waterman-Storer, C.M., and Cohan, C.S., 2002, Focal loss of actin bundles causes microtubule redistribution and growth cone turning, *J. Cell Biol.* **157:** 839–849.

Zhou, F.Q., Zhou, J., Dedhar, S., Wu, Y.H., and Snider, W.D., 2004, NGF-induced axon growth is mediated by localized inactivation of GSK-3β and functions of the microtubule plus end binding protein APC, *Neuron* **42:** 897–912.

Zumbrunn, J., Kinoshita, K., Hyman, A.A., and Nathke, I.S., 2001, Binding of the adenomatous polyposis coli protein to microtubules increases microtubule stability and is regulated by GSK3 beta phosphorylation, *Curr. Biol.* **11:** 44–49..

4
Small GTPases: Mechanisms Linking Membrane Traffic to Cytoskeleton During Neuritogenesis

IVAN DE CURTIS AND SARA CORBETTA

4.1. Introduction

During neurite extension, the protrusive activity of the neuronal growth cone is mediated by the coordination of membrane traffic, adhesion to the cell substrate, and reorganization of the cytoskeleton that includes the dynamic rearrangement of actin filaments at the peripheral region of the growth cone. The small GTPases of the Rho family are important regulators of the actin cytoskeleton (Hall, 1998) and are critical for the regulation of axonal and dendritic extension. In particular, studies in different organisms and studies with primary neurons have shown that Rac acts as a regulator of process outgrowth and axonal guidance (Luo, 2000), by stimulating actin polymerization at the growth cone edge surface. On the other hand, the extension of long axons and branched dendrites requires the addition of membrane to the plasmalemma, to support the extraordinary increase of the neuronal surface occurring during neuritogenesis. Therefore, an important contribution of membrane trafficking to neuritogenesis must be accounted for during neuronal differentiation. *Trans*-Golgi network-derived constitutive exocytosis and membrane recycling after endocytosis are possible sources of membrane addition to the growing tips of neurites.

Traffic within the cell is characterized by the flow of membrane among different compartments and is believed to be mediated by populations of vesicles budding from distinct donor compartments and fusing with specific acceptor compartments. Membrane trafficking is regulated by members of two families of small GTPases, the ADP ribosylation factor (Arf) and Rab families. A third group of molecules, the SNAREs (soluble N-ethyl maleimide-sensitive factor attachment protein receptors) are probably catalyzing membrane fusion, and major progresses have been made to unravel their function (Ungermann and Langosch, 2005). Much less is known about the mechanisms required before fusion to target and tether the vesicular membrane to the membrane of specific compartments (Whyte and Munro, 2002). In particular, the details of the molecular mechanisms that direct membrane traffic at sites of adhesion and of actin rearrangement in migrating growth cones are largely unknown. Other chapters of this book are dealing with different aspects of the molecular events involved in membrane trafficking in developing neurons. The aim of this chapter is to present the available evidence and some hypotheses on the mechanisms that may link membrane trafficking to cytoskeletal rearrangements at the growth cone. Findings have shown that a number of multidomain adaptor proteins characterized by an Arf GTPase activating protein (ArfGAP) domain interact with actin-regulating and integrin-binding proteins and are able to affect Rac-mediated protrusive activity and cell motility. Some of these proteins have been shown to localize at endocytic compartments and to have a role in regulating endocytosis. Given the participation of Arf proteins in membrane traffic, one appealing hypothesis is that these ArfGAPs act as molecular devices that coordinate membrane traffic and cytoskeletal reorganization at growth cones during neurite development.

4.2. Membrane Traffic and Actin Dynamics at the Growth Cone

Neurite extension is driven by the protrusive activity at the edge of the growth cone, where continuous remodeling of actin and adhesive contacts occurs. Actin dynamics during growth cone navigation evolve into stabilization of the cytoskeleton and neurite elongation (Tanaka and Sabry, 1995). Development of dendrites and axon is the base for the establishment of neuronal shape and polarity, and it requires new membrane addition to both types of extending processes. Several evidences support a fundamental role of membrane traffic in the outgrowth of neurites. Plasma membrane expansion necessary to neurite elongation is mediated by constant membrane addition and retrieval in growth cones (Cheng and Reese, 1987; Dailey and Bridgman, 1993; Igarashi et al., 1996). Contribution to membrane expansion during neuritogenesis may come from two major intracellular sources: newly synthesized membrane from the exocytic secretory pathway, ultimately deriving from membranes outing from the *trans*-face of the Golgi apparatus, and membrane from endocytic compartment(s), recycled after internalization from the plasma membrane. Here, evidence will be presented supporting the implication of both newly made and recycling membrane in neurite extension.

4.2.1. *Role of Exocytosis in Neurite Extension*

Exocytosis plays an essential part in axonal and dendritic outgrowth (Bradke and Dotti, 1997; Futerman and Banker, 1996). Many observations indicate that the expansion of the plasma membrane in the growing neurite occurs by exocytic incorporation of plasma membrane precursor vesicles from the Golgi into the cell surface, primarily at the growth cone (Craig et al., 1995; Pfenninger and Maylié-Pfenninger, 1981). Moreover in isolated growth cones, membrane expansion from fusion with the plasma membrane of internal vesicles may be induced by Ca^{2+} influx (Lockerbie et al., 1991). Detailed characterization of vesicles containing newly synthesized lipids localized at the growth cone confirms their identity as precursor vesicles. These vesicles cluster to form distinct, dynamic organelles specialized for plasmalemmal expansion in the growth cone (Pfenninger and Friedman, 1993; Pfenninger and Maylié-Pfenninger, 1981). Inhibition of microtubule dynamic instability does not prevent the delivery of new membrane to the growth cone, but it strongly inhibits the insertion of vesicles into the plasma membrane, suggesting that local destabilization of microtubules is required to supply membrane to the growth cone (Zakharenko and Popov, 1998).

One open question is whether the same or different exocytic routes are accountable for the respective outgrowth of axons and dendrites. In this direction, the tetanus neurotoxin-insensitive vesicle-associated membrane protein (TI-VAMP) defines a network of tubulo-vesicular structures present both at the leading edge of elongating dendrites and axons of immature hippocampal neurons in culture. TI-VAMP is essential for neurite outgrowth in PC12 cells and for the outgrowth of both dendrites and axons in primary neuronal cultures, suggesting that a common exocytic mechanism that relies on TI-VAMP mediates both axonal and dendritic outgrowth in developing neurons (Martinez-Arca et al., 2001).

4.2.1.1. Links Between Traffic from the Golgi Apparatus and the Actin Cytoskeleton

A number of regulators of the actin cytoskeleton involved in neurite outgrowth have been found associated to the Golgi. Among these proteins, the finding that LIM kinase-1 is enriched in the Golgi apparatus and in growth cones raises the interesting possibility that this kinase regulates the trafficking of Golgi-derived vesicles to the tips of neurites. Overexpression of LIM kinase-1 accelerates axon formation and increases the accumulation at growth cones of growth factor receptors, cell adhesion molecules and Par proteins (Rosso et al., 2004), important for the polarity of different mammalian cell types including neurons (Ohno, 2001).

The serine/threonine kinase Cdk5 (Cyclin-dependent kinase-5)-p35 is important during brain development to regulate neuronal migration and neurite extension. Cdk5-p35 is concentrated at the leading edge of cortical neurons axonal growth cones, where it colocalizes with Rac and PAK, with p35 interacting directly with Rac in a GTP-dependent manner (Nikolic et al., 1998; Rashid et al., 2001). The Cdk5 kinase induces PAK1 hyperphosphorylation on threonine-212, which results in downregulation of PAK1, a serine/threonine kinase that controls microtubule dynamics as well as the organization of F-actin microfilaments downstream of the Rho GTPases Rac and Cdc42. Therefore, the p35/Cdk5-dependent modification of PAK1 is likely to affect the dynamics of actin and microtubule organization in neurons, thus promoting neuronal migration and neurite outgrowth. Cdk5 kinase and Cdk5-dependent phosphorylation of PAK have been found associated with Golgi membranes. Antisense oligonucleotide suppression of Cdk5 in young cultured neurons blocks the formation of membrane vesicles from the Golgi apparatus suggesting a role for Cdk5 in membrane traffic during neurite outgrowth (Paglini et al., 2001). Moreover, the interaction of Cdk5 with a number of endocytic vesicle-associated proteins implicates this kinase in directing endocytic traffic in neurons (Smith and Tsai, 2002).

4.2.2. Role of Endocytosis and Membrane Recycling in Neurite Extension

It has been demonstrated that nonneuronal cells internalize and recycle membrane from the cell surface rapidly and that rapid recycling involves passage through an endocytic organelle that has been identified as the sorting endosome by colocalization with internalized transferrin and low-density lipoprotein (Hao and Maxfield, 2000). These results imply that the membrane internalization rate can be very high, with a $t_{1/2}$ as short as 5–10 min. This indicates that large amounts of recycling membrane are available for polarized delivery during migratory events. Similarly, by using the membrane-impermeable fluorescent dye FM1-43, it has been shown that constitutive endocytosis accounts for a large amount of membrane retrieved at the tip of neurites: the equivalent of the entire growth cone surface area can be internalized within 30 min (Diefenbach et al., 1999).

It has been hypothesized that membrane internalized from the cell surface by endocytic processes is recycled to the leading edge of migrating cells to contribute to the extension of the cell front (Bretscher, 1996). Consistent with this model, recycling transferrin receptors and low-density lipoprotein receptors are distributed to the cell front of migrating fibroblasts and to Rac-induced ruffles (Bretscher and Aguado-Velasco, 1998a; Hopkins et al., 1994). Thus, the reinsertion of internalized membranes at the surface of a neuron may be directed to the sites of protrusion near or at the growth cone during its navigation in response to motogenic stimuli (Bretscher and Aguado-Velasco, 1998b).

Growth cones are sites of intense endocytosis and require recycling of membrane back to the surface to maintain equilibrium. The recycling endosomes present in axons and dendrites of developing hippocampal neurons (Prekeris et al., 1999) may represent a pool of membrane that can be quickly added/removed for growth cone-mediated neurite extension or retraction, respectively (Craig et al., 1995; Diefenbach et al., 1999). Inhibition of vesicular transport from the cell soma to axons and dendrites by microtubules depolymerization with nocodazole results in almost complete block of axonal and dendritic endosomal trafficking in hippocampal neurons, with strikingly large accumulations of endosomal markers toward the distal end of the dendrites, indicating that perturbation of endosomal traffic affects dendritic stability (Zakharenko and Popov, 1998).

The involvement in neuritogenesis of proteins with an established role in synaptic function and endocytosis has already been demonstrated. Two examples are amphiphysin-I, an Src homology type-3 (SH3) domain-containing cytosolic protein enriched in axon terminals, and dynamin-1, a GTPase that plays a role in early steps of receptor-mediated endocytosis. The two proteins bind to each other and participate in synaptic vesicle endocytosis. Downregulation of either protein in cultured developing hippocampal neurons by antisense oligonucleotides causes the collapse of growth cones and a severe inhibition of neurite outgrowth and axon formation, suggesting that both proteins are necessary for normal neuronal morphogenesis (Mundigl et al., 1998; Torre et al., 1994).

4.3. Regulation of Neurite Extension by Arf and Rab Proteins

While our knowledge of the molecular machinery underlying the propulsive mechanism driven by actin and mediated by Rho family GTPases has increased dramatically (Borisy and Svitkina, 2000; Hall, 1998), it is still unclear how membrane traffic is incorporated into the extension process. Progress in this direction, however, comes from studies of Rab and Arf family GTPases, two classes of proteins implicated in the regulation of different aspects of membrane traffic.

4.3.1. Arf GTPases as Regulators of Membrane Traffic

There are six mammalian Arf proteins that have been structurally grouped into three classes: class I (Arf1, Arf2, and Arf3), class II (Arf4 and Arf5), and class III (Arf6). Of these, Arf1 and Arf6 are the most extensively studied and have been implicated at multiple sites as regulators of membrane traffic (Nie et al., 2003; Takai et al., 2001). Arf proteins cycle between the GTP- and GDP-bound forms with the help of specific GAPs and guanosine nucleotide exchanging factors (GEFs) (Donaldson and Jackson, 2000). Arf proteins interact with at least three different types of effectors: (1) structural proteins, the vesicle coat proteins; (2) lipid-metabolizing enzymes; and (3) other proteins that bind to Arf-GTP but whose functions are less defined.

Members of the Arf family are essential players in membrane trafficking, which subserves constitutive protein transport along exocytic and endocytic pathways within eukaryotic cells. In this respect, the molecular details of Arf1 function have been better elucidated compared to other members of the family. The engagement of Arf1 in the formation and budding of vesicles from the Golgi through interaction with coat proteins is well established (Spang, 2002). Based on several studies on this protein, the prevailing model suggests that cycles of Arf1 activation and inactivation are directly linked to coat protein recruitment and dissociation at the membrane of the Golgi apparatus, a process required for the budding of vesicles carrying cargo molecules among Golgi subcompartments along the exocytic pathway.

Based on the fact that Arf1 regulates specifically the formation of vesicles within the Golgi compartment (Roth, 1999), and given the specific localization of Arf6 at endosomes and plasma membrane, one could speculate that Arf6 would analogously regulate vesicle formation between endosomes and the plasma membrane. The subcellular localization of Arf6 is unique when compared to class I and class II Arfs. A number of studies indicate that Arf6 is involved in the regulation of receptor-mediated endocytosis, and in the recycling of endocytosed membrane back to the cell surface (Altschuler et al., 1999; D'Souza-Schorey et al., 1995, 1998; Radhakrishna and Donaldson, 1997; Zhang et al., 1998). Arf6 is involved also in Rac1-dependent remodeling of the cytoskeleton at the plasma membrane, and it especially localizes to membrane ruffles in spreading cells (D'Souza-Schorey et al., 1997; Radhakrishna and Donaldson, 1997; Song et al., 1998). The functional link between Arf6 and Rac1 is supported by the colocalization of the two GTPases on Arf6 recycling endosomes and at the plasma membrane, where Rac1-stimulated ruffling is blocked by the GTP binding-defective N27-Arf6 mutant (Radhakrishna et al., 1999). The data on Arf6 involvement in vesicle trafficking and the association with Rac1 have led to the suggestion that the ability of Arf6 to influence Rac1-mediated lamellipodia formation depends, in part, on Arf6-mediated regulation of Rac1 targeting to the plasma membrane. In addition, it has been shown that two GEFs for Arf6, Arf nucleotide-binding site opener (ARNO) and EFA6, modulate growth factor- and protein kinase C-mediated cytoskeletal reorganization through activation of Arf6 (Franco et al., 1999; Frank et al., 1998), while a dominant negative mutant of ARNO inhibits both Arf6 translocation and cortical actin formation induced by growth factors (Venkateswarlu and Cullen, 2000). The proposed cross talk between Arf6 and Rac1 pathways is further supported by the finding that the cytoskeletal remodeling by EFA6 can be blocked by coexpression of a dominant negative mutant of either Arf6 or Rac1 (Franco et al., 1999).

It should be considered that more than one source of recycling membrane is probably involved in the development of axons and dendrites. In this direction both Arf6-dependent and independent endocytic pathways have been identified in neurons, which may contribute to neurite formation (Kanaani et al., 2004).

4.3.1.1. Contribution of Arf-Mediated Constitutive Exocytosis to Neuritogenesis

In neurons, treatment with the fungal metabolite brefeldin A (BFA) induces disruption of the Golgi complex including the trans-Golgi network and tubulation of endosomes (Cid-Arregui et al., 1995). BFA interferes with membrane transport between endoplasmic reticulum and Golgi, and between Golgi and plasma membrane, by inhibiting the activity of GEFs required for the activation of Arf1 and its recruitment from cytosol to membranes (Chardin et al., 1996; Donaldson et al., 1992b; Helms and Rothman, 1992; Morinaga et al., 1996; Randazzo et al., 1993; Tsai et al., 1996), thus potently inhibiting Arf function (Donaldson et al., 1991, 1992a; Klausner et al., 1992; Lippincott-Schwartz et al., 1991). To examine the role of the availability of membranes during the development of polarity and axonal elongation, Jareb and Banker (1997) have treated hippocampal neurons with BFA. Treatment of neurons with BFA inhibits axonal growth within 0.5 h and prevents the formation of an axon in unpolarized cells indicating that the availability of membrane components of Golgi-derived vesicles is required for axonal growth and hence the development of polarity. The rapid arrest of axonal extension by BFA in rat dorsal root ganglia neurons has been ascribed to inhibition of Arf proteins that are constituents of axonal growth cones (Hess et al., 1999), indicating that Arf proteins participate not only in constitutive membrane traffic within the cell body but also in membrane dynamics within growing axon endings.

Arf1 is mainly localized to the Golgi complex and is a common regulator of non-clathrin and clathrin coat recruitment, while Arf6 (the most divergent Arf with respect to primary sequence) links endocytic/exocytic events to the organization of the actin cytoskeleton, and does not localize at the Golgi (Chavrier and Goud, 1999; Takai et al., 2001). In general, the GEFs for Arf6 are insensitive to BFA (Cavenagh et al., 1996; Jackson and Casanova, 2000; Peters et al., 1995; Radhakrishna et al., 1996). For this reason, the effects of BFA are generally considered Arf6-independent. On the other hand, the identification of BIG2 (brefeldin A-inhibited guanine nucleotide-exchange protein 2) as a BFA-sensitive GEF expressed in neural tissue (Togawa et al., 1999) that localizes and affects the organization of the recycling endosomes through activating class I Arfs (Shin et al., 2004) indicates that this drug also affects the recycling pathways.

4.3.1.2. Contribution of Arf-Mediated Endocytic Membrane Recycling in Neuritogenesis

In nonneuronal cells, Arf6 regulates endocytic traffic, and affects cortical actin organization (Boshans et al., 2000; D'Souza-Schorey et al., 1997; Frank et al., 1998; Radhakrishna and Donaldson, 1997; Radhakrishna et al., 1996, 1999), and cell adhesion (Palacios et al., 2001; Santy and Casanova, 2001). Arf GTPases activate the lipid-modifying enzymes phospholipase D and phosphatidyl-inositol-4-phosphate 5-kinase (PI(4)P5-K) (Exton, 1997; Honda et al., 1999) and indirectly Rac1 (Santy and Casanova, 2001). Also the Arf6 GEF ARNO modulates cell migration (Santy and Casanova, 2001). Therefore ARNO and Arf6 emerge as possible key players in controlling actin dynamics and membrane trafficking in cells.

The implication of Arf proteins in the regulation of neurite extension of *Aplysia* motor neurons is confirmed by the GEF activity-dependent enhancement of neuritogenesis observed following overexpression of the mammalian Arf GEFs msec7-1 and ARNO (Huh et al., 2003). ARNO is a BFA-insensitive GEF that catalyzes exchange on ARF1 and ARF6 (Chardin et al., 1996; Frank et al., 1998). Biochemical studies have revealed high expression of both Arf6 and ARNO proteins in the developing hippocampus (Hernandez-Deviez et al., 2002; Suzuki et al., 2001, 2002). Overexpression of inactive ARNO and Arf6 causes increased branching of dendritic processes in embryonic rat hippocampal neurons, an effect reversed by Rac1 expression (Hernandez-Deviez et al., 2002). The two mutants enhance axonal extension and branching (Hernandez-Deviez et al., 2004), an effect abrogated by coexpression of constitutively active Arf6. Accordingly, the overexpression of a GEF-defective mutant of EFA6A enhances the formation of dendrites in hippocampal neurons (Sakagami et al., 2004).

On the other hand, we have shown strong inhibition of neurite extension in chick retinal neurons expressing either a constitutively active or a dominant negative mutant of Arf6 (Albertinazzi et al., 2003). In retinal neurons, these effects are stronger for the Arf6 mutants when compared to the corresponding mutants for Arf1 and Arf5, two Arfs implicated in membrane traffic through the Golgi apparatus, suggesting that trafficking of recycling membranes plays an important regulatory role in retinal neurite extension. The apparent contrast between the results obtained by expressing the dominant negative Arf6 mutant in retinal neurons and in hippocampal neurons may be explained by differences in the two types of neurons. Moreover a major experimental difference is the time of transfection: retinal neurons are transfected before plating on the substrate, while hippocampal neurons are transfected after 5 days in culture, once neurites are already established. The inhibitory effects of both constitutively active and dominant negative Arf6 mutants in retinal neurons suggest that a cycling GTPase is critical for the dynamic events at the onset of neurite extension.

The search for effectors of Arf6 during neurite extension has revealed that PI(4)P5-K-α plays a role in neurite extension and branching downstream of Arf6 (Hernandez-Deviez et al., 2004). PI(4)P5-K-α is implicated in membrane ruffling (Honda et al., 1999) and translocates to ruffling membranes with Arf6, where active Arf6 and phosphatidic acid produced by Phospholipase D act synergistically to enhance phosphatidyl-inositol-4,5-phosphate (PIP2) production by activating PI(4)P5-K-α. Since phospholipase D itself is an effector and PIP2 is an activator of Arf proteins, it is tempting to speculate that the local phospholipid metabolism regulated by Arf6 plays a crucial role in actin-mediated protrusion at the growth cone. One contribution of Arf6 in the reorganization of the actin cytoskeleton at the growth cone edge may come from the stimulation of a number of actin-binding proteins by PIP2 (Yin and Janmey, 2003). Inactive ARNO or Arf6 mutants deplete the actin-binding protein Mena, a member of the Ena/VASP family (Krause et al., 2003), from the growth cone edge, again indicating that the effects of Arf6 on neuritogenesis are mediated by changes in cytoskeletal dynamics.

4.3.1.3. Arf6 Controls Axonal Guidance at the Midline by Modulating Endocytosis of Adhesive Cues

In bilaterally organized animals, axons need to cross the midline to move to the contralateral body side. Axons first translate attractive extracellular signals that guide them to the nervous system midline (Mueller, 1999), while after crossing the midline the growth cones change behavior and continue to grow away from the midline to avoid crossing again. Extracellular netrins mediate attraction of growth cones toward the midline (Serafini et al., 1996). In *Drosophila* netrin mutants most axons fail to cross the midline (Mitchell et al., 1996). A study has shown that the *Drosophila* Arf6-GEF Schizo is required for midline crossing (Onel et al., 2004). Schizo acts by regulating the expression of the axon repulsive protein Slit at the surface of midline glial cells. The mutation of this Arf6-GEF leads to increased expression of the repulsive ligand, possibly due to the reduction of its Arf6-dependent internalization. A similar phenotype is obtained by expression of dominant negative form of Arf6. The balance of attractive (netrin) and repulsive (Slit) cues may therefore be regulated by controlling membrane dynamics.

4.3.2. Regulation of Neurite Development by Rab Proteins

Rab proteins and their effectors are highly compartmentalized in organelles to coordinate successive steps of vesicular transport including vesicle formation, motility, and tethering of vesicles to their target compartment (Zerial and

McBride, 2001). Rab5 is involved in early endocytic traffic in neurons (de Hoop et al., 1994), and its overexpression in cultured hippocampal neurons induces the formation of abnormal endosomes in both the somatodendritic and axonal domains. Another member of the family, Rab8, regulates polarized membrane transport to the dendrites. In nonneuronal BHK cells, expression of wild type or constitutively activated Rab8 induces a dramatic change in cell morphology, with extension of long processes resulting from the reorganization of actin and microtubules, and newly synthesized Golgi-derived proteins preferentially delivered into the outgrowths (Peranen et al., 1996). Therefore, Rab8 may provide a link between the machinery responsible for cell protrusion and polarized biosynthetic membrane traffic. In hippocampal neurons, suppression of the Rab8 expression by sequence-specific antisense oligonucleotides results in the blockage of morphological maturation (Huber et al., 1995). This impairment is due to inhibition of membrane traffic, as shown by a dramatic reduction in the number of vesicles undergoing anterograde transport in Rab8-depleted neurons. Moreover, Rab8 regulates the post-Golgi trafficking of rhodopsin-bearing vesicles in retinal photoreceptors. In these specialized neurons, ezrin/moesin and Rac1 colocalize with Rab8 on the transport vesicles at sites of fusion with the neuronal plasma membrane, suggesting that cytoskeletal elements act in concert with Rab8 to regulate tethering and fusion of rhodopsin-bearing vesicles (Deretic et al., 2004).

Rab proteins are believed to be important players also for the regulation of the transition of vesicular traffic between microtubules and actin filaments at the periphery of growth cones. Rab27a and its Rabphilin-like effector protein recruit myosin-V to melanosomes and appear to serve as membrane anchor. Myosin-V is a motor for short-range transport of vesicles in the actin-rich cortex of axons and dendrites that forms a complex with the microtubule-based motor kinesin (Huang et al., 1999). This heteromotor complex allows long-range movement of different kinds of vesicles on microtubules along axons and dendrites and short-range movement on actin filaments in dendritic spines and axon terminals (Langford, 2002). Distinct Rab GTPases could be involved in the recruitment of myosin-V to distinct types of carrier vesicles to be delivered to the neurite tips.

4.4. Coupling Arf and Rho Signaling Pathways at the Growth Cone

4.4.1. ArfGAP Proteins

In fully differentiated neurons, networks including large scaffolding proteins specifically localized in the presynaptic active zone link synaptic constituents to the underlying cytoskeleton and regulate synaptic-specific membrane traffic (Zhen and Jin, 2004). Likewise, one could envisage that during neuritogenesis, multimolecular complexes are required to coordinate exocytic events from the Golgi and the recycling compartment with those required for cytoskeletal rearrangement at the edge of migrating growth cones. These mechanisms have to regulate and coordinate the cross talk between different small GTPases that control distinct cellular processes during neuritogenesis. In this respect, the available results indicate that Arf6 functions in concert with Rac1 to regulate the actin-mediated protrusive activity at the cell edge. Evidence suggesting a connection between Arf6 and Rac1 pathways is the finding that the ArfGAPs of the GIT (G protein-coupled receptor kinase interactor) family identified by a number of laboratories including ours, serve as linkers between the focal adhesions and the Rac/Cdc42 pathway by assembling complexes including effectors and GEFs for Rho family GTPases (Bagrodia et al., 1999; Di Cesare et al., 2000; Premont et al., 1998; Turner et al., 1999). GIT proteins represent likely candidates for coordinating Arf6-mediated trafficking events with Rac-mediated cytoskeletal reorganization.

GIT1/p95-APP1/Cat1 is one of the two members of the GIT family of ArfGAPs (de Curtis, 2001; Donaldson and Jackson, 2000) that are able to interact with the PIX exchanging factors for Rac and Cdc42 (Manser et al., 1998), and with the focal

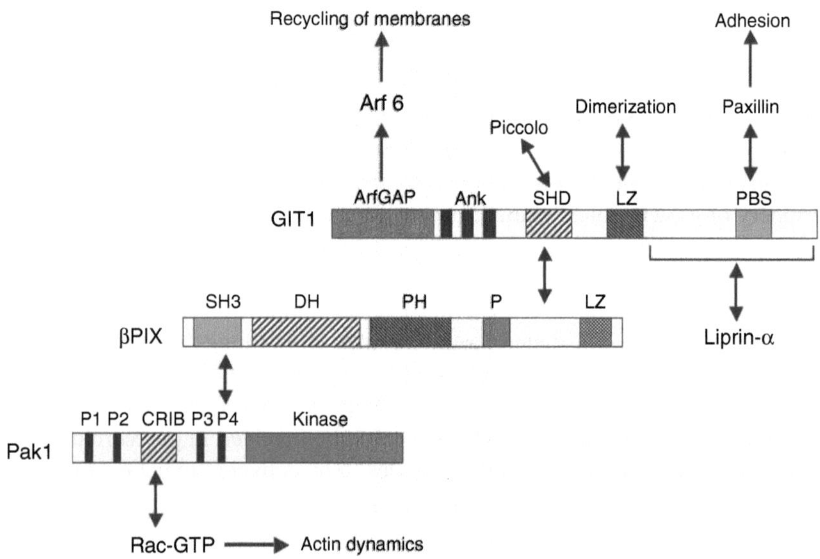

FIGURE 4.1. Model for the intermolecular interactions and functional connections identified for GIT proteins. Double-ended arrows point to some of the identified direct intermolecular interactions relevant to the issues discussed in this chapter; single-ended arrows indicate functional connections. Ank = ankyrin repeats; CRIB = Cdc42/Rac interactive binding motif; DH = Dbl homology; LZ = leucine zipper; P, P1-P4 = proline-rich regions; PBS = paxillin binding subdomain; PH = pleckstrin homology; SH3 = Src homology type-3; SHD = Spa2 homology domain. GIT family of ArfGAP proteins interacts with PIX and PAK (Bagrodia et al., 1999; Turner et al., 1999; Di Cesare et al., 2000), which mediate the interaction of the complex with active Rac and Cdc42. The complex may regulate actin remodeling at the cell surface by controlling Rac/Cdc42 activity via PAK and PIX. The interaction of GIT proteins with paxillin (Turner et al., 1999) functionally links the complex to integrin-mediated adhesion. Moreover, GIT proteins may be localized to endosomal compartments by PIX-dependent mechanisms.

adhesion protein Paxillin (Turner et al., 1999) (Figure 4.1). On top of their ability to interact with proteins involved in cell adhesion and actin organization, GIT proteins can affect endocytosis (Claing et al., 2000). By forming a stable complex with PIX that binds PAK via the SH3 domain (Manser et al., 1998), GIT1 can interact with Rac and Cdc42 GTPases in a GTP-dependent manner (Di Cesare et al., 2000) (Figure 4.1). The amino-terminal ArfGAP domain of GIT1 accelerates the hydrolysis of Arf proteins that have very slow intrinsic rates of GTP hydrolysis (Welsh et al., 1994), and *in vitro* and *in vivo* studies indicate that GIT proteins specifically regulate the activity of Arf6 (Albertinazzi et al., 2003; Vitale et al., 2000).

Given the proposed role of Arf6 in membrane recycling, one attractive hypothesis is that the ArfGAP activity of these proteins is required for the regulation of Arf-mediated membrane recycling to or near the sites of protrusion during neurite extension (Figure 4.2). In this direction, expression of the constitutively active, ArfGAP-resistant L67-Arf6 mutant prevents recycling to the plasma membrane of internalized membranes, which accumulate in PIP2-positive actin-coated vacuoles (Brown et al., 2001). These data corroborate the requirement of GTP hydrolysis and therefore of Arf6 GAPs for the recycling of endocytosed membrane back to the cell surface. One intriguing hypothesis is that Arf6 and GIT1 are required to assemble a so far undefined coat on the membrane of an endocytic compartment, for the formation and budding of vesicles to be recycled along the neurite, paraphrasing the role of Arf1 and its ArfGAP in the Golgi (Goldberg, 1999). A short variant of GIT2 has been shown to specifically affect Arf1-mediated distribution of the Golgi coatomer protein β-COP in nonneuronal cells (Mazaki et al., 2001), while no effect has been detected on either Arf1 or β-COP distribution by overexpressing the GIT1 constructs in retinal neurons

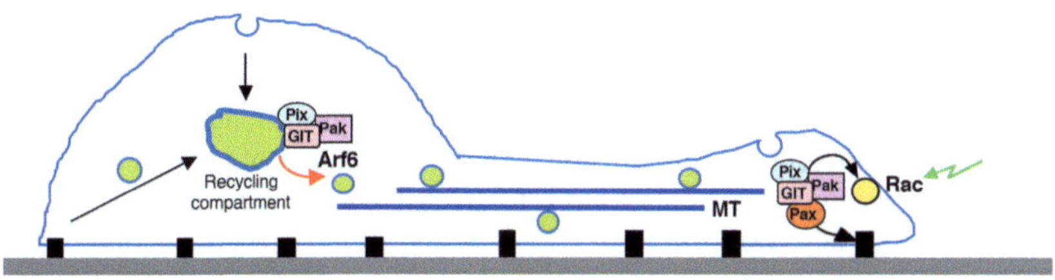

FIGURE 4.2. Model for the role of the GIT complex during neurite extension. The GIT1 complex colocalizes with Arf6 in the recycling endosomal compartment via the interaction with PIX. There the internalized membranes converge before recycling. According to the proposed model, Arf6 and the ArfGAP activity of GIT1 are required for the formation of recycling vesicles. Once formed, vesicles may travel along microtubules to reach the growth cone, where they can be recruited to Rac-enriched sites near or at the leading edge of the growth cone. By recruiting paxillin to the complex, GIT proteins induce the redistribution of paxillin from established adhesion sites to newly forming sites at the growth cone. Here, paxillin may contribute to the formation of membrane protrusions, by participating in the formation of new adhesive complexes where paxillin is required for the anchorage of Rac-induced actin filaments to sites of adhesion to the substrate. MT = microtubules; Pax = paxillin.

(Albertinazzi et al., 2003). These findings and the localization of GIT1 mutants at Arf6-positive endocytic structures in neurons support a specific role of this ArfGAP in the regulation of Arf6 *in vivo*.

4.4.2. GIT Proteins in Neurite Outgrowth and Synapse Formation

GIT proteins are strong candidates to connect the action of Rho GTPases on cytoskeleton with that of Arf GTPases on membrane traffic. The dissection of these multidomain proteins has been useful in identifying distinct functions for individual domains. GIT proteins interact with paxillin via their carboxy-terminal region (Turner et al., 1999), important to enhance protrusion and neurite extension in a Rac- and Arf6-dependent manner (Albertinazzi et al., 2003; Di Cesare et al., 2000). GIT1 overexpression causes relocalization of paxillin from focal adhesions to large perinuclear vesicles, and stimulates cell motility (Zhao et al., 2000), indicating a link between paxillin and membrane recycling and implicating the GIT-paxillin complexes in the regulation of protrusion. Moreover, inhibition of the interaction between paxillin and GIT2/PKL prevents the formation of lamellipodia (Turner et al., 1999).

GIT1 participates in different events during neuronal differentiation including neurite development and synaptogenesis (Zhang et al., 2003, 2005). We have shown a specific functional connection between Arf6 and GIT1 during Rac3-mediated neurite extension and branching. ArfGAP mutants of GIT1 as well as Arf6 mutants affecting the GDP/GTP cycle inhibit neurite extension from retinal neurons (Albertinazzi et al., 2003). In cotransfected neurons, Arf6 and the GIT1 ArfGAP-defective mutants accumulate specifically at large structures positive for Rab11, a functional marker for the perinuclear endocytic recycling compartment (Ren et al., 1998; Ullrich et al., 1996). In nonneuronal cells Arf6 colocalizes with Rab11 at pericentriolar recycling endosomes, and with a minor pool of Rab11 at the cell periphery (Powelka et al., 2004), while in retinal neurons endogenous Rab11 is distributed in a cytoplasmic punctated pattern along neurites and at growth cones (Albertinazzi et al., 2003). The accumulation of Arf6 in neurons is induced also by the expression of the GTP hydrolysis-defective mutant L67Arf6 (Albertinazzi et al., 2003). In both cases the intracellular accumulation can be explained by the inhibition of the conversion of Arf6-GTP to Arf6-GDP. One interpretation of these results is that GIT1 ArfGAP mutants negatively affect membrane recycling along neurites by interfering with Arf6 function, thus leading to membrane accumulation in an abnormal perinuclear recycling compartment in the soma, with consequent inhibition of neurite extension. Moreover, the inhibition of Rac3-enhanced neuritogenesis in retinal neurons by the ArfGAP mutants of GIT1 points to a functional link between Arf6 and Rac during neurite development.

The published data support the requirement of the Spa2 homology domain (SHD) domain of GIT1 responsible for PIX binding (Figure 4.1) in the localization of the GIT1 complex at the endocytic Rab11-positive compartment. Expression of a form of GIT1 lacking the SHD domain, and therefore unable to bind PIX, induces the formation of long, branched neurites (Albertinazzi et al.,

2003). Altogether, the data suggest a model in which the effects of the GIT1 complex on neurite extension may result from an equilibrium between the ability of the complex to interact with endocytic membranes through the SHD–PIX interaction, and the protrusive activity mediated by the interaction of the carboxy-terminal domain of GIT1 with paxillin, to promote protrusion at the growth cones that would be supported by maintaining Rac activity via the PIX-PAK complex (Figure 4.2). According to this hypothesis, the recruitment of both structural (paxillin) and signaling (PAK and PIX) proteins to the leading edge of migrating growth cones, a step that favors the formation of new adhesive contacts at these sites, is not mediated by simple cytoplasmic diffusion but rather through these ArfGAPs.

The kinase PAK is a downstream effector of Rac and Cdc42. In PC12 cells overexpression of PAK induces a Rac phenotype, consisting in enhanced lamellipodia formation along nerve growth factor (NGF)-induced neurites and at growth cones, via the amino-terminal noncatalytic domain of PAK. These effects require Rac activity, since PAK-induced lamellipodia are prevented by expression of the dominant-negative N17Rac1. PAK-induced lamellipodia require the interaction via a proline-rich region in the amino-terminal part of PAK with the SH3 domain of βPIX (Obermeier et al., 1998) (Figure 4.1). The guanin nucleotide exchanging activity of βPIX is required to enhance lamellipodial activity. Therefore, the localization of the GIT-βPIX-PAK complex at the plasma membrane of extending neurites appears to be important to maintain the activation of Rac and to recruit the machinery required for actin-driven membrane protrusion.

Beside the role in neurite extension, GIT proteins also contribute to synaptogenesis. GIT1 is enriched in pre- and post-synaptic terminals of hippocampal neurons and the interaction of GIT1 with PIX is necessary for the formation of synapses as shown by the reduction in synapses following disruption of this interaction (Zhang et al., 2003). Piccolo is a large protein implicated in the organization of molecular networks and actin cytoskeleton at the active zone of presynaptic densities, where synaptic vesicles dock and fuse for neurotransmitter release. Piccolo interacts directly with both GIT1 and GIT2 via their SHD domain (Kim et al., 2003). A large number of dense-core vesicles positive for Piccolo and several other components of the presynaptic active zone have been identified along axons and in growth cones of developing hippocampal neurons (Shapira et al., 2003). These vesicles are molecularly distinct from the synaptic vesicles present in the same neurons and are believed to represent transport vesicles necessary for the assembly of active zones in developing neurons. Given the direct interaction of Piccolo with GIT1, and the colocalization of the two proteins at presynaptic sites in mature hippocampal cultures (Ko et al., 2003), it will be interesting to find out whether GIT1 is among the components of these precursor vesicles in nonmature neurons, where GIT1 shows a similar punctate distribution. Piccolo, by acting as a specific membrane docking site for the GIT-PIX complex and for other components of the active zone, may represent a point of convergence of the integration of Rho and Arf signals, required for the formation of the large molecular networks constituting presynaptic densities.

GIT1 interacts directly with liprin-α via the carboxy-terminal paxillin-binding site. Liprin-α is a scaffolding protein regulating pre- and postsynaptic differentiation in neurons and adhesion in nonneuronal cells. In mature neurons, the two proteins colocalize both at postsynaptic densities and presynaptic active zones (Ko et al., 2003). They interact with the LAR (leukocyte common antigen-related) family of receptor protein tyrosine phosphatases (Serra-Pages et al., 1995) and are required for postsynaptic targeting of AMPA (α-amino-3-hydroxy-5-methyl-4-isoxazole propionic acid) receptors (Wyszynski et al., 2002) and for the development of the presynaptic active zone (Schoch et al., 2002). Interference with the GIT1–liprin-α interaction selectively reduces dendritic clustering of AMPA receptors in cultured neurons. GIT1 is strongly associated to a membrane fraction obtained from embryonic brain (Botrugno et al., 2005), and immunofluorescence in hippocampal neurons reveals the presence along neurites of GIT1-positive structures that are distinct from EEA1-positive endocytic structures and partially overlapping with transferrin receptor-positive vesicles (Ko et al., 2003), while in immature hippocampal neurons, GIT1 colocalizes with liprin-α at the tips of growth cones, suggesting a role of the GIT1–liprin-α complex during neuritogenesis.

4.4.3. Mechanisms of Delivery

Microtubules are considered responsible for long-distance membrane traffic in neurons, whereas actin filaments are responsible for relatively short distance trafficking. Distinct classes of motor proteins have been implicated in the transport along these two cytoskeletal systems: kinesins and myosins, respectively. Several microtubule-dependent motor proteins have been identified and characterized. The kinesin superfamily (KIFs) includes several motors for anterograde transport along the microtubules to their plus ends (Miki et al., 2005). It has been shown that KIF3 is engaged in fast axonal transport to convey membranous components important for neurite extension, as demonstrated by the fact that microinjection of function-blocking antibodies against KIF3 into superior cervical ganglion neurons blocks fast axonal transport and causes inhibition of neurite extension (Takeda et al., 2000). This result is consistent with the knowledge that fast axonal transport is required for the advancement of the growth cone (Martenson et al., 1993). Dynamic analysis in living neurons has shown that KIF1A, a neuron-specific KIF1/Unc104 family of plus-end–directed kinesin motors (Bloom, 2001; Okada et al., 1995), mediates neuronal anterograde transport at a high velocity and processivity (Lee et al., 2003). Liprin-α interacts with KIF1A (Figure 4.3A), suggesting that it works as a KIF1A receptor by linking KIF1A to various liprin-α–associated proteins for their transport in neurons (Shin et al., 2003) (Figure 4.3B). Miller et al. (2005) have shown that liprin-α promotes the delivery of synaptic material by a direct increase in kinesin-1 processivity and an indirect suppression of dynein activation. Live observation in *Drosophila* liprin-α mutant axons, showing synaptic strength and morphology defects that may be due to a failure in the delivery of synaptic-vesicle precursors, has demonstrated a role of this scaffolding protein in the regulation of bidirectional transport.

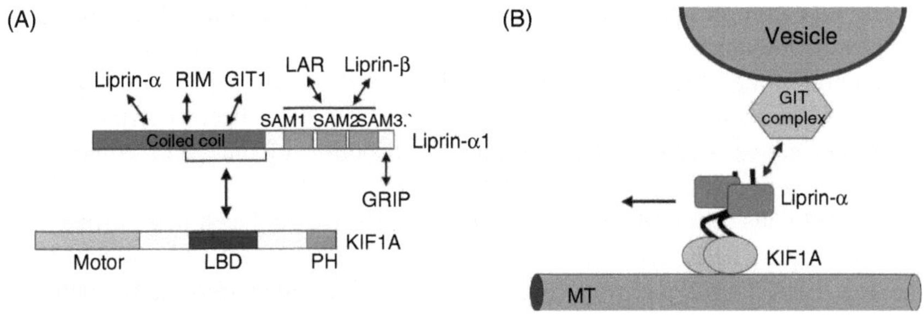

FIGURE 4.3. Characterization of the interaction of KIF1A with liprin-α. (A) Domain organization of liprin-α1 and KIF1A. Double-ended arrows point to identified direct intermolecular interactions. The identified interactions with liprin-α proteins include: a second liprin-α polypeptide for homodimerization; liprin; the GIT proteins; RIM, an active zone protein that was identified as a putative effector for the synaptic vesicle protein Rab3A (Wang et al., 1997); GRIP/ABP family of multi-PDZ proteins (Wyszynski et al., 2002); Leukocyte-common antigen related (LAR) tyrosine phosphatase receptors (Serra-Pages et al., 1995); and the kinesin motor protein KIF1A (Shin et al., 2003). (B) KIF1A is a plus-end–directed kinesin-like motor that mediates neuronal anterograde transport of large protein complexes including liprin-α and GIT1. Given the localization of GIT proteins at endocytic recycling compartments, this may represent a mechanism for the delivery of recycling membrane to the growth cone. Motor = kinesin-like motor domain; LBD = liprin-binding domain; MT = microtubule; PH = pleckstrin homology domain; SAM = sterile alpha motif.

4.4.4. Mechanisms of Targeting to the Plasma Membrane

How are recycling or exocytic membranes targeted to the growth cone? Different possibilities exist for the mechanisms driving the targeting of membranes at sites of protrusion. One hypothesis is that specific molecules localized at the protruding edge of growth cones act as docking sites for vesicles from the exocytic and recycling pathways transported to the tip of neurites along microtubules and actin filaments. A number of candidates have been identified that may serve as docking sites for recycling membranes, including the exocyst complex (Figure 4.4A) that is considered a signal for targeting transport vesicles originating from the Golgi or the endocytic recycling compartments to the plasma membrane (Finger and Novick 1998; Zhang et al., 2004). Impairment of neurite outgrowth by dominant-negative forms of Sec10 and Sec8 in PC12 cells has suggested a role for this complex in neurite extension (Vega and Hsu, 2001). In this direction, it has been demonstrated that Rab11 binds proteins of the exocyst complex in a GTP-dependent manner, indicating that components of the exocyst are Rab11 effectors, and that recycling vesicles may be targeted via this interaction (Beronja et al., 2005; Wu et al., 2005). In *Drosophila*, mutation of the exocyst component Sec5 disrupts neuronal membrane traffic and inhibits neurite extension (Murthy et al., 2003). Loss of function of Sec5, Sec6, and Sec15 in epithelial cells ends up in the inhibition of membrane delivery to the plasma membrane and accumulation of cadherins in enlarged Rab11 recycling endosomes (Langevin et al., 2005).

The GIT complex may contribute to the docking of recycling membranes at the plasma membrane in different ways. GIT1 can localize to membranes via the interaction of the PIX/PAK complex with Rac-GTP (Figure 4.4B) following activation of Rac by motogenic signals. In this direction, it has been shown that constitutively active V12Rac1 induces the redistribution of the GIT1 complex at lamellipodia in nonneuronal cells (Matafora et al., 2001).

In PC12 cells, basic fibroblast growth factor (bFGF)-induced neurite-like outgrowths require ERK-dependent phosphorylation of βPIX, which is mediated via the Ras/ERK/PAK2 pathway (Figure 4.4B). ERK interacts directly with the PAK2 kinase expressed in these cells, leading to activation and translocation at sites of protrusion of the PAK2-βPIX complex. A mutant form of βPIX in which the major phosphorylation sites (S525, T526) are replaced by alanines shows inhibition of neurite extension possibly due to a defect in targeting membranes at the cell periphery (Shin et al., 2002). Further analysis by the same laboratory has shown that phosphorylation of βPIX is responsible for FGF-induced activation of Rac1 needed for protrusive activity during neurite extension. A GEF-inactive mutant of βPIX induces retraction of preexisting neurites, suggesting that it competes with endogenous βPIX at the plasma membrane possibly by preventing the activation of Rac required for cytoskeletal remodeling during neurite extension (Shin et al., 2004).

Yoshii et al. (1999) have shown that PIX can be activated by signaling cascades from the platelet-derived growth factor (PDGF) receptor and EphB2 receptor and from integrin-induced signaling through phosphatidylinositol 3-kinase (PI3-kinase). Under these conditions, the GIT complex may be recruited at membranes by association of PIX with the p85 regulatory subunit of PI3-kinase (Figure 4.4C), with PAK locally stimulating the exchange factor activity of PIX (Daniels et al., 1999).

Another way to recruit the GIT complex at the cell periphery is via the direct interaction of βPIX with the PDZ domains of Scribble (Figure 4.4D). Scribble recruits PIX and GIT1 at the plasma membrane of PC12 cells following secretagogue-induced stimulation (Audebert et al., 2004). The relevance of this interaction for trafficking events is demonstrated by the finding that mutations of Scribble sequestering PIX in the cytosol impair Ca^{2+}-dependent exocytosis. PIX itself increases secretion, while a catalytically inactive mutant defective in GEF activity reduces it, implicating PIX in the exocytic pathway in PC12 cells. Therefore, Scribble may act as one component of a membrane anchor for the PIX/GIT1 complex regulating exocytic events, by establishing a spatial landmark at the plasma membrane, where Rac and Arf signaling pathways meet at sites of vesicle docking and fusion.

4.5. Concluding Remarks

Given the dynamic nature of the processes involved in axonal and dendritic development, it is not surprising that the underlying molecular mechanisms are highly

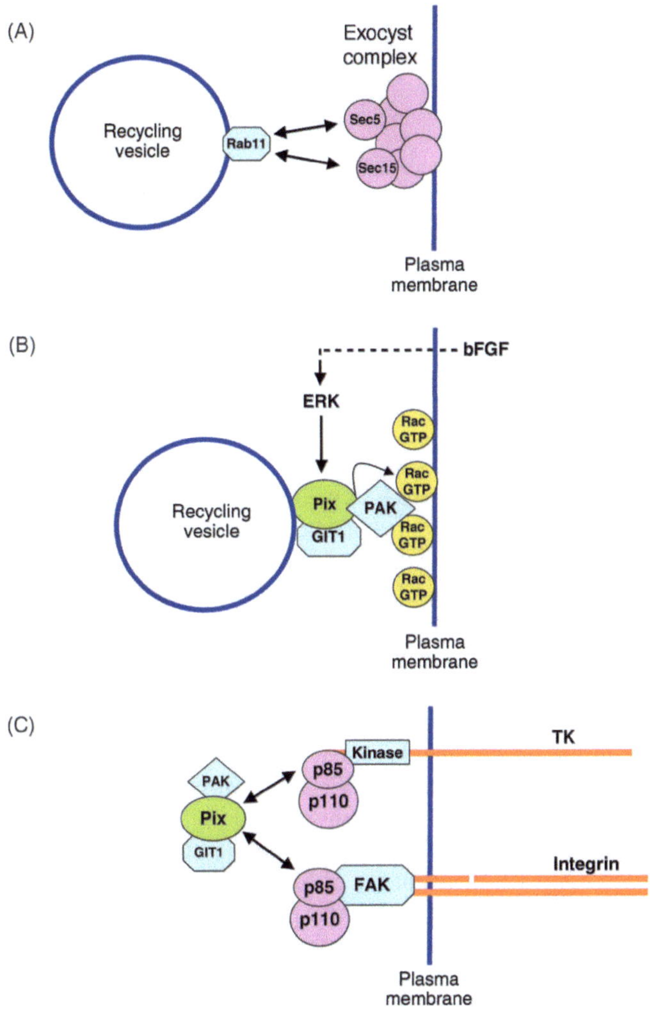

FIGURE 4.4. Mechanisms for targeting of recycling membranes to the plasma membrane. Models describing some of the mechanisms that may account for the targeting of recycling membranes and/or protein complexes to the front of neurites to regulate extension (see the text for more details). (A) Rab11-positive recycling membranes may be targeted at the plasma membrane of the growth cone by interacting with components of the exocyst complex. (B) The GIT complex may contribute to the docking of membrane vesicles originating from the recycling compartment via the interaction of the PIX/PAK complex with Rac activated by motogenic signals. In this direction, bFGF-induced neurite outgrowth has been shown to require ERK-dependent phosphorylation of βPIX, via the Ras/ERK/PAK2 pathway (Shin et al., 2002). (C) Alternatively, the GIT complex may be recruited to the plasma membrane by association of PIX with the p85 regulatory subunit of PI3-kinase (Yoshii et al., 1999) that is recruited at the plasma membrane by different mechanisms, including tyrosine kinase (TK) receptor- and integrin-mediated mechanisms.

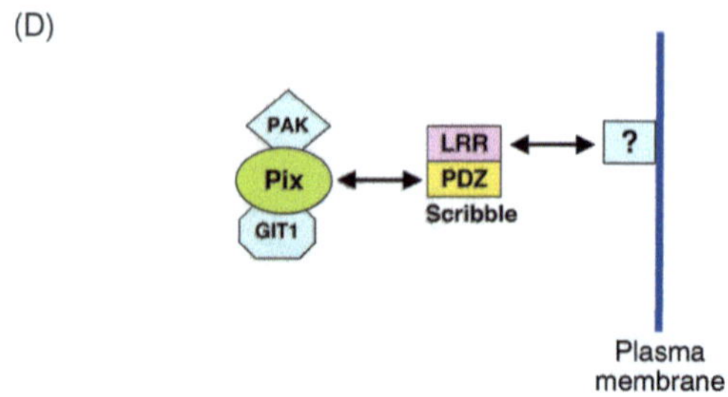

FIGURE 4.4. (*continued*) (D) The GIT complex can be recruited at the neuronal plasma membrane by the direct interaction of βPIX with the PDZ domains of the cytoplasmic protein Scribble (Audebert et al., 2004) that is recruited at the plasma membrane of the growth cone by so far undefined mechanisms. LRR = leucine rich repeat.

complex. Growth cones are sites of intense endocytosis and require an equivalent membrane flow back to the surface to maintain equilibrium. The endocytic and exocytic compartments represent dynamic reservoirs of mobile membrane to quickly respond to extracellular stimuli, leading to growth cone-mediated neurite extension and retraction. The available data on the existence of dynamic recycling endosomes in axons and dendrites of developing neurons indicate that an important contribution to neurite progression comes from endocytosed, recycled membranes at the growth cone. The identification of proteins that can assemble molecular networks by functionally linking to different types of GTPases promotes them as interesting candidates, and provides a framework to test models and hypotheses on how the action of distinct classes of small GTPases may be linked together to contribute to different dynamic processes underlying neurite extension. Some of these multimolecular adaptor proteins already have a defined role in the assembly and maintenance of functional synapses. One possibility is that proteins like GIT, PIX, and liprins may perform different functions during the life of the neuron.

Several issues remain to be explored, including the molecular characterization of the intracellular compartment(s) involved, a better definition of the targeting mechanisms that bring about the coordination between membrane traffic with the adhesive and protrusive processes at the edge of the migrating growth cone, the mechanisms responsible for the regulation of these highly dynamic processes, and the identification of other components of the molecular machinery linking the different traffic pathways considered to the cytoskeletal devices contributing to neurite elongation or retraction. Given the complexity of these processes, integration and merging of the information from different areas of neuronal cell biology will be essential for progress in our understanding of the events underlying growth cone behavior.

Acknowledgments. The financial support of Telethon-Italy (grant n. GGP05051) is gratefully acknowledged.

References

Albertinazzi, C., Za, L., Paris, S., and de Curtis, I., 2003, ADP-ribosylation factor 6 and a functional PIX/p95–APP1 complex are required for Rac1B-mediated neurite outgrowth, *Mol. Biol. Cell* **14:** 1295–1307.

Altschuler, Y., Liu, S., Katz, L., Tang, K., Hardy, S., Brodsky, F., et al., 1999, ADP-ribosylation factor 6 and endocytosis at the apical surface of Madin-Darby canine kidney cells, *J. Cell Biol.* **147:** 7–12.

Audebert, S., Navarro, C., Nourry, C., Chasserot-Golaz, S., Lecine, P., Bellaiche, Y., et al., 2004, Mammalian Scribble forms a tight complex with the betaPIX exchange factor, *Curr. Biol.* **14:** 987–995.

Bagrodia, S., Bailey, D., Lenard, Z., Hart, M., Guan, J.L., Premont, R.T., et al., 1999, A tyrosine-phosphorylated protein that binds to an important regulatory region on the cool family of p21–activated kinase-binding proteins, *J. Biol. Chem.* **274:** 22393–22400.

Beronja, S., Laprise, P., Papoulas, O., Pellikka, M., Sisson, J., and Tepass, U., 2005, Essential function of Drosophila Sec6 in apical exocytosis of epithelial photoreceptor cells, *J. Cell Biol.* **169:** 635–646.

Bloom, G.S., 2001, The UNC-104/KIF1 family of kinesins, *Curr. Opin. Cell Biol.* **13:** 36–40.

Borisy, G.G., and Svitkina, T.M., 2000, Actin machinery: Pushing the envelope, *Curr. Opin. Cell Biol.* **12:** 104–112.

Boshans, R.L., Szanto, S., van Aelst, L., and D'Souza-Schorey, C., 2000, ADP-ribosylation factor 6 regulates actin cytoskeleton remodeling in coordination with Rac1 and RhoA, *Mol. Cell. Biol.* **20:** 3685–3694.

Botrugno, O.A., Paris, S., Za, L., Gualdoni, S., Cattaneo, A., Bachi, A., et al., 2006, Characterization of the endogenous GIT1–βPIX complex, and identification of its association to membranes, *Eur. J. Cell Biol.* **85:** 35–46.

Bradke, F., and Dotti, C.G., 1997, Neuronal polarity: Vectorial cytoplasmic flow precedes axon formation, *Neuron* **19:** 1175–1186.

Bretscher, M.S., 1996, Getting membrane flow and the cytoskeleton to cooperate in moving cells, *Cell* **87:** 601–606.

Bretscher, M.S., and Aguado-Velasco, C., 1998a, EGF induces recycling membrane to form ruffles, *Curr. Biol.* **8:** 721–724.

Bretscher, M.S., and Aguado-Velasco, C., 1998b, Membrane traffic during cell locomotion, *Curr. Opin. Cell Biol.* **10:** 537–541.

Brown, F.D., Rozelle, A.L., Yin, H.L., Balla, T., and Donaldson, J.G., 2001, Phosphatidylinositol 4,5–bisphosphate and Arf6–regulated membrane traffic, *J. Cell Biol.* **54:** 1007–1017.

Cavenagh, M.M., Whitney, J.A., Carroll, K., Zhang, C., Boman, A.L., Rosenwald, A.G., et al., 1996, Intracellular distribution of Arf proteins in mammalian cells. Arf6 is uniquely localized to the plasma membrane, *J. Biol. Chem.* **271:** 21767–21774.

Chardin, P., Paris, S., Antonny, B., Robineau, S., Beraud-Dufour, S., Jackson, C.L., et al., 1996, A human exchange factor for ARF contains Sec7– and pleckstrin-homology domains, *Nature* **384:** 481–484.

Chavrier, P., and Goud, B., 1999, The role of ARF and Rab GTPases in membrane transport, *Curr. Opin. Cell Biol.* **11:** 466–475.

Cheng, T.P., and Reese, T.S., 1987, Recycling of plasmalemma in chick tectal growth cones, *J. Neurosci.* **7:** 1752–1759.

Cid-Arregui, A., Parton, R.G., Simons, K., and Dotti, C.G., 1995, Nocodazole-dependent transport, and brefeldin A-sensitive processing and sorting, of newly synthesized membrane proteins in cultured neurons, *J. Neurosci.* **15:** 4259–4269.

Claing, A., Perry, S.J., Achiriloaie, M., Walker, J.K., Albanesi, J.P., Lefkowitz, R.J., et al., 2000, Multiple endocytic pathways of G protein-coupled receptors delineated by GIT1 sensitivity, *Proc. Natl. Acad. Sci. USA* **97:** 1119–11124.

Craig, A.M., Wyborski, R.J., and Banker, G., 1995, Preferential addition of newly synthesized membrane protein at axonal growth cones, *Nature* **375:** 592–594.

D'Souza-Schorey, C., Li, G., Colombo, M.I., and Stahl, P.D., 1995, A regulatory role for ARF6 in receptor-mediated endocytosis, *Science* **267:** 1175–1178.

D'Souza-Schorey, C., Boshans, R.L., McDonough, M., Stahl, P.D., and Van Aelst, L., 1997, A role for POR1, a Rac1–interacting protein, in ARF6–mediated cytoskeletal rearrangements, *EMBO J.* **16:** 5445–5454.

D'Souza-Schorey, C., van Donselaar, E., Hsu, V.W., Yang, C., Stahl, P.D., and Peters, P.J., 1998, ARF6 targets recycling vesicles to the plasma membrane: Insights from an ultrastructural investigation, *J. Cell Biol.* **140:** 603–616.

Dailey, M.E., and Bridgman, P.C., 1993, Vacuole dynamics in growth cones: Correlated EM and video observations, *J. Neurosci.* **13:** 3375–3393.

Daniels, R.H., Zenke, F.T., and Bokoch, G.M., 1999, AlphaPix stimulates p21–activated kinase activity through exchange factor-dependent and -independent mechanisms, *J. Biol. Chem.* **274:** 6047–6050.

de Curtis, I., 2001, Cell migration: GAPs between membrane traffic and the cytoskeleton, *EMBO Rep.* **2:** 277–281.

de Hoop, M.J., Huber, L.A., Stenmark, H., Williamson, E., Zerial, M., Parton, R.G., et al., 1994, The involvement of the small GTP-binding protein Rab5a in neuronal endocytosis, *Neuron* **13:** 11–22.

Deretic, D., Traverso, V., Parkins, N., Jackson, F., Rodriguez de Turco, E.B., and Ransom, N., 2004, Phosphoinositides, ezrin/moesin, and rac1 regulate fusion of rhodopsin transport carriers in retinal photoreceptors, *Mol. Biol. Cell* **15:** 359–370.

Di Cesare, A., Paris, S., Albertinazzi, C., Dariozzi, S., Andersen, J., Mann, M., et al., 2000, P95–APP1 links membrane transport to Rac-mediated reorganization of actin, *Nature Cell Biol.* **2:** 521–530.

Diefenbach, T.J., Guthrie, P.B., Stier, H., Billups, B., and Kater, S.B., 1999, Membrane recycling in the neuronal growth cone revealed by FM1–43 labeling, *J. Neurosci.* **19:** 9436–9444.

Donaldson, J.G., and Jackson, C.L., 2000, Regulators and effectors of the ARF GTPases, *Curr. Opin. Cell Biol.* **12:** 475–482.

Donaldson, J.G., Kahn, R.A., Lippincott-Schwartz, J., and Klausner, R.D., 1991, Binding of ARF and beta-COP to Golgi membranes: Possible regulation by a trimeric G protein, *Science* **254:** 1197–1199.

Donaldson, J.G., Cassel, D., Kahn, R.A., and Klausner, R.D., 1992a, ADP-ribosylation factor, a small GTP-binding protein, is required for binding of the coatomer protein beta-COP to Golgi membranes, *Proc. Natl. Acad. Sci. USA* **189:** 6408–6412.

Donaldson, J.G., Finazzi, D., and Klausner, R.D., 1992b, Brefeldin A inhibits Golgi membrane-catalysed exchange of guanine nucleotide onto ARF protein, *Nature* **360:** 350–352.

Exton, J.H., 1997, Regulation of phosphoinositide phospholipases by G-proteins, *Adv. Exp. Med. Biol.* **400A:** 3–8.

Finger, F.P., and Novick, P., 1998, Spatial regulation of exocytosis: Lessons from yeast, *J. Cell Biol.* **142:** 609–612.

Franco, M., Peters, P.J., Boretto, J., van Donselaar, E., Neri, A., D'Souza-Schorey, C., et al., 1999, EFA6, a sec7 domain-containing exchange factor for ARF6, coordinates membrane recycling and actin cytoskeleton organization, *EMBO J.* **18:** 1480–1491.

Frank, S., Upender, S., Hansen, S.H., and Casanova, J.E., 1998, ARNO is a guanine nucleotide exchange factor for ADP-ribosylation factor 6, *J. Biol. Chem.* **273:** 23–27.

Futerman, A.H., and Banker, G.A., 1996, The economics of neurite outgrowth-the addition of new membrane to growing axons, *Trends Neurosci.* **19:** 144–149.

Goldberg, J., 1999, Structural and functional analysis of the ARF1–ARFGAP complex reveals a role for coatomer in GTP hydrolysis, *Cell* **96:** 893–902.

Hall, A., 1998, Rho GTPases and the actin cytoskeleton, *Science* **279:** 509–514.

Hao, M., and Maxfield, F.R., 2000, Characterization of rapid membrane internalization and recycling, *J. Biol. Chem.* **75:** 15279–15286.

Helms, J.B., and Rothman, J.E., 1992, Inhibition by brefeldin A of a Golgi membrane enzyme that catalyses exchange of guanine nucleotide bound to ARF, *Nature* **360:** 352–354.

Hernandez-Deviez, D.J., Casanova, J.E., and Wilson, J.M., 2002, Regulation of dendritic development by the ARF exchange factor ARNO, *Nat. Neurosci.* **5:** 623–624.

Hernandez-Deviez, D.J., Roth, M.G., Casanova, J.E., and Wilson, J.M., 2004, ARNO and ARF6 regulate axonal elongation and branching through downstream activation of phosphatidylinositol 4–phosphate 5–kinase alpha, *Mol. Biol. Cell* **15:** 111–120.

Hess, D.T., Smith, D.S., Patterson, S.I., Kahn, R.A., Skene, J.H., and Norden, J.J., 1999, Rapid arrest of axon elongation by brefeldin A: A role for the small GTP-binding protein ARF in neuronal growth cones, *J. Neurobiol.* **38:** 105–115.

Honda, A., Nogami, M., Yokozeki, T., Yamazaki, M., Nakamura, M., Watanabe, H., et al., 1999, Phosphatidyl-inositol 4–phosphate 5–kinase alpha is a downstream effector of the small G protein ARF6 in membrane ruffle formation, *Cell* **99:** 521–532.

Hopkins, C.R., Gibson, A., Shipman, M., Strickland, D.K., and Trowbridge, I.S., 1994, In migrating fibroblasts, recycling receptors are concentrated in narrow tubules in the pericentriolar area, and then routed to the plasma membrane of the leading lamella, *J. Cell Biol.* **125:** 1265–1274.

Huang, J.D., Brady, S.T., Richards, B.W., Stenolen, D., Resau, J.H., Copeland, N.G., et al., 1999, Direct interaction of microtubule- and actin-based transport motors, *Nature* **397:** 267–270.

Huber, L.A., Dupree, P., and Dotti, C.G., 1995, A deficiency of the small GTPase rab8 inhibits membrane traffic in developing neurons, *Mol. Cell Biol.* **15:** 918–924.

Huh, M., Han, J.H., Lim, C.S., Lee, S.H., Kim, S., Kim, E., et al., 2003, Regulation of neuritogenesis and synaptic transmission by msec7–1, a guanine nucleotide exchange factor, in cultured Aplysia neurons, *J. Neurochem.* **85:** 282–285.

Igarashi, M., Kozaki, S., Terakawa, S., Kawano, S., Ide, C., and Komiya, Y., 1996, Growth cone collapse and inhibition of neurite growth by Botulinum neurotoxin C1: A t-SNARE is involved in axonal growth, *J. Cell Biol.* **134:** 205–215.

Jackson, C.L., and Casanova, J.E., 2000, Turning on ARF: The Sec7 family of guanine-nucleotide-exchange factors, *Trends Cell Biol.* **10:** 60–67.

Jareb, M., and Banker, G., 1997, Inhibition of axonal growth by brefeldin A in hippocampal neurons in culture, *J. Neurosci.* **17:** 8955–8963.

Kanaani, J., Diacovo, M.J., El-Husseini, A.-D., Bredt, D.S., and Baekkeskov, S., 2004, Palmitoylation controls trafficking of GAD65 from Golgi membranes to axon-specific endosomes and a Rab5a-dependent pathway to presynaptic clusters, *J. Cell Sci.* **117:** 2001–2013.

Kim, S., Ko, J., Shin, H., Lee, J.R., Lim, C., Han, J.H., et al., 2003, The GIT family of proteins forms multimers and associates with the presynaptic cytomatrix protein Piccolo, *J. Biol. Chem.* **278:** 6291–6300.
Klausner, R.D., Donaldson, J.G., and Lippincott-Schwartz, J., 1992, Brefeldin A: Insights into the control of membrane traffic and organelle structure, *J. Cell Biol.* **116:** 1071–1080.
Ko, J., Kim, S., Valtschanoff, J.G., Shin, H., Lee, J.R., Sheng, M., et al., 2003, Interaction between liprin-alpha and GIT1 is required for AMPA receptor targeting, *J. Neurosci.* **23:** 1667–1677.
Krause, M., Dent, E.W., Bear, J.E., Loureiro, J.J., and Gertler, F.B., 2003, Ena/VASP proteins: Regulators of the actin cytoskeleton and cell migration, *Annu. Rev. Cell Dev. Biol.* **19:** 541–564.
Langevin, J., Morgan, M.J., Sibarita, J.B., Aresta, S., Murthy, M., Schwarz, T., et al., 2005, Drosophila exocyst components Sec5, Sec6, and Sec15 regulate DE-Cadherin trafficking from recycling endosomes to the plasma membrane, *Dev. Cell* **9:** 355–376.
Langford, G.M., 2002, Myosin-V, a versatile motor for short-range vesicle transport, *Traffic* **3:** 859–865.
Lee, J.-R., Shin, H., Ko, J., Choi, J., Lee, H., and Kim, E., 2003, Characterization of the movement of the kinesin motor KIF1A in living cultured neurons, *J. Biol. Chem.* **278:** 2624–2629.
Lippincott-Schwartz, J., Yuan, L., Tipper, C., Amherdt, M., Orci, L., and Klausner, R.D., 1991, Brefeldin A's effects on endosomes, lysosomes, and the TGN suggest a general mechanism for regulating organelle structure and membrane traffic, *Cell* **67:** 601–616.
Lockerbie, R.O., Miller, V.E., and Pfenninger, K.H., 1991, Regulated plasmalemmal expansion in nerve growth cones, *J. Cell Biol.* **112:** 1215–1227.
Luo, L., 2000, Rho GTPases in neuronal morphogenesis, *Nat. Rev. Neurosci.* **1:** 173–180.
Manser, E., Loo, T.H., Koh, C.G., Zhao, Z.S., Chen, X.Q., Tan, L., et al., 1998, PAK kinases are directly coupled to the PIX family of nucleotide exchange factors, *Mol. Cell* **1:** 183–192.
Martenson, C., Stone, K., Reedy, M., and Sheetz, M., 1993, Fast axonal transport is required for growth cone advance, *Nature* **366:** 66–69.
Martinez-Arca, S., Coco, S., Mainguy, G., Schenk, U., Alberts, P., Bouille, P., et al., 2001, A common exocytotic mechanism mediates axonal and dendritic outgrowth, *J. Neurosci.* **21:** 3830–3838.
Matafora, V., Paris, S., Dariozzi, S., and de Curtis, I., 2001, Molecular mechanisms regulating the subcellular localization of p95–APP1 between the endosomal recycling compartment and sites of actin organization at the cell surface, *J. Cell Sci.* **114:** 4509–4520.
Mazaki, Y., Hashimoto, S., Okawa, K., Tsubouchi, A., Nakamura, K., Yagi, R., et al., 2001, An ADP-ribosylation factor GTPase-activating protein Git2–short/KIAA0148 is involved in subcellular localization of paxillin and actin cytoskeletal organization, *Mol. Biol. Cell* **12:** 645–662.
Miki, H., Okada, Y., and Hirokawa, N., 2005, Analysis of the kinesin superfamily: Insights into structure and function, *Trends Cell Biol.* **15:** 467–476.
Miller, K.E., DeProto, J., Kaufmann, N., Patel, B.N., Duckworth, A., and Van Vactor, D., 2005, Direct observation demonstrates that Liprin-alpha is required for trafficking of synaptic vesicles, *Curr. Biol.* **15:** 684–689.
Mitchell, K.J., Doyle, J.L., Serafini, T., Kennedy, T.E., Tessier-Lavigne, M., Goodman, C.S., et al., 1996, Genetic analysis of Netrin genes in Drosophila: Netrins guide CNS commissural axons and peripheral motor axons, *Neuron* **17:** 203–215.

Morinaga, N., Tsai, S.C., Moss, J., and Vaughan, M., 1996, Isolation of a brefeldin A-inhibited guanine nucleotide-exchange protein for ADP ribosylation factor (ARF) 1 and ARF3 that contains a Sec7–like domain, *Proc. Natl. Acad. Sci. USA* **93:** 12856–12860.

Mueller, B.K., 1999, Growth cone guidance: First steps towards a deeper understanding, *Annu. Rev. Neurosci.* **22:** 351–388.

Mundigl, O., Ochoa, G.C., David, C., Slepnev, V.I., Kabanov, A., and De Camilli, P., 1998, Amphiphysin I antisense oligonucleotides inhibit neurite outgrowth in cultured hippocampal neurons, *J. Neurosci.* **18:** 93–103.

Murthy, M., Garza, D., Scheller, R.H., and Schwarz, T.L., 2003, Mutations in the exocyst component Sec5 disrupt neuronal membrane traffic, but neurotransmitter release persists, *Neuron* **37:** 433–447.

Nie, Z., Hirsch, D.S., and Randazzo, P.A., 2003, Arf and its many interactors, *Curr. Opin. Cell Biol.* **15:** 396–404.

Nikolic, M., Chou, M.M., Lu, W., Mayer, B.J., and Tsai, L.H., 1998, The p35/Cdk5 kinase is a neuron-specific Rac effector that inhibits Pak1 activity, *Nature* **395:** 194–198.

Obermeier, A., Ahmed, S., Manser, E., Yen, S.C., Hall, C., and Lim, L., 1998, PAK promotes morphological changes by acting upstream of Rac, *EMBO J.* **17:** 4328–4339.

Ohno, S., 2001, Intercellular junctions and cellular polarity: The Par-aPKC complex, a conserved core cassette playing fundamental roles in cell polarity, *Curr. Opin. Cell Biol.* **13:** 641–648.

Okada, Y., Yamazaki, H., Sekine-Aizawa, Y., and Hirokawa, N., 1995, The neuron-specific kinesin superfamily protein KIF1A is a unique monomeric motor for anterograde axonal transport of synaptic vesicle precursors, *Cell* **81:** 769–780.

Onel, S., Bolke, L., and Klambt, C., 2004, The Drosophila ARF6–GEF Schizo controls commissure formation by regulating Slit, *Development* **131:** 2587–2594.

Paglini, G., Peris, L., Diez-Guerra, J., Quiroga, S., and Caceres, A., 2001, The Cdk5–p35 kinase associates with the Golgi apparatus and regulates membrane traffic, *EMBO Rep.* **2:** 1139–1144.

Palacios, F., Price, L., Schweitzer, J., Collard, J.G., and D'Souza-Schorey, C., 2001, An essential role for ARF6–regulated membrane traffic in adherens junction turnover and epithelial cell migration, *EMBO J.* **20:** 4973–4986.

Peranen, J., Auvinen, P., Virta, H., Wepf, R., and Simons, K., 1996, Rab8 promotes polarized membrane transport through reorganization of actin and microtubules in fibroblasts, *J. Cell Biol.* **135:** 153–167.

Peters, P.J., Hsu, V.W., Ooi, C.E., Finazzi, D., Teal, S.B., Oorschot, V., et al., 1995, Overexpression of wild-type and mutant ARF1 and ARF6: Distinct perturbations of nonoverlapping membrane compartments, *J. Cell Biol.* **128:** 1003–1017.

Pfenninger, K.H., and Friedman, L.B., 1993, Sites of plasmalemmal expansion in growth cones, *Dev. Brain Res.* **71:** 181–192.

Pfenninger, K.H., and Maylié-Pfenninger, M.-F., 1981, Lectin labeling of sprouting neurons. II. Relative movement and appearance of glycoconjugates during plasmalemmal expansion, *J. Cell Biol.* **89:** 547–559.

Powelka, A.M., Sun, J., Li, J., Gao, M., Shaw, L.M., Sonnenberg, A., et al., 2004, Stimulation-dependent recycling of integrin beta1 regulated by ARF6 and Rab11, *Traffic* **5:** 20–36.

Prekeris, R., Foletti, D.L., and Scheller, R.H., 1999, Dynamics of tubulovesicular recycling endosomes in hippocampal neurons, *J. Neurosci.* **19:** 10324–10337.

Premont, R.T., Claing, A., Vitale, N., Freeman, J.L., Pitcher, J.A., Patton, W.A., et al., 1998, Beta2–Adrenergic receptor regulation by GIT1, a G protein-coupled receptor kinase-associated ADP ribosylation factor GTPase-activating protein, *Proc. Natl. Acad. Sci. USA* **95:** 14082–14087.

Radhakrishna, H., and Donaldson, J.G., 1997, ADP-ribosylation factor 6 regulates a novel plasma membrane recycling pathway, *J. Cell Biol.* **139:** 49–61.

Radhakrishna, H., Klausner, R.D., and Donaldson, J.G., 1996, Aluminum fluoride stimulates surface protrusions in cells overexpressing the ARF6 GTPase, *J. Cell Biol.* **134:** 935–947.

Radhakrishna, H., Al-Awar, O., Khachikian, Z., and Donaldson, J.G., 1999, ARF6 requirement for Rac ruffling suggests a role for membrane trafficking in cortical actin rearrangements, *J. Cell Sci.* **112:** 855–866.

Randazzo, P.A., Yang, Y.C., Rulka, C., and Kahn, R.A., 1993, Activation of ADP-ribosylation factor by Golgi membranes. Evidence for a brefeldin A- and protease-sensitive activating factor on Golgi membranes, *J. Biol. Chem.* **268:** 9555–9563.

Rashid, T., Banerjee, M., and Nikolic, M., 2001, Phosphorylation of Pak1 by the p35/Cdk5 kinase affects neuronal morphology, *J. Biol. Chem.* **276:** 49043–49052.

Ren, M., Xu, G., Zeng, J., De Lemos-Chiarandini, C., Adesnik, M., and Sabatini, D.D., 1998, Hydrolysis of GTP on rab11 is required for the direct delivery of transferrin from the pericentriolar recycling compartment to the cell surface but not from sorting endosomes, *Proc. Natl. Acad. Sci. USA* **95:** 6187–6192.

Rosso, S., Bollati, F., Bisbal, M., Peretti, D., Sumi, T., Nakamura, T., et al., 2004, LIMK1 regulates Golgi dynamics, traffic of Golgi-derived vesicles, and process extension in primary cultured neurons, *Mol. Biol. Cell* **15:** 3433–3449.

Roth, M.G., 1999, Snapshots of ARF1: Implications for mechanisms of activation and inactivation, *Cell* **97:** 149–152.

Sakagami, H., Matsuya, S., Nishimura, H., Suzuki, R., and Kondo, H., 2004, Somatodendritic localization of the mRNA for EFA6A, a guanine nucleotide exchange protein for ARF6, in rat hippocampus and its involvement in dendritic formation, *Eur. J. Neurosci.* **19:** 863–870.

Santy, L.C., and Casanova, J.E., 2001, Activation of ARF6 by ARNO stimulates epithelial cell migration through downstream activation of both Rac1 and phospholipase D, *J. Cell Biol.* **154:** 599–610.

Schoch, S., Castillo, P.E., Jo, T., Mukherjee, K., Geppert, M., Wang, Y., et al., 2002, RIM1 forms a protein scaffold for regulating neurotransmitter release at the active zone, *Nature* **415:** 321–326.

Serafini, T., Colamarino, S.A., Leonardo, E.D., Wang, H., Beddington, R., Skarnes, W.C., et al., 1996, Netrin-1 is required for commissural axon guidance in developing vertebrate nervous system, *Cell* **87:** 1001–1014.

Serra-Pages, C., Kedersha, N.L., Fazikas, L., Medley, Q., Debant, A., and Streuli, M., 1995, The LAR transmembrane protein tyrosine phosphatase and a coiled-coil LAR-interacting protein co-localize at focal adhesions, *EMBO J.* **14:** 2827–2838.

Shapira, M., Zhai, R.G., Dresbach, T., Bresler, T., Torres, V.I., Gundelfinger, E.D., et al., 2003, Unitary assembly of presynaptic active zones from Piccolo-Bassoon transport vesicles, *Neuron* **38:** 237–252.

Shin, E.Y., Shin, K.S., Lee, C.S., Woo, K.N., Quan, S.H., Soung, N.K., et al., 2002, Phosphorylation of p85 beta PIX, a Rac/Cdc42–specific guanine nucleotide exchange factor, via the Ras/ERK/PAK2 pathway is required for basic fibroblast growth factor-induced neurite outgrowth, *J. Biol. Chem.* **277:** 44417–44430.

Shin, H., Wyszynski, M., Huh, K.H., Valtschanoff, J.G., Lee, J.R., Ko, J., et al., 2003, Association of the kinesin motor KIF1A with the multimodular protein liprin-alpha, *J. Biol. Chem.* **278:** 11393–11401.

Shin, H.-W., Morinaga, N., Noda, M., and Nakayama, K., 2004, BIG2, a guanine nucleotide exchange factor for ADP-ribosylation factors: Its localization to recycling endosomes and implication in the endosome integrity, *Mol. Biol. Cell* **15:** 5283–5294.

Smith, D.S., and Tsai, L.H., 2002, Cdk5 behind the wheel: A role in trafficking and transport? *Trends Cell Biol.* **12:** 28–36.

Song, J., Khachikian, Z., Radhakrishna, H., and Donaldson, J.G., 1998, Localization of endogenous ARF6 to sites of cortical actin rearrangement and involvement of ARF6 in cell spreading, *J. Cell Sci.* **111:** 2257–2267.

Spang, A., 2002, ARF1 regulatory factors and COPI vesicle formation, *Curr. Opin. Cell Biol.* **14:** 423–427.

Suzuki, I., Owada, Y., Suzuki, R., Yoshimoto, T., and Kondo, H., 2001, Localization of mRNAs for six ARFs (ADP-ribosylation factors) in the brain of developing and adult rats and changes in the expression in the hypoglossal nucleus after its axotomy, *Brain Res. Mol. Brain Res.* **88:** 124–134.

Suzuki, I., Owada, Y., Suzuki, R., Yoshimoto, T., and Kondo, H., 2002, Localization of mRNAs for subfamily of guanine nucleotide-exchange proteins (GEP) for ARFs (ADP-ribosylation factors) in the brain of developing and mature rats under normal and postaxotomy conditions, *Brain Res. Mol. Brain Res.* **98:** 41–50.

Takai, Y., Sasaki, T., and Matozaki, T., 2001, Small GTP-Binding Proteins, *Physiol. Rev.* **81:** 153–208.

Takeda, S., Yamazaki, H., Seog, D.H., Kanai, Y., Terada, S., and Hirokawa, N., 2000, Kinesin superfamily protein 3 (KIF3) motor transports fodrin-associating vesicles important for neurite building, *J. Cell Biol.* **148:** 1255–1265.

Tanaka, E., and Sabry, J., 1995, Making the connection: Cytoskeletal rearrangements during growth cone guidance, *Cell* **83:** 171–176.

Togawa, A., Morinaga, N., Ogasawara, M., Moss, J., and Vaughan, M., 1999, Purification and cloning of a brefeldin A-inhibited guanine nucleotide-exchange protein for ADP-ribosylation factors, *J. Biol. Chem.* **274:** 12308–12315.

Torre, E., McNiven, M.A., and Urrutia, R., 1994, Dynamin 1 antisense oligonucleotide treatment prevents neurite formation in cultured hippocampal neurons, *J. Biol. Chem.* **269:** 32411–32417.

Tsai, S.C., Adamik, R., Moss, J., and Vaughan, M., 1996, Purification and characterization of a guanine nucleotide-exchange protein for ADP-ribosylation factor from spleen cytosol, *Proc. Natl. Acad. Sci. USA* **93:** 305–309.

Turner, C.E., Brown, M.C., Perrotta, J.A., Riedy, M.C., Nikolopoulos, S.N., McDonald, A.R., et al., 1999, Paxillin LD4 motif binds PAK and PIX through a novel 95–kD ankyrin repeat, ARF-GAP protein: A role in cytoskeletal remodelling, *J. Cell Biol.* **145:** 851–863.

Ullrich, O., Reinsch, S., Urbe, S., Zerial, M., and Parton, R.G., 1996, Rab11 regulates recycling through the pericentriolar recycling endosome, *J. Cell Biol.* **135:** 913–924.

Ungermann, C., and Langosch, D., 2005, Functions of SNAREs in intracellular membrane fusion and lipid bilayer mixing, *J. Cell Sci.* **118:** 3819–3828.

Vega, I.E., and Hsu, S.C., 2001, The exocyst complex associates with microtubules to mediate vesicle targeting and neurite outgrowth, *J. Neurosci.* **21:** 3839–3848.

Venkateswarlu, K., and Cullen, P.J., 2000, Signalling via ADP-ribosylation factor 6 lies downstream of phosphatidylinositide 3–kinase, *Biochem. J.* **345:** 719–724.

Vitale, N., Patton, W.A., Moss, J., Vaughan, M., Lefkowitz, R.J., and Premont, R.T., 2000, GIT proteins, a novel family of phosphatidylinositol 3,4,5–trisphosphate-stimulated GTPase-activating proteins for ARF6, *J. Biol. Chem.* **275:** 13901–13906.

Wang, Y., Okamoto, M., Schmitz, F., Hofman, K., and Südhof, T.C., 1997, RIM is a putative Rab3A-effector in regulating synaptic vesicle fusion, *Nature* **388:** 593–598.

Welsh, C.F., Moss, J., and Vaughan, M., 1994, Isolation of recombinant ADP-ribosylation factor 6, an approximately 20–kDa guanine nucleotide-binding protein, in an activated GTP-bound state, *J. Biol. Chem.* **1269:** 15583–15587.

Whyte, J.R.C., and Munro, S., 2002, Vesicle tethering complexes in membrane traffic, *J. Cell Sci.* **115:** 2627–2637.

Wu, S., Mehta, S.Q., Pichaud, F., Bellen, H.J., and Quiocho, F.A., 2005, Sec15 interacts with Rab11 via a novel domain and affects Rab11 localization in vivo, *Nat. Struct. Mol. Biol.* **12:** 879–885.

Wyszynski, M., Kim, E., Dunah, A.W., Passafaro, M., Valtschanoff, J.G., Serra-Pages, C., et al., 2002, Interaction Between GRIP and liprin-a/SYD2 required for AMPA receptor targeting, *Neuron* **34:** 39–52.

Yin, H.L., and Janmey, P.A., 2003, Phosphoinositide regulation of the actin cytoskeleton, *Annu. Rev. Physiol.* **65:** 761–789.

Yoshii, S., Tanaka, M., Otsuki, Y., Wang, D.Y., Guo, R.J., Zhu, Y., et al., 1999, AlphaPIX nucleotide exchange factor is activated by interaction with phosphatidylinositol 3–kinase, *Oncogene* **18:** 5680–5690.

Zakharenko, S., and Popov, S., 1998, Dynamics of axonal microtubules regulate the topology of new membrane insertion into the growing neuritis, *J. Cell Biol.* **143:** 1077–1086.

Zerial, M., and McBride, H., 2001, Rab proteins as membrane organizers, *Nat. Rev. Mol. Cell. Biol.* **2:** 107–117.

Zhang, Q., Cox, D., Tseng, C.C., Donaldson, J.G., and Greenberg, S., 1998, A requirement for ARF6 in Fc receptor-mediated phagocytosis in macrophages, *J. Biol. Chem.* **273:** 19977–19981.

Zhang, H., Webb, D.J., Asmussen, H., Horwitz, A.F., 2003, Synapse formation is regulated by the signaling adaptor GIT1, *J. Cell Biol.* **161:** 131–142.

Zhang, H., Webb, D.J., Asmussen, H., Niu, S., and Horwitz, A.F., 2005, A GIT1/PIX/Rac/PAK signaling module regulates spine morphogenesis and synapse formation through MLC, *J. Neurosci.* **25:** 3379–3388.

Zhang, X.M., Ellis, S., Sriratana, A., Mitchell, C.A., and Rowe, T., 2004, Sec15 is an effector for the Rab11 GTPase in mammalian cells, *J. Biol. Chem.* **279:** 43027–43034.

Zhao, Z.S., Manser, E., Loo, T.H., and Lim, L., 2000, Coupling of PAK-interacting exchange factor PIX to GIT1 promotes focal complex disassembly, *Mol. Cell Biol.* **20:** 6354–6363.

Zhen, M., and Jin, Y., 2004, Presynaptic terminal differentiation: Transport and assembly, *Curr. Opin. Neurobiol.* **14:** 280–287.

5 Exocytic Mechanisms for Axonal and Dendritic Growth

THIERRY GALLI AND PHILIPP ALBERTS

5.1. Introduction

The structure of the nervous system is of great complexity. A large number of distinct neuronal cell types interact in a sophisticated network that runs over long distances. At the same time, communication points between neurons are not static but plastic during both development and adulthood. These fascinating properties underline the construction of the nervous system and its capacity to learn and memorize.

Mature neurons are highly polarized cells with two main compartments with distinct processes called neurite: the axon which contains the presynaptic boutons from where neurotransmitters are released and the somato-dendritic domain which contains the postsynaptic receptors. The specific contacts between axons and dendrites, so called synapses, are the basis of communication in the brain and with the other organs. This is the result of the growth of both axons and dendrite toward their specific target along specific pathways during development. The formation of neuronal cell contacts has to occur at the right place and the right time, and, during its journey, the growing neurite has to ignore a number of cues that are recognized by others as their target. Thus, neurons have to be equipped with a highly sensitive apparatus that reads and interprets cues and that allows advance while preventing the premature establishment of contacts. Several studies on the role of attractive and repulsive cues and their effects on the neuronal cytoskeleton have led to an impressive knowledge as to how these molecules might work during brain development. In contrast, the role of membrane trafficking and exocytosis during neuronal morphogenesis has received little attention. It is intuitively obvious that delivery of membrane and exocytosis allow the neuron to grow, but this process has to be tightly connected to the signaling molecules that guide axons and dendrites for site-directed growth to occur during brain development.

The question of exocytic mechanisms for axonal and dendritic growth can be split into two: (1) what is the basic function of exocytosis in neurite outgrowth or, in other words, to which extent is it similar to exocytosis involved in other cell growth mechanism? and (2) what are the specific regulations of exocytosis in neuritogenesis? The purpose of this chapter is to address the role of exocytosis in neuritogenesis and its regulation.

5.2. Basic Mechanism of Exocytosis is Conserved from Yeast to Neurons

The outgrowth of axons and dendrites of neurons is accompanied by a major increase in surface area of the cell. For example, an axon growing at a rate of 20–50 μm/h has to incorporate a membrane area corresponding to the size of its cell body every 2 h (Futerman and Banker, 1996). Neurite outgrowth resembles other cell growth processes that require exocytosis like the sustained polar growth at the tips of pollen tubes (120 μm/h) (Cole et al., 2005), the hyphae elongation of the yeast *Candida albicans* (20 μm/h) (Hausauer et al., 2005), and the growth of the budding yeast *Saccharomyces cerevisiae* (*S. cerevisiae*).

Schekman and coworkers took advantage of the fact that *S. cerevisiae* yeast cells defective in secretion accumulate newly synthesized proteins and lipids, while plasma membrane growth ceases, inducing an increase in cell density to isolate mutants and identify important gene products. Temperature sensitive mutant cells were selected by simple density gradient centrifugation separating dense and therefore secretion defective cells from nondefective cells. This approach led to the identification of 23 distinct gene products necessary for secretion and thus cell growth in yeast, called Sec 1–23 (Novick et al., 1980). As we will see later on, the molecules involved in exocytosis mediating neurite outgrowth are often the orthologues of the ones involved in growth and secretion in yeast. Further important progresses were made by Rothman and colleagues using their intra-Golgi transport *in vitro* assay with the discovery of numerous key molecules in membrane trafficking. Transport between Golgi membranes is mediated by vesicular intermediates (Balch et al., 1984) and requires *N*-ethyl-maleimide-sensitive fusion (NSF) protein (for N-ethyl maleimide sensitive factor), the mammalian homologue to the sec18 gene (Block and Rothman, 1992), and α-, β-, and γ-SNAP (for soluble NSF attachment proteins), the orthologues of sec17 (Clary et al., 1990). Salt stripped Golgi membranes would bind NSF only when purified α- or β-SNAP was present indicating the existence of a membrane integral SNAP receptor (SNARE) protein. Purified NSF and SNAP bound to the SNARE could be extracted from Golgi membranes with detergent as a stable 20S particle. Three proteins were isolated from bovine brain as SNAP receptors, which had been previously purified and cloned as abundant transmembrane proteins of the presynaptic terminal: synaptobrevin-2, syntaxin-1, and SNAP25 (Söllner et al., 1993b) (Figure 5.1). Further characterization of the 20S particle revealed that all three SNARE proteins identified from bovine brain were present in a single 20S particle together with NSF and SNAP and that ATPase activity of NSF dissociates the complex (Söllner et al., 1993a). Intriguingly, synaptobrevin localizes to synaptic vesicles, whereas syntaxin and SNAP25 are expressed on the synaptic plasma membrane (Baumert et al., 1989; Bennett et al., 1992; Oyler et al., 1992), thus complex formation between these proteins should occur when the synaptic vesicle has approached the presynaptic membrane (Rothman, 1994; Rothman and Warren, 1994). An essential role for SNARE proteins in neurotransmitter release was proposed at about the same time from the characterization of the molecular mechanism of action of clostridial neurotoxins, the most potent blockers of neurotransmitter release. These toxins carry a protease that cleaves the synaptic SNARE proteins synaptobrevin-2, SNAP25, or syntaxin-1 (Blasi et al., 1993a, 1993b; Schiavo et al., 1992). Taking all these observations into consideration, Rothman and colleagues suggested a molecular mechanism, which would explain vesicle docking and fusion in molecular terms, the so-called *SNARE Hypothesis* (Rothman and Warren, 1994) (Figure 5.1). Each transport step within the eukaryotic cell is mediated by an original pair of SNAREs which reside on the vesicle (the v-SNARE) and the target membrane (the t-SNARE). Complex formation between cognate v- and t-SNAREs would provide the specificity of vesicular docking to the target membrane. The predictions of the SNARE hypothesis as they were formulated in 1993 did not hold true in all details, but the SNARE hypothesis provided the conceptual framework that stimulated and guided studies that aimed at understanding the molecular mechanisms of membrane trafficking. SNARE proteins are indeed necessary and sufficient to fuse artificial membranes *in vitro* (Weber et al., 1998) and cells engineered to express the correct set of cognate SNARE proteins on their plasma membrane facing the extracellular medium fuse with each other (Hu et al., 2003). Yet, in mutants of synaptobrevin in mice, fly, and nematodes, spontaneous release of neurotransmitter is reduced, but clearly detectable, in contrast to evoked neurotransmitter release, which is essentially abrogated (Deitcher et al., 1998; Nonet et al., 1998; Schoch et al., 2001a; Washbourne et al., 2001). A similar result was obtained in mice mutant for SNAP25 (Washbourne et al., 2002). These results raised the question of whether SNARE-independent membrane fusion events might occur in the cell (Martinez-Arca et al., 2000a). We will see later on that the exocytosis mediating neuritogenesis is yet another mechanism resistant to certain neurotoxins.

In conclusion, SNARE proteins fulfill the structural requirements for proteins involved in membrane fusion events, they are necessary for membrane fusion in a large number of trafficking events and mediate cell growth in the budding yeast.

The regulation of membrane fusion is in part encoded in the structure of SNARE proteins. A number of SNAREs have N-terminal extensions, which participate directly in the kinetics of membrane fusion by negatively regulating SNARE complex formation (Filippini et al., 2001). The negative regulation of SNARE complex formation by N-terminal extensions is best understood for the synaptic SNARE syntaxin-1. The N-terminal domain of syntaxin-1 folds back on the SNARE motif, thus leading to a so called closed conformation in which the SNARE motif is not accessible for SNARE complex formation (Dulubova et al., 1999). Removal of the N-terminal extension greatly enhances SNARE complex formation and fusion efficiency mediated by syntaxin-1 (Parlati et al., 1999). A protein called n-Sec1/Munc18, which binds to syntaxin-1, plays an essential role in activating syntaxin-1 for SNARE complex formation *in vivo*. Munc18 is a member of a family of proteins found from yeast to mammals called SM proteins

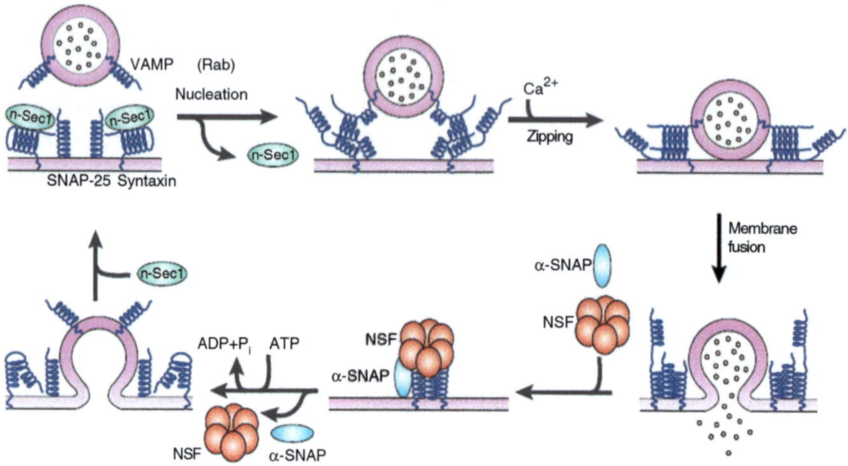

FIGURE 5.1. The SNARE cycle in exocytosis syntaxin is bound to n-Sec1 before formation of the core complex. Rab proteins might facilitate the dissociation of n-Sec1 from syntaxin, allowing subsequent binding (nucleation) between the three neuronal SNAREs, syntaxin, SNAP-25, and VAMP (for simplicity, only one coil is drawn for SNAP25). Ca^{2+} triggers the full zipping of the coiled-coil complex, which results in membrane fusion and release of vesicle contents. After the fusion event, recruitment of α-SNAP and NSF from the cytoplasm and subsequent hydrolysis of ATP by NSF causes dissociation of the SNARE complex. Syntaxin, VAMP, and SNAP25 are then free for recycling and another round of exocytosis. NSF = *N*-ethyl-maleimide-sensitive fusion protein; SNAP25 = 25 kDa synaptosome-associated protein; SNARE = soluble NSF attachment protein receptor; VAMP = vesicle-associated membrane protein. (Reproduced with permission from Chen, Y.A., and Scheller, R.H., 2001, SNARE-mediated membrane fusion, *Nat. Rev. Mol. Cell Biol.* **2:** 98–106. Copyright 2001 Macmillan Magazines Ltd.)

(for Sec1/Munc18), which were shown to be crucial in diverse membrane fusion including exocytosis (Jahn et al., 2003) (Figure 5.1).

The molecular nature of docking factors that act upstream of SNARE proteins was also first described in yeast. The importance of tethering factors *in vivo* became clear when the gene products of Sec 3, 5, 6, 8, 10, 15, together with Exo70 and Exo84, where shown to form a multisubunit tethering complex called the exocyst (Novick et al., 1980; TerBush et al., 1996). A mammalian homologue of the exocyst has been isolated, which plays a role similar to the yeast exocyst in recruiting exocytotic vesicles (Grindstaff et al., 1998). Tethering factors, which are recruited by the action of Rab proteins, have been detected in a number of trafficking steps (Zerial and McBride, 2001) (Figure 5.1). Rab proteins are also involved in the organization of organelle transport along the cytoskeleton, highlighting the importance of Rab proteins in a number of membrane trafficking steps (Hammer and Wu, 2002). As we will see later, SNAREs, the exocyst, and Rab proteins were shown to be involved in neuritogenesis, further establishing the functional analogy between the basic mechanism of yeast growth and neuritogenesis.

5.3. Exocytosis in Neurite Outgrowth: A Basic Mechanism Involving Endosomes

In the course of their differentiation, neurons extend axons and dendrites that elongate along specific paths within the developing brain. The extension of axons and dendrites implies the targeted delivery of membranes to allow the net increase in surface during neurite outgrowth. Brefeldin A, a fungal toxin that blocks post-Golgi trafficking of membrane carriers, inhibits neurite outgrowth suggesting a role of post-Golgi membrane trafficking in this phenomenon (Jareb and Banker, 1997). Newly synthesized Neural-glial cell adhesion molecule (NgCAM) is delivered and inserted in the plasma membrane at the distal end of the growing axon (Vogt et al., 1996). Similar results were obtained when CD8 alpha, a nonneuronal protein, was expressed in neurons in culture and found to be inserted at the tip of the growing axons (Craig et al., 1995). These studies suggest that newly synthesized proteins and lipids may be transported and inserted into the membrane at the sites where growth actually takes place. A study on NgCAM demonstrated that transport carriers delivering newly synthesized molecules and endosomal carriers are not necessarily different. Newly synthesized NgCAM is first transported to the dendritic plasma membrane followed by immediate uptake and delivery to the axon in endosomal carriers (Wisco et al., 2003), thus the biosynthetic route to the plasma membrane could involve endosomes. Furthermore, endosomal membranes are observed as soon as the axon starts to emerge (de Anda et al., 2005) suggesting that the endosomal contribution to neuritogenesis is required early.

As discussed previously, the clostridial neurotoxins cleave the key components of the synaptic exocytic membrane fusion machinery. The inactivation of the neuronal plasma membrane target SNARE composed of SNAP25 and syntaxin-1, by botulinum neurotoxin (BoNT) A and C1 leads to a pronounced reduction in axonal outgrowth (Grosse et al., 1999; Osen-Sand et al., 1996), thus suggesting the involvement of exocytosis in neurite outgrowth. A role for exocytosis in neurite outgrowth *in vivo* was also suggested by the *Dorosophila* mutant of sec5, a subunit of the exocyst complex involved in tethering of exocytotic vesicles, because neurites stop growing as soon as the maternal, cytosolic pool of sec5 is exhausted (Murthy et al., 2003). In contrast, tetanus neurotoxin (TeNT)-mediated cleavage of the synaptic v-SNARE synaptobrevin-2 has no effect on neurite outgrowth in spite of complete abolition of neurotransmitter release (Grosse et al., 1999; Osen-Sand et al., 1996). These findings were confirmed by analyzing knock out animals for synaptobrevin-2, which showed a severe impairment in transmitter release but apparently a normal development of the brain (Schoch et al., 2001b). Thus, distinct types of vesicles with apparently different targeting mechanisms are implicated in neurotransmitter release and neuritogenesis. Similarly, mutants for Munc18 and Munc 13–1/2, two partners of syntaxin-1, showed severe defects in neurotransmitter release without any apparent impairment of the brain development (Aravamudan et al., 1999; Verhage et al., 2000). The precise role of SNAP-25 is less clear because the knock out of SNAP-25 shows normal brain development (Washbourne et al., 2002), whereas the previously mentioned experiments using clostridial NTs suggested a role of this molecule in neurite outgrowth. This discrepancy could be owing to the expression of SNAP23, a close homologue that may, at least partially, be able to functionally replace SNAP25 in the SNAP25 knockout mice (Chieregatti et al., 2004). In any case, these results clearly indicate that the membrane trafficking pathway mediating neurite outgrowth is different from the one involved in neurotransmitter release.

Intrigued by the resistance of neurite outgrowth to TeNT, our group has showed that the vesicular SNARE tetanus NT insensitive-VAMP [TI-VAMP, also called syntobrevin-like gene 1 and VAMP7 (D'Esposito et al., 1996)], which is resistant to clostricial NTs (Galli et al., 1998), defines a new type of vesicles that is different from synaptic vesicles and transferrin receptor positive recycling endosomes/early endosomes in neuronal cells. In young neurons developing in culture, TI-VAMP accumulates at the leading edge of growing axons and dendrites, consistent with a role for this protein in neurite outgrowth (Coco et al., 1999). Further studies based on the overexpression of fragments of TI-VAMP (Martinez-Arca et al., 2000a, 2001) and RNAi silencing of TI-VAMP's expression (Alberts et al., 2003) showed that TI-VAMP mediates neurite outgrowth in PC12 cells and in neurons in culture (Figure 5.2).

No true homologue of TI-VAMP exists in yeast, whereas TI-VAMP homologues are found in higher unicellular and multicellular organisms like *Dictyostelium*, *Caenorhabditis elegans*, *Drosophila*, and humans (Filippini et al., 2001). The structure of TI-VAMP is different from the classical structure of synaptobrevin in that it carries an N-terminal extension in addition to the SNARE motif and the transmembrane domain (Galli et al., 1998). This N-terminal extension, called the Longin domain, is found in a number of v-SNAREs from yeast to human and its sequence shows a high degree of conservation between different v-SNAREs carrying this extension (Filippini et al., 2001). The Longin domain of TI-VAMP and another v-SNARE called Ykt6 were shown to negatively regulate SNARE complex formation (Martinez-Arca et al., 2000b, 2003b; Tochio et al., 2001) similar to the structurally unrelated N-terminal extensions of syntaxin-1 (Gonzalez et al., 2001; Tochio et al., 2001). In addition to its role as negative regulator in SNARE complex formation, the Longin domain of TI-VAMP binds to the adaptor complex AP3, which is important for the localization of this protein to late endosomal compartments in nonneuronal cells (Martinez-Arca et al., 2003b). This regulation does not affect the localization and function of TI-VAMP in immature neurons but does affect TI-VAMP localization to the synapse in the mature brain (Rachel Rudge, Thierry Galli, unpublished observations).

Although a number of studies attempted to elucidate the membrane trafficking pathway(s) mediated by TI-VAMP, its exact nature is still unclear. TI-VAMP localizes to late endosomes/multivesicular bodies and to vesicles scattered throughout the cytoplasm (Advani et al., 1999; Coco et al., 1999; Martinez-Arca

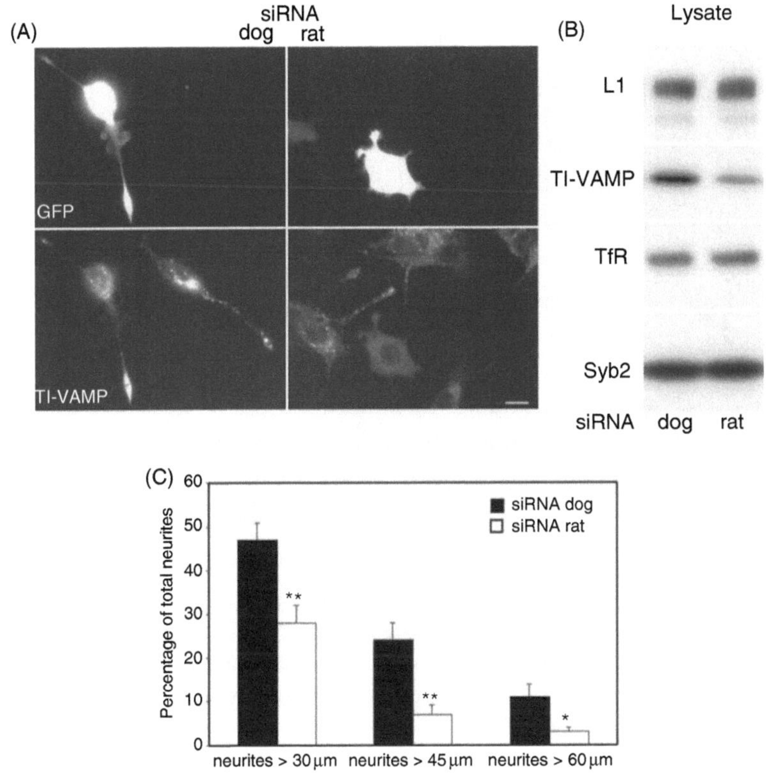

FIGURE 5.2. Silencing of TI-VAMP expression impairs neurite outgrowth in PC12 cells and neurons. (A) PC12 cells were transfected with siRNAr or siRNAd combined with an expression plasmiden coding GFP. After 72 h, differentiation was induced by treatment with staurosporine for 24 h. SiRNAr silenced the expression of TI-VAMP and inhibited neurite outgrowth (GFP-positive cell, right), whereas siRNAd had no effect (GFP-positive cell, left) (bar, 5 μm). (B) Lysates of cells corresponding to A were analyzed by Western blotting for the expression levels of TI-VAMP, L1, transferring receptor (TfR), and Syb 2. (C) The neurite length of GFP-positive cells transfected with siRNAr or siRNAd was quantified. Bars represent the percentage of the total number of neurites >30, 45, and 60 μm from two independent experiments ($^*p < 0.05$; $^{**}p < 0.01$).

et al., 2003a,b). It forms complexes with SNARE proteins functioning at the plasma membrane and in endosomal compartments (Alberts et al., 2003; Bogdanovic et al., 2000, 2002; Galli et al., 1998; Martinez-Arca et al., 2000b, 2003a). Function blocking assays have implicated TI-VAMP in the degradative pathway of the EGF receptor and fusion of late endosomes/lysosomes (Advani et al., 1999; Ward et al., 2000), but this function might be redundant with endobrevin/VAMP8, another endosomal v-SNARE (Antonin et al., 2000). Another series of studies implicated TI-VAMP in an exocytotic pathway. In epithelial cells, TI-VAMP plays a role in transport to the apical plasma membrane, a pathway

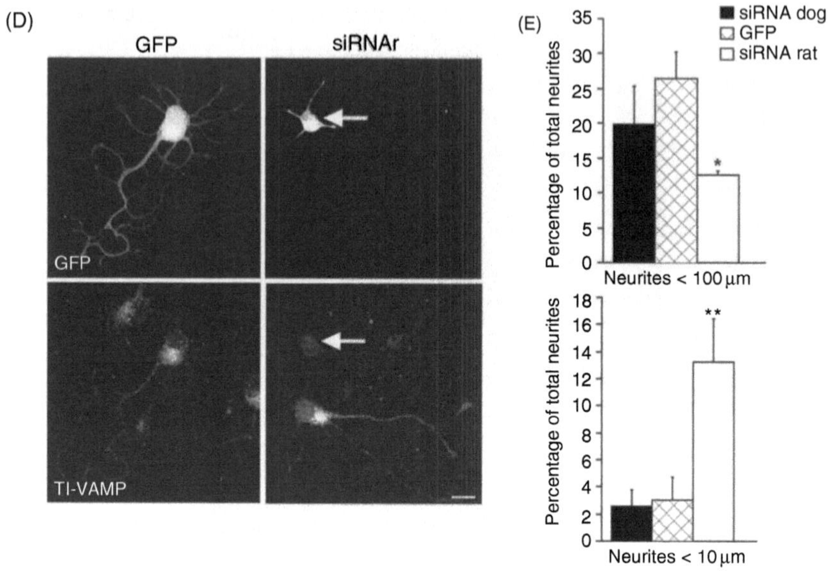

FIGURE 5.2. (*continued*) (D) TI-VAMP expression in hippocampal neurons was silenced by transfecting with siRNA and small amounts of EGFP vector. Three days after plating, TI-VAMP expression in EGFP-positive cells was assessed by immunofluorescence with mAb 158.2. As in PC12 cells, siRNAr decreased expression levels of TI-VAMP and inhibited neurite outgrowth (GFP-positive cell, right) compared with control cells (GFP-positive cell, left) (bar, 10 μm). (E) Quantification of axon lengths in neurons cotransfected with siRNAd or siRNAr and EGFP or transfected with EGFP alone is shown. Bars represent the percentage of total axons >100 μ (top) or <10 μm(bottom) from three independent experiments ($^{*}p < 0.02$ with respect to GFP control; $^{**}p < 0.01$ with respect to siRNAd control). [Published in Alberts et al. (2003).]

insensitive to TeNT (Galli et al., 1998; Ikonen et al., 1995; Lafont et al., 1999). TI-VAMP is also involved in the degranulation of mastocytes (Hibi et al., 2000), in late endosomal exocytosis in phagocytosis in macrophages (Braun et al., 2004). The relationship between the late endosomal nature of the nonneuronal TI-VAMP compartment and its role in neurite extension in neurons is still unclear. TI-VAMP mediates the release of so-called secretory lysosomes, a TeNT resistant pathway (Andrews, 2000) in fibroblasts (Rao et al., 2004). The exocytosis of secretory lysosomes has been involved in membrane repair (Reddy et al., 2001). It is tempting to speculate that membrane repair and cell outgrowth could depend on the same exocytic pathway. Both membrane repair and cell outgrowth require massive transport to the plasma membrane. Moreover, several fly mutants show impairment in late endosome function coupled with abnormal neurite outgrowth. The case of spinster is particularly interesting because spinster localizes to a late endosomal compartment and late endosomes show altered size and distribution in spinster mutant flies. This phenotype is associated with a very pronounced synaptic overgrowth (200% increase in synaptic bouton number at the neuromuscular junction), and an enhanced Transforming Growth Factor -α (TGF-α) signaling (Sweeney and Davis, 2002). Synaptic overgrowth of this range has also been observed in the case of overexpression of an ubiquitin hydrolase (DiAntonio et al., 2001) and of a dominant negative NSF (Stewart et al., 2002). The reason for the apparent discrepancy of TI-VAMP function in either endosomal or exocytotic trafficking is not understood but may underline the fact that some cells like developing neurons or phagocyting macrophages show an important and fast growth of their surface area, whereas other cell types grow more slowly. Altogether, these lines of evidence suggest that TI-VAMP, the neuronal late endosomal compartment, play key roles in neurite outgrowth.

5.4. Regulation of Exocytosis Mediating Neurite Outgrowth

To enable axons and dendrites to locate their appropriate synaptic partners and therefore to enable the correct wiring of the nervous sytem, the delivery of membrane to the axon and dendrites has to be integrated into the demands of directed axonal and dendritic growth. This plasticity in neurite outgrowth is exemplified by the fact that the extent of axonal outgrowth of a neuron in culture can be drastically modified by the substrate that is presented to the neuron [see Figure 5.3; (Lemmon et al., 1989)] and signaling cues exist, which preferentially stimulate either axonal or dendritic outgrowth (Higgins et al., 1997; Prochiantz, 1995). Thus, as shown in Figure 5.3, axonal outgrowth is selectively stimulated by a member of the IgCAM superfamily of cell adhesion molecules called L1, whereas dendritic outgrowth is not affected compared to the control substrate poly-L-lysine.

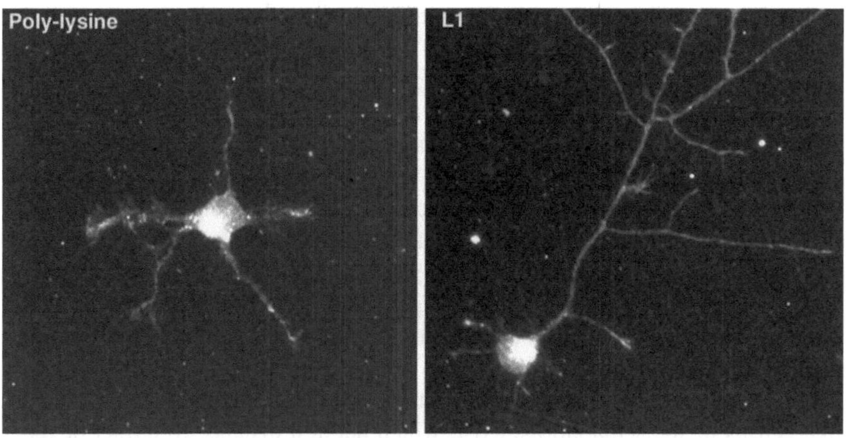

FIGURE 5.3. L1 stimulates axonal outgrowth in hippocampal neurons. Poly-L-lysine–coated coverslips were incubated with anti-human Fc antibodies followed by incubation with (L1) or without (poly-lysine) L1-Fc fragment. Hippocampal neurons seeded at low density were kept in culture for 24 h and analyzed by immunofluorescence with pAb to L1.

How do guidance cues, e.g., the ligation of cell adhesion molecules, translate into directed growth of the axon? Directed axonal navigation is mediated by growth cones, which are specialized, highly motile structures found at the tip of developing axons. Long-range or short-range guidance cues, which can be attractive or repulsive, are recognized by receptors expressed on the growth cone and provoke the classical forms of growth cone behavior such as advance, turning, withdrawal, and target recognition (Suter and Forscher, 1998). The growth cone allows for the establishment of the complex pattern of neuronal connections by probing and sensing the environment and by translating the information received into motile behavior (Tessier-Lavigne and Goodman, 1996).

The cytoskeleton of growth cones is intimately involved in the axonal response to extracellular cues as drugs interfering with actin dynamics were shown to render neurons "blind" *in vivo* resulting in axons that bypass and extend beyond normal synaptic partners (Bentley and Toroian-Raymond, 1986; Chien et al., 1993; Kaufmann et al., 1998). In growth cones, F-actin undergoes a retrograde flow from the peripheral to the central region. This so-called retrograde actin flow is the result of constant *de novo* polymerization of actin at the leading edge of the growth cone and retrograde transport of actin filaments mediated by myosin motors (Forscher and Smith, 1988; Lin and Forscher, 1995; Lin et al., 1996). In order for the growth cone to be able to advance, the retrograde actin flow has to be coupled to the substrate. This substrate-cytoskeletal coupling is mediated by surface receptors that link the underlying cytoskeleton to the extracellular matrix and ultimately allows the growth cone to generate force and pull forwards. Thus, cell adhesion molecules of the integrin, cadherin, or IgCAM family play essential roles in growth cone movement. They all can be linked to the underlying actin-cytoskeleton via distinct adaptor proteins in a stimulus-dependent manner, which is provided by ligation of the respective receptor with its extracellular ligand (Suter and Forscher, 1998). For the growth cone to advance, substrate-cytoskeletal coupling has to be under tight control. In fact, a gradient of strong adhesion at the leading front and weak adhesion at the rear edge of the growth cone has to exist to allow the cytoskeletal machinery to pull the growth cone forward. Such a gradient of adhesiveness has been detected for several adhesion molecules in agreement with the predictions of the substrate-cytoskeletal coupling model of growth cone motility (Kamiguchi and Lemmon, 2000; Kamiguchi and Yoshihara, 2001; Schmidt et al., 1995). The signaling cascades initiated by CAMs allow for the site-directed rearrangement of cytoskeletal dynamics and the resulting change in growth cone motility (Challacombe et al., 1996; Dent and Kalil, 2001; Suter et al., 1998). Whereas adhesion molecules exert their function via direct cell–cell or cell–matrix interaction, other target-derived secreted factors like netrins or the semaphorins exist, which act at a distance to attract or repel axons. Secretion establishes a gradient and thus attracts or repels axons or dendrites along the gradient if the target neurite expresses the respective receptor. Secreted guidance cues induce motile behavior of the axon by regulating the growth cone cytoskeleton similar to what was described previously for CAMs

(Dickson et al., 2002). A physical and functional interaction between the receptor for semaphorin-3a (Sema3a), Neuropillin, and the cell adhesion molecule L1 has been observed. Whereas binding of Sema3a to its receptor induces growth cone turning away from a Sema3a gradient in wild type neurons, repulsion is not observed in L1 knockout cells (Castellani et al., 2000, 2002, 2004). Thus, adhesion molecules can be functionally linked to soluble guidance molecules and their receptors, illustrating the high plasticity by which growth cones can respond to guidance cues.

In our studies, we have demonstrated that TI-VAMP expression is necessary for neurite outgrowth and adhesion mediated by L1. Furthermore, L1-mediated adhesion controls TI-VAMP-mediated trafficking via an actin-based mechanism (Alberts et al., 2006). L1 is cell adhesion molecule of the IgCAM family expressed mainly in brain, where it plays a crucial role in development. A human neurological disorder known as MASA syndrome is caused by mutations in the L1 gene (Brummendorf et al., 1998). The disease is characterized by malformations of the brain, particularly of the cortico-spinal tract. Similarly, mice mutant for L1 show defects in corticospinal axon guidance (Cohen et al., 1998; Dahme et al., 1997). We have obtained the following evidence: (i) TI-VAMP mediates a recycling pathway in neuronal cells and inhibiting this pathway by TI-VAMP–specific RNAi strongly inhibits neurite outgrowth in PC12 cells; (ii) TI-VAMP colocalizes with L1 *in situ* in the developing brain, PC12 cells, and neurons, and endocytosed L1 is localized to the TI-VAMP compartment; (iii) impaired TI-VAMP expression affects the surface expression of L1 and reduces the binding of L1-coated beads to PC12 cells; (iv) the TI-VAMP expression in growth cones depends on actin dynamics and L1-dependent adhesive contacts induce local clustering of TI-VAMP in an actin-dependent manner, suggesting that L1-mediated adhesion controls the anchoring of TI-VAMP containing vesicles (Alberts et al., 2003, and our unpublished observations). Therefore, this report demonstrates that a cross talk exists between TI-VAMP–dependent trafficking and L1-mediated adhesion and between L1-induced signaling and TI-VAMP dynamics (Alberts et al., 2003). It is not yet known how the TI-VAMP membrane trafficking pathway is regulated, or how it is coordinated with the cytoskeletal dynamics that guide neurite outgrowth from the growth cone. We have found that TI-VAMP, but not synaptobrevin-2, concentrates in the peripheral, F-actin–rich region of the growth cones of hippocampal neurons in primary culture. Its accumulation correlates strictly with and depends on the presence of F-actin as shown by the effect of actin sequestering drugs (Alberts et al., 2006). GTPases of the Rho/Rac/Cdc42 subfamily are of crucial importance in the control of cell polarity and signal-dependent actin remodeling in neurons and other cell types (Dickson, 2002; Li et al., 2000; Luo et al., 1996; Schwamborn and Puschel, 2004; Van Aelst and Cline, 2004). Expression of a dominant-positive mutant of Cdc42 stimulates formation of F-actin– and TI-VAMP–rich filopodia outside the growth cone, an effect not seen when expressing Rac1 or RhoA dominant-positive mutants. Furthermore, we found that Cdc42 directly activates actin-dependent exocytosis of TI-VAMP–containing vesicles by imaging pHLuorin-tagged TI-VAMP (Alberts et al., 2006). Collectively, these data indicate that Cdc42 and regulated assembly of the F-actin network control the accumulation and exocytosis of TI-VAMP–containing membrane vesicles in growth cones and suggest that membrane trafficking and actin remodeling are coordinated in neurite outgrowth.

Neuronal L1 harbors an additional short exon that yields a tyrosine-based motif YRSLE. This motif mediates binding to the AP2 adaptor complex and clathrin dependent endocytosis and accelerates the recycling of neuronal L1 (Kamiguchi et al., 1998). Thus, a role for intracellular trafficking in L1 function, especially in the course of neuronal development, can be anticipated. In analogy to migrating cells, it was suggested that L1 trafficking in the axonal growth cone establishes a dynamic gradient of L1 adhesiveness necessary to move along the substrate (Kamiguchi and Lemmon, 2000). Such a gradient was detected and its maintenance was dependent on endocytic trafficking (Kamiguchi and Yoshihara, 2001). Our results suggest that the TI-VAMP–dependent intracellular trafficking of L1 may be necessary to stabilize and regulate adhesive contacts of neuronal cells. Thus, the TI-VAMP compartment could correspond to a specialized recycling endosome to which L1 would be targeted following internalization and from which they would then be recycled to the plasma membrane. Additionally, the YRSLE motif important for L1 trafficking is conserved in other members of the IgCAM family like NrCAM and neurofascin suggesting that these molecules might also travel through the TI-VAMP compartment (Kayyem et al., 1992; Volkmer et al., 1992).

Since decreased neurite outgrowth in TI-VAMP–depleted cells was observed in PC12 cells grown on collagen or hippocampal neurons grown on poly-L-lysine (Alberts et al., 2003), neurite extension should not directly depend on homophilic L1-ligation in these systems. In addition to homophilic L1-interactions, a wide spectrum of heterophilic interactions of L1 in *cis* and *trans* with other members of the IgCAM family, receptors of the integrin family or receptors for secreted guidance cues have been reported (Brummendorf et al., 1998; Castellani et al., 2000). Other IgCAMs have been shown to cooperate in *cis* in promoting axonal outgrowth (Buchstaller et al., 1996) suggesting important functions in neurite outgrowth for the observed *cis*-interactions of L1 with other members of the IgCAM family. Moreover, L1 was shown to enhance β1 integrin-mediated cell migration and this promoting effect seems to depend on L1-mediated endocytosis of β1 integrin (Thelen et al., 2002). Therefore, interfering with the TI-VAMP–dependent trafficking of L1 would be expected to affect functional interactions of L1 with other molecules that promote neurite outgrowth. Clearly, the exact consequences of inhibiting TI-VAMP–expression for neurite outgrowth will not be understood until a more complete picture of the cargo proteins of the TI-VAMP compartment exists.

How does L1-dependent signaling control membrane trafficking? Activation of L1 by homophilic interaction stimulates a signaling pathway involving p60Src, and activation of Src seems to play an essential role in L1-dependent neurite outgrowth (Ignelzi et al., 1994; Schaefer et al., 1999; Schmid et al., 2000). This view is in agreement with studies on apCAM, a cell adhesion molecule of the IgCAM family found in Aplysia. Binding of apCAM-coated beads to growth cones induces local activation of Src kinases (Suter and Forscher, 2001). P60Src was proposed to phosphorylate L1 on tyrosine 1176 of the YRSLE motif, which inhibits L1-binding to the AP2 adaptor complex thus inhibiting its endocytosis (Schaefer et al., 2002). It can be speculated that TI-VAMP–mediated transport of L1 to sites of newly formed contacts enhances homophilic L1-ligation and therefore L1-induced activation of Src kinases. Thus, a dynamic equilibrium might exist between TI-VAMP–dependent L1-delivery and signaling events regulating the stability or removal of L1 from the plasma membrane. P60Src has been localized to endosomes in fibroblasts (Kaplan et al., 1992) and a study showed that p60Src activity on endosomes is necessary for the stable association of endosomes with actin cables via a RhoD/hDia2C complex (Gasman et al., 2003). Thus, Src activity could serve as an integrator for coordinated delivery of L1-containing, TI-VAMP–positive endosomal structures to L1 junctions and L1 stability at the plasma membrane, which would provide the cell with a highly dynamic regulatory mechanism to control adhesion and growth cone movement.

Other molecules are likely to regulate the exocytosis involved in neuritogenesis. This is the case of scaffolding proteins like ankyrin and ezrin, which interact directly with L1, and could play a role in recruiting the TI-VAMP–compartment to L1-junctions (Davis and Bennett, 1994; Dickson et al., 2002). Ankyrin-B is a scaffolding protein that links L1 to the spectrin-based actin cytoskeleton via dual interaction with L1 and spectrin (Davis and Bennett, 1994). Similar to ankyrin, ezrin interacts directly with L1 and is known to serve as a membrane-cytoskeleton linker (Dickson et al., 2002). Ezrin or other members of the ezrin-radixin-moesin (ERM) family of proteins are good candidates to mediate early events in L1-dependent adhesion, since they localize to actin rich, peripheral regions in growth cones (Paglini et al., 1998). Furthermore, Rab proteins play an important role in vesicle targeting and association of organelles with the cytoskeleton (Hammer and Wu, 2002) and in site-directed exocytosis involved in neuritogenesis. Rab5 is expressed in neurons and localizes to multilamellar, endocytic organelles different from synaptic vesicles in brain (Ikin et al., 1996), and its activity is involved in the regulation of actin-based membrane motility, particularly in highly dynamic processes of neurite outgrowth (Sabo et al., 2003). Rab8 has also been linked to both regulation and polarized membrane traffic and actin (Hattula et al., 2002) in neuronal cell morphogenesis (Hattula and Peranen, 2000). Future studies should test to what extent the recruitment of TI-VAMP vesicles to L1-dependent contacts depends on any of the proteins mentioned earlier or whether an as yet unknown mechanism is recruiting and stabilizing TI-VAMP vesicles.

5.5. Conclusions

In conclusion, we would like to propose a model in which the restricted activity of signaling pathways including Src and Cdc42 within the growth cone mediates the polarized accumulation and exocytosis of TI-VAMP in growth cones through its regulatory action on the actin cytoskeleton. The importance of spatially restricted Cdc42 activation for directed cellular movement was demonstrated before as both dominant-positive and dominant-negative Cdc42 mutants inhibit cellular polarization and directed migration of macrophages and astrocytes (Allen et al., 1998; Etienne-Manneville and Hall, 2001). We do not know the molecular cascade leading to the concentration of TI-VAMP in the growth cone and the stimulation of its exocytosis by Cdc42 in a physiological context. Signaling events induced by adhesion molecules like L1 would activate Cdc42 leading to the actin-dependent accumulation and exocytosis of TI-VAMP (Figure 5.4). This mechanism would be similar to that described for phagocytosis in macrophages, another cellular process that also requires localized TI-VAMP trafficking (Braun et al., 2004) upon receptor and Cdc42 activation (Hoppe and Swanson, 2004) and actin remodeling (May and Machesky, 2001). Thus, it is tempting to speculate that TI-VAMP–mediated exocytosis and actin dynamics are tightly coupled to direct the formation of cell protrusions not only in neurons and macrophages but also in other cell types as a more general mechanism for the formation of directed cellular extensions.

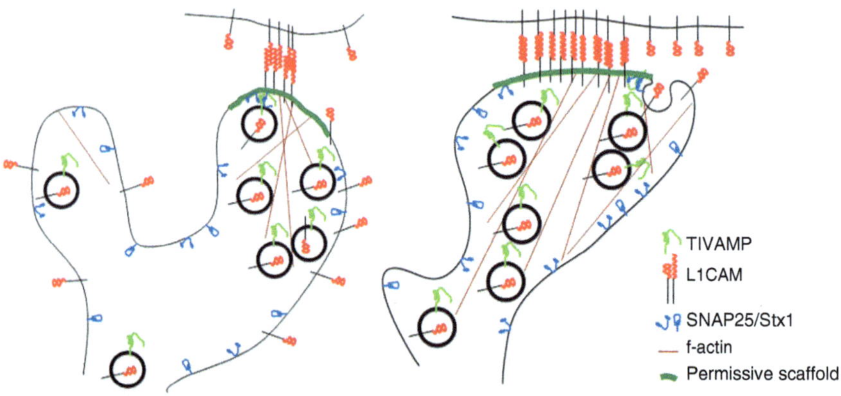

FIGURE 5.4. Model for the role of TI-VAMP in neuritogenesis. (A) Homophilic binding between L1 expressed on the growth cone of an axon (bottom) and an adjacent cell (top) leads to conformational change and activation of the L1 molecule. Activation of L1 leads to an assembly of a molecular scaffold including the exocyst, perhaps via Src-activation, Cdc42 activation and reorganization of the actin cytoskeleton. The establishment of this spatial cue allows for the recruitment and accumulation of TI-VAMP–positive vesicles containing L1 at sites of contact formation. SNARE complex formation between TI-VAMP and the t-SNARE SNAP25/Stx 1 mediates exocytosis of L1 molecules, which will reinforce the initial contact and lead to stable contact formation. (B) Stable contact formation will reinforce L1-induced signaling and thus recruitment of TI-VAMP vesicles. In turn, enhanced recruitment of TI-VAMP vesicles will stimulate growth toward the attractive substrate. Therefore, a positive feedback mechanism based on L1-signaling and TI-VAMP–mediated trafficking could result in directed growth of the axon or stable target recognition.

Furthermore, we speculate that TI-VAMP could be envisioned as a new target for axonal regeneration. Activating TI-VAMP as seen by expressing dominant positive Cdc42 should enhance axonal outgrowth. Finding therapeutic ways to stimulate axonal regeneration is still an important goal for which fundamental work has to be carried out. The problem is clearly more complex than anticipated as growth factors, guidance molecules, artificial substrates, and electrical stimulations despite positive effects are not enough to mediate satisfactory restoration. Finding drugs capable of stimulating TI-VAMP–dependent exocytosis may be a way to open new therapeutic strategies.

Acknowledgments. Work in TG's group was supported in part by grants from INSERM (Avenir Program), the European Commission ("Signalling and Traffic" STREP 503229), the Association Française contre les Myopathies, the Ministère de la Recherche (ACI-BDP), the Fondation pour la Recherche Médicale, and the Fondation pour la Recherche sur le Cerveau.

References

Advani, R.J., Yang, B., Prekeris, R., Lee, K.C., Klumperman, J., and Scheller, R.H., 1999, VAMP-7 mediates vesicular transport from endosomes to lysosomes, *J. Cell Biol.* **146:** 765–775.

Alberts, P., Rudge, R., Hinners, I., Muzerelle, A., MartinezArca, S., Irinopoulou, T., et al., 2003, Cross talk between tetanus neurotoxin-insensitive vesicle-associated membrane protein-mediated transport and L1-mediated adhesion, *Mol. Biol. Cell* **14:** 4207–4220.

Alberts, P., Rudge, R., Irinopoulou, T., Danglot, L., Gauthier-Rouviere, C., Galli, T., 2006, Cdc42 and actin control polarized expression of TI-VAMP vesicles to neuronal growth cones and their fusion with the plasma membrane, *Mol. Biol. Cell.* **17:** 1194–203

Allen, W.E., Zicha, D., Ridley, A.J., and Jones, G.E., 1998, A role for Cdc42 in macrophage chemotaxis, *J. Cell Biol.* **141:** 1147–1157.

Andrews, N.W., 2000, Regulated secretion of conventional lysosomes, *Trends Cell Biol.* **10:** 316–321.

Antonin, W., Holroyd, C., Fasshauer, D., Pabst, S., vonMollard, G.F., and Jahn, R., 2000, A SNARE complex mediating fusion of late endosomes defines conserved properties of SNARE structure and function, *EMBO J.* **19:** 6453–6464.

Aravamudan, B., Fergestad, T., Davis, W.S., Rodesch, C.K., and Broadie, K., 1999, Drosophila Unc-13 is essential for synaptic transmission, *Nat. Neurosci.* **2:** 965–971.

Balch, W.E., Dunphy, W.G., Braell, W.A., and Rothman, J.E., 1984, Reconstitution of the transport of protein between successive compartments of the Golgi measured by the coupled incorporation of N-acetylglucosamine, *Cell* **39:** 405–416.

Baumert, M., Maycox, P.R., Navone, F., De Camilli, P., and Jahn, R., 1989, Synaptobrevin: An integral membrane protein of 18,000 daltons present in small synaptic vesicle of rat brain, *EMBO J.* **8:** 379–384.

Bennett, M.K., Calakos, N., and Scheller, R.H., 1992, Syntaxin: A synaptic protein implicated in docking of synaptic vesicles at presynaptic active zones, *Science* **257:** 255–259.

Bentley, D., and Toroian-Raymond, A., 1986, Disoriented pathfinding by pioneer neurone growth cones deprived of filopodia by cytochalasin treatment, *Nature* **323:** 712–715.

Blasi, J., Chapman, E.R., Link, E., Binz, T., Yamasaki, S., De Camilli, P., et al., 1993a, Botulinum neurotoxin A selectively cleaves the synaptic protein SNAP-25, *Nature* **365:** 160–163.

Blasi, J., Chapman, E.R., Yamasaki, S., Binz, T., Niemann, H., and Jahn, R., 1993b, Botulinum neurotoxin C1 blocks neurotransmitter release by means of cleaving HPC-1/syntaxin, *EMBO J.* **12:** 4821–4828.

Block, M.R., and Rothman, J.E., 1992, Purification of N-ethylmaleimide-sensitive fusion protein, *Methods Enzymol* **219:** 300–309.

Bogdanovic, A., Bruckert, F., Morio, T., and Satre, M., 2000, A syntaxin 7 homologue is present in Dictyostelium discoideum endosomes and controls their homotypic fusion, *J. Biol. Chem.* **275:** 36691–36697.

Bogdanovic, A., Bennett, N., Kieffer, S., Louwagie, M., Morio, T., Garin, J., et al., 2002, Syntaxin 7, Syntaxin 8, Vti1 and VAMP7 form an active SNARE complex for early macropinocytic compartment fusion in Dictyostelium discoideum, *Biochem. J.* **15:** 29–39.

Braun, V., Fraisier, V., Raposo, G., Hurbain, I., Sibarita, J.B., Chavrier, P., et al., 2004, TI-VAMP/VAMP7 is required for optimal phagocytosis of opsonised particles in macrophages, *EMBO J.* **23:** 4166–4176.

Brummendorf, T., Kenwrick, S., and Rathjen, F.G., 1998, Neural cell recognition molecule L1: From cell biology to human hereditary brain malformations, *Curr. Opin. Neurobiol.* **8:** 87–97.

Buchstaller, A., Kunz, S., Berger, P., Kunz, B., Ziegler, U., Rader, C., et al., 1996, Cell adhesion molecules NgCAM and axonin-1 form heterodimers in the neuronal membrane and cooperate in neurite outgrowth promotion, *J. Cell Biol.* **135:** 1593–1607.

Castellani, V., Chedotal, A., Schachner, M., Faivre-Sarrailh, C., and Rougon, G., 2000, Analysis of the L1-deficient mouse phenotype reveals cross-talk between Sema3A and L1 signaling pathways in axonal guidance, *Neuron* **27:** 237–249.

Castellani, V., De Angelis, E., Kenwrick, S., and Rougon, G., 2002, Cis and trans interactions of L1 with neuropilin-1 control axonal responses to semaphorin 3A, *EMBO J.* **21:** 6348–6357.

Castellani, V., Falk, J., and Rougon, G., 2004, Semaphorin3A-induced receptor endocytosis during axon guidance responses is mediated by L1 CAM, *Mol. Cell. Neurosci.* **26:** 89–100.

Challacombe, J.F., Snow, D.M., and Letourneau, P.C., 1996, Actin filament bundles are required for microtubule reorientation during growth cone turning to avoid an inhibitory guidance cue, *J. Cell Sci.* **109:** 2031–2040.

Chien, C.B., Rosenthal, D.E., Harris, W.A., and Holt, C.E., 1993, Navigational errors made by growth cones without filopodia in the embryonic Xenopus brain, *Neuron* **11:** 237–251.

Chieregatti, E., Chicka, M.C., Chapman, E.R., and Baldini, G., 2004, SNAP-23 functions in docking/fusion of granules at low Ca2+, *Mol. Biol. Cell* **15:** 1918–1930.

Clary, D.O., Griff, I.C., and Rothman, J.E., 1990, SNAPs, a family of NSF attachment proteins involved in intracellular membrane fusion in animals and yeast, *Cell* **61:** 709–721.

Coco, S., Raposo, G., Martinez, S., Fontaine, J.J., Takamori, S., Zahraoui, A., et al., 1999, Subcellular localization of tetanus neurotoxin-insensitive vesicle-associated membrane protein (VAMP)/VAMP7 in neuronal cells: Evidence for a novel membrane compartment, *J. Neurosci.* **19:** 9803–9812.

Cohen, N.R., Taylor, J.S., Scott, L.B., Guillery, R.W., Soriano, P., and Furley, A.J., 1998, Errors in corticospinal axon guidance in mice lacking the neural cell adhesion molecule L1, *Curr. Biol.* **8:** 26–33.

Cole, R.A., Synek, L., Zarsky, V., and Fowler, J.E., 2005, SEC8, a subunit of the putative Arabidopsis exocyst complex, facilitates pollen germination and competitive pollen tube growth, *Plant Physiol.* **138:** 2005–2018.

Craig, A.M., Wyborski, R.J., and Banker, G., 1995, Preferential addition of newly synthesized membrane protein at axonal growth cones, *Nature* **375:** 592–594.

D'Esposito, M., Ciccodicola, A., Gianfrancesco, F., Esposito, T., Flagiello, L., Mazzarella, R., et al., 1996, A synaptobrevin-like gene in the Xq28 pseudoautosomal region undergoes X inactivation, *Nat. Genet.* **13:** 227–229.

Dahme, M., Bartsch, U., Martini, R., Anliker, B., Schachner, M., and Mantei, N., 1997, Disruption of the mouse L1 gene leads to malformations of the nervous system, *Nat. Genet.* **17:** 346–349.

Davis, J.Q., and Bennett, V., 1994, Ankyrin binding activity shared by the neurofascin/L1/NrCAM family of nervous system cell adhesion molecules, *J. Biol. Chem.* **269:** 27163–27166.

de Anda, F.C., Pollarolo, G., Da Silva, J.S., Camoletto, P.G., Feiguin, F., and Dotti, C.G., 2005, Centrosome localization determines neuronal polarity, *Nature* **436:** 704–708.

Deitcher, D.L., Ueda, A., Stewart, B.A., Burgess, R.W., Kidokoro, Y., and Schwarz, T.L., 1998, Distinct requirements for evoked and spontaneous release of neurotransmitter are revealed by mutations in the Drosophila gene neuronal-synaptobrevin, *J. Neurosci.* **18:** 2028–2039.

Dent, E.W., and Kalil, K., 2001, Axon branching requires interactions between dynamic microtubules and actin filaments, *J. Neurosci.* **21:** 9757–9769.

DiAntonio, A., Haghighi, A.P., Portman, S.L., Lee, J.D., Amaranto, A.M., and Goodman, C.S., 2001, Ubiquitination-dependent mechanisms regulate synaptic growth and function, *Nature* **412:** 449–452.

Dickson, B.J., 2002, Molecular mechanisms of axon guidance, *Science* **298:** 1959–1964.

Dickson, T.C., Mintz, C.D., Benson, D.L., and Salton, S.R., 2002, Functional binding interaction identified between the axonal CAM L1 and members of the ERM family, *J. Cell Biol.* **157:** 1105–1112.

Dulubova, I., Sugita, S., Hill, S., Hosaka, M., Fernandez, I., Sudhof, T.C., et al., 1999, A conformational switch in syntaxin during exocytosis: Role of munc18, *EMBO J.* **18:** 4372–4382.

Etienne-Manneville, S., and Hall, A., 2001, Integrin-mediated activation of Cdc42 controls cell polarity in migrating astrocytes through PKCzeta, *Cell* **106:** 489–498.

Filippini, F., Rossi, V., Galli, T., Budillon, A., D'Urso, M., and D'Esposito, M., 2001, Longins: A new evolutionary conserved VAMP family sharing a novel SNARE domain, *Trends Biochem. Sci.* **26:** 407–409.

Forscher, P., and Smith, S.J., 1988, Actions of cytochalasins on the organization of actin filaments and microtubules in a neuronal growth cone, *J. Cell Biol.* **107:** 1505–1516.

Futerman, A.H., and Banker, G.A., 1996, The economics of neurite outgrowth–the addition of new membrane to growing axons, *Trends. Neurosci.* **19:** 144–149.

Galli, T., Zahraoui, A., Vaidyanathan, V.V., Raposo, G., Tian, J.M., Karin, M., et al., 1998, A novel tetanus neurotoxin-insensitive vesicle-associated membrane protein in SNARE complexes of the apical plasma membrane of epithelial cells, *Mol. Biol. Cell* **9:** 1437–1448.

Gasman, S., Kalaidzidis, Y., and Zerial, M., 2003, RhoD regulates endosome dynamics through Diaphanous-related Formin and Src tyrosine kinase, *Nat. Cell Biol.* **5:** 195–204.

Gonzalez, L.C., Weis, W.I., and Scheller, R.H., 2001, A novel SNARE N-terminal domain revealed by the crystal structure of Sec22b, *J. Biol. Chem.* **276:** 24203–24211.

Grindstaff, K.K., Yeaman, C., Anandasabapathy, N., Hsu, S.C., RodriguezBoulan, E., Scheller, R.H., et al., 1998, Sec6/8 complex is recruited to cell-cell contacts and specifies transport vesicle delivery to the basal-lateral membrane in epithelial cells, *Cell* **93:** 731–740.

Grosse, G., Grosse, J., Tapp, R., Kuchinke, J., Gorsleben, M., Fetter, I., et al., 1999, SNAP-25 requirement for dendritic growth of hippocampal neurons, *J. Neurosci. Res.* **56:** 539–546.

Hammer, J.A., III, and Wu, X.S., 2002, Rabs grab motors: Defining the connections between Rab GTPases and motor proteins, *Curr. Opin. Cell Biol.* **14:** 69–75.
Hattula, K., and Peranen, J., 2000, FIP-2, a coiled-coil protein, links Huntingtin to Rab8 and modulates cellular morphogenesis, *Curr. Biol.* **10:** 1603–1606.
Hattula, K., Furuhjelm, J., Arffman, A., and Peranen, J., 2002, A Rab8-specific GDP/GTP exchange factor is involved in actin remodeling and polarized membrane transport, *Mol. Biol. Cell* **13:** 3268–3280.
Hausauer, D.L., Gerami-Nejad, M., Kistler-Anderson, C., and Gale, C.A., 2005, Hyphal guidance and invasive growth in Candida albicans require the Ras-like GTPase Rsr1p and its GTPase-activating protein Bud2p, *Eukaryot. Cell* **4:** 1273–1286.
Hibi, T., Hirashima, N., and Nakanishi, M., 2000, Rat basophilic leukemia cells express syntaxin-3 and VAMP-7 in granule membranes, *Biochem. Biophys. Res. Commun.* **271:** 36–41.
Higgins, D., Burack, M., Lein, P., and Banker, G., 1997, Mechanisms of neuronal polarity, *Curr. Opin. Neurobiol.* **7:** 599–604.
Hoppe, A.D., and Swanson, J.A., 2004, Cdc42, Rac1, and Rac2 display distinct patterns of activation during phagocytosis, *Mol. Biol. Cell* **15:** 3509–3519.
Hu, C., Ahmed, M., Melia, T.J., Sollner, T.H., Mayer, T., and Rothman, J.E., 2003, Fusion of cells by flipped SNAREs, *Science* **300:** 1745–1749.
Ignelzi, M.A., Jr., Miller, D.R., Soriano, P., and Maness, P.F., 1994, Impaired neurite outgrowth of src-minus cerebellar neurons on the cell adhesion molecule L1, *Neuron* **12:** 873–884.
Ikin, A.F., Annaert, W.G., Takei, K., De Camilli, P., Jahn, R., Greengard, P., et al., 1996, Alzheimer amyloid protein precursor is localized in nerve terminal preparations to Rab5-containing vesicular organelles distinct from those implicated in the synaptic vesicle pathway, *J. Biol. Chem.* **271:** 31783–31786.
Ikonen, E., Tagaya, M., Ullrich, O., Montecucco, C., and Simons, K., 1995, Different requirements for NSF, SNAP, and rab proteins in apical and basolateral transport in MDCK cells, *Cell* **81:** 571–580.
Jahn, R., Lang, T., and Sudhof, T.C., 2003, Membrane fusion, *Cell.* **112:** 519–533.
Jareb, M., and Banker, G., 1997, Inhibition of axonal growth by brefeldin A in hippocampal neurons in culture, *J. Neurosci.* **17:** 8955–8963.
Kamiguchi, H., and Lemmon, V., 2000, Recycling of the cell adhesion molecule L1 in axonal growth cones, *J. Neurosci.* **20:** 3676–3686.
Kamiguchi, H., and Yoshihara, F., 2001, The role of endocytic l1 trafficking in polarized adhesion and migration of nerve growth cones, *J. Neurosci.* **21:** 9194–9203.
Kamiguchi, H., Long, K.E., Pendergast, M., Schaefer, A.W., Rapoport, I., Kirchhausen, T., et al., 1998, The neural cell adhesion molecule L1 interacts with the AP-2 adaptor and is endocytosed via the clathrin-mediated pathway, *J. Neurosci.* **18:** 5311–5321.
Kaplan, K.B., Swedlow, J.R., Varmus, H.E., and Morgan, D.O., 1992, Association of p60c-src with endosomal membranes in mammalian fibroblasts, *J. Cell Biol.* **118:** 321–333.
Kaufmann, N., Wills, Z.P., and Van Vactor, D., 1998, Drosophila Rac1 controls motor axon guidance, *Development* **125:** 453–461.
Kayyem, J.F., Roman, J.M., de la Rosa, E.J., Schwarz, U., and Dreyer, W.J., 1992, Bravo/Nr-CAM is closely related to the cell adhesion molecules L1 and Ng-CAM and has a similar heterodimer structure, *J. Cell Biol.* **118:** 1259–1270.
Lafont, F., Verkade, P., Galli, T., Wimmer, C., Louvard, D., and Simons, K., 1999, Raft association of SNAP receptors acting in apical trafficking in Madin-Darby canine kidney cells, *Proc. Nat. Acad. Sci. USA* **96:** 3734–3738.

Lemmon, V., Farr, K.L., and Lagenaur, C., 1989, L1-mediated axon outgrowth occurs via a homophilic binding mechanism, *Neuron* **2:** 1597–1603.
Li, Z., Van Aelst, L., and Cline, H.T., 2000, Rho GTPases regulate distinct aspects of dendritic arbor growth in Xenopus central neurons in vivo, *Nat. Neurosci.* **3:** 217–225.
Lin, C.H., and Forscher, P., 1995, Growth cone advance is inversely proportional to retrograde F-actin flow, *Neuron* **14:** 763–771.
Lin, C.H., Espreafico, E.M., Mooseker, M.S., and Forscher, P., 1996, Myosin drives retrograde F-actin flow in neuronal growth cones, *Neuron* **16:** 769–782.
Luo, L., Hensch, T.K., Ackerman, L., Barbel, S., Jan, L.Y., and Jan, Y.N., 1996, Differential effects of the Rac GTPase on Purkinje cell axons and dendritic trunks and spines, *Nature* **379:** 837–840.
Martinez-Arca, S., Alberts, P., and Galli, T., 2000a, Clostridial neurotoxin-insensitive vesicular SNAREs in exocytosis and endocytosis, *Biol. Cell* **92:** 449–453.
Martinez-Arca, S., Alberts, P., Zahraoui, A., Louvard, D., and Galli, T., 2000b, Role of tetanus neurotoxin insensitive vesicle-associated membrane protein (TI-VAMP) in vesicular transport mediating neurite outgrowth, *J. Cell Biol.* **149:** 889–899.
Martinez-Arca, S., Coco, S., Mainguy, G., Schenk, U., Alberts, P., Bouille, P., et al., 2001, A common exocytotic mechanism mediates axonal and dendritic outgrowth, *J. Neurosci.* **21:** 3830–3838.
Martinez-Arca, S., Proux-Gillardeaux, V., Alberts, P., Louvard, D., and Galli, T., 2003a, Ectopic expression of syntaxin 1 in the ER redirects TI-VAMP- and cellubrevin-containing vesicles, *J. Cell Sci.* **116:** 2805–2816.
Martinez-Arca, S., Rudge, R., Vacca, M., Raposo, G., Camonis, J., Proux-Gillardeaux, V., et al., 2003b, A dual mechanism controlling the localization and function of exocytic v-SNAREs, *Proc. Natl. Acad. Sci. USA* **100:** 9011–9016.
May, R.C., and Machesky, L.M., 2001, Phagocytosis and the actin cytoskeleton, *J. Cell Sci.* **114:** 1061–1077.
Murthy, M., Garza, D., Scheller, R.H., and Schwarz, T.L., 2003, Mutations in the exocyst component Sec5 disrupt neuronal membrane traffic, but neurotransmitter release persists, *Neuron* **37:** 433–447.
Nonet, M.L., Saifee, O., Zhao, H.J., Rand, J.B., and Wei, L.P., 1998, Synaptic transmission deficits in *Caenorhabditis elegans* synaptobrevin mutants, *J. Neurosci.* **18:** 70–80.
Novick, P., Field, C., and Schekman, R., 1980, Identification of 23 complementation groups required for post-translational events in the yeast secretory pathway, *Cell* **21:** 205–215.
Osen-Sand, A., Staple, J.K., Naldi, E., Schiavo, G., Rossetto, O., Petitpierre, S., et al., 1996, Common and distinct fusion proteins in axonal growth and transmitter release, *J. Comp. Neurol.* **367:** 222–234.
Oyler, G.A., Polli, J.W., Higgins, G.A., Wilson, M.C., and Billingsley, M.L., 1992, Distribution and expression of SNAP-25 immunoreactivity in rat brain, rat PC-12 cells and human SMS-KCNR neuroblastoma cells, *Dev. Brain Res.* **65:** 133–146.
Paglini, G., Kunda, P., Quiroga, S., Kosik, K., and Caceres, A., 1998, Suppression of radixin and moesin alters growth cone morphology, motility, and process formation in primary cultured neurons, *J. Cell Biol.* **143:** 443–455.
Parlati, F., Weber, T., McNew, J.A., Westermann, B., Sollner, T.H., and Rothman, J.E., 1999, Rapid and efficient fusion of phospholipid vesicles by the alpha-helical core of a SNARE complex in the absence of an N-terminal regulatory domain, *Proc. Natl. Acad. Sci. USA* **96:** 12565–12570.
Prochiantz, A., 1995, Neuronal polarity: Giving neurons heads and tails, *Neuron* **15:** 743–746.

Rao, S.K., Huynh, C., Proux-Gillardeaux, V., Galli, T., and Andrews, N.W., 2004, Identification of SNAREs involved in synaptotagmin VII-regulated lysosomal exocytosis, *J. Biol. Chem.* **279:** 20471–20479.
Reddy, A., Caler, E.V., and Andrews, N.W., 2001, Plasma membrane repair is mediated by Ca2+-regulated exocytosis of lysosomes, *Cell* **106:** 157–169.
Rothman, J.E., 1994, Mechanisms of intracellular protein transport, *Nature* **372:** 55–63.
Rothman, J.E., and Warren, G., 1994, Implication of the SNARE hypothesis for intracellular membrane topology and dynamics, *Curr. Biol.* **4:** 220–233.
Sabo, S.L., Ikin, A.F., Buxbaum, J.D., and Greengard, P., 2003, The amyloid precursor protein and its regulatory protein, FE65, in growth cones and synapses in vitro and in vivo, *J. Neurosci.* **23:** 5407–5415.
Schaefer, A.W., Kamiguchi, H., Wong, E.V., Beach, C.M., Landreth, G., and Lemmon, V., 1999, Activation of the MAPK signal cascade by the neural cell adhesion molecule L1 requires L1 internalization, *J. Biol. Chem.* **274:** 37965–37973.
Schaefer, A.W., Kamei, Y., Kamiguchi, H., Wong, E.V., Rapoport, I., Kirchhausen, T., et al., 2002, L1 endocytosis is controlled by a phosphorylation-dephosphorylation cycle stimulated by outside-in signaling by L1, *J. Cell Biol.* **157:** 1223–1232.
Schiavo, G., Benfenati, F., Poulain, B., Rossetto, O., Polverino de Laureto, P., DasGupta, B.R. et al., 1992, Tetanus and botulinum-B neurotoxins block neurotransmitter release by proteolytic cleavage of synaptobrevin, *Nature* **359:** 832–835.
Schmid, R.S., Pruitt, W.M., and Maness, P.F., 2000, A MAP kinase-signaling pathway mediates neurite outgrowth on L1 and requires Src-dependent endocytosis, *J. Neurosci.* **20:** 4177–4188.
Schmidt, C.E., Dai, J., Lauffenburger, D.A., Sheetz, M.P., and Horwitz, A.F., 1995, Integrin-cytoskeletal interactions in neuronal growth cones, *J. Neurosci.* **15:** 3400–3407.
Schoch, S., Cibelli, G., Magin, A., Steinmuller, L., and Thiel, G., 2001a, Modular structure of cAMP response element binding protein 2 (CREB2), *Neurochem. Int.* **38:** 601–608.
Schoch, S., Deak, F., Konigstorfer, A., Mozhayeva, M., Sara, Y., Sudhof, T.C., et al., 2001b, SNARE function analyzed in synaptobrevin/VAMP knockout mice, *Science* **294:** 1117–1122.
Schwamborn, J.C., and Puschel, A.W., 2004, The sequential activity of the GTPases Rap1B and Cdc42 determines neuronal polarity, *Nat. Neurosci.* **7:** 923–929.
Söllner, T., Bennett, M.K., Whiteheart, S.W., Scheller, R.H., and Rothman, J.E., 1993a, A protein assembly-disassembly pathway in vitro that may correspond to sequential steps of synaptic vesicle docking, activation, and fusion, *Cell* **75:** 409–418.
Söllner, T., Whiteheart, S.W., Brunner, M., Erdjument-Bromage, H., Geromanos, S., Tempst, P., et al., 1993b, SNAP receptors implicated in vesicle targeting and fusion, *Nature* **362:** 318–324.
Stewart, B.A., Mohtashami, M., Rivlin, P., Deitcher, D.L., Trimble, W.S., and Boulianne, G.L., 2002, Dominant-negative NSF2 disrupts the structure and function of Drosophila neuromuscular synapses, *J. Neurobiol.* **51:** 261–271.
Suter, D.M., and Forscher, P., 1998, An emerging link between cytoskeletal dynamics and cell adhesion molecules in growth cone guidance, *Curr. Opin. Neurobiol.* **8:** 106–116.
Suter, D.M., and Forscher, P., 2001, Transmission of growth cone traction force through apCAM-cytoskeletal linkages is regulated by Src family tyrosine kinase activity, *J. Cell Biol.* **155:** 427–438.
Suter, D.M., Errante, L.D., Belotserkovsky, V., and Forscher, P., 1998, The Ig superfamily cell adhesion molecule, apCAM, mediates growth cone steering by substrate-cytoskeletal coupling, *J. Cell Biol.* **141:** 227–240.

Sweeney, S.T., and Davis, G.W., 2002, Unrestricted synaptic growth in spinster-a late endosomal protein implicated in TGF-beta-mediated synaptic growth regulation, *Neuron* **36:** 403–416.

TerBush, D.R., Maurice, T., Roth, D., and Novick, P., 1996, The Exocyst is a multiprotein complex required for exocytosis in *Saccharomyces cerevisiae*, *EMBO J.* **15:** 6483–6494.

Tessier-Lavigne, M., and Goodman, C.S., 1996, The molecular biology of axon guidance, *Science* **274:** 1123–1133.

Thelen, K., Kedar, V., Panicker, A.K., Schmid, R.S., Midkiff, B.R., and Maness, P.F., 2002, The neural cell adhesion molecule L1 potentiates integrin-dependent cell migration to extracellular matrix proteins, *J. Neurosci.* **22:** 4918–4931.

Tochio, H., Tsui, M.M.K., Banfield, D.K., and Zhang, M.J., 2001. An autoinhibitory mechanism for nonsyntaxin SNARE proteins revealed by the structure of Ykt6p, *Science* **293:** 698–702.

Van Aelst, L., and Cline, H.T., 2004, Rho GTPases and activity-dependent dendrite development, *Curr. Opin. Neurobiol.* **14:** 297–304.

Verhage, M., Maia, A.S., Plomp, J.J., Brussaard, A.B., Heeroma, J.H., Vermeer, H., et al., 2000, Synaptic assembly of the brain in the absence of neurotransmitter secretion, *Science* **287:** 864–869.

Vogt, L., Giger, R.J., Ziegler, U., Kunz, B., Buchstaller, A., Hermens, W., et al., 1996, Continuous renewal of the axonal pathway sensor apparatus by insertion of new sensor molecules into the growth cone membrane, *Curr. Biol.* **6:** 1153–1158.

Volkmer, H., Hassel, B., Wolff, J.M., Frank, R., and Rathjen, F.G., 1992, Structure of the axonal surface recognition molecule neurofascin and its relationship to a neural subgroup of the immunoglobulin superfamily, *J. Cell Biol.* **118:** 149–161.

Ward, D.M., Pevsner, J., Scullion, M.A., Vaughn, M., and Kaplan, J., 2000, Syntaxin 7 and VAMP-7 are soluble N-ethylmaleimide-sensitive factor attachment protein receptors required for late endosome-lysosome and homotypic lysosome fusion in alveolar macrophages, *Mol. Biol. Cell* **11:** 2327–2333.

Washbourne, P., Cansino, V., Mathews, J.R., Graham, M., Burgoyne, R.D., and Wilson, M.C., 2001, Cysteine residues of SNAP-25 are required for SNARE disassembly and exocytosis, but not for membrane targeting, *Biochem. J.* **357:** 625–634.

Washbourne, P., Thompson, P.M., Carta, M., Costa, E.T., Mathews, J.R., Lopez-Bendito, G., et al., 2002, Genetic ablation of the t-SNARE SNAP-25 distinguishes mechanisms of neuroexocytosis, *Nat. Neurosci.* **5:** 19–26.

Weber, T., Zemelman, B.V., McNew, J.A., Westermann, B., Gmachl, M., Parlati, F., et al., 1998, SNAREpins: Minimal machinery for membrane fusion, *Cell* **92:** 759–772.

Wisco, D., Anderson, E.D., Chang, M.C., Norden, C., Boiko, T., Folsch, H., et al., 2003, Uncovering multiple axonal targeting pathways in hippocampal neurons, *J. Cell Biol.* **162:** 1317–28.

Zerial, M., and McBride, H., 2001, Rab proteins as membrane organizers, *Nat. Rev. Mol. Cell Biol.* **2:** 107–117.

6
Role of the Golgi Apparatus During Axon Formation

ALFREDO CÁCERES, GABRIELA PAGLINI, SANTIAGO QUIROGA,
AND ADRIANA FERREIRA

6.1. Summary

During recent years, a large body of evidence has accumulated regarding cellular events and potential mechanisms underlying the development of specialized cellular projections, namely axons and dendrites in mammalian neurons. Most of these studies have focused on the regulation of cytoskeletal organization and function during different stages of neuronal polarization. Much less effort has been invested in analyzing how cytoskeletal assembly and dynamics are coordinately coupled with membrane addition at sites of active growth or with membrane formation and/or exit from the Golgi complex. In this chapter, we shall review current evidence concerning the participation of the Golgi apparatus in axon specification, one of the earliest events in neuronal polarization.

6.2. Relationship Between Golgi Functioning and Axon Formation

The ability of cells to polarize is crucial for many biological activities such as the functioning the nervous system. Indeed, neurons are among the best examples of highly differentiated and polarized cell types, typically extending a single, thin, long axon, which send signals, and several shorter and thicker dendrites, which receive them. A wealth of data (Goda and Davis, 2003; Govek et al., 2005; Jan and Jan, 2003; Wiggin et al., 2005) has identified a number of gene products having capacity to impose cellular asymmetry, through their ability to form dynamic multiprotein complexes. Such polarity proteins participate in neurite outgrowth, axon specification, elongation, dendritic formation and branching, as well as synaptogenesis, by signaling to the actin and microtubule cytoskeletons.

The initial event in establishing a polarized neuron is the specification of a single axon. Proteins that are crucial for neuronal polarity determination have been studied extensively in cultured pyramidal neurons from the embryonic rodent hippocampus; in this system, cellular asymmetry develops in the absence of extracellular spatial cues (Craig and Banker, 1994). Initially there is no morphological polarity, as neurons elaborate lamellipodia (stage 1), and then short neurites (stage 2). Later on, one of these neurites forms the axon by extending a large and highly dynamic growth cone with labile actin cytoskeleton (stage 3) (Bradke and Dotti, 1997; Chuang et al., 2005; Kunda et al., 2001), leaving the rest of the neurites to form dendrites (stage 4), and setting the stage for synaptogenesis (stage 5) (Craig and Banker, 1994). Current evidence favors the view that many polarity proteins exert their functions, acting at sites of active growth, such as the axonal growth cone, where they regulate cytoskeletal assembly and dynamics (Govek et al., 2005; Jan and Jan, 2003; Wiggin et al., 2005). During recent years, it has also become clear that for an axon to become specified, changes in cytoskeletal organization and dynamics need to be coordinately coupled with the addition of new membrane at sites of active growth. Thus, the growth of a single axon from the neuronal cell body during differentiation involves a substantial and selective increase in its surface area (Futerman and Banker, 1996) and also the targeted delivery to the growth cone of regulatory membrane constituents, such as tyrosine kinase receptors (neurotrophin receptors) or polarity proteins (e.g. Par3/Par6) essentially required for driving/triggering polarization (Jan and Jan, 2003; Wiggin et al., 2005). It is also widely accepted that most of the membranes needed to elongate the growing axon derived from "constitutive" exocytosis from the *trans*-Golgi network (TGN) and that the nature of transport carriers and the mechanisms of their formation may be similar to those described in other polarized cells (Tang, 2001; Horton and Ehlers, 2003b). In this chapter, we shall review current evidence regarding the link between axonal growth and the supply of membrane materials produced in the cell body, with particular emphasis to the role of the Golgi apparatus.

6.3. Origin of Vesicles at the Axonal Growth Cone

One of the hallmarks of axon outgrowth is membrane expansion due to the incorporation/addition of vesicles to the neuronal plasma membrane. Most of the addition of this new membrane occurs at the axonal growth cone of differentiating neurons (Craig et al., 1995; Laurino et al., 2005; Pfenninger and Friedman, 1993; Pfenninger et al., 2003) by incorporation of small TGN-derived vesicles (plasmalemmal precursor vesicles—PPVs) that fuse with the plasma membrane of the growing tip (Paglini et al., 2001; Pfenninger et al., 2003). In cultured hippocampal pyramidal neurons, this phenomenon, termed expansive regulated exocytosis (Chieregatti and Meldolesi, 2005), is locally regulated by IGF1 through the rapid activation of the IRS2/PI3K/Akt signaling pathway (Laurino et al., 2005; Pfenninger et al., 2003). This implies that membrane components assembled at the Golgi apparatus are packed in vesicles and selectively shuttled to the axonal growth cone by axoplasmic transport where PPVs accumulate at sites of insertion, until they fuse with the plasma membrane after an appropriate specific stimuli (e.g., IGF1 in cultured hippocampal neurons) (Laurino et al., 2005).

Coincident with this live imaging of hippocampal neurons before and during polarization has revealed that overall anterograde membrane trafficking, including post-Golgi vesicles or tubulo-vesicular organelles, as well as mitochondria, all appear polarized in a subpopulation of stage 2 cells (e.g., neurons displaying at least one minor neurite with a large growth cone) (Bradke and Dotti, 1997; Kunda et al., 2001) or to the axon of most stage 3 cells (Bradke and Dotti, 1997; Paglini et al., 2001; Pfenninger et al., 2003).

The importance of Golgi-derived membrane elements in axon outgrowth and specification has also been examined by evaluating the effects of brefeldin-A (BFA) in cultured hippocampal neurons (Jareb and Banker, 1997). Treatment with BFA (1 μg/ml) produces a rapid and selective inhibition of axonal growth. Thus, in stage 3 hippocampal neurons axon elongation stops within 30 min after the initiation of BFA treatment, and shortly thereafter the axon begins to retract. Moreover in unpolarized cells (stage 2 neurons), BFA prevents the initiation of axon outgrowth that marks the initial establishment of morphological polarity (Jareb and Banker, 1997). The time course of inhibition of either axon outgrowth or elongation is consistent with the fastest rates of vesicle transport from the endoplasmic reticulum (ER) to the cell surface (Wieland et al., 1987; Young et al., 1992), further suggesting that blockade of axonal elongation is causally related with depletion of Golgi-derived vesicles, and hence PPVs, at the growth cone (Jareb and Banker, 1997; Paglini et al., 2001; Pfenninger et al., 2003).

TGN-derived axonal PPVs have different morphologies. For example, amyloid precursor protein (APP) appears to be transported in elongated tubules (as long as 10 μm) that move extremely fast (on average 4.5 μm/s) and over long distances; by contrast, L1- or synaptophysin-transporting structures are more vesicular and moved slower and over shorter distances only (Kaether et al., 2000; Rosso et al., 2004; Silverman et al., 2001). The transport of PPVs depends on the activity of microtubular-based motors (Hirokawa and Takemura, 2005). Different TGN-derived PPVs have been described based on the type of motor and "cargo" proteins. PPVs containing a β-subunit variant of the IGF1 receptor designated as βgc (Mascotti et al., 1997; Quiroga et al., 1995) are transported toward the axonal growth cone in a distinct vesicle population by the central motor domain protein KIF2 (Morfini et al., 1997; Pfenninger et al., 2003). Other membrane proteins are exported from the TGN to the growth cone in different PPVs as suggested by the reports showing that axonal transport of APP and GAP43 is kinesin-, rather than KIF2-dependent (Ferreira et al., 1992, 1993; Kamal et al., 2000; Morfini et al., 1997), while that of L1-containing or fodrin-containing vesicles has been linked to the KIF4 (Peretti et al., 2000) or KIF3 (Takeda et al., 2000) motors respectively. A study form our laboratory also indicates that axonal TGN-derived vesicles differ in the composition of membrane-associated and actin-regulatory proteins. For example, βgc- or TrkB-containing vesicles associate with the polarity proteins Par3–Par6, as well as cortactin, PAK1, phosphorylated LIMK1 and phosphorylated cofilin; by contrast, synaptophysin-containing vesicles lack all these proteins but contain actin and Arp 2/3 (Rosso et al., 2004).

6.4. Positioning of the Golgi Apparatus and Axon Specification

In several cells, such as fibroblasts, endothelial cells, T cells, astrocytes, and neuroblasts, the microtubule-organizing center (MTOC) and the Golgi apparatus are reoriented to a position between the leading edge and the nucleus during polarized migration (Etienne-Manneville and Hall, 2001; Gomes et al., 2005; Gotlieb et al., 1981; Gregory et al., 1988; Gundersen and Bulinski, 1988; Kupfer et al., 1982, 1983; Palazzo et al., 2001). It is believed that the reorientation of both organelles gives the cells an overall polarity that may be required for establishing a polarized microtubule network as well as the directed and polarized flow of membrane precursors and polarity proteins. In accordance with this idea, several studies have examined whether or not a relationship exists between the localization of the MTOC–Golgi apparatus complex and the site of axon emergence. Surprisingly, and despite considerable efforts, a clear picture has not yet emerged.

Studies carried out by C. Dotti and G. Banker in the early 1990s revealed that cultured hippocampal neurons contained a single Golgi apparatus and MTOC, which were frequently localized together in close proximity to a shallow indentation of the cells nucleus (Dotti and Banker, 1991). In this study no correlation was found between the position of the MTOC–Golgi apparatus and the site of origin of the axon, both in fixed and live hippocampal pyramidal cell that develop in culture. Thus, when outgrowth was followed by time-lapse microscopy and the location of the Golgi complex and MTOC determined at the exact moment that the axon emerged, no correlation was apparent (Dotti and Banker, 1991). It was then concluded that the localization of these cellular constituents might not play a major role in determining which of the neurites (minor processes) initially extended by stage 2 hippocampal neurons becomes the definitive axon. Another study has reported that the Golgi localization in the cell soma of cultured hippocampal neurons has no spatial relationship with the site of axon origin, either during the early stages of axon specification or at later time points (Horton et al., 2005).

Similarly, no correlation was found between the position of the TGN and the site of axon origin during the development and maturation of rat neocortical neurons *in vitro* (Lowenstein et al., 1994). These authors showed that before and during axon formation, TGN38 (a marker of the TGN) immunoreactivity labels several vesicles dispersed throughout the cell body cytoplasm that later merge into a single structure located at the base of a thick MAP2-immunopositive neurite, presumably a dendrite. In the majority of the neurons, TGN38 immunoreactivity was located within 45° of the major MAP2-immunoreactive process. Thus, in cortical neurons the distribution of TGN38 immunoreactivity does not polarize to the axon but rather localize within a single, usually the major apical dendrite. The presence of Golgi cisternae and/or Golgi posts localized to a single major dendritic processes have also been detected in cultured hippocampal pyramidal neurons after staining for GM1 130 (Horton and Ehlers, 2003a,b, 2004) or transfection with GFP-GalT2 or YFP-SialT2 (Rosso et al., 2004). Moreover, these authors have demonstrated that Golgi polarization to a single dendrite precedes and is required for asymmetric dendritic growth (Horton et al., 2005).

The lack of correlation between the position of the MTOC–Golgi apparatus with either the site of axon origin and/or with events leading to axon specification/outgrowth is consistent with observations showing that: (1) All minor processes from stage 2 or stage 3 cells have the potential to become axons. (2) It is possible to induce multiple axons, after treatment with cytochalasin D (Kunda et al., 2001) or overexpression of Tiam1 (Kunda et al., 2001) or Tctex-1 (Chuang et al., 2001), even though neurons still display a single randomly localized/oriented MTOC–Golgi apparatus (Bollati and Cáceres, unpublished observations). Consistent with this, it is possible to induce a second or even a third or fourth axon sprout in stage 3 polarized neurons by either the local addition of laminin or cytochalasin D to the growth cone of a minor neurite; we have found no correlation between the position of the Golgi apparatus labeled with GFP-GalT2 or YFP-SialT2, and the site of emergence of the second axon in neurons locally treated with cytochalasin D or laminin (Bollati, Jausoro, and Cáceres, unpublished observations). (3) A lag of at least 10–12 h exist between the formation of minor neurites and the emergence of the first axon; in fact, a "tug of war" has been described among minor neurites during the transition between stage 2 and 3 in cultured hippocampal pyramidal neurons; a dramatic example of such situation is observed in a time-lapse sequence published by Goslin and Banker (1990). Thus, taken together these observations favor the view that minor neurites may compete for the acquisition of some rate-limiting factor required for driving/initiating axon outgrowth. These factors may include regulatory enzymes, such as PI3K (Shi et al., 2003), microtubule-assembly promoting factors, such as CRMP-2 (Fukata et al., 2002), guanosine-nucleotide exchange factors, such a Tiam1 (Kunda et al., 2001), or Golgi-derived membrane proteins, such as receptors for trophic factors (BDNF, NGF, IGF1), or membrane-associated polarity proteins such as Par3/Par6 (Rosso et al., 2004; Shi et al., 2003).

By contrast, several other reports have demonstrated a clear relationship between the positioning of the MTOC–Golgi apparatus and the site of axon emergence. For example, cultured cerebellar granule neurons develop their characteristic axonal and dendritic morphologies in a series of discrete temporal steps highly similar to those observed *in situ*. Initially they extend a single neurite, which is then followed by a second one emerging from the opposite pole; both neurites will become axons to generate a bipolar morphology (Zmuda and Rivas, 1998). When the Golgi apparatus was labeled with the fluorescent lipid analogue, C5-DMB-Ceramide, or by indirect immunofluorescence using antibodies against α-mannosidase II, these authors found that the Golgi was first positioned at the site of initial axonal extension and later was repositioned to the site of secondary axon outgrowth. Labeling of the MTOC with an antibody against γ-tubulin revealed that this organelle colocalizes with the Golgi apparatus at all stages of differentiation and also appeared to be repositioned to the base of the newly emerging axon during the transition from a unipolar to a bipolar morphology. As the second axon elongates and the neurons acquired their bipolar state, or at later stages when distinct axonal and somatodendritic domains had been established, the Golgi was not consistently positioned at the base of either axons or dendrites and was most often found at sites on the plasma membrane from which no processes originated (Zmuda and Rivas, 1998).

The possibility that a similar phenomenon may be occurring in other developing neurons, including cultured hippocampal pyramidal neurons, has been reexamined by Dotti and coworkers (de Anda et al., 2005). For such a purpose, these authors analyzed the morphological differentiation of recent postmitotic hippocampal neurons; instead of using the standard 18-day-old rat embryos, hippocampi were taken from 16-day-old rat embryos, therefore increasing the number of neurons, which are close to their last mitosis. Under these conditions, they observe that the MTOC, the Golgi apparatus, and endosomes localize/accumulate (cluster) close to the area where the first neurite emerge, which will later become the axon; the positioning of the MTOC–Golgi apparatus-endosomes and the site of axon origin is opposite to the plane of the last division both in culture hippocampal pyramidal neurons developing *in vitro* and also in *Drosophila melanogaster* neurons that develop *in situ*. To prove a causal relationship between the positioning/activities of the centrosome and the specification of an axon, and/or the site of axon origin, and/or the number of axons that an hippocampal neuron extend, neurons were generated by blocking cytokinesis with cytochalasin D and then analyzing axon formation in neurons displaying more than one centrosome; neurons showing two centrosomes also display two long axon-like neurites that emerge from the poles where the centrosomes are located. Taken together, these observations provide intriguing evidence favoring a major role for the asymmetric positioning of the centrosome-Golgi apparatus in axon specification.

However, and in the absence of detailed mechanistic evidence, it is at present safe to propose that MTOC–Golgi apparatus positioning has an instructive facilitatory role, yet not essentially required, for axon formation/specification. Thus, an appropriate positioning (e.g., opposite to the cleavage furrow) of the MTOC–Golgi apparatus complex may accelerate the events leading to axon formation/specification by driving polarized cytoskeletal assembly/dynamics and Golgi-derived membrane trafficking to a particular region of the cell. Under this context, rate-limiting factors might accumulate in a selected region of the cortex by events promoted by the MTOC–Golgi apparatus activity. Such a phenomenon may involve a microtubule-capture machinery similar to that described in budding yeast or migrating fibroblasts or astrocytes. However, when neurite outgrowth is initiated several hours after the last mitosis, and therefore in the absence of a cleavage furrow (e.g., E18 hippocampal pyramidal neurons), the positioning of the centrosome-Golgi complex may not be crucially required to determine initial neuronal polarization. Accumulation of polarity proteins and/or rate limiting factors may occur by random events at the cell periphery (e.g., growth cones of minor neurites) until they reach a threshold level required for triggering axon formation. In the absence of cell division orientation mechanisms, axon formation may proceed at a much slower rate and explain the prolonged "tug of war" among minor neurites described in stage 2 hippocampal neurons obtained from E18 embryos. Environmental factors may accelerate/promote axon formation by locally activating randomly distributed polarity proteins. A more detailed comprehension of the biological significance of MTOC–Golgi apparatus positioning during initial neuronal polarization would certainly require the identification of proteins (or motifs) that mediate and/or control "steps" such as sorting into carrier vesicles, fission, and exit from the Golgi apparatus (see in a later section).

6.5. Regulation of Golgi Functioning During Axon Formation

6.5.1. Rho GTPases, Downstream Effectors, and the Trafficking of Golgi-Derived Vesicles

In mammalian cells, the small GTP-binding ADP-ribosylation factor 1 (ARF1) acts as a master regulator of Golgi structure and function through the recruitment and activation of various downstream effectors, including members of the Rho family of small GTPases (Chen et al., 2004; Dubois et al., 2005; Fucini et al., 2002). It is likely that a major action of these proteins is the regulation of actin organization and dynamics at the Golgi apparatus (Stamnes, 2002). In favor of this view, a series of experimental data have established that Rho family members and several of their effector proteins localized to vesicular compartments, where they regulate not only actin dynamic but also several transport steps (Cerione, 2004; Ridley, 2001). This includes the early demonstration that Cdc42 localizes primarily to the Golgi apparatus and that in epithelial cells it is involved in the exit and sorting of proteins from the TGN (Cohen et al., 2001; Kroschewski et al., 1999; Musch et al., 2001), as well as the more recent data showing that active Cdc42 participates in ER to Golgi transport by regulating dynein recruitment to COP1-coated vesicles (Chen et al., 2005), in Golgi to ER transport by inducing relocation of N-WASP and Arp3 to the lateral rims of Golgi cisternae (Luna et al., 2002; Matas et al., 2004), and that Golgi positioning during polarized growth is also dependent on Cdc42 signaling (Cau and Hall, 2005; Etienne-Manneville and Hall, 2001). Additional studies have identified Citron-N, a Rho-binding protein, ROCK-II, LIMK1, and cofilin in the Golgi apparatus (Camera et al., 2003; Rosso et al., 2004). Besides, PAK4, an effector for Cdc42, not only localizes to the BFA-sensitive compartment of the Golgi (Abo et al., 1998) but also interacts specifically with LIMK1 stimulating its ability to phosphorylate cofilin (Dan et al., 2001).

Many of these proteins have also been identified in the neuronal Golgi apparatus. This includes, among others: Cdc42 (Paglini et al., 2001a), Rac (Paglini et al., 2001a), the p21-activated kinases PAK1 and PAK4 (Paglini et al., 2001a; Qu et al., 2003; Rosso et al., 2004), Citron-N, ROCK-II, and Profilin IIa (Camera et al., 2003), LIMK1 (Camera et al., 2003; Rosso et al., 2004), cofilin and phospho-cofilin (Rosso et al., 2004), cyclin-dependent kinase 5 (Cdk5), the Cdk5 activator p35 (Paglini and Cáceres, 2001; Paglini et al., 2001a), and a Cdk5/p35-regulated kinase (Kesavapany et al., 2003). Unfortunately, few studies have examined their role during development. Despite that, the available information suggests that some of them are required for maintaining Golgi architecture, while others for regulating Golgi-to-ER transport, Golgi exit and/or post-Golgi trafficking (see in a later section).

6.5.1.1. Citron-N

Antisense suppression of the brain-specific Rho-binding protein Citron-N or expression of a mutant form lacking Rho-binding activity results in Golgi dispersion; inhibition of Rho functioning after treatment with C3 toxin also results in Golgi fragmentation (Camera et al., 2003). To prove that Citron-N is involved in the local regulation of actin assembly by Rho recruitment, these authors showed that Golgi disruption induced by actin depolymerizing drugs (e.g., cytochalasin D) was prevented by overexpressing Citron-N. Additional experiments demonstrated that two downstream Rho effectors, namely ROCK-II and profilin IIa (PIIa), are not only involved in Citron-N mediated actin recruitment to the Golgi apparatus but also in the maintenance of its architecture (Camera et al., 2003).

6.5.1.2. LIMK1-Cofilin

LIMK1, a major effector of Rho-ROCK-II and Cdc42 signaling pathways, is present at the Golgi apparatus (Camera et al., 2003; Rosso et al., 2004). Studies from our laboratory have shown that LIMK1 is enriched in this organelle, and its association with Golgi membranes is dependent on the integrity of the LIM domain (Rosso et al., 2004). LIMK1 is activated by phosphorylation at Thr508, by the action of PAK1 and PAK4, two kinases that also localize to the Golgi apparatus and have been implicated in neuronal development. Active LIMK1 phosphorylates actin-depolymerizing factor (ADF) and cofilin at a single site (Ser3) inhibiting its binding to actin monomers and actin-depolymerizing activity (Sarmiere and Bamburg, 2004). Active-phosphorylated LIMK1 was detected in the Golgi apparatus of young (stage 2–3) and more mature (stage 4–5) cultured hippocampal pyramidal neurons (Rosso et al., 2004); p-LIMK1 punctate immunolabeling was also found along neurites; these vesicle-like organelles were more abundant in axons and their growth cones, than in minor neurites (Rosso et al., 2004). Cofilin and p-cofilin were also detected in the Golgi apparatus of developing neurons, with overexpression of LIMK1 increasing p-cofilin and F-actin staining of the Golgi, and expression of a kinase dead mutant having the opposite effect.

LIMK1 appears to participate in several different aspects of Golgi functioning, all of which involve regulation of cofilin activity by phosphorylation. For example, the long-term expression (more than a day) of the kinase defective mutant of LIMK1 or a constitutively active form of cofilin (S3A mutant) produce Golgi fragmentation and dramatically accelerates cytochalasin-induced Golgi disruption; by contrast, overexpression of wild-type (wt)LIMK1 or treatment with the actin stabilizer Jasplakinolide make the Golgi more compact preventing the effects of the actin depolymerizing agent (Rosso et al., 2004). Taken together, these observations showed for the first time that regulation of actin turnover is critically involved in maintaining neuronal Golgi architecture. Since these effects are similar to those observed after suppression of Citron-N or Rho (Camera et al., 2003), LIMK1 and cofilin may be part and act downstream of the Citron-N-Rho-ROCK-II-signaling pathway involved in the maintenance of Golgi organization and dynamics.

Golgi-to-ER-transport, which is crucial for proper Golgi functioning, requires actin filaments (Valderrama et al., 2000, 2001) and myosin motors (presumably myosin II) to move transport carriers in the retrograde direction (Duran et al., 2003). The observation that expression of the kinase dead mutant of LIMK1 (LIMK1-kd) or S3A cofilin accelerates BFA-induced Golgi dispersal suggests that cofilin activity may regulate the formation of the actin tracks used by myosin motors during Golgi-to-ER transport (Jausoro, Paglini, Bamburg, Ferreira, and Cáceres, unpublished observations). In favor of this, we observed that S3A-cofilin induced acceleration of BFA-mediated Golgi disassembly was prevented in neurons treated with a broad spectrum myosin inhibitor (BDM) or ML7 (an inhibitor of myosin II light chain) or after transfection with a cofilin mutant lacking severing activity. Whether severing leads to an increase in F-actin resulting from additional elongation from the newly generated barbed ends remains to be established. In any case, these observations suggest that both dynamizing and severing properties of cofilin (Sarmiere and Bamburg, 2004) contribute to the regulation of Golgi organization.

LIMK1 and cofilin also have a more subtle involvement in the regulation of Golgi morphology and dynamics. Thus, LIMK1 activation and hence cofilin phosphorylation-inactivation are paralleled by a reduction in the number, length, and rate of elongation/retraction of Golgi-derived membrane tubules; by contrast, activation of cofilin dramatically increases the length and elongation rate of these tubules (Rosso et al., 2004). In many occasions, these tubules penetrate more than 30 μm within a thicker dendritic-like neurite (Figure 4: Rosso et al., 2004; Figure 1: Horton et al., 2005). Thus, cofilin activity may be responsible for determining the presence of elongated Golgi cisternae and/or Golgi tubules and/or Golgi posts in dendrites, an event important for polarized asymmetric dendritic growth (Horton et al., 2005).

Aside from all this, LIMK1 and cofilin are involved in regulating post-Golgi trafficking, in a manner that appears to be important for axon formation. Thus, immunofluorescence and subcellular fractionation experiments indicate that active LIMK1, PAK1, and the polarity proteins Par3/Par6, which are crucially required for axon specification in hippocampal neurons (Nishimura et al., 2005; Shi et al., 2003; Wiggin et al., 2005) colocalized in the same type of post-Golgi transport carrier (Rosso et al., 2004). Overexpression of LIMK1 accelerates axonal formation promoting the accumulation of Par3/Par6 and p-Akt at the axonal growth cone (Rosso et al., 2004). This function of LIMK1 involves cofilin phosphorylation since it is prevented by coexpression of S3A-cofilin and is not observed in neurons overexpressing LIMK1-kd. Importantly, blockade of axon formation and reduced accumulation of Par3/Par6 at growth cones were observed after expression of the Δ-LIM mutant (loss of Golgi localization) of wt-LIMK1 or the LIM domain alone (Rosso et al., 2004). Immunoisolation experiments allow us to precipitate Par3-containing vesicles and show that they contain Par6, PAK1, activated LIMK1, phospho-cofilin, cortactin, and actin (Rosso et al., 2004; our unpublished observations). These results suggest that by inactivating cofilin at the Golgi apparatus, LIMK1 promotes the trafficking of a selected population of Golgi-derived vesicles capable of interacting with Par3–Par6 proteins. Two integral membrane proteins, the βgc subunit of the IGF1 receptor (Laurino et al., 2005; Mascotti et al., 1997; Pfenninger et al., 2003) and the cell adhesion molecule L1 (Itoh et al., 2004; Kamiguchi et al., 1998; Peretti et al., 2000), both of which are involved in axon formation, are also present in these vesicles.

As mentioned earlier, synaptophysin-containing vesicles have low amounts of associated p-LIMK1 and p-cofilin suggesting that activation–deactivation of LIMK1 and cofilin may differentially regulate the exit and/or trafficking of post-Golgi vesicles. In accordance with this, overexpression of LIMK1-kd or S3A cofilin increases synaptophysin-GFP labeling in axonal shafts and growth cones, while dramatically reducing Par3/Par6, L1, and βgc at these locations with most of the fluorescent signals found within the Golgi region. These effects are dependent on the Golgi localization of LIMK1 (Rosso et al., 2004).

All these observations are in line with studies in nonneuronal cells suggesting distinct roles for actin in the formation of different classes of Golgi-derived vesicles (Fucini et al., 2000; Heimann et al., 1999; Musch et al., 2001; Rozelle et al., 2000). The recent demonstration of β/γ-actin on vesicles budding from the Golgi by immunoelectron microscopy (Valderrama et al., 2000) and the existence of different actin-binding proteins, including multiple nonmuscle myosins, on distinct populations of Golgi-derived vesicles (Heimann et al., 1999; Rosso et al., 2004) support this idea.

The mechanisms by which LIMK1 participates in post-Golgi trafficking are not known. One possibility is that LIMK1 and cofilin are part of the machinery regulating scission events at the TGN. Proteins, such as dynamin and cortactin, might provide dynamic components to the actin-membrane matrix, thereby producing vesicle scission by severing membrane vesicle necks (Cao et al., 2005). By regulating actin assembly, actin filament length, and/or turnover, LIMK1 and cofilin may contribute to this essential Golgi function. A GTPase-deficient dynamin2-GFP prevents vesiculation and induces the formation of long membrane tubules terminated with a prominent bulb (Orth and McNiven, 2003), resembling those observed after expression of LIMK1-kd or S3A cofilin. Additional studies are now required to further elucidate the role of LIMK1 and cofilin in membrane protrusion, tubulation, and scission at the Golgi apparatus.

6.5.1.3. Cyclin-Dependent Kinase 5

The Cdk5 along with p35, its neuron specific activator, are required for proper brain development, including neuronal migration, axon outgrowth, and synapse formation (Cruz and Tsai, 2004; Dhavan and Tsai, 2001; Paglini and Cáceres, 2001). During axon extension, the Cdk5-p35 kinase becomes enriched in axonal growth cones (Paglini et al., 1998) and at the Golgi apparatus (Paglini et al., 2001a) as well as in post-Golgi transport vesicles trafficking in the anterograde direction (Paglini et al., 2001a; see also Morfini et al., 2004). Within the Golgi complex, active Cdk5-p35 kinase associates with a detergent insoluble fraction containing actin; besides, immunoprecipitation and biochemical experiments showed an interaction among Cdk-p35, Cdc42 and PAK proteins in Golgi membranes, and the Cdk5-p35 dependent phosphorylation of two PAK protein species (Paglini et al., 2001a). These observations are in good agreement with studies showing that Cdk5-p35 kinase forms large macromolecular complexes (Lee et al., 1996), interacts with components of the actin subcortical cytoskeleton (Paglini et al., 1998), and phosphorylates and inactivates PAK1 by associating with GTP-bound Rac (Dhavan and Tsai, 2001; Nikolic et al., 1998). The localization of Cdk5-p35 at the Golgi apparatus appears to be functionally relevant, since antisense suppression of either Cdk5 or p35 or treatment with olomoucine (a Cdk5 kinase inhibitor) in cultured hippocampal neurons result in the disappearance of PPVs from growth cones, a decrease of transport vesicles along axonal shafts paralleled by their accumulation in the Golgi region (Paglini et al., 2001a). This phenotype suggests that this kinase participates in early steps of the secretory pathway, regulating either vesicle formation of budding from the Golgi (Paglini et al., 2001a) and therefore could be required for the proper delivery of PPVs to the growth cone.

The Cdk5-p35 kinase also regulates post-Golgi membrane trafficking by controlling kinesin-mediated transport along neurites and delivery of vesicles and/or PPVs to the growth cone plasma membrane (Morfini et al., 2004; Ratner et al., 1998). Thus, inhibition of Cdk5-p35 kinase activity with pharmacological agents reduced rates of anterograde but not retrograde transport of kinesin-driven motility (Ratner et al., 1998). In subsequent studies, these authors described a pathway linking Cdk5 and protein phosphatase 1 (PP1) to glycogen synthase kinase-3 (GSK3). Inhibition of Cdk5 leads to activation of PP1 and GSK3, resulting in phosphorylation of kinesin light chains (KLCs) with the subsequent release of kinesin from membrane bound organelles (Morfini et al., 2002, 2004). According to this model, Cdk5-p35 activity during vesicle transport along axonal shafts would prevent the early release of transport carriers before reaching delivery sites (e.g., growth cones); local inactivation of Cdk5 followed by activation of GSK3 at growth cones would in turn promote the release of PPVs from the kinesin motor and their delivery to sites of membrane addition.

6.5.1.4. Protein Kinase D1

Protein kinase D1 (PKD1) is a member of a novel family of dyacylglycerol (DAG)-stimulated Ser/Thr kinases. It has been implicated in many cellular functions including the regulation of Golgi organization and plasma membrane directed transport (Lijedahl et al., 2001; Van Lint et al., 2002). PKD binds primarily to the TGN through its first cysteine-rich domain in a DAG-dependent manner (Baron and Malhotra, 2000). Suppression of PKD1 activity inhibits polarized membrane delivery in migrating fibroblasts (Prigozhina and Waterman-Storer, 2004) and a membrane fission pathway specifically involved in the transport of cargo carrying basolateral sorting signals in MDCK cells (Yeaman et al., 2004). Overexpression of PKD kinase-defective mutants also leads to extensive tubulation of the TGN, with cargo-containing vesicles failing to detach from the TGN. Conversely, overexpression of PKD1 over activates the fission reaction leading to Golgi fragmentation, an event inhibited by expression of a kinase-defective mutant of PKD1. Because of these functions and because PKD1 is expressed in the nervous system (Cabrera-Poch et al., 2004), including cultured hippocampal pyramidal neurons (Horton et al., 2005; our unpublished observations), it is likely that this kinase may regulate Golgi function during neuronal development, including polarization. In favor of this, a study has demonstrated that expression of a kinase dead mutant of PKD1 (PKD1-kd) significantly and rapidly inhibits dendritic growth and polarized secretory traffic toward major apical dendrites (Horton et al., 2005); by contrast, axonal outgrowth or elongation was not affected by overexpressing PKD1-kd. As in the case of polarized epithelial cells (e.g., MDCK cells) where PKD1 is selectively involved in the trafficking and sorting of basolateral but not apical membranes proteins, in neurons it may only regulate the sorting and/or delivery of dendritic, but no axonal, membrane proteins. Futures studies should address this and related issues as well as identify the signaling pathways involved in PKD1 activation and the targets of PKD1 action during neuronal development.

6.5.1.5. Rab Proteins

The SNAREs and Rabs are large protein families capable of conferring specificity and directionality to membrane trafficking by ensuring the fidelity of fusion events (Grote and Novick, 1999; Pfeffer, 1996; Rothman and Warren, 1994). In mammalian cells, more than 30 Rab family members have been identified, with each member localizing to a particular compartment of the exocytic or endocytic pathways and regulating a specific transport step (Simons and Zerial, 1993). Several of these proteins, including Rab2, Rab3a, and Rab5, have been identified in neurons (Ayala et al., 1990; de Hoop et al., 1994; Fischer et al., 1991; Krijnse-Locker et al., 1995), and Rab8 directly implicated in the regulation of trafficking between the Golgi apparatus and the plasma membrane in developing neurons. Rab8 is selectively found in dendrites of mature cultured hippocampal neurons while distributing to both axons and dendrites in younger cells (de Hoop et al., 1994). Subsequent work showed that treatment of cultured neurons with antisense oligonucleotides specific for Rab8 inhibits neurite outgrowth and axonal elongation (Huber et al., 1995). Besides, using video-enhanced microscopy they observed an impairment of both anterograde and retrograde transport in Rab8-suppressed neurons as well as an accumulation of Bodipy-ceramide labeled vesicles in and around the Golgi region. Since Rab8 was not found in growth cones, these results were interpreted as indicative of an important role for this protein in vesicle formation, budding from the TGN and/or binding to motor proteins

(Huber et al., 1995). While potentially important, no further studies have addressed the role of other Rab proteins, or the mechanisms involved in the regulation of Golgi functioning during axonal formation. In this regard, it will be of considerable interest to begin analyzing the relationship between Rabs and golgins during neuronal development.

Golgins are a family of coiled-coil proteins that function in a variety of membrane–membrane and membrane–cytoskeleton tethering events at the Golgi apparatus; they are regulated by small GTPases of the Rab and Arl families (Barr and Short, 2003). While the function of golgins in neurons has not been explored in detail, a study showed that overexpression GRASP65 (Barr et al., 1997), a golgin that interacts with GM-130 (also known as golgin-95) inhibits polarized dendritic growth in 7–8 DIV cultured hippocampal neurons (Horton et al., 2005).

6.6. Signaling to the Golgi Apparatus During Axon Formation

The trafficking of PPVs from the Golgi apparatus to sites of membrane addition needs to be coordinately coupled with cytoskeletal assembly during axon outgrowth and subsequent elongation (Dent and Gertler, 2003). Therefore, it is likely that pathways regulating microtubule/actin organization and dynamics at the axonal growth may also affect at the same time Golgi functioning. Unfortunately, little is known about this regulation but a few of the available examples may help illustrate this issue.

In nonneuronal cells, Rho GTPases and the cytoskeleton regulate the distribution of Par proteins, and positive feedback loops linking PI(3,4,5)P3 with Rho-GTPases and/or polymerized actin have been shown to be important for neutrophil polarity during chemotaxis (Wang et al., 2002; Weiner et al., 2002). Similarly, in cultured hippocampal neurons, a signal transduction pathway linking growth factor tyrosine kinase receptors (e.g., BDNF, NGF, or IGF1 receptors), activation of PI 3-kinase, Par proteins, and atypical protein kinase C is crucially involved in axonal specification (Shi et al., 2003) at least in part by activating Rac activity by regulating the guanosine-nucleotide exchange factor, Tiam1/2 (Nishimura et al., 2005; Wiggin et al., 2005). Activation of Rac may lead to PAK1/4 mediated phosphorylation of LIMK1 at the Golgi apparatus and polarized trafficking of Par3/Par6 and/or βgc and/or TrkB containing vesicles to the growth cone of the nascent axon (Rosso et al., 2004). This could set a positive feedback loop (Shi et al., 2003) involving two different cellular locations and linking Golgi-derived membrane trafficking with growth cone cytoskeletal assembly and membrane addition. Transport of PPVs from the TGN to the axonal growth cones in developing neurons seems to be a regulated phenomenon, as suggested by the faster transport of βgc containing IGF1 receptors to the axonal growth cone of cultured hippocampal neurons after stimulation with brain derived neurotrophic factor (BDNF) or NGF in PC12 cells (Mascotti et al., 1997; Pfenninger et al., 2003).

6.7. Future Perspectives

One of the major activities of developing neurons is the transport of new membrane from the Golgi apparatus to the growing axon. The information reviewed in this article clearly indicates that the Golgi apparatus has an active and important role in axon specification, outgrowth, and elongation, still the study of its function during neuronal polarization is in its infancy. Much remains to be learned. First of all, precise knowledge on the identity and function of the players involved in regulating Golgi function during axon and/or dendritic formation is highly required. Second, signaling pathways, as well as feedback loops, involved in coordinating events at the growth cone and the Golgi apparatus should be identified; along this line, the concept of constitutive versus regulated exocytosis, especially during polarized growth should be revisited. Finally, mechanisms controlling fission and sorting events at the neuronal Golgi apparatus should be elucidated.

Acknowledgments. The work of the authors has been supported by Fogarty International Research Collaboration Awards (FIRCA) to AF and SQ and to Karl Pfenninger and SQ, a grant from the National Scientific Agency form Argentina awarded to AC, SQ, and GP, and a grant from the Howard Hughes Medical Institute (HHMI 75197–553201) awarded under the International Research Scholar Program to AC.

References

Abo, A., Qu, J., Cammarano, M., Dan, C., Fritsch, A., Baud, A., et al., 1998, PAK4, a novel effector for Cdc42Hs, is implicated in the reorganization of the actin cytoskeleton and in the formation of filopodia, *EMBO J.* **22:** 6527–6540.

Ayala, J., Touchot, N., Zahraoui, A., Tavitian, A., and Prochiantz, A., 1990, The product of rab2, a small GTP binding protein, increases neuronal adhesion, and neurite growth in vitro, *Neuron* **4:** 797–805.

Baron, C., and Malhotra, V., 2000, Role of dyacylglycerol in PKD recruitment to the TGN and protein transport to the plasma membrane, *Science* **295:** 325–328.

Barr, F.A., and Short, B., 2003, Golgins in the structure and dynamics of the Golgi apparatus, *Curr. Opin. Cell Biol.* **15:** 405–413.

Barr, F., Puype. M., Vandekerckhove, J., and Warren, G., 1997, GRASP65, a protein involved in the stacking of Golgi cisternae, *Cell* **91:** 253–262.

Bradke, F., and Dotti, C.G., 1997, Neuronal polarity: Vectorial cytoplasmic flow precedes axon formation, *Neuron* **19:** 1175–1186.

Cabrera-Poch, N., Sanchez-Ruiloba, L., Rodriguez-Martinez, M., and Iglesias, T., 2004, Lipid raft disruption triggers protein kinase C and Src-dependent protein kinase D activation and Kidins220 phosphorylation in neuronal cells, *J. Biol. Chem.* **279:** 28592–28602.

Camera, P., Santos Da Silva, J., Griffiths, G., Giuffrida, M., Ferrara, L., Schubert, V., et al., 2003, Citron-N is a neuronal Rho-associated protein involved in Golgi organization through actin cytoskeletal organization, *Nat. Cell Biol.* **5:** 1071–1078.

Cao, H., Weller, S., Orth, J.D., Chen, J., Huang, B., Chen, J.L., et al., 2005, Actin and Arf1-dependent recruitment of a cortactin-dynamin complex to the Golgi regulates post-Golgi transport, *Nat. Cell Biol.* **7:** 483–492.

Cau, J., and Hall, A., 2005, Cdc42 controls the polarity of the actin and microtubule cytoskeletons through two distinct signal transduction pathways, *J. Cell Sci.* **118:** 2579–2587.

Cerione, R., 2004, Cdc42: new roads to travel, *Trends Cell Biol.* **14:** 127–132.

Chen, J., Lacomis, L., Erdjument-Bromage, H., Tempst, P., and Stamnes, M., 2004, Cytosol-derived proteins are sufficient for Arp2/3 recruitment and ARF/coatomer-dependent actin polymerization on Golgi membranes, *FEBS Lett.* **566:** 281–286.

Chen, J., Fucini, R., Lacomis, L., Erdjument-Bromage, H., Tempst, P., and Stamnes, M., 2005, Coatomer-bound Cdc42 regulates dynein recruitment to COPI vesicles, *J. Cell Biol.* **169:** 383–389.

Chieregatti, E., and Meldolesi, J., 2005, Regulated exocytosis: new organelles for non-secretory purposes, *Nat. Rev. Mol. Cell Biol.* **6:** 181–187.

Chuang, J., Yen, T., Bollati, F., Conde, C., Canavosio, F., Cáceres, A., et al., 2005, The dynein light chain Tctex-1 has a dynein-independent role in actin remodeling during neurite outgrowth, *Dev. Cell* **9:** 75–86.

Cohen, D., Musch, A., and Rodriguez-Boulan, E., 2001, Selective control of basolateral membrane protein polarity by Cdc42, *Traffic* **2:** 556–564.

Craig, A., and Banker, G., 1994, Neuronal polarity, *Ann. Rev. Neurosci.* **17:** 267–310.

Craig, A., Wyborski, R, and Banker, G., 1995, Preferential addition of newly synthesized membrane protein at axonal growth cones, *Nature* **375:** 592–594.

Cruz, J., Tsai, L.-H., 2004, A Jekyll and Hyde kinase: Roles for Cdk5 in brain development and disease, *Curr. Opin. Neurobiol.* **14:** 390–394.

Dan, C., Kelly, A., Bernard, O., and Minden, A., 2001, Cytoskeletal changes regulated by the PAK4 serine-threonine kinase are mediated by LIMK1 and cofilin, *J. Biol. Chem.* **276:** 32115–32121.

de Anda, F., Pollarolo, G., Da Silva, J., Camoletto, P., Feiguin, F., and Dotti, C., 2005, Centrosome localization determines neuronal polarity, *Nature* **436:** 704–708.

Dent, E., and Gertler, F., 2003, Cytoskeletal dynamics and transport in growth cone motility and axon guidance, *Neuron* **40:** 209–227.

De Hoop, M.J., Huber, L., Stenmark, H., Williamson, E., Zerial, M., Parton, R., et al., 1994, Rab5 involvement in axonal and dendritic endocytosis, *Neuron* **13:** 11–22.

Dhavan, R., and Tsai, L-H., 2001, A decade of Cdk5, *Nature Rev.* **2:** 749–759.

Dotti, C.G., and Banker, G., 1991, Intracellular organization of hippocampal neurons during the development of neuronal polarity, *J. Cell Sci. Suppl.* **15:** 75–84.

Dubois, T., Paleotti, O., Mironov, A., Fraisier, V., Stradal, T., De Matteis, M., et al., 2005, Golgi-localized GAP for Cdc42 functions downstream of ARF1 to control Arp2/3 complex and F-actin dynamics, *Nat. Cell Biol.* **7:** 353–364.

Duran, J., Valderrama, F., Castel, S., Magdalena, J., Tomas, M., Hosoya, H., et al., 2003, Myosin motors and not actin comets are mediators of the actin-based Golgi-to-endoplasmic reticulum protein transport, *Mol. Biol. Cell.* **14:** 445–459.

Etienne-Manneville, S., and Hall, A., 2001, Integrin-mediated activation of Cdc42 controls cell polarity in migrating astrocytes through PKCzeta, *Cell* **106:** 489–498.

Ferreira, A., Niclas, J., Vale, R., Banker, G., and Kosik, K., 1992, Suppression of kinesin expression in cultured hippocampal neurons using antisense oligonucleotides, *J. Cell Biol.* **117:** 595–606.

Ferreira, A., Cáceres, A., and Kosik, K., 1993, Intraneuronal compartments of the amyloid precursor protein, *J. Neurosci.* **13:** 3112–3123.

Fischer von Mollard, G., Sudhof, T., and Jahn, R., 1991, A small GTP-binding protein dissociates from synaptic vesicles during exocytosis, *Nature* **349:** 79–81.

Fucini, R., Navarrete, A., Vadakkan, C., Lacomis, L., Erdjument-Bromage, H., Tempts, P., et al., 2000, Activated ADP-ribosylation factor assembles distinct pools of actin in Golgi membranes, *J. Biol. Chem.* **275:** 18824–18829.

Fucini, R., Chen, J., Sharma, C., Kessels, M., and Stamnes, M., 2002, Golgi vesicle proteins are linked to the assembly of an actin complex defined by mAbp1, *Mol. Biol. Cell* **13:** 621–631.

Fukata, Y., Itoh, T., Kimura, T., Menager, C., Nishimura, T., Shiromizu, T., et al., 2002, CRMP2 binds to tubulin heterodimers to promote microtubule assembly, *Nat. Cell Biol.* **4:** 583–591.

Futerman, A., and Banker, G., 1996, The economics of neurite outgrowth-the addition of new membrane to growing axons, *Trends Neurosci.* **19:** 144–149.

Goda, Y., and Davis, G., 2003, Mechanisms of synapse assembly and disassembly, *Neuron* **40:** 243–264.

Gomes, E.R., Jani S., Gundersen G.G., 2005, Nuclear movement regulated by Cdc42, MRCK, myosin, and actin flow establishes MTOC polarization in migrating cells, *Cell* **121:** 451–463.

Goslin, K., and Banker, G., 1990, Rapid changes in the distribution of GAP-43 correlate with the expression of neuronal polarity during normal development and under experimental conditions, *J. Cell Biol.* **110:** 1319–1331.

Gotlieb, A.I., May, L.M., Subrahmanyan, L., and Kalnins, V.I., 1981, Distribution of microtubule organizing centers in migrating sheets of endothelial cells, *J. Cell Biol.* **91:** 589–594.

Govek, E., Newey, S., and VanAelst, L., 2005, Role of the GTPases in neuronal development, *Gene Dev.* **19:** 1–49.

Gregory, W.A., Edmondson, J.C., Hatten, M.E., and Mason, C.A., 1988, Cytology and neuron-glial apposition of migrating cerebellar granule cells in vitro, *J. Neurosci.* **8:** 1728–1738.

Grote, E., and Novick, P., 1999, Promiscuity in Rab-SNARE interactions, *Mol. Biol. Cell.* **10:** 4149–4161.

Gundersen, G.G., and Bulinski, J.C., 1988, Selective stabilization of microtubules oriented toward the direction of cell migration, *Proc. Natl. Acad. Sci. USA* **85:** 5946–5950.

Heimann, K., Percival, J., Weiberger, R., Gunning, P., and Stow, J., 1999, Specific isoforms of actin-binding protein on distinct populations of Golgi-derived vesicles, *J. Biol. Chem.* **274:** 10743–10750.

Hirokawa, N., and Takemura, R., 2005, Molecular motors and mechanisms of directional transport in neurons, *Nat. Rev. Neurosci.* **6:** 201–214.

Horton, A., and Ehlers, M., 2003a, Dual modes of endoplasmic reticulum-to-Golgi transport in dendrites revealed by live-cell imaging, *J. Neurosci.* **23:** 6188–6199.

Horton A., and Ehlers, M., 2003b, Neuronal polarity and trafficking, *Neuron* **40:** 277–295.

Horton, A, and Ehlers, M., 2004, Secretory trafficking in neuronal dendrites, *Nat. Cell Biol.* **6:** 585–591.

Horton, A., Racz, B., Monson, E., Lin, A., Weinberg, R., and Ehlers, M., 2005, Polarized secretory trafficking directs cargo for asymmetric dendrite growth and morphogenesis, *Neuron* **48:** 757–771.

Huber, L., Dupree, P., and Dotti, C., 1995, A deficiency of the small GTPase rab8 inhibits membrane traffic in developing neurons, *Mol. Cell. Biol.* **15:** 918–924.

Itoh, K., Cheng, L., Kamei, Y., Fushiki, S., Kamiguchi, H., Gutwein, P., et al., 2004, Brain development in mice lacking L1-L1 homophilic adhesion, *J. Cell Biol.* **165:** 145–154.

Jan, J., and Jan, L., 2003, The control of dendritic development, *Neuron* **40:** 229–242.

Jareb, M., and Banker, G., 1997, Inhibition of axonal growth by brefeldin A in hippocampal neurons in culture, *J. Neurosci.* **17:** 8955–8963.

Kaether, C., Skehel, P., and Dotti, C., 2000, Axonal membrane proteins are transported in distinct carriers: A two-color video microscopy study in cultured hippocampal neurons, *Mol. Biol. Cell* **11:** 1213–1224.

Kamal, A., Stokin, G., Yang, Z., Xia, C., and Goldstein, L.S., 2000, Axonal transport of amyloid precursor protein is mediated by direct binding to the kinesin light chain subunit of kinesin-I, *Neuron* **28:** 449–459.

Kamiguchi, H., Hlavin, M., Yamasaki, M., and Lemmon, V., 1998, Adhesion molecules and inherited diseases of the human nervous system, *Annu. Rev. Neurosci.* **21:** 97–125.

Kesavapany, S., Lau, K., Ackerley, S., Banner, J., Shemilt, J., Cooper, J., et al., 2003, Identification of a novel, membrane-associated neuronal kinase, cyclin-dependent kinase 5/p35-regulated kinase, *J. Neurosci.* **23:** 4975–4983.

Krijnse-Locker, J., Parton, R., Fuller, S., Griffiths, G., and Dotti, C., 1995, The organization of the endoplasmic reticulum and the intermediate compartment in cultured rat hippocampal neurons, *Mol. Biol. Cell.* **6:** 1315–1332.

Kroschewski, R., Hall, A., and Mellman, I., 1999, Cdc42 controls secretion and endocytic transport to the basolateral plasma membrane of MDCK cells, *Nat. Cell Biol.* **1:** 8–13.

Kupfer, A., Louvard, D., and Singer, S.J., 1982, Polarization of the Golgi apparatus and the microtubule-organizing center in cultured fibroblasts at the edge of an experimental wound, *Proc. Natl. Acad. Sci. USA* **79:** 2603–2607.

Kupfer, A., Dennert, G., and Singer, S.J., 1983, Polarization of the Golgi apparatus and the microtubule-organizing center within cloned natural killer cells bound to their targets, *Proc. Natl. Acad. Sci. USA* **80:** 7224–7228.

Kunda, P., Paglini, G., Kosik, K., Quiroga, S., and Cáceres, A., 2001, Evidence for the involvement of Tiam-1 in axon formation, *J. Neurosci.* **21:** 2361–2372.

Laurino, L., Xiaoxin X., de la Houssaye B., Sosa L., Dupras S., Cáceres A., et al., 2005, PI3K activation by IGF-1 is essential for the regulation of membrane expansion at the nerve growth cone, *J. Cell Sci.* **118:** 3653–3662.

Lee, K., Hrosales, J., Tang, D., and Wang, J., 1996, Interaction of cyclin-dependent kinase 5 (Cdk5) and neuronal Cdk5 activator in bovine brain, *J. Biol. Chem.* **271:** 423–426.

Lijedahl, M., Maeda, Y., Colanzi, A., Ayala, I., Van Lint, J., and Malhotra, V., 2001, Protein kinase D regulates the fission of cell surface destined transport carriers from the trans-Golgi network, *Cell* **104:** 409–420.

Lowenstein, P., Morrison, E., Bain, D., Shering, A., Banting, G., Douglas, P., et al., 1994, Polarized distribution of the trans-Golgi network marker TGN38 during the in vitro development of neocortical neurons: Effects of nocodazole and brefeldin A, *Eur. J. Neurosci.* **6:** 1453–65.

Luna, A., Matas, O., Martinez-Menarguez, J., Mato, E., Duran, J., Ballesta, J., et al., 2002, Regulation of protein transport from the Golgi complex to the endoplasmic reticulum by CDC42 and N-WASP, *Mol. Biol. Cell.* **13:** 866–879.

Mascotti, F., Cáceres, A., Pfenninger, K., and Quiroga, S., 1997, Expression and distribution of IGF-1 receptors containing a beta-subunit variant (betagc) in developing neurons, *J. Neurosci.* **17:** 1447–1459.

Matas, O., Martinez-Menarguez, J., and Egea, G., 2004, Association of Cdc42/N-WASP/Arp2/3 signaling pathway with Golgi membranes, *Traffic* **5:** 838–46.

Morfini, G., Rosa, A., Quiroga, S., Kosik, K., and Cáceres, A., 1997, Suppression of KIF2 alters the distribution of a growth cone non-synaptic membrane receptor and inhibits neurite outgrowth, *J. Cell Biol.* **138:** 657–669.

Morfini, G., Szebenyi, G., Elluru, R., Ratner, N., and Brady, S.T., 2002, Glycogen synthase kinase-3 phosphorylates kinesin light chains and negatively regulates kinesin-based motility, *EMBO J.* **23:** 281–293.

Morfini, G., Szebenyi, G., Brown, H., Pant, H., Pigino, G., DeBoer, S., et al., 2004, A novel CDK5-dependent pathway for regulating GSK3 activity and kinesin-driven motility in neurons, *EMBO J.* **23:** 2235–2245.

Musch, A., Cohen, D., Kreitzer, G., and Rodriguez-Boulan, E., 2001, Cdc42 regulates the exit of apical and basolateral proteins from the trans-Golgi network, *EMBO J.* **20:** 2171–2179.

Orth, J., and McNiven, M., 2003, Dynamin at the actin-membrane interface, *Curr. Opin. Cell Biol.* **15:** 31–39.

Nikolic, M., Chou, M., Lu, W., Mayer, B., and Tsai, L-H., 1998, The p35/Cdk5 kinase is a neuron specific Rac effector and inhibits PAK1 activity, *Nature* **395:** 194–198.

Nishimura, T., Yamaguchi, T., Katsukiro, K., Yoshizawa, M., Nabeshima, Y., Ohno, S., et al., 2005, Par6-Par3 mediates Cdc42-induced Rac activation through the Rac GEF STEF/Tiam1, *Nat. Cell Biol.* **7:** 270–277.

Paglini, G., and Cáceres, A., 2001, The role of cdk5-p35 kinase in neuronal development, *Eur. J. Biochem.* **268:** 1528–1533.

Paglini, G., Pigino, G., Morfini, G., Kunda, P., Maccioni, R., Quiroga, S., et al., 1998, Evidence for the participation of the neuron-specific activator p35 during laminin-enhanced axonal growth, *J. Neurosci.* **18:** 9858–9869.

Paglini, G., Peris, L., Diez-Guerra, J., Quiroga, S., and Cáceres, A., 2001a, The Cdk5-p35 kinase associates with the Golgi apparatus and regulates membrane traffic, *EMBO Rep.* **2:** 1139–1144.

Palazzo, A.F., Joseph, H.L., Chen, Y.J., Dujardin, D.L., Alberts, A.S., Pfister, K.K., et al., 2001, Cdc42, dynein, and dynactin regulate MTOC reorientation independent of Rho-regulated microtubule stabilization, *Curr. Biol.* **11:** 1536–1541.

Pfeffer, S., 1996, Transport vesicle docking: SNAREs and associates, *Annu. Rev. Cell Dev. Biol.* **12:** 441–461.

Pfenninger, K.H., and Friedman, L.B., 1993, Sites of plasmalemmal expansion in growth cones, *Dev. Brain Res.* **71:** 181–192.

Pfenninger, K.H., Laurino, L., Peretti, D., Wang, X., Rosso, S., Morfini, G., et al., 2003, Regulation of membrane expansion at the nerve growth cone, *J. Cell Sci.* **16:** 1209–1217.

Peretti, D., Peris, L., Rosso, S., Quiroga, S., and Cáceres, A., 2000, Evidence for the involvement of KIF4 in the anterograde transport of L1-containing vesicles, *J. Cell Biol.* **149:** 141–152.

Prigozhina, N., and Waterman-Storer, C., 2004, Protein kinase D-mediated anterograde membrane trafficking is required for fibroblast motility, *Curr. Biol.* **14:** 88–98.

Qu, J., Li, X., Novitch, B., Zheng, Y., Kohn, M., Xie, J., et al., 2003, PAK4 kinase is essential for embryonic viability and for proper neuronal development, *Mol. Cell. Biol.* **20:** 7122–7133.

Quiroga, S., Garofalo, R., and Pfenninger, K., 1995, Insulin-like growth factor I receptors of fetal brain are enriched in nerve growth cones and contain a beta-subunit variant, *Proc. Natl. Acad. Sci. USA* **92:** 4309–4312.

Ratner, N., Bloom, G., and Brady, S.T., 1998, A role for Cdk5 kinase in fast anterograde axonal transport: novel effects of olomoucine and the APC tumor suppressor protein, *J. Neurosci.* **18:** 7717–7726.

Ridley, A., 2001, Rho proteins: Linking signaling with membrane trafficking, *Traffic* **2:** 303–310.

Rosso, S., Bollati, F., Bisbal, M., Peretti, D., Sumi, T., Nakamura, T., et al., 2004, LIMK1 regulates Golgi dynamics, traffic of Golgi-derived vesicles, and process extension in primary cultured neurons, *Mol. Biol. Cell* **15:** 3433–3449.

Rothman, J.E., and Warren, G., 1994, Implications of the SNARE hypothesis for intracellular membrane topology and dynamics, *Curr. Biol.* **4:** 220–233.

Rozelle, A.L., Machesky, L., Yamamoto, Y., Driessens, M., Insall, R., Roth, M., et al., 2000, Phosphatidylinositol 4, 5 bisphosphate induces actin-based movements of raft-enriched vesicles through WASP-Arp2/3, *Curr. Biol.* **10:** 311–320.

Sarmiere, P., and Bamburg, J., 2004, Regulation of the neuronal actin cytoskeleton by ADF/cofilin, *J. Neurobiol.* **58:** 103–117.

Shi, S., Jan, L.Y., Jan, N.Y., 2003, Hippocampal neuronal polarity specified by spatially localized mPar3.mPar6 and PI30kinase activity, *Cell* **112:** 63–75.

Silverman, M., Kaech, S., Jareb, M., Burack, M., Vogt, L., Sonderegger, P., et al., 2001, Sorting and directed transport of membrane proteins during development of hippocampal neurons in culture, *Proc. Natl. Acad. Sci. USA* **98:** 7051–7057.

Simons, K., and Zerial, M., 1993, Rab proteins and the road maps for intracellular transport, *Neuron* **11:** 789–799.

Stamnes, M., 2002, Regulating the actin cytoskeleton during vesicular transport, *Curr. Opin. Cell Biol.* **14:** 428–433.

Takeda, S., Yamazaki, H., Seog, D., Kanai, Y., Terada, S., and Hirokawa, N., 2000, Kinesin superfamily protein 3 (KIF3) motor transports fodrin-associating vesicles important for neurite building, *J. Cell Biol.* **148:** 1255–1265.

Tang, B., 2001, Protein trafficking mechanisms associated with neurite outgrowth and polarized sorting in neurons, *J. Neurochem.* **79:** 923–930.

Valderrama, F., Luna, A., Babia, T., Martinez-Menarguez, J., Ballesta, J., Barth, H., et al., 2000, The Golgi-associated COPI-coated buds and vesicles contain beta/gamma –actin, *Proc. Natl. Acad. Sci. USA* **97:** 1560–1565.

Valderrama, F., Duran, J., Babia, T., Barth, H., Renau-Piqueras, J., and Egea, G., 2001, Actin microfilaments facilitate the retrograde transport from the Golgi complex to the endoplasmic reticulum in mammalian cells, *Traffic* **2:** 717–726.

Van Lint, J., Rykx, A., Maeda, Y., Vantus, T., Sturany, S., Malhotra, V., et al., 2002, Protein kinase D: an intracellular traffic regulator on the move, *Trends Cell Biol.* **12:** 193–200.

Wang, F., Herzmark, P., Weiner, O.D., Srinivasan, S., Servant, G., and Bourne, H.R., 2002, Lipids products of PI(3)Ks maintain persistent cell polarity and directed motility in neutrophils, *Nat. Cell Biol.* **4:** 509–512.

Weiner, O.D., Neilsen, P., Prestwich, G., Kirschner, M., Cantley, L., and Bourne, H.R., 2002, A PtdInsP3- and Rho GTPase-mediated positive feedback loop regulates neutrophil polarity, *Nat. Cell Biol.* **4:** 509–512.

Wieland, F., Gleason, M., Serafini, T., and Rothman, J., 1987, The rate of bulk flow from the endoplasmic reticulum to the cell surface. *Cell* **50:** 289–300.

Wiggin, R., Fawcett, J., and Pawson, T., 2005, Polarity proteins in axon specification and synaptogenesis, *Dev. Cell* **8:** 803–816.

Yeaman, C., Ayala, I., Wright, J., Bard, F., Bossard, C., Ang, A., et al., 2004, Protein kinase D regulates basolateral membrane protein exit from the trans-Golgi network, *Nat. Cell Biol.* **6:** 106–112.

Young, Jr., W., Lutz, M., and Blackburn, W., 1992, Endogenous glycosphingolipids move to the cell surface at a rate consistent with bulk flow estimates, *J. Biol. Chem.* **267:** 12011–12015.

Zmuda, J., and Rivas, R., 1998, The Golgi apparatus and the centrosome are localized to the sites of newly emerging axons in cerebellar granule neurons in vitro, *Cell Motil. Cytosk.* **41:** 18–38.

7
Focal Adhesion Kinase in Neuritogenesis

DARIO BONANOMI AND FLAVIA VALTORTA

7.1. Summary

The transduction of extracellular signals through adhesion and guidance receptors into cytoskeleton reorganization underlies axon outgrowth and pathfinding decisions. The cytoplasmic protein tyrosine kinase *focal adhesion kinase* (FAK), a central regulator of cell motility phenomena during morphogenesis, is highly expressed in the developing nervous system and selectively enriched in axonal growth cones. Novel studies implicate FAK in the control of neuronal migration and axon behavior during nervous system development. By scaffolding receptor-proximal multiprotein complexes, FAK guarantees the integration of adhesive signals derived from the extracellular matrix with guidance cues and growth factors at the level of the developing axons.

7.2. Cell Adhesion Devices: General Considerations

The considerable expansion of the number of genes encoding for adhesion and growth factor receptors during evolution underlies the need for a tight regulation of the interactions between cells and their environment to allow the development and maintenance of the high complexity of vertebrates, in particular for the elaboration of sophisticated structures such as the nervous system (Hynes, 1999). Cell adhesion receptors interact with either other receptors expressed on the surface of neighboring cells or ligands associated with the extracellular matrix and impinge on intracellular organization and cell motility through the connection of their cytoplasmic domains with the cytoskeleton.

Since integrins, the major adhesion receptors for the extracellular matrix, have short cytoplasmic domains and do not possess enzymatic activity, their signaling mechanisms depend on their ability of scaffolding large adhesion/signaling complexes linked to the actin cytoskeleton termed *focal contacts* (Figure 7.2A and B). In addition to integrins, focal contacts also contain cytoskeletal actin-binding proteins, such as vinculin, α-actinin, and talin, low levels of nonenzymatic proteins with regulative functions, such as paxillin and tensin, as well as several enzymes which associate with the actin cytoskeleton during cell adhesion, including FAK, protein kinase C, Src tyrosine kinases, and Rho GTPases (Yamada and Geiger, 1997).

Cytoskeletal molecules of adhesion complexes act as sites for the localization and recruitment of the components of a number of signaling cascades. Indeed, the sequential recruitment of molecular components is a key process for the formation and function of adhesion complexes. The activation of receptors at specific sites on the plasma membrane triggers a cascade of molecular events leading to the phosphorylation of neighboring molecules, which in turn fosters the gathering of new components at adhesion sites. Once assembled, these adhesion/transduction complexes activate several cytoplasmic signaling pathways regulating a variety of cell processes such as survival, proliferation, differentiation, and migration. Signal transduction downstream of adhesion complexes is further enhanced by multifunctional enzymes, such as FAK, whose binding to a number of signaling molecules and cytoskeletal proteins is modulated by the phosphorylation state of the protein itself and by its kinase activity.

The large focal contacts typical of fibroblasts are absent in neuronal cells, which *in vitro* display smaller adhesion structures with distinct composition named *point contacts*. Point contacts are somewhat similar to podosomes of highly motile transformed cells (Arregui et al, 1994; Nermut et al., 1991) and are present together with focal contacts in migrating astrocytes (Tawil et al., 1993). Therefore, focal contacts and point contacts may coexist in the same cell, being involved in specific and perhaps overlapping processes. In glial cells, the early, rapid stages of cell spreading appear to be mediated by point contacts, whereas the further stabilization of cell adhesion requires the formation of focal contacts (Tawil et al., 1993). At variance, point contacts are the only adhesive structures described in growth cones, where they appear as small β1 integrin-containing adhesion sites localized at the cell-substratum interface, defined by the presence of cytoskeleton-associated molecules, such as vinculin, talin, and paxillin, and signaling molecules including FAK and Rho-family GTPases (Renaudin et al., 1999).

It is tempting to speculate that at the level of the neuronal growth cone the presence of small structures with rapid turnover instead of the large and more stable focal contacts is related to the high mobility of this compartment during neurite outgrowth (see in a later section). Moreover, the association of signaling molecules with point contacts suggests that they may contribute to the regulation of integrin-mediated adhesion in motility and pathfinding.

7.3. Molecular Principles of Growth Cone Motility

Adhesion receptors create a crucial connection between the extracellular matrix and the cytoskeleton, generating the traction forces required for cell motility (Hynes, 1992). Cell adhesion to the substrate requires a tight regulation in order for cell migration to occur. Exceedingly strong adhesion does not allow the cell to move, while weak adhesion will not produce enough strength to pull the cell forward (Schwarzbauer, 1997). Cells can sense changes in substrate stiffness and modulate accordingly the strength of the connection between extracellular matrix, integrins, and cytoskeleton. The strengthening of adhesive connections depends on integrin clustering and subsequent recruitment of actin-associated proteins as well as tyrosine phosphorylation events (Chouquet et al., 1997). The important contribution of tyrosine kinases to the regulation of the interaction between integrins and the cytoskeleton during cell motility is confirmed by the impaired migration of Src-null cells, which correlates with the strengthening of adhesive connections in response to substrate stiffness (Felsenfeld et al., 1999).

Similarly to the migration of nonneuronal cells, neurite extension requires extracellular stimuli coming from the matrix or target cells to be translated into a substantial reorganization of the cytoskeleton. These changes mainly occur in the growth cone, the specialized compartment at the tip of neurites, which integrates positive and negative signals from the extracellular space. The elaboration of these signals leads to rapid rearrangements of both the actin cytoskeleton and microtubules within the growth cone, accounting for the changes in morphology and directionality required during neurite outgrowth (Tanaka and Sabry, 1999).

In neuronal growth cones, microtubules are organized in bundles restricted to the central domain, while F-actin creates a meshwork in the lamellar distal domain and enters filopodial structures which originate at the periphery. Growth cone dynamics largely depend on actomyosin contractility within the distal domain and appear to be characterized by three main processes with distinct kinetics: (i) net F-actin assembly in the peripheral area; (ii) constant retrograde flow of the F-actin meshwork driven by myosin motors associated with microfilaments and counterbalanced by dynein connecting the microfilament and microtubule systems; and (iii) F-actin disassembly in the proximal area and its recycling in the transition zone, where F-actin and microtubules partially overlap (Ahmad et al., 2000; Lin et al.,1994; Suter and Forscher, 1998). Similar processes are observed to occur in lamellipodia of numerous migratory cells, and their modulation affects the migration rate (Mitchison and Cramer, 1996).

In order for neurite outgrowth to occur, microtubules have to extend into the peripheral, actin-rich domain of the growth cone, passing through a corridor in which the F-actin flow is attenuated (Lin and Forscher, 1995). As the growth cone moves on an attractive substrate, adhesion receptors stabilize the actin cytoskeleton, thus reducing the F-actin flow. This generates tension between microfilaments and microtubules, which allows the latter to extend into the distal domain, thus moving the growth cone forward (Suter and Forscher, 1998).

A growing list of molecules operate in the formation of the molecular clutch involved in substrate-cytoskeleton coupling during growth cone migration (Valtorta and Leoni, 1999). Extracellular matrix components produce variable effects on the rate and directionality of neurite extension (Kuhn et al., 1995). Growth cone attraction by netrin-1, a soluble laminin homologue, is converted into repulsion through the engagement of $\alpha6\beta1$ integrin by laminin-1 (Hopker et al., 1999). Vinculin depletion in PC12 cells drammatically impairs neurite extension due to filopodia and lamellipodia destabilization (Varnum-Finney and Reichardt, 1994).

Cell adhesion receptors of the immunoglobuline superfamily (IgCAM) mediate, through both homophilic and heterophilic interactions, several aspects of neuronal differentiation such as neurite outgrowth, axonal fasciculation, and guidance. apCAM, the *Aplysia* homologue of vertebrate neural cell adhesion molecule (NCAM), has been reported to control the functional substrate-cytoskeleton coupling through the recruitment of Src-family tyrosine kinases, resulting in attenuation of F-actin flow and generation of the traction force required for growth cone movement (Suter and Forscher, 2001; Suter et al., 1998). NCAM and L1, another member of the IgCAM superfamily, are enriched in neuronal projections and growth cones where they promote neurite extension through the regulation of adhesion and signaling events mediated by the tyrosine kinases c-Fyn and Src, respectively (Beggs et al., 1994; Doherty et al., 1990; Ignelzi et al., 1994). Tyrosine phosphorylation of cytoskeleton-associated proteins is a common mechanism for the release of adhesive interactions both in cell migration and neurite extension (Crowley and Horwitz, 1995; Wu and Goldberg, 1993). Thus, signaling cascades modulating tyrosine phosphorylation levels are crucial for the control of the stability of adhesion structures during cell motility.

Accumulating evidences indicate that the members of the Rho GTPase family play a crucial role in the regulation of neurite extension and growth cone motility by coupling the stimulation of membrane receptors by extracellular ligands to actin cytoskeleton dynamics within the growth cone (Mueller, 1999). Involvement of the Rho GTPases Rac, Cdc42, and Rho in cell motility and cytoskeleton remodeling has been largely investigated in nonneuronal cells. In fibroblasts, RhoA induces the formation of actin stess fibers and the stabilization of cell adhesion during migration. Rac is essential for lamellipodia protrusion and cell progression, whereas Cdc42 is implicated in filopodia extension and cell polarization. In addition, all these proteins are required for focal adhesion assembly (Hall, 1998; Nobes and Hall, 1999).

7.4. Domains, Activation and Signaling

7.4.1. Structural Organization

The tyrosine kinase FAK was originally identified as a 125 kDa protein highly enriched at focal adhesions and undergoing prominent tyrosine phosphorylation on either fibroblast transformation by the viral oncogene pp60v-Src or interaction with the extracellular matrix (Hanks et al., 1992; Schaller et al., 1992). FAK homologues with high sequence similarity have been identified in humans, mice, chicken, and Xenopus.

In both normal and transformed cells, FAK signaling can promote cell motility (Mitra et al., 2005) and the formation of podosomes and invadopodia, which lead to cell invasiveness (Hauck et al., 2002; Hsia et al., 2003). Consistently, FAK expression is increased in many malignant human cancers (Cance et al., 2000).

Although FAK contains sequences common to all nonreceptor tyrosine kinases, its structural organization is unique (Figure 7.1A). Its kinase domain is flanked by two large amino (N)-terminal and carboxy (C)-terminal domains, which do not have homologues in other tyrosine kinases. Moreover, FAK does not possess acylation or myristoylation sites for the association with the plasma membrane, nor SH2 and SH3 (Src homology 2–3) domains for protein–protein

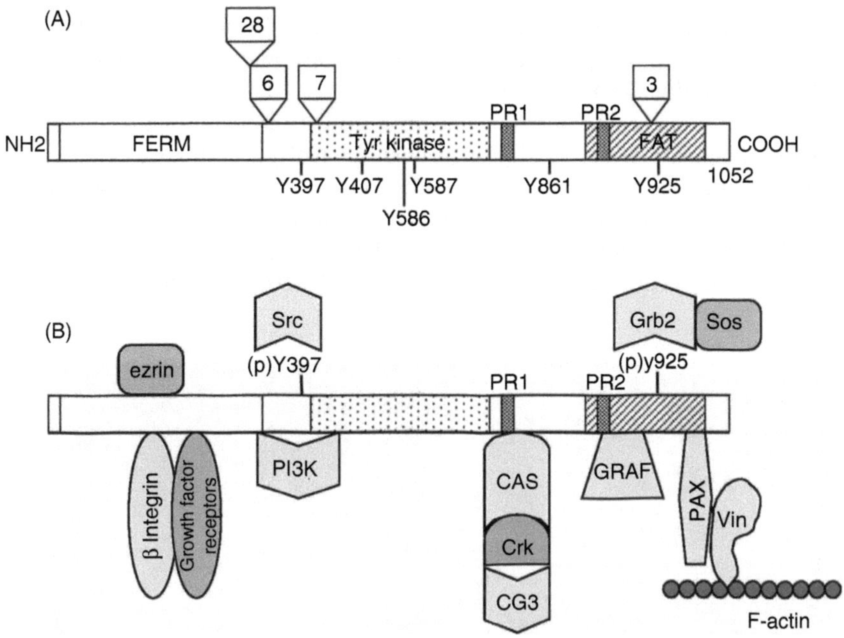

FIGURE 7.1. Structure and molecular interactions of FAK. (A) The main bar represents FAK peptide sequence without the alternative spliced exons, which are indicated by the smaller rectangles, in which the corresponding number of amino acids inserted by each exon is reported. FAK autophosphorylation site (Y397) and tyrosine residues phosphorylated by Src-family kinases are indicated. FERM, protein 4.1/ezrin/radixin/moesin domain; PR1 and PR2 = C-terminal proline-rich sequences; FAT = focal adhesion targeting sequence; Tyr kinase = FAK catalitic domain. (B) FAK associates with a number of cytoskeletal and signaling molecules fostering the assembly of adhesion/signaling macromolecular complexes. Src kinases bind autophosphorylated Y397 and phosphorylate multiple tyrosines of FAK, including (p)Y925, which is required for the interaction with Grb2/Sos and activation of Ras-dependent MAPK cascade. p130Cas SH3 domain binds the first FAK proline-rich sequence and recruits the molecular adapter Crk. The interaction with paxillin indirectly connects FAK to F-actin. The PI3K regulative subunit p85 binds through its SH2 domain to (p)Y397. The SH3 domain GRAF binds to the second proline-rich sequence of FAK. FAK N-terminal domain interacts with ezrin, integrin β1 and growth factor receptors such as EGFR and PDGFR. NH2 = N-terminus; COOH = C-terminus; Pax = paxillin; Vin = vinculin; CAS = p130Cas; PR1 and PR2 = first and second prolin-rich sequences respectively; PI3K = PI3-kinase. (Original by D. Bonanomi and F. Valtorta)

interactions. However, its C-terminal region displays two proline-rich sequences (711–741 and 861–882), which are putative ligands for SH3 domains of molecules such as p130Cas and Graf (GTPase regulator associated with FAK) (Schaller et al., 1992). The C-terminal domain also contains a 159 aa sequence named FAT (focal adhesion targeting), required for FAK localization to focal adhesion sites (Hildebrand et al., 1993) (Figure 7.2C). This effect is likely to be accomplished through binding of the FAT domain to integrin-associated proteins such as paxillin and talin (Hayashi et al., 2002). Moreover, the FAT domain also binds directly to an activator of Rho-family GTPases known as p190 RhoGEF, and FAK-mediated phosphorylation of p190 RhoGEF might provide a direct link to RhoA activation (Zhai et al., 2003).

The N-terminal region of FAK contains a FERM (protein 4.1, ezrin, radixin, moesin) domain, a widespread protein module involved in linking cytoplasmic proteins to the plasma membrane (Girault et al., 1999b). The FERM domain of FAK mediates association with receptor tyrosine kinases, such as epidermal growth factor receptor (EGFR) and platelet-derived growth factor receptor (PDGFR) (Sieg et al., 2000). In addition, the FAK N-terminal domain can directly

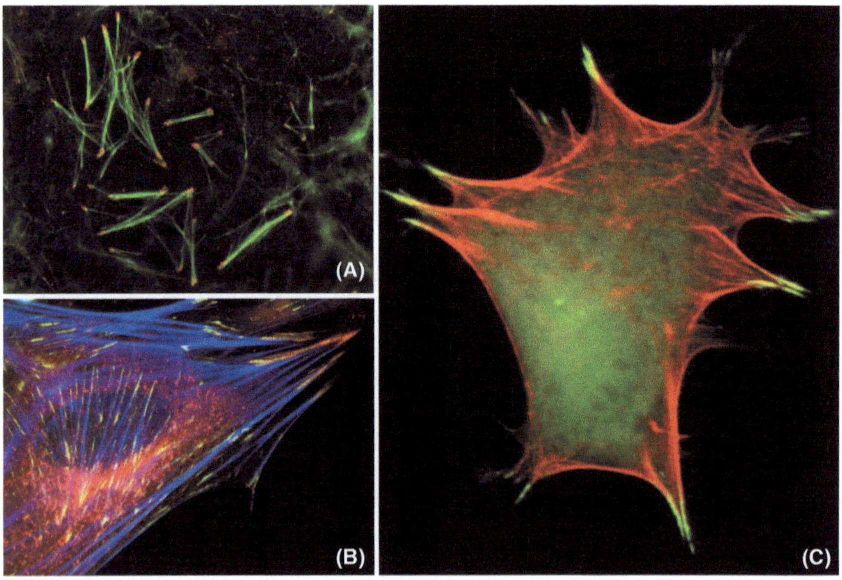

FIGURE 7.2. Targeting of FAK in nonneuronal cells. (A) COS-7 cells stained for FAK (red) and F-actin (green). (B) Chicken embryo fibroblast expressing a chimera of FAK fused to enhanced green fluorescent protein (green) and double stained for vinculin (red) and F-actin (blue). FAK is associated with focal adhesions underlined by vinculin staining at the edge of actin stress fibers. (C) Glial cell expressing the FAT domain of FAK fused to enhanced green fluorescent protein (green) and stained for F-actin (red). The FAT domain is responsible for the selective targeting of the chimera to focal adhesions. (original by D. Bonanomi and F. Valtorta)

bind the β integrin cytoplasmic domain (Eliceiri et al., 2002; Schaller et al., 1995). Lys152 within the FERM domain can be posttranslationally modified by sumoylation, which consists in the addition of a small ubiquitin-related molecule in most instances associated with nuclear import of the protein. Sumoylation appears to promote catalytic activation and nuclear translocation of FAK (Kadare et al., 2003), thus providing a potential link between focal adhesions and gene transcription regulated by FAK signaling (McKean et al., 2003).

7.4.2. Splicing Variants

Although FAK is encoded by a single gene, both the use of internal promoters and the occurrence of alternative splicing generate multiple isoforms of the protein (Burgaya and Girault, 1996). A truncated isoform of 42–44 kDa named FRNK ($p125^{FAK}$-related nonkinase), which lacks the catalitic domain, is autonomously expressed in fibroblasts through the use of an internal promoter (Schaller et al., 1993). FRNK, which contains the FAT region, competes with FAK for the same targeting site at focal adhesions. Thus, it acts as an endogenous regulator of FAK, transiently blocking the formation of focal contacts in cells grown on fibronectin (Richardson and Parsons, 1996). Since FRNK is not expressed in all cell types, other still uncharacterized regulators of FAK localization might exist (Andre and Becker-Andre, 1993).

Numerous FAK isoforms are selectively enriched in neuronal cells (Figure 7.1A). The isoform termed FAK^{+}, which contains a three-amino acids insertion (Pro-Trp-Arg) approximately in the middle of the FAT sequence, is expressed at very low levels in nonneuronal cells while is highly expressed in the central nervous systems (Burgaya and Girault, 1996). Other two FAK isoforms preferentially expressed in the rat brain contain alternative exons which encode for peptides of 28, 6, and 7 amino acids surrounding the autophosphorylation site (Tyr-397). At least in the rat striatum, exon 6 is always found in association with exon 7, while exon 28 is found only in combination with the others (Burgaya et al., 1997). The presence of additional exons does not prevent the targeting of FAK neuronal isoforms to adhesion sites in both nonneuronal and neuronal cells (Contestabile et al., 2003; Toutant et al., 2000).

A 110 kDa tyrosine kinase named PYK2 (proline-rich tyrosine kinase 2)/CAKβ (cell-adhesion kinase β), displaying 45% sequence identity to FAK, has been included together with FAK in a new subfamily of nonreceptor tyrosine kinases. PYK2 has been reported to undergo autophosphorylation on the increase of intracellular calcium levels (Lev et al., 1995). The C-terminal region of PYK2 can be expressed alone by the use of an internal promoter. Moreover, a PYK2 isoform lacking 42 amino acids in the C-terminal domain is abundantly expressed in hematopoietic cells, while the full-length isoform is predominant in the brain (Xiong et al., 1998).

7.4.3. Mechanisms of FAK Activation

FAK represents a point of convergence of the signaling pathways activated by integrins, oncogenes, and mitogenic neuropeptides (Zachary and Rozengurt,

1992). Indeed, both integrin-dependent adhesion to the substrate and cell transformation by v-Src increase FAK phosphorylation and subsequently its kinase activity (Guan and Schalloway, 1992). In addition, FAK is activated on stimulation of G-protein–coupled receptors by mitogenic peptides, such as bombesin and vasopressin, and phospholipids, including lysophosphatidic acid (Rodriguez-Fernandez and Rozengurt, 1998).

Both integrin engagement and G-protein–coupled receptor stimulation promote FAK activation, leading to its phosphorylation on six tyrosine residues (Hanks and Polte, 1997).

The catalytic activity of FAK appears to be downregulated through an autoinhibitory intramolecular interaction between its N-terminal and kinase domains (Cooper et al., 2003). The evidence that incubation of a GST fusion protein containing the β1 integrin cytoplasmic domain with recombinant FAK stimulates its kinase activity *in vitro* (Cooper et al., 2003) suggests that such binding may displace the inhibitory intramolecular interaction of FAK, thus leading to its activation. These findings support a two-step model for FAK activation by integrins in cell adhesion (Cooper et al., 2003). On integrin-mediated cell adhesion, FAK becomes recruited to focal contacts through interaction of its FAT domain with the integrin-associated proteins paxillin and talin. At these complexes, the autoinhibition within FAK is released by interaction of its N-terminus with the β1 integrin cytoplasmic domain. FAK activation is followed by autophosphorylation on Tyr-397, which is located in a motif (pTyr-Ala-Glu-Ile) corresponding to the consensus sequence for the binding of the SH2 domain of Src kinases (Cobb et al., 1994). Once recruited by FAK, Src undergoes conformational activation resulting in the phosphorylation of FAK on several tyrosine residues: Tyr-407 in the N-terminal domain, Tyr-576,577 in the activation loop of the catalitic domain, and Tyr-861,925 in the C-terminal domain (Hanks and Polte, 1997). Phosphorylation on Tyr-407,576,577 produces a massive stimulation of FAK kinase activity *in vitro* (Calalb et al., 1995). Since an inactivating mutation in the autophosphorylation site inhibits FAK phosphorylation of its downstream target paxillin, Src recruitment is likely to be essential for FAK-mediated signaling (Shen and Schaller, 1999).

FAK activation can also occur in an integrin-independent manner mediated by binding of actin- and membrane-associated adaptors, such as ezrin, to the FERM domain of FAK (Poullet et al., 2001).

Ectopic expression of the neuronal isoforms FAK^{+}, $FAK^{+}6,7$, and $FAK^{+}6,7,28$ in COS-7 cells has revealed that the presence of exons 6 and 7 either alone or in combination with exon 28 promote FAK autophosphorylation, which is not affected by the expression of the PWR insertion alone (Burgaya et al., 1997). Moreover, exons 6 and 7 prevent FAK phosphorylation by Src-family kinases (Toutant et al., 2000). Autophosphorylation of FAK^{+} has been shown to occur intermolecularly, whereas that of $FAK^{+}6,7$ and $FAK^{+}7$ is predominantly intramolecular (Toutant et al., 2002). Therefore, the presence of neuro-specific exons leads to critical modifications in the biochemical properties of FAK isoforms, suggesting that specific mechanisms of FAK activation and signaling might take place in neurons.

7.4.4. *FAK as a Molecular Adapter*

FAK activation is necessary for the recruitment of Src kinases and components of the Ras/mitogen-activated-protein kinase (MAPK) pathway at focal adhesions (Miyamoto et al., 1995). Therefore, FAK seems to act as an adapter for the assembly of a multimolecular scaffold for intracellular signaling downstream of adhesion receptors (Figure 7.1B). FAK phosphorylation by Src kinases creates binding sites for SH2 containing proteins, such as Grb2, which in turn activate Ras-mediated extracellular-signal-regulated kinase (ERK)/MAPK signaling (Schlaepfer et al., 1994). FAK appears to be crucial in mediating the cross talk between adhesion and growth factor receptors in the context of the MAPK pathway but less important for the direct pathway of MAPK activation by integrins (Renshaw et al., 1999).

Several important signaling events mediated by FAK are due to its interaction with the molecular adapter p130Cas, which through its SH3 domain binds to both FAK C-terminal proline-rich sequences. Binding of p130Cas to FAK is linked to enhanced Src-operated phosphorylation of p130Cas on multiple tyrosines, which creates sites for the SH2-based recruitment of the Crk adaptor protein. Crk, in turn, interacts with the guanidine nucleotide exchanging factor CG3, eventually leading to activation of the JNK/MAPK cascade implicated in the transcriptional control of cell proliferation (Giancotti and Ruoslahti, 1999; Schlaepfer et al., 1994). Signaling downstream of p130Cas also results in increased Rac activity and enhanced cell motility (Cho and Klemke, 2002; Hsia et al., 2003). Moreover, the FAK/p130Cas complex plays a key role in mediating cell survival signaling from the extracellular matrix and thus in preventing apoptosis due to cell adhesion loss (Almeida et al., 2000).

7.5. FAK Controls Adhesion and Cell Motility

7.5.1. *Cell Migration in FAK-Null Mice*

The functional characterization of FAK has been fostered a great deal by the study of mice in which its gene had been knocked out. This ablation results in a lethal phenotype in the early stages of embryonic development (E8.0–8.5) due to dramatic defects in the morphogenesis of the axial mesoderm and cardiovascular system (Furuta et al., 1995). These abnormalities are attributed to the impaired migration of mesodermic fak$^{-/-}$ cells and alterations in the anchorage, migration, and proliferation during the formation of endothelial tissues. Since the phenotype of fak$^{-/-}$ embryos strongly resembles that reported for fibronectin knockout mice (George et al., 1993), the observed morphogenetic and migratory defects might reflect a general impairment of the integrin-dependent adhesion to the extracellular matrix. The molecular basis of these abnormalities have been investigated through the analysis of fak$^{-/-}$ fibroblasts (Ilic et al., 1995). FAK-defective fibroblasts show a remarkably reduced migration rate compared to the wild-type cells. Unexpectedly, mutant cells display an increased number of focal adhesions, thus suggesting that FAK is involved in the control of the balance between assembly and destruction of these structures during cell migration rather than in their formation. Consistently, cells injected with the FAT domain alone, which displaces FAK from focal contacts, are defective in cell migration without showing any reduction in the number of focal adhesions (Gilmore and Romer, 1996).

The current model for interpreting these data holds that the ablation of FAK, slowing down the turnover of focal contacts, causes a net increase in their number, hence reducing the probability of the release of adhesive interactions during cell migration. The systematic analysis of fibroblasts lacking focal contact-associated proteins, such as paxillin, p130Cas, and Src family kinases, has revealed similarities with FAK-null cells: accumulation of immature focal contacts, reduced rate of focal contact disassembly, and refractory migration defects (Webb et al., 2004). This comprehensive study showed that focal contact disassembly requires FAK phosphorylation on Tyr-397, Src kinase activity, and paxillin phosphoryation at two distinct sites (Tyr-31 and Tyr-118).

The recent generation of a conditional fak knockout mouse allowed investigation of FAK in brain development, as yet precluded by the early embryonic lethality of $fak^{-/-}$ embryos which die prior to extensive development of the nervous system. Tissue-specific deletion of fak from neuronal and glial precursors of the dorsal telencephalon resulted in severe cortical dysplasia resembling type II lissencephaly, as seen in congenital muscular dystrophy, linked to local disruption of the cortical basement membrane (Beggs et al., 2003). Abnormalities were still present following selective ablation of fak from meningeal fibroblasts, which are essential for basal lamina organization and secretion of matrix components, but absent following neuron-specific deletion of the gene. Consistently, fak-deficient meningeal fibroblasts grown in culture exhibited impaired basal lamina organization. These results establish FAK as an essential component of the bidirectional signaling cascade linking the extracellular matrix with the cytoskeleton, regulating laminin polymerization and organization of the basal lamina at the interface between radial glial endfeet and meningeal fibroblasts.

7.5.2. Scaffolding the Motility Machinery with FAK

Several of the effects of FAK on cell motility seem to depend on its ability to act as a platform for the assembly of a signaling scaffold at adhesion sites. Mutagenesis strategy has revealed that the kinase activity, the autophosphorylation site, and the first proline-rich region (711–741) of FAK are all needed for optimal rescuing of the migratory defects of $fak^{-/-}$ cells grown on fibronectin (Sieg et al., 1999). FAK overexpression enhances the migration of CHO cells independently of FAK kinase activity but requiring its autophosphorylation and subsequent association with Src kinases (Cary et al., 1996). The study of the viral Src (v-Src) mutant, a constitutively active variant of c-Src, has allowed to shed light on the molecular mechanisms used by FAK to regulate focal adhesion turnover during cell motility (Fincham and Frame, 1998; Guan and Schalloway, 1992). v-Src moves to the cell periphery, where it is recruited at focal adhesions through its association with FAK. Phosphorylating FAK, v-Src drives the loss of the FAK/v-Src interaction, thus increasing FAK degradation. This leads to focal contact disassembly, hence fostering cell migration.

An intriguing mechanism by which FAK might control focal contact turnover relies on its association with a complex consisting of the Ca^{2+}-dependent protease calpain-2 and p42ERK/MAPK (Carragher et al., 2003). Once recruited at focal contacts via interaction with FAK, ERK-activated calpain-2 cleaves several focal adhesion components, including FAK, thus promoting disassembly of adhesion complexes during migration and transformation. In addition to controlling intracellular proteolytic events, FAK influences the expression and activity of matrix metalloproteases (MMPs), proteolytic enzymes that degrade the extracellular matrix and have important roles in tissue remodeling and metastasis (Hauck et al., 2001; Hsia et al., 2003; Wu et al., 2005). The control exerted by FAK on MMPs appears to contribute to the effects of FAK on cell motility and tumor progression. FAK can regulate local endocytosis and surface expression of the membrane-anchored MT1-MMP at invasive membrane protrusions (i.e., podosomes) of transformed fibroblasts (Wu et al., 2005). This regulatory mechanism relies on the scaffolding role of FAK to bind endophilin A2, a key component of the endocytotic machinery (Brodin et al., 2000) at the second Pro-rich motif and Src at (auto)phosphorylated Tyr-397. Src-mediated phosphorylation of endophilin A2 on a critical tyrosine (Tyr-315) reduces its association with dynamin and attenuates MT1-MMP endocytosis, thus leading to increased surface expression of MT1-MMP and extracellular matrix degradation at podosomes. The control of MMP functions by FAK might also operate at growth cones during neurite elongation. There is increasing evidence for the physiological involvement of MMPs in axon outgrowth and guidance (Vaillant et al., 2003; Webber et al., 2002; Yang et al., 2005).

The interaction between p130Cas and the FAK/c-Src complex mediates FAK-dependent cell migration through a signaling cascade independent of MAPK activation. SH2 domain containing proteins, such as Crk, PI3-kinase (PI3K), and Src kinases, recruited by the p130Cas/FAK/c-Src complex, might be the effectors responsible for the transmission of migratory signals downstream of integrins (Cary et al., 1998). Moreover, it has been shown that on localization at membrane ruffles, the p130Cas-Crk complex activates the small GTPase Rac, thus enhancing cell migration (Klemke et al., 1998).

The proline-rich sequences of FAK bind the SH3 domain of GRAF (GTPase regulator associated with FAK), a GAP for Rho and Cdc42, ultimately leading to the inhibition of their modulating activity on actin dynamics. This GRAF-mediated negative regulation of Rho might account for some of the known effects produced by FAK on the regulation of the cytoskeleton and cell morphology (Schoenwaelder and Burridge, 1999). Rho is constitutively activated in $fak^{-/-}$ cells, leading to impairment of focal adhesion turnover (Ren et al., 2000). A specific inhibitor for Rho-associated kinase promotes the spreading of $fak^{-/-}$ cells through the reorganization of the actin cytoskeleton and focal adhesions. Besides, Rho-associated kinase mediates the recruitment of myosin light chain kinase at cell adhesion sites, thus enhancing actomyosin-generated contractile forces which prevent cell spreading and migration (Chen et al., 2002). FAK can influence the activity of Cdc42 also through the binding and phosphorylation of the Cdc42 effector N-WASP (neuronal Wiskott-Aldrich syndrome protein), which is linked to regulation of the actin cytoskeleton via activation of the ARP2/3 complex. By associating with Cdc42-activated N-WASP FAK controls its intracellular localization and promotes cell motility (Wu et al., 2004).

FAK autophosphorylation site has been shown to bind the SH2 domain of the adaptor protein Grb7, mediating the transmission of migratory signals. Grb7 is a member of a family of pleckstrin homology (PH) domain-containing adaptor molecules which exhibits a high homology to Mig10, a protein involved in neuronal cell migration during *Caenorhabditis elegans* development. Fibroblasts migration on integrin-engaging substrates is enhanced by Grb7 overexpression but reduced by the disassembly of the FAK-Grb7 complex (Han and Guan, 1999). The interaction of the Grb7 PH domain with phosphoinositides plays a role in the stimulation of cell migration by Grb7, consistent with its requirement for Grb7 phosphorylation by FAK (Shen et al., 2002).

FAK has been reported to participate in integrin-stimulated motility mediated by the Etk/BMX tyrosine kinase. Etk is a member of the Btk family of tyrosine kinases that are expressed at high levels in highly motile cells, including metastatic carcinoma cell lines. The interaction between the FERM domain of FAK and the PH domain of Etk modulates the activation of the latter by proteins of the extracellular matrix and appears to be required for efficient cell migration induced by both kinases (Chen et al., 2001).

fak$^{-/-}$ cells are refractory to motility signals induced by the growth factors platelet-derived growth factor (PDGF) and epidermal growth factor (EGF), whereas stable reconstitution of FAK expression restores PDGF and EGF receptor-dependent migration. An efficient growth factor-stimulated migration requires FAK to be targeted to focal adhesions, where it connects growth factor receptors and integrins, binding the former through its N-terminal domain and the latter through its C-terminal domain (Sieg et al., 2000). In this manner, FAK promotes cell migration by integrating signals from adhesion and growth factor receptors. However, distinct components are required for FAK-mediated cell motility depending on whether it is stimulated by integrins or by growth factor receptors. FAK kinase activity and the assembly of the FAK/p130Cas complex are dispensable when cell migration is stimulated by EGF and PDGF receptors (Sieg et al., 2000) but essential when it is stimulated by the extracellular matrix (Cary et al., 1998; Sieg et al., 1999). This difference could be explained by the fact that integrins lack kinase activity to phosporylate and hence activate FAK, while growth factor receptors possess tyrosine-kinase activity.

In addition to interacting with integrins and growth factor receptors, in epithelial cells FAK appears to be constitutively associated with the receptor tyrosine-kinase EphA2 at the level of focal contacts (Miao et al., 2000). The stimulation of Eph receptors by their ligands, the ephrins, is involved in the reduction of the intermingling between adjacent cell populations during the formation of tissue borders and in the transmission of repulsive, contact-dependent migratory signals (Mellitzer et al., 1999). In the central nervous system, Eph receptors are implicated in cortex patterning as well as axonal fasciculation and branching. Moreover, Eph receptor stimulation elicits neuronal growth cone collapse through cytoskeletal disassembly (Mueller, 1999). The activation of EphA2 by ephrin-A1 downregulates integrin signaling and enhances rapid association of SHP2 tyrosine-phosphatase with the receptor, followed by transient FAK dephosphorylation and disassembly of the EphA2/FAK complex. These events inhibit integrin-dependent cell adhesion, focal contact turnover, and cell migration (Miao et al., 2000). However, under different experimental conditions, EphA2 signaling can induce cell adhesion and actin cytoskeletal assembly in a FAK-dependent and p130Cas-dependent manner (Carter et al., 2002).

The tuning of FAK activity through repeated cycles of phosphorylation and dephosphorylation seems to be crucial in order to link this kinase to the control of cell migration. Overexpression of the tyrosine-phosphatase PTEN, which dephosphorylates FAK, leads to a robust inhibition of integrin-stimulated migration. PTEN overexpressing cells show a reduced number of focal contacts and severe alteration of the actin cytoskeleton, which can be rescued by FAK overexpression (Tamura et al., 1998). As reported previously, FAK is a substrate for SHP2 phosphatase, which participates in signal transduction downstream of a number of growth factor receptors, playing an important role in the control of cellular architecture and cell migration. Fibroblasts expressing an inactive form of SHP2 exhibit increased levels of FAK phosphorylation in addition to decreased motility and increased number of focal contacts, a phenotype resembling that displayed by fak$^{-/-}$ cells (Yu et al., 1998). A similar phenotype has been observed in fibroblasts lacking the protein tyrosine-phosphatase (PTP)-PEST (Angers-Loustau et al., 1999), indicating that an excess of FAK phosphorylation leads to the inactivation of the kinase, possibly due to its degradation (see earlier). Moreover, it has been shown that PTP α controls cell migration and morphology through the modulation of the kinetics and efficiency of FAK autophosphorylation on Tyr-397 (Zeng et al., 2003).

7.5.3. FAK Organizes Microtubule Networks During Cell Motility

Although the analysis of mice bearing a targeted deletion of fak in cortical neurons did not reveal a cell-autonomous defect in neuronal migration, FAK appears to control neuronal positioning during neocortex development through an effect on nuclear translocation (i.e., nucleokinesis), a well established behavior of migrating neurons that is critically dependent on microtubule organization (Xie et al., 2003). Phosphorylation of FAK on Ser-732 by the serine–threonine kinase Cdk5 is critical for the organization of a small network of centrosome-associated microtubules arranged in a fork-like structure that extends around the nucleus. The microtubule fork is thought to be essential for achieving the tight association between the centrosome and the nucleus required for proper nucleokinesis (Nikolic, 2004; Xie et al., 2003). In neurons devoid of Cdk5 or expressing a nonphosphorylatable FAK Ser-732-Ala mutant, the microtubule fork is disrupted and nuclear morphology and translocation are impaired. Moreover, expression of the FAK-Ser-732Ala mutant in the developing cortex with an in utero electroporation technique results in neuronal positioning defects *in vivo* (Xie et al., 2003).

The role of microtubules during cell migration is not limited to nuclear translocation. Coordinated regulation of microtubule structures and microfilaments is crucial in cell motility. Local stabilization of microtubules is required during several cell motility phenomena, including neurite outgrowth (Gundersen and Cook, 1999). Microtubule stabilization at the leading edge of migrating cells is accomplished through posttranslational modifications of α-tubulin controlled by RhoA GTPase and its effector protein diaphanous (mDia) (Palazzo et al., 2001).

A connection exists between FAK activation by integrins and the RhoA-mediated generation of stable microtubules, and these signaling events require lipid rafts containing the ganglioside GM1 (Palazzo et al., 2004). FAK-deficient cells have less stable microtubules and show misdistributed GM1. At the leading edge of migrating cells, activation of FAK by integrins appears to be necessary for the organization of GM1-containing plasma membrane microdomains, which in turn are required for RhoA-mDia-dependent microtubule stabilization. The molecular link between integrins/FAK and GM1-positive microdomains is as yet unknown. However, this connection appears to be highly specific since the distribution of lipid raft markers other than GM1 (e.g., cholesterol, caveolin, CD44) is unaffected in fak$^{-/-}$ cells.

7.6. FAK Controls Axon Extension and Guidance

7.6.1. FAK Expression and Regulation in the Nervous System

Although the fak gene is ubiquitously expressed in all cells of vertebrates, it reaches the highest levels of expression in the brain with particular enrichments in cerebral and cerebellar cortex and the hippocampus (Burgaya et al., 1995). While FAK levels in the brain are maximal at the end of embryonic development, PYK2 levels increase progressively after birth (Menegon et al., 1999). In contrast, the levels of the neuronal isoform FAK^{+} are stable, thus during development they increase relatively to FAK. FAK^{+}6,7 is the predominant isoform in the adult brain (Toutant et al., 2000). In the brain, this isoform is regulated by neurotransmitters (glutamate and acetylcholine), depolarization (Derkinderen et al., 1998; Siciliano et al., 1996), and lipid extracellular messenger such as endocannabinoids and lysophosphatidic acid (LPA) (Derkinderen et al., 1996, 1998). Extracellular messengers activating FAK were shown not to affect PYK2 phosphorylation (Derkinderen et al., 1998). Thus, due to their differential distribution (Menegon et al., 1999) and regulation in the brain, FAK and PYK2 are likely to exert distinct, although possibly overlapping, functions in the nervous system.

Molecular events induced by FAK and PYK2, such as Src-family kinase and ERK signaling activation, might be important for synaptic plasticity in the brain

(Girault et al., 1999a). FAK is enriched in the neuronal growth cones, where it colocalizes with F-actin reaching the distal most lamellipodial region, from which PYK2 is excluded (Menegon et al., 1999) (Figure 7.3). Within the growth cone, FAK is associated with Fyn and NCAM (Beggs et al., 1997) in point contacts, where it colocalizes with paxillin and integrin β1 (de Curtis and Malanchini, 1997). The colocalization of FAK and vinculin has been reported by some authors (Renaudin et al., 1999; Stevens et al., 1996) but not confirmed by others (Contestabile et al., 2003). This discrepancy possibly reflects changes in FAK localization and/or molecular interactions during different states of growth cone motility. *In vivo* imaging of a fluorescent chimera of FAK^{+} expressed in developing hippocampal neurons allowed to study protein dynamics within the growth cones (Contestabile et al., 2003). FAK^{+} was shown to undergo rapid relocalization from a cytosolic pool to membrane ruffles and concentrate at the tip of cytochalasin D-induced protrusions generated by mibrotubule invasion into the peripheral actin-rich domain of the growth cone. Since both membrane ruffles and leading edges of protrusions are sites of rapid F-actin turnover, these findings support a role for FAK in the regulation of F-actin dynamics in the growth cone.

First clues as to the involvement of FAK in neurite extension have been gained through the use of rat pheochromocytoma PC12 and human neuroblastoma SH-SY5Y cells, two commonly employed cell models to study neuronal differentiation on stimulation with growth factors. FAK and PYK2 were shown to associate with integrins and growth factor receptors clustered at adhesion sites which participate in neurite extension. Costimulation of both integrins and growth factor

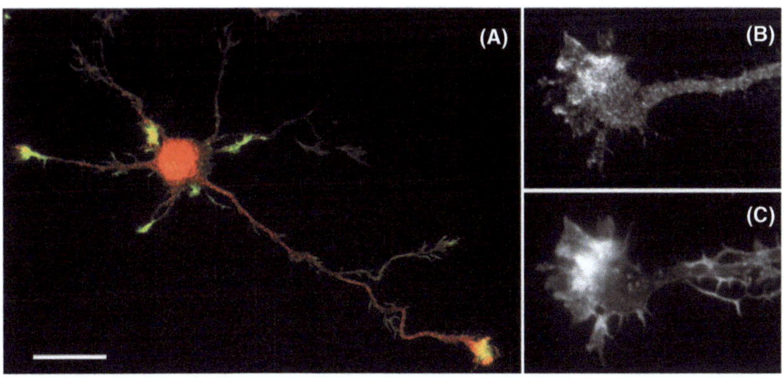

FIGURE 7.3. FAK localizes at the growth cone of developing neurons. (A) Embryonic rat hippocampal neuron developing in culture double stained for FAK (red) and F-actin (green). Note the enrichment of FAK in the growth cone associated with the longest neurite, probably destined to become the future axon. (B and C) High magnification of a growth cone showing FAK (B) and F-actin (C) staining. FAK displays a punctuate staining pattern throughout the growth cone, due to its localization at point contacts, and overlaps with F-actin enrichments in the peripheral domain. Bar, 10 μm. Reproduced with permission from Contestabile et al. (2003).

receptors elicits FAK and PYK2 signaling cascades responsible for the control of neurite formation without requiring the activation of ERK/MAP kinases (Ivankovic-Dikic et al., 2000). In SH-SY5Y cells, the insulin-like growth factor-I (IGF-I) triggers PI3K-mediated FAK phosphorylation, which in turn promotes the assembly of adhesion structures required for neurite extension (Kim and Feldman, 1998). In the same cell line, the kinase inhibitor K-252a promotes neurite extension by fostering FAK phosphorylation (Maroney et al., 1995).

LPA, which enhances FAK activation and Rho-dependent acto-myosin contractility, also induces neurite retraction and growth cone collapse in PC12 cells (Jalink et al., 1994). Consistently, bradykinin, which leads to growth cone collapse, stimulates Ca^{2+}-dependent PYK2 phosphorylation in differentiated PC12 cells (Park et al., 2000). Thus, a tight control of FAK and PYK2 phosphorylation seems to be essential for growth cone stability and motility, while loss of regulation and massive phosphorylation lead to growth cone arrest, detachment from the substrate, and collapse.

7.6.2. FAK Is a Receptor-Proximal Regulator of Axon Elongation and Guidance

The direct involvement of FAK-mediated signaling in neurite outgrowth during nervous system development is now supported by compelling evidences.

Postnatal ablation of fak in Purkinje cells at the time when they are collateralizing their axons and refining their synaptic connections results in increased branching of axons and, consequently, enhanced synaptogenesis. An increase in the total length of the axonal arbor and the number of branchpoints was also observed in cultured hippocampal neurons following conditional *fak* deletion or expression of a dominant-negative mutant, as a result of both enhanced formation and reduced pruning of axonal branches. Nevertheless, axons lacking FAK grow at abnormally low speed. Thus, FAK acts as a negative regulator of axonal branch formation, and this effect is mediated, in part, through the recruitment of modulators of Rho-family GTPases, including the RhoA activator p190RhoGEF (Rico et al., 2004).

The complex interactions between FAK and Rho are thought to underlie both cytoskeletal remodeling and focal adhesion turnover. In analogy to what has been observed in nonneuronal cells, loss of this regulatory loop might produce aberrantly stable adhesion structures in growth cones, thus reducing their motility and preventing their retraction. In addition, FAK might be implicated in the control of axonal arborization by mediating signaling cascades initiated by soluble factors, including neurotrophins and guidance cues, membrane-bound ligands, such as ephrins, and adhesion-receptors including integrins, which have been shown to regulate axonal branching and synaptogenesis (Bagri et al., 2003; Gao et al., 1999; Markus et al., 2002; Murai et al., 2003; Wang et al., 1999). Misinterpretation of guidance cues might also underlie the severe disorientation of dendritic processes of $fak^{-/-}$ pyramidal neurons, which display increased branch complexity primarily affecting the apical dendrite (Beggs et al., 2003).

The possibility that FAK mediates signaling downstream of membrane receptors of factors involved in the control of axon pathfinding and outhgrowth has been directly addressed by investigating the cascade of events involved in netrin-mediated axon guidance (Li et al., 2004; Liu et al., 2004; Ren et al., 2004). Netrins are prototypical soluble guidance cues that promote axon outgrowth and guide the direction of axon pathfinding in many regions of the nervous system (Tessier-Lavigne and Goodman, 1996). DCC (deleted in colorectal cancer) is a functional receptor for netrin and mediates both the axon growth and the chemoattractive function of netrin (Culotti and Merz, 1998). In addition, the association of DCC with the UNC-5 receptor mediates growth cone repulsion (Hong et al., 1999). Remarkably, with the exception of ephrins, whose receptors (Ephs) are tyrosine kinases, all other neuronal guidance cues, including netrins, function through receptors devoid of enzymatic activity. Biochemical evidences have been provided for the direct association of FAK with the intracellular domain of DCC under physiological conditions. Stimulation of neurons with netrin promotes the interaction between DCC and a Src family kinase (identified as Fyn by Liu et al., 2004), which in turn activates FAK through the phosphorylation of Tyr-576 and Tyr 577 and phosphorylates the cytotail of DCC on Tyr-1420, which is critical for downstream signaling to modify growth cone behavior (Li et al., 2004). Although FAK is not directly responsible for DCC phosphorylation, its binding to the receptor is required for Src-mediated Tyr-1420 phosphorylation, consistent with a scaffolding role of FAK in recruiting Src to DCC. This model is supported by both the rapid netrin-induced enhancement of FAK (auto)phosphorylation on Tyr-397, which creates a docking site for Src kinase, and the abolishment of DCC tyrosine phosphorylation by mutation of this residue (Ren et al., 2004).

Genetic approaches and the use of dominant-negative mutants have been exploited to investigate the functional role of the Src/FAK association with DCC in netrin signaling events controlling growth cone motility. Interference with Src function blocks netrin-induced neurite outgrowth and growth cone attraction (Li et al., 2004; Liu et al., 2004). The stimulatory effect of netrin on axon outgrowth is significantly reduced in cortical neurons lacking *fak*. Importantly, *fak*-null neurons do not present a general impairment in axon outgrowth, since in the absence of netrin they extend neurites at a rate comparable with that of wild-type neurons (Liu et al., 2004). The involvement of FAK in netrin-induced attraction was shown by the loss of responsiveness to the cue in both cortical axons from FAK-null mice and spinal cord neurons expressing a dominant-negative FAK mutant (Liu et al., 2004). In addition, expression of a FAK mutant lacking the phosphorylation site Tyr-861, which is maximally phosphorylated on netrin stimulation, but not the expression of FAK mutated on Tyr-397, prevents attractive turning of growth cone from *Xenopus* spinal neurons toward a source of netrin (Ren et al., 2004).

Tyrosine phosphorylation of FAK was induced by netrin in neurons devoid of DCC derived from the cortex, but not in neurons derived from the spinal cord, indicating the presence of other receptors that are functionally redundant with DCC in cortical neurons (Liu et al., 2004; Ren et al., 2004). It is therefore possible that integrins contribute in part to netrin-induced FAK phosphorylation. Several guidance cues with neurite-promoting activities, including netrins (Serafini et al., 1994) and semaphorins (Pasterkamp et al., 2003), contain integrin-binding motifs. Therefore, canonical guidance cue receptors might cross talk with integrin-mediated signaling to mediate the effects of these factors on axon extension and pathfinding. Membrane-anchored semaphorin-7A binds to β1-containing integrins rather than a member of the canonical plexin receptor family and enhances axon outgrowth in a MAPK-dependent signaling mechanism which involves the activation of FAK (Pasterkamp et al., 2003).

References

Ahmad, F.J., Hughey, J., Wittmann, T., Hyman, A., Greaser, M., and Baas, P.W., 2000, Motor proteins regulate force interactions between microtubules and microfilaments in the axon, *Nat. Cell Biol.* **2:** 276–280.

Almeida, E.A., Ilic, D., Han, Q., Hauck, C.R., Jin, F., Kawakatsu, H., et al., 2000, Matrix survival signaling: From fibronectin via focal adhesion kinase to c-Jun NH(2)-terminal kinase, *J. Cell Biol.* **149:** 741–754.

Andre, E., and Becker-Andre, M., 1993, Expression of an N-terminally truncated form of human focal adhesion kinase in brain, *Biochem. Biophys. Res. Commun.* **190:** 140–147.

Angers-Loustau, A., Cote, J.F., Charest, A., Dowbenko, D., Spencer, S., Lasky, L.A., et al., 1999, Protein tyrosine phosphatase-PEST regulates focal adhesion disassembly, migration, and cytokinesis in fibroblasts, *J. Cell Biol.* **144:** 1019–1031.

Arregui, C.O., Carbonetto, S., and McKerracher, L., 1994, Characterization of neural cell adhesion sites: Point contacts are the sites of interaction between integrins and the cytoskeleton in PC12 cells, *J. Neurosci.* **14:** 6967–6977.

Bagri, A., Cheng, H.J., Yaron, A., Pleasure, S.J., and Tessier-Lavigne, M., 2003, Stereotyped pruning of long hippocampal axon branches triggered by retraction inducers of the semaphorin family, *Cell* **11:** 285–299.

Beggs, H.E., Soriano, P., and Maness, P.F., 1994, NCAM-dependent neurite outgrowth is inhibited in neurons from Fyn-minus mice, *J. Cell Biol.* **127:** 825–833.

Beggs, H.E., Baragona, S.C., Hemperly, J.J., and Maness, P.F., 1997, NCAM140 interacts with the focal adhesion kinase p125(fak) and the SRC-related tyrosine kinase p59(fyn), *J. Biol. Chem.* **272:** 8310–8319.

Beggs, H.E., Schahin-Reed, D., Zang, K., Goebbels, S., Nave, K.A., Gorski, J., et al., 2003, FAK deficiency in cells contributing to the basal lamina results in cortical abnormalities resembling congenital muscular dystrophies, *Neuron* **40:** 501–514.

Brodin, L., Low, P., and Shupliakov, O., 2000, Sequential steps in clathrin-mediated synaptic vesicle endocytosis, *Curr. Opin. Neurobiol.* **10:** 312–320.

Burgaya, F., and Girault, J.A., 1996, Cloning of focal adhesion kinase, pp125FAK, from rat brain reveals multiple transcripts with different patterns of expression, *Brain Res. Mol. Brain Res.* **37:** 63–73.

Burgaya, F., Menegon, A., Menegoz, M., Valtorta, F., and Girault, J.A., 1995, Focal adhesion kinase in rat central nervous system, *Eur. J. Neurosci.* **7:** 1810–1821.

Burgaya, F., Toutant, M., Studler, J.M., Costa, A., Le Bert, M., Gelman, M., et al., 1997, Alternatively spliced focal adhesion kinase in rat brain with increased autophosphorylation activity, *J. Biol. Chem.* **272:** 28720–28725.

Calalb, M.B., Polte, T.R., and Hanks, S.K., 1995, Tyrosine phosphorylation of focal adhesion kinase at sites in the catalytic domain regulates kinase activity: A role for Src family kinases, *Mol. Cell. Biol.* **15:** 954–963.

Cance, W.G., Harris, J.E., Iacocca, M.V., Roche, E., Yang, X., Chang, J., et al., 2000, Immunohistochemical analyses of focal adhesion kinase expression in benign and malignant human breast and colon tissues: Correlation with preinvasive and invasive phenotypes, *Clin. Cancer Res.* **6:** 2417–2423.

Carragher, N.O., Westhoff, M.A., Fincham, V.J., Schaller, M.D., and Frame, M.C., 2003, A novel role for FAK as a protease-targeting adaptor protein: Regulation by p42 ERK and Src, *Curr. Biol.* **13:** 1442–1450.

Carter, N., Nakamoto, T., Hirai, H., and Hunter, T., 2002, EphrinA1-induced cytoskeletal re-organization requires FAK and p130(cas), *Nat. Cell Biol.* **4:** 565–573.

Cary, L.A., Chang, J.F., and Guan, J.L., 1996, Stimulation of cell migration by overexpression of focal adhesion kinase and its association with Src and Fyn, *J. Cell. Sci.* **109 (Pt. 7):** 1787–1794.

Cary, L.A., Han, D.C., Polte, T.R., Hanks, S.K., and Guan, J.L., 1998, Identification of p130Cas as a mediator of focal adhesion kinase-promoted cell migration, *J. Cell Biol.* **140:** 211–221.

Chen, R., Kim, O., Li, M., Xiong, X., Guan, J.L., Kung, H.J., et al., 2001, Regulation of the PH-domain-containing tyrosine kinase Etk by focal adhesion kinase through the FERM domain, *Nat. Cell Biol.* **3:** 439–444.

Chen, B.H., Tzen, J.T., Bresnick, A.R., and Chen, H.C., 2002, Roles of Rho-associated kinase and myosin light chain kinase in morphological and migratory defects of focal adhesion kinase-null cells, *J. Biol. Chem.* **277:** 33857–33863.

Cho, S.Y., and Klemke, R.L., 2002, Purification of pseudopodia from polarized cells reveals redistribution and activation of Rac through assembly of a CAS/Crk scaffold, *J. Cell Biol.* **156:** 725–736.

Cobb, B.S., Schaller, M.D., Leu, T.H., and Parsons, J.T., 1994, Stable association of pp60src and pp59fyn with the focal adhesion-associated protein tyrosine kinase, pp125FAK, *Mol. Cell. Biol.* **14:** 147–155.

Contestabile, A., Bonanomi, D., Burgaya, F., Girault, J.A., and Valtorta, F., 2003, Localization of focal adhesion kinase isoforms in cells of the central nervous system, *Int. J. Dev. Neurosci.* **21:** 83–93.

Cooper, L.A., Shen, T.L., and Guan, J.L., 2003, Regulation of focal adhesion kinase by its amino-terminal domain through an autoinhibitory interaction, *Mol. Cell. Biol.* **23:** 8030–8041.

Crowley, E., and Horwitz, A.F., 1995, Tyrosine phosphorylation and cytoskeletal tension regulate the release of fibroblast adhesions, *J. Cell Biol.* **131:** 525–537.

Culotti, J.G., and Merz, D.C., 1998, DCC and netrins, *Curr. Opin. Cell Biol.* **10:** 609–613.

de Curtis, I., and Malanchini, B., 1997, Integrin-mediated tyrosine phosphorylation and redistribution of paxillin during neuronal adhesion, *Exp. Cell Res.* **230:** 233–243.

Derkinderen, P., Toutant, M., Burgaya, F., Le Bert, M., Siciliano, J.C., de Franciscis, V., et al., 1996, Regulation of a neuronal form of focal adhesion kinase by anandamide, *Science* **273:** 1719–1722.

Derkinderen, P., Siciliano, J., Toutant, M., and Girault, J.A., 1998, Differential regulation of FAK+ and PYK2/Cakbeta, two related tyrosine kinases, in rat hippocampal slices: Effects of LPA, carbachol, depolarization and hyperosmolarity, *Eur. J. Neurosci.* **10:** 1667–1675.

Doherty, P., Fruns, M., Seaton, P., Dickson, G., Barton, C.H., Sears, T.A., et al., 1990, A threshold effect of the major isoforms of NCAM on neurite outgrowth, *Nature* **343:** 464–466.

Eliceiri, B.P., Puente, X.S., Hood, J.D., Stupack, D.G., Schlaepfer, D.D., Huang, X.Z., et al., 2002, Src-mediated coupling of focal adhesion kinase to integrin alpha(v)beta5 in vascular endothelial growth factor signaling, *J. Cell Biol.* **157:** 149–160.

Felsenfeld, D.P., Schwartzberg, P.L., Venegas, A., Tse, R., and Sheetz, M.P., 1999, Selective regulation of integrin–cytoskeleton interactions by the tyrosine kinase Src, *Nat. Cell Biol.* **1:** 200–206.

Fincham, V.J., and Frame, M.C., 1998, The catalytic activity of Src is dispensable for translocation to focal adhesions but controls the turnover of these structures during cell motility, *EMBO J.* **17:** 81–92.

Furuta, Y., Ilic, D., Kanazawa, S., Takeda, N., Yamamoto, T., and Aizawa, S., 1995, Mesodermal defect in late phase of gastrulation by a targeted mutation of focal adhesion kinase, FAK, *Oncogene* **11:** 1989–1995.

Gao, P.P., Yue, Y., Cerretti, D.P., Dreyfus, C., and Zhou, R., 1999, Ephrin-dependent growth and pruning of hippocampal axons, *Proc. Natl. Acad. Sci. USA* **96:** 4073–4077.

George, E.L., Georges-Labouesse, E.N., Patel-King, R.S., Rayburn, H., and Hynes, R.O., 1993, Defects in mesoderm, neural tube and vascular development in mouse embryos lacking fibronectin, *Development* **119:** 1079–1091.

Giancotti, F.G., and Ruoslahti, E., 1999, Integrin signaling, *Science* **285:** 1028–1032.

Gilmore, A.P., and Romer, L.H., 1996, Inhibition of focal adhesion kinase (FAK) signaling in focal adhesions decreases cell motility and proliferation, *Mol. Biol. Cell* **7:** 1209–1224.

Girault, J.A., Costa, A., Derkinderen, P., Studler, J.M., and Toutant, M., 1999a, FAK and PYK2/CAKbeta in the nervous system: A link between neuronal activity, plasticity and survival? *Trends Neurosci.* **22:** 257–263.

Girault, J.A., Labesse, G., Mornon, J.P., and Callebaut, I., 1999b, The N-termini of FAK and JAKs contain divergent band 4.1 domains, *Trends Biochem. Sci.* **24:** 54–57.

Guan, J.L., and Shalloway, D., 1992, Regulation of focal adhesion-associated protein tyrosine kinase by both cellular adhesion and oncogenic transformation, *Nature* **358:** 690–692.

Gundersen, G.G., and Cook, T.A., 1999, Microtubules and signal transduction, *Curr. Opin. Cell Biol.* **11:** 81–94.

Hall, A., 1998, Rho GTPases and the actin cytoskeleton, *Science* **279:** 509–514.

Han, D.C., and Guan, J.L., 1999, Association of focal adhesion kinase with Grb7 and its role in cell migration, *J. Biol. Chem.* **274:** 24425–24430.

Hanks, S.K., and Polte, T.R., 1997, Signaling through focal adhesion kinase, *Bioessays* **19:** 137–145.

Hanks, S.K., Calalb, M.B., Harper, M.C., and Patel, S.K., 1992, Focal adhesion protein-tyrosine kinase phosphorylated in response to cell attachment to fibronectin, *Proc. Natl. Acad. Sci. USA* **89:** 8487–8491.

Hauck, C.R., Sieg, D.J., Hsia, D.A., Loftus, J.C., Gaarde, W.A., Monia, B.P., et al., 2001, Inhibition of focal adhesion kinase expression or activity disrupts epidermal growth factor-stimulated signaling promoting the migration of invasive human carcinoma cells, *Cancer Res.* **61:** 7079–7090.

Hauck, C.R., Hsia, D.A., Ilic, D., and Schlaepfer, D.D., 2002, v-Src SH3-enhanced interaction with focal adhesion kinase at beta 1 integrin-containing invadopodia promotes cell invasion, *J. Biol. Chem.* **277:** 12487–12490.

Hayashi, I., Vuori, K., and Liddington, R.C., 2002, The focal adhesion targeting (FAT) region of focal adhesion kinase is a four-helix bundle that binds paxillin, *Nat. Struct. Biol.* **9:** 101–106.

Hildebrand, J.D., Schaller, M.D., and Parsons, J.T., 1993, Identification of sequences required for the efficient localization of the focal adhesion kinase, pp125FAK, to cellular focal adhesions, *J. Cell Biol.* **123:** 993–1005.

Hong, K., Hinck, L., Nishiyama, M., Poo, M.M., Tessier-Lavigne, M., and Stein, E., 1999, A ligand-gated association between cytoplasmic domains of UNC5 and DCC family receptors converts netrin-induced growth cone attraction to repulsion, *Cell* **97:** 927–941.

Hopker, V.H., Shewan, D., Tessier-Lavigne, M., Poo, M., and Holt, C., 1999, Growth-cone attraction to netrin-1 is converted to repulsion by laminin-1, *Nature* **401:** 69–73.

Hsia, D.A., Mitra, S.K., Hauck, C.R., Streblow, D.N., Nelson, J.A., Ilic, D., et al., 2003, Differential regulation of cell motility and invasion by FAK, *J. Cell Biol.* **160:** 753–767.

Hynes, R.O., 1992, Integrins: Versatility, modulation, and signaling in cell adhesion, *Cell* **69:** 11–25.

Hynes, R.O., 1999, Cell adhesion: Old and new questions, *Trends Cell Biol.* **9:** M33–7.

Ignelzi, M.A., Jr., Miller, D.R., Soriano, P., and Maness, P.F., 1994, Impaired neurite outgrowth of src-minus cerebellar neurons on the cell adhesion molecule L1, *Neuron* **12:** 873–884.

Ilic, D., Furuta, Y., Kanazawa, S., Takeda, N., Sobue, K., Nakatsuji, N., et al., 1995, Reduced cell motility and enhanced focal adhesion contact formation in cells from FAK-deficient mice, *Nature* **377:** 539–544.

Ivankovic-Dikic, I., Gronroos, E., Blaukat, A., Barth, B.U., and Dikic, I., 2000, Pyk2 and FAK regulate neurite outgrowth induced by growth factors and integrins, *Nat. Cell Biol.* **2:** 574–581.

Jalink, K., van Corven, E.J., Hengeveld, T., Morii, N., Narumiya, S., and Moolenaar, W.H., 1994, Inhibition of lysophosphatidate- and thrombin-induced neurite retraction and neuronal cell rounding by ADP ribosylation of the small GTP-binding protein Rho, *J. Cell Biol.* **126:** 801–810.

Kadare, G., Toutant, M., Formstecher, E., Corvol, J.C., Carnaud, M., Boutterin, M.C., et al., 2003, PIAS1-mediated sumoylation of focal adhesion kinase activates its autophosphorylation, *J. Biol. Chem.* **278:** 47434–47440.

Kim, B., and Feldman, E.L., 1998, Differential regulation of focal adhesion kinase and mitogen-activated protein kinase tyrosine phosphorylation during insulin-like growth factor-I-mediated cytoskeletal reorganization, *J. Neurochem.* **71:** 1333–1336.

Klemke, R.L., Leng, J., Molander, R., Brooks, P.C., Vuori, K., and Cheresh, D.A., 1998, CAS/Crk coupling serves as a "molecular switch" for induction of cell migration, *J. Cell Biol.* **140:** 961–972.

Kuhn, T.B., Schmidt, M.F., and Kater, S.B., 1995, Laminin and fibronectin guideposts signal sustained but opposite effects to passing growth cones, *Neuron* **14:** 275–285.

Lev, S., Moreno, H., Martinez, R., Canoll, P., Peles, E., Musacchio, J.M., et al., 1995, Protein tyrosine kinase PYK2 involved in Ca(2+)-induced regulation of ion channel and MAP kinase functions, *Nature* **376:** 737–745.

Li, W., Lee, J., Vikis, H.G., Lee, S.H., Liu, G., Aurandt, J., et al., 2004, Activation of FAK and Src are receptor-proximal events required for netrin signaling, *Nat. Neurosci.* **7:** 1213–1221.

Lin, C.H., and Forscher, P., 1995, Growth cone advance is inversely proportional to retrograde F-actin flow, *Neuron* **14:** 763–771.

Lin, C.H., Thompson, C.A., and Forscher, P., 1994, Cytoskeletal reorganization underlying growth cone motility, *Curr. Opin. Neurobiol.* **4:** 640–647.

Liu, G., Beggs, H., Jurgensen, C., Park, H.T., Tang, H., Gorski, J., et al., 2004, Netrin requires focal adhesion kinase and Src family kinases for axon outgrowth and attraction, *Nat. Neurosci.* **7:** 1222–1232.

Markus, A., Patel, T.D., and Snider, W.D., 2002, Neurotrophic factors and axonal growth, *Curr. Opin. Neurobiol.* **12:** 523–531.

Maroney, A.C., Lipfert, L., Forbes, M.E., Glicksman, M.A., Neff, N.T., Siman, R., et al., 1995, K-252a induces tyrosine phosphorylation of the focal adhesion kinase and neurite outgrowth in human neuroblastoma SH-SY5Y cells, *J. Neurochem.* **64:** 540–549.

McKean, D.M., Sisbarro, L., Ilic, D., Kaplan-Alburquerque, N., Nemenoff, R., Weiser-Evans, M., et al., 2003, FAK induces expression of Prx1 to promote tenascin-C-dependent fibroblast migration, *J. Cell Biol.* **161:** 393–402.

Mellitzer, G., Xu, Q., and Wilkinson, D.G., 1999, Eph receptors and ephrins restrict cell intermingling and communication, *Nature* **400:** 77–81.

Menegon, A., Burgaya, F., Baudot, P., Dunlap, D.D., Girault, J.A., and Valtorta, F., 1999, FAK+ and PYK2/CAKbeta, two related tyrosine kinases highly expressed in the central nervous system: Similarities and differences in the expression pattern, *Eur. J. Neurosci.* **11:** 3777–3788.

Miao, H., Burnett, E., Kinch, M., Simon, E., and Wang, B., 2000, Activation of EphA2 kinase suppresses integrin function and causes focal-adhesion-kinase dephosphorylation, *Nat. Cell Biol.* **2:** 62–69.

Mitchison, T.J., and Cramer, L.P., 1996, Actin-based cell motility and cell locomotion, *Cell* **84:** 371–379.

Mitra, S.K., Hanson, D.A., and Schlaepfer, D.D., 2005, Focal adhesion kinase: In command and control of cell motility, *Nat. Rev. Mol. Cell Biol.* **6:** 56–68.

Miyamoto, S., Teramoto, H., Coso, O.A., Gutkind, J.S., Burbelo, P.D., Akiyama, S.K., et al., 1995, Integrin function: Molecular hierarchies of cytoskeletal and signaling molecules, *J. Cell Biol.* **131:** 791–805.

Mueller, B.K., 1999, Growth cone guidance: First steps towards a deeper understanding, *Annu. Rev. Neurosci.* **22:** 351–388.

Murai, K.K., Nguyen, L.N., Irie, F., Yamaguchi, Y., and Pasquale, E.B., 2003, Control of hippocampal dendritic spine morphology through ephrin-A3/EphA4 signaling, *Nat. Neurosci.* **6:** 153–160.

Nermut, M.V., Eason, P., Hirst, E.M., and Kellie, S., 1991, Cell/substratum adhesions in RSV-transformed rat fibroblasts, *Exp. Cell Res.* **193:** 382–397.

Nikolic, M., 2004, The molecular mystery of neuronal migration: FAK and Cdk5, *Trends Cell Biol.* **14:** 1–5.

Nobes, C.D., and Hall, A., 1999, Rho GTPases control polarity, protrusion, and adhesion during cell movement, *J. Cell Biol.* **144:** 1235–1244.

Palazzo, A.F., Cook, T.A., Alberts, A.S., and Gundersen, G.G., 2001, mDia mediates Rho-regulated formation and orientation of stable microtubules, *Nat. Cell Biol.* **3:** 723–729.

Palazzo, A.F., Eng, C.H., Schlaepfer, D.D., Marcantonio, E.E., and Gundersen, G.G., 2004, Localized stabilization of microtubules by integrin- and FAK-facilitated Rho signaling, *Science* **303:** 836–839.

Park, S.Y., Avraham, H., and Avraham, S., 2000, Characterization of the tyrosine kinases RAFTK/Pyk2 and FAK in nerve growth factor-induced neuronal differentiation, *J. Biol. Chem.* **275:** 19768–19777.

Pasterkamp, R.J., and Kolodkin, A.L., 2003, Semaphorin junction: Making tracks toward neural connectivity, *Curr. Opin. Neurobiol.* **13:** 79–89.

Pasterkamp, R.J., Peschon, J.J., Spriggs, M.K., and Kolodkin, A.L., 2003, Semaphorin 7A promotes axon outgrowth through integrins and MAPKs, *Nature* **424:** 398–405.

Poullet, P., Gautreau, A., Kadare, G., Girault, J.A., Louvard, D., and Arpin, M., 2001, Ezrin interacts with focal adhesion kinase and induces its activation independently of cell-matrix adhesion, *J. Biol. Chem.* **276:** 37686–37691.

Ren, X.D., Kiosses, W.B., Sieg, D.J., Otey, C.A., Schlaepfer, D.D., and Schwartz, M.A., 2000, Focal adhesion kinase suppresses Rho activity to promote focal adhesion turnover, *J. Cell. Sci.* **113:** 3673–3678.

Ren, X.R., Ming, G.L., Xie, Y., Hong, Y., Sun, D.M., Zhao, Z.Q., et al., 2004, Focal adhesion kinase in netrin-1 signaling, *Nat. Neurosci.* **7:** 1204–1212.

Renaudin, A., Lehmann, M., Girault, J., and McKerracher, L., 1999, Organization of point contacts in neuronal growth cones, *J. Neurosci. Res.* **55:** 458–471.

Renshaw, M.W., Price, L.S., and Schwartz, M.A., 1999, Focal adhesion kinase mediates the integrin signaling requirement for growth factor activation of MAP kinase, *J. Cell Biol.* **147:** 611–618.

Richardson, A., and Parsons, T., 1996, A mechanism for regulation of the adhesion-associated proteintyrosine kinase pp125FAK, *Nature* **380:** 538–540.

Rico, B., Beggs, H.E., Schahin-Reed, D., Kimes, N., Schmidt, A., and Reichardt, L.F., 2004, Control of axonal branching and synapse formation by focal adhesion kinase, *Nat. Neurosci.* **7:** 1059–1069.

Rodriguez-Fernandez, J.L., and Rozengurt, E., 1998, Bombesin, vasopressin, lysophosphatidic acid, and sphingosylphosphorylcholine induce focal adhesion kinase activation in intact Swiss 3T3 cells, *J. Biol. Chem.* **273:** 19321–19328.

Schaller, M.D., Borgman, C.A., Cobb, B.S., Vines, R.R., Reynolds, A.B., and Parsons, J.T., 1992, pp125FAK a structurally distinctive protein-tyrosine kinase associated with focal adhesions, *Proc. Natl. Acad. Sci. USA* **89:** 5192–5196.

Schaller, M.D., Borgman, C.A., and Parsons, J.T., 1993, Autonomous expression of a non-catalytic domain of the focal adhesion-associated protein tyrosine kinase pp125FAK, *Mol. Cell. Biol.* **13:** 785–791.

Schaller, M.D., Otey, C.A., Hildebrand, J.D., and Parsons, J.T., 1995, Focal adhesion kinase and paxillin bind to peptides mimicking beta integrin cytoplasmic domains, *J. Cell Biol.* **130:** 1181–1187.

Schlaepfer, D.D., Hanks, S.K., Hunter, T., and van der Geer, P., 1994, Integrin-mediated signal transduction linked to Ras pathway by GRB2 binding to focal adhesion kinase, *Nature* **372:** 786–791.

Schoenwaelder, S.M., and Burridge, K., 1999, Bidirectional signaling between the cytoskeleton and integrins, *Curr. Opin. Cell Biol.* **11:** 274–286.

Schwarzbauer, J.E., 1997, Cell migration: May the force be with you, *Curr. Biol.* **7:** R292–R294.

Shen, T.L., Han, D.C., and Guan, J.L., 2002, Association of Grb7 with phosphoinositides and its role in the regulation of cell migration, *J. Biol. Chem.* **277:** 29069–29077.

Shen, Y., and Schaller, M.D., 1999, Focal adhesion targeting: The critical determinant of FAK regulation and substrate phosphorylation, *Mol. Biol. Cell* **10:** 2507–2518.

Siciliano, J.C., Toutant, M., Derkinderen, P., Sasaki, T., and Girault, J.A., 1996, Differential regulation of proline-rich tyrosine kinase 2/cell adhesion kinase beta (PYK2/CAKbeta) and pp125(FAK) by glutamate and depolarization in rat hippocampus, *J. Biol. Chem.* **271:** 28942–28946.

Sieg, D.J., Hauck, C.R., and Schlaepfer, D.D., 1999, Required role of focal adhesion kinase (FAK) for integrin-stimulated cell migration, *J. Cell. Sci.* **112:** 2677–2691.

Sieg, D.J., Hauck, C.R., Ilic, D., Klingbeil, C.K., Schaefer, E., Damsky, C.H., et al., 2000, FAK integrates growth-factor and integrin signals to promote cell migration, *Nat. Cell Biol.* **2:** 249–256.

Stevens, G.R., Zhang, C., Berg, M.M., Lambert, M.P., Barber, K., Cantallops, I., et al., 1996, CNS neuronal focal adhesion kinase forms clusters that co-localize with vinculin, *J. Neurosci. Res.* **46:** 445–455.

Suter, D.M., and Forscher, P., 1998, An emerging link between cytoskeletal dynamics and cell adhesion molecules in growth cone guidance, *Curr. Opin. Neurobiol.* **8:** 106–116.

Suter, D.M., and Forscher, P., 2001, Transmission of growth cone traction force through apCAM-cytoskeletal linkages is regulated by Src family tyrosine kinase activity, *J. Cell Biol.* **155:** 427–438.

Suter, D.M., Errante, L.D., Belotserkovsky, V., and Forscher, P., 1998, The Ig superfamily cell adhesion molecule, apCAM, mediates growth cone steering by substrate-cytoskeletal coupling, *J. Cell Biol.* **141:** 227–240.

Tamura, M., Gu, J., Matsumoto, K., Aota, S., Parsons, R., and Yamada, K.M., 1998, Inhibition of cell migration, spreading, and focal adhesions by tumor suppressor PTEN, *Science* **280:** 1614–1617.

Tanaka, E., and Sabry, J., 1995, Making the connection: Cytoskeletal rearrangements during growth cone guidance, *Cell* **83:** 171–176.

Tawil, N., Wilson, P., and Carbonetto, S., 1993, Integrins in point contacts mediate cell spreading: Factors that regulate integrin accumulation in point contacts vs. focal contacts, *J. Cell Biol.* **120:** 261–271.

Tessier-Lavigne, M., and Goodman, C.S., 1996, The molecular biology of axon guidance, *Science* **274:** 1123–1133.

Toutant, M., Studler, J.M., Burgaya, F., Costa, A., Ezan, P., Gelman, M., et al., 2000, Autophosphorylation of Tyr397 and its phosphorylation by Src-family kinases are altered in focal-adhesion-kinase neuronal isoforms, *Biochem. J.* **348:** 119–128.

Toutant, M., Costa, A., Studler, J.M., Kadare, G., Carnaud, M., and Girault, J.A., 2002, Alternative splicing controls the mechanisms of FAK autophosphorylation, *Mol. Cell. Biol.* **22:** 7731–7743.

Vaillant, C., Meissirel, C., Mutin, M., Belin, M.F., Lund, L.R., and Thomasset, N., 2003, MMP-9 deficiency affects axonal outgrowth, migration, and apoptosis in the developing cerebellum, *Mol. Cell. Neurosci.* **24:** 395–408.

Valtorta, F., and Leoni, C., 1999, Molecular mechanisms of neurite extension, *Philos. Trans. R. Soc. Lond. B. Biol. Sci.* **354:** 387–394.

Varnum-Finney, B., and Reichardt, L.F., 1994, Vinculin-deficient PC12 cell lines extend unstable lamellipodia and filopodia and have a reduced rate of neurite outgrowth, *J. Cell Biol.* **127:** 1071–1084.

Wang, K.H., Brose, K., Arnott, D., Kidd, T., Goodman, C.S., Henzel, W., et al., 1999, Biochemical purification of a mammalian slit protein as a positive regulator of sensory axon elongation and branching, *Cell* **96:** 771–784.

Webb, D.J., Donais, K., Whitmore, L.A., Thomas, S.M., Turner, C.E., Parsons, J.T., et al., 2004, FAK-Src signalling through paxillin, ERK and MLCK regulates adhesion disassembly, *Nat. Cell Biol.* **6:** 154–161.

Webber, C.A., Hocking, J.C., Yong, V.W., Stange, C.L., and McFarlane, S., 2002, Metalloproteases and guidance of retinal axons in the developing visual system, *J. Neurosci.* **22:** 8091–8100.

Wu, D.Y., and Goldberg, D.J., 1993, Regulated tyrosine phosphorylation at the tips of growth cone filopodia, *J. Cell Biol.* **123:** 653–664.

Wu, X., Suetsugu, S., Cooper, L.A., Takenawa, T., and Guan, J.L., 2004, Focal adhesion kinase regulation of N-WASP subcellular localization and function, *J. Biol. Chem.* **279:** 9565–9576.

Wu, X., Gan, B., Yoo, Y., and Guan, J.L., 2005, FAK-Mediated src phosphorylation of endophilin A2 inhibits endocytosis of MT1-MMP and promotes ECM degradation, *Dev. Cell.* **9:** 185–196.

Xie, Z., Sanada, K., Samuels, B.A., Shih, H., and Tsai, L.H., 2003, Serine 732 phosphorylation of FAK by Cdk5 is important for microtubule organization, nuclear movement, and neuronal migration, *Cell* **114:** 469–482.

Xiong, W.C., Macklem, M., and Parsons, J.T., 1998, Expression and characterization of splice variants of PYK2, a focal adhesion kinase-related protein, *J. Cell. Sci.* **111:** 1981–1991.

Yamada, K.M., and Geiger, B., 1997, Molecular interactions in cell adhesion complexes, *Curr. Opin. Cell Biol.* **9:** 76–85.

Yang, P., Baker, K.A., and Hagg, T., 2005, A disintegrin and metalloprotease 21 (ADAM21) is associated with neurogenesis and axonal growth in developing and adult rodent CNS, *J. Comp. Neurol.* **490:** 163–179.

Yu, D.H., Qu, C.K., Henegariu, O., Lu, X., and Feng, G.S., 1998, Protein-tyrosine phosphatase Shp-2 regulates cell spreading, migration, and focal adhesion, *J. Biol. Chem.* **273:** 21125–21131.

Zachary, I., and Rozengurt, E., 1992, Focal adhesion kinase (p125FAK): A point of convergence in the action of neuropeptides, integrins, and oncogenes, *Cell* **71:** 891–894.

Zeng, L., Si, X., Yu, W.P., Le, H.T., Ng, K.P., Teng, R.M., et al., 2003, PTP alpha regulates integrin-stimulated FAK autophosphorylation and cytoskeletal rearrangement in cell spreading and migration, *J. Cell Biol.* **160:** 137–146.

Zhai, J., Lin, H., Nie, Z., Wu, J., Canete-Soler, R., Schlaepfer, W.W., et al., 2003, Direct interaction of focal adhesion kinase with p190RhoGEF, *J. Biol. Chem.* **278:** 24865–24873.

8
Regulation of Neuronal Morphogenesis by Abl Family Kinases

HAMEEDA SULTANA AND ANTHONY J. KOLESKE

8.1. Summary

Genetic studies indicate that Abl family kinases are required for proper axon and dendrite formation in developing metazoan nervous systems. Biochemical experiments suggest that Abl family kinases relay signals from guidance, growth factor, and adhesion receptors to promote cytoskeletal rearrangements in developing neurons. This chapter focuses on the recent advances in our understanding of how Abl family kinases regulate neuronal morphogenesis in developing metazoan nervous systems.

8.2. Introduction

The proper function of the nervous system depends on the complex architecture of neuronal networks. The elaborate morphology of neurons allows them to form thousands of specific synaptic connections with other neurons. Neurons start out as simple spherical neuroblast. Extracellular cues induce the formation of cylindrical extensions of the cell body and regulate the extension and branching of these processes into complex axon and dendrite arbors. These morphogenetic processes require dramatic cytoskeletal rearrangements that are coordinated by a variety of signaling and cytoskeletal regulatory proteins. Among these regulatory proteins are the Abl family nonreceptor tyrosine kinases that serve as essential mediators of guidance and adhesion receptor signaling in developing nervous systems (for structural organization of Abl family kinases, see Figure 8.1). This chapter will review the mechanisms by which Abl family kinases regulate neuronal morphogenesis.

8.3. Regulation of neuronal morphogenesis by Abl family kinases

Genetic studies in flies and mice indicate that Abl family kinases regulate axon guidance and dendrite formation in the developing nervous system.

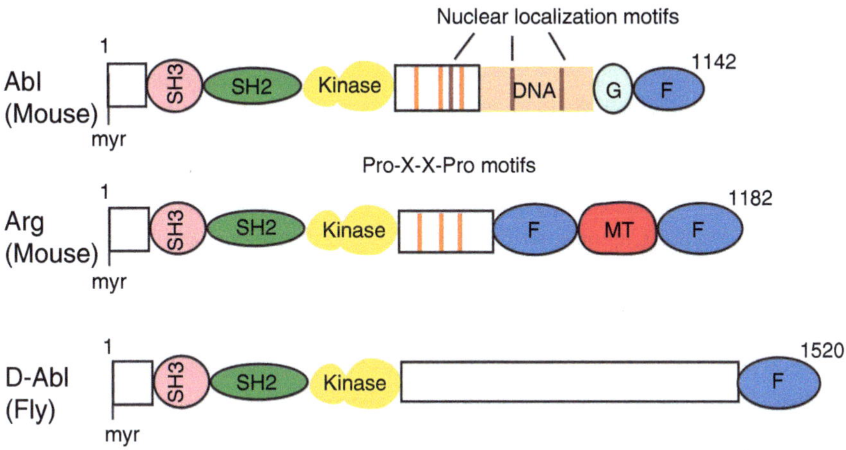

FIGURE 8.1. Structural organization of Abl family kinases. This diagram shows the domain organization of Abl, Arg, and D-Abl. Abl family kinases share extensive sequence similarity and conservation in their domain organization. After one of several alternatively spliced N-termini, each Abl family kinase contains Src homology-3 (SH3), Src homology-2 (SH2) protein–protein interaction domains followed by a kinase domain. The Abl family kinases contain three conserved Pro-X-X-Pro motifs immediately C-terminal to the kinase domain that serves as prospective binding sites for the SH3 domain containing proteins (Pendergast, 2002). The C-terminal halves of Abl family kinases contain multiple domains that interact directly with cytoskeletal proteins. Abl contains binding domains for G- and (F-) actin (McWhirter and Wang, 1993; Van Etten et al., 1994). Arg contains two F-actin–binding domains (F) and a microtubule-binding domain (MT) that may be used to directly regulate the cytoskeletal structure (McWhirter and Wang, 1993; Van Etten et al., 1994). Sequence alignment suggests that D-Abl also contains an F-acting–binding domain (F), although this has not been demonstrated experimentally. In addition to the cytoskeletal regulatory regions, Abl contains a DNA binding motif (Miao and Wang, 1996) and three nuclear localization sequences (NLS) (Wen et al., 1996) that are important for Abl functions in the nucleus.

8.3.1. D-Abl Regulates Axon Guidance at the Drosophila Midline

Drosophila (D)-Abl, the single Abl family kinase in flies, is abundant in the nervous system, where it localizes to developing axons (Bennett and Hoffmann, 1992; Gertler et al., 1989). Genetic studies indicate that D-Abl (encoded by the *abl* gene) is required to regulate midline crossing of CNS axons and motor axon guidance during fly embryogenesis (Bashaw et al., 2000; Comer et al., 1998; Forsthoefel et al., 2005; Gertler et al., 1989; Hsouna et al., 2003; Lanier and Gertler, 2000; Liebl et al., 2000; Wills et al., 1999a, 2002).

The loss of D-Abl function results in ectopic crossing of the midline by longitudinal axons (Figure 8.2A) (Forsthoefel et al., 2005; Wills et al., 2002). However, D-Abl overexpression also promotes ectopic midline crossing (Bashaw et al.,

2000). These observations indicate that D-Abl activity must be tightly balanced to maintain an optimal state for avoidance of the midline. Longitudinal axons are prevented from crossing the midline through the actions of three roundabout (*robo, robo-3*) receptors, which function as receptors for the midline chemorepellent Slit. The ectopic midline crossing phenotypes of *robo, robo-2/+,+* and *robo,robo-3/+,+* heterozygotes are dramatically enhanced by heterozygosity for an *abl* mutation (Figure 8.2A) (Hsouna et al., 2003; Wills et al., 2002). Mutations in a cytoskeletal regulatory protein (Capulet), an Ena/VASP protein (Enabled), and a Rho guanine nucleotide exchange factor (Trio), also modulate *Abl*-dependent midline phenotypes (Forsthoefel et al., 2005; Gertler et al., 1989, 1995; Liebl et al., 2000; Wills et al., 1999a, 2002). Together, these observations support a model in which D-Abl acts downstream of Robo receptors to coordinate the cytoskeletal rearrangements required for avoidance of the midline.

In addition to its role as a transducer of robo signaling, D-Abl also appears to act as a feedback regulator of Robo activity. D-Abl phosphorylates the Robo receptor at a single tyrosine residue in conserved cytoplasmic domain 1 (Bashaw et al., 2000). A Robo tyrosine to phenylalanine substitution mutant that cannot be phosphorylated on this site has an enhanced ability to prevent midline crossing (Bashaw et al., 2000). By modulating *robo* function, *Abl* may allow the receptor to respond to a wide range of Slit concentrations (Willis et al., 2002).

D-Abl also interacts functionally with frazzled (*fra*), the *Drosophila* receptor for the midline attractant Netrin, to regulate the formation of commissural axon bundles (Figure 8.2B). The loss of D-Abl function enhances the *fra* phenotype, leading to an increase in the number of thin or missing commissures in *fra/fra; abl/abl* mutants (Figure 8.2B) (Forsthoefel et al., 2005). D-Abl can bind directly to the *fra* cytoplasmic tail, suggesting that it might transduce signals from *fra* in response to ligand binding (Forsthoefel et al., 2005). Coexpression of D-Abl with fra promotes tyrosine phosphorylation of the Rho family GTPase regulator Trio in S2 cells. As outlined below, Rho family GTPases are important regulators of axon and dendrite morphogenesis. A Fra:D-Abl pathway might act through Trio to regulate the activity of Rho family GTPases during commissural axon outgrowth.

8.3.2. D-Abl Regulates Motor Axon Guidance

In the normal course of fly development, the b subset of intersegmental neurons (ISNb) defasciculates from the intersegmental nerve (ISN) shortly after it leaves the ventral nerve cord and branches off to innervate its target muscles. The guidance of ISNs to their muscle targets requires D-Abl. In *abl* mutant flies, these neurons defasciculate normally from the ISN but stop short of reaching their final targets (Wills et al., 1999a). Overexpression of *Abl* leads to a complete bypass phenotype in which the ISNb neurons do not defasciculate and enter the muscle, continuing instead along with the other ISN neurons (Wills et al., 1999b). This phenotype is shared with mutants in the Dlar receptor tyrosine phosphatase (Wills et al., 1999b). Thus, Dlar and D-Abl appear to act antagonistically to modulate ISNb axon guidance.

D-Abl and Dlar might regulate ISNb axon guidance by regulating the phosphorylation of the Enabled (Ena) protein. Ena can be phosphorylated by D-Abl and dephosphorylated by Dlar *in vitro*. In *ena* mutant flies, ISNb neurons exhibit a bypass phenotype similar to that observed upon D-Abl overexpression. Thus, D-Abl phosphorylation appears to inhibit Ena function and Dlar counteracts this inhibition. Mammalian Ena/VASP family proteins (Lebrand et al., 2004) promote the formation of filopodia. In flies, Ena may act similarly to promote the formation of growth cone filopodia in growth cones. D-Abl may inhibit this process by restricting the formation of filopodia.

8.3.3. D-Abl Restricts Dendrite Branching in Peripheral Neurons

D-Abl has been shown to regulate dendrite formation in neurons of the *Drosophila* peripheral nervous system. Dendritic arborization (DA) sensory neurons, have slightly more branches in *abl* null embryos (Li et al., 2005). Overexpression of D-Abl in these neurons leads to a more simplified dendrite arbor structure. Overexpression of a kinase-inactive D-Abl point mutant does not affect dendrite morphology (Li et al., 2005). These data suggest that D-Abl acts to restrict dendrite branch formation *in vivo*. The *ena* mutant flies exhibit reductions in the dendrite arbors of DA neurons. These data indicate that D-Abl and Ena have antagonistic roles in the formation of dendrite branches on DA neurons just as they have been shown to play antagonistic roles in motor axon guidance (Li et al., 2005).

8.3.4. Abl and Arg Are Required for Dendrite Branch Maintenance in Mice

Genetic studies in mice have shown that Abl and Arg are required for dendrite branch maintenance in mice (Moresco et al., 2005). The dendrite arbors of cortical layer 5 pyramidal neurons develop normally in *arg*$^{-/-}$ mice through postnatal week 3, but dendrite branches are not stabilized, leading to a reduction in dendrite arbor by early adulthood (Moresco et al., 2005). More severe dendrite atrophy is observed in the brains of mice that lack both Abl and Arg (Figure 8.2C). These data suggest that Abl and Arg may have overlapping function in the regulation of cortical dendrite maintenance. Loss of Arg function leads to a similar dendrite atrophy in hippocampal CA1 pyramidal neurons after postnatal day 18 (A.J.K., unpublished data).

Abl and Arg localize to both axons and dendrites in cultured cortical neurons. It is not clear where Abl and Arg act in axons or dendrites to promote dendrite stabilization. In cultured cortical neurons, Abl promotes neurite elongation and branching (Jones et al., 2004; Woodring et al., 2002; Zukerberg et al., 2000), wherease Arg promotes axon and dendrite branching (Moresco et al., 2005). The observation that Arg is required for adhesion-dependent stimulation of neurite dynamics suggests that Arg might act directly within neurons to regulate dendrite arbor maintenance.

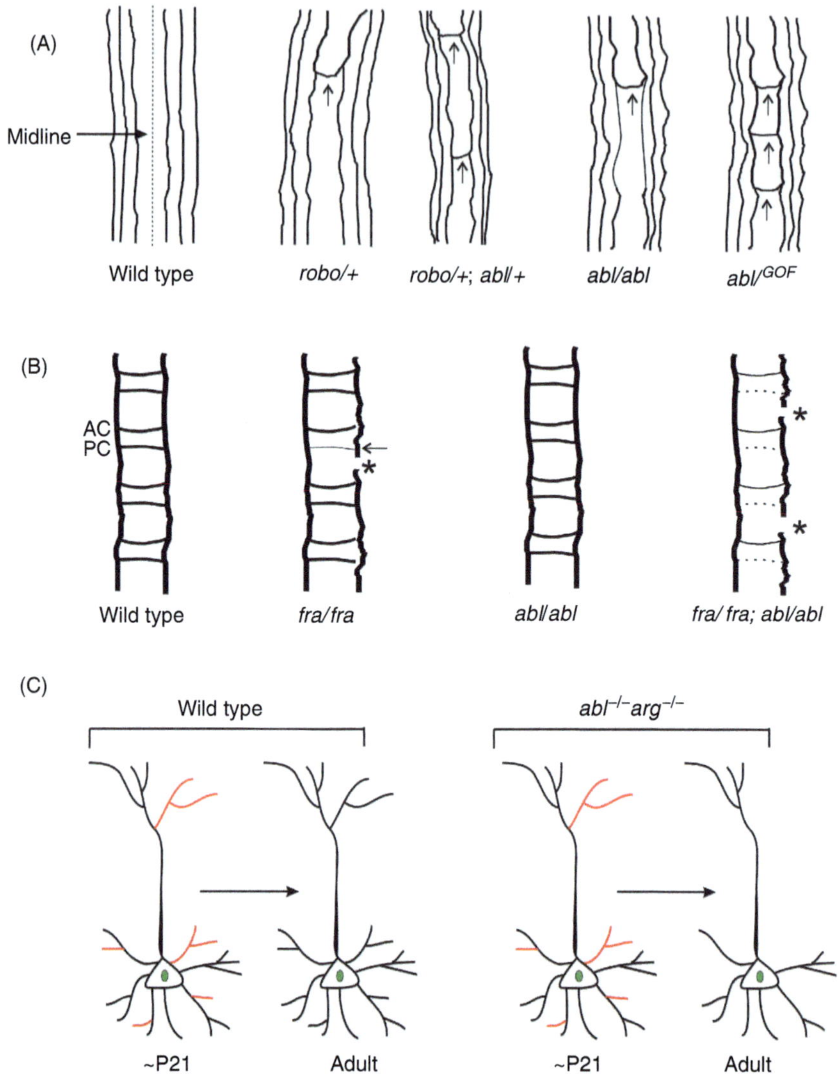

FIGURE 8.2. Abl family kinases regulate neuronal morphogenesis. (A) D-Abl regulates midline crossing. The schematic shows the midline axon guidance phenotypes of the longitudinal axon bundles in stages 16–17 of the *Drosophila* embryos. Wild-type embryos have three longitudinal axon bundles on each side of the midline that do not cross the midline. Heterozygosity for the midline repellent receptor robo (*robo/+*) leads to ectopic midline crossing of longitudinal axons (arrow) and this phenotype is enhanced by heterozygosity for D-Abl (*robo/+; abl/+*). Both *abl* homozygous mutants and flies overexpressing wild-type D-Abl (*abl*^{GOF}) show ectopic midline crossing phenotypes. (B) Abl regulates commissure formation. In wild-type embryos, CNS axons form two commissures per segment (AC and PC). Homozygous *fra/fra* mutants exhibit only subtle disruption of the commissural axon bundles with thin or missing commissures (arrows) and breaks in

The fact that developing dendrite arbors become stabilized by synaptic activity (Lohmann et al., 2002; McAllister, 2000; Redmond et al., 2002; Vaillant et al., 2002; Whitford et al., 2002) suggests an alternative scenario in which Abl family kinases indirectly control dendrite branch maintenance by regulating the formation of afferents that innervate the cortical dendrite arbor. The finding that the two major axon tracts that innervate the cortex (the corpus callosum and the internal capsule) are normal in *arg*$^{-/-}$ and *abl*$^{-/-}$ *arg*$^{-/-}$ brains argues that Abl family kinases are not required for the general targeting of axons to their specific brain region (Moresco et al., 2005). It remains possible that Abl family kinases are required for the fine targeting of afferent axons to their dendritic contacts.

A third possibility is that Abl and Arg influence dendrite branch stability by regulating the efficiency of neurotransmitter release at excitatory synapses. Hippocampal slices from *arg*$^{-/-}$ and *abl*$^{-/-}$ mice exhibit defects in paired-pulse facilitation, a form of synaptic plasticity that measures the efficiency of presynaptic neurotransmitter release following two closely paired stimuli. Reductions in synaptic efficiency could lead to a failure of dendrite branch maintenance over time. Future studies of dendrite development in mice bearing inducible Abl/Arg transgenes or brain region-specific knockouts of Abl and Arg should clarify when and where these kinases act to control dendrite branch maintenance *in vivo*.

8.3.5. How Can Abl Family Kinases Regulate so Many Different Processes?

Clearly, Abl family kinases regulate a diverse set of biological processes in developing nervous systems. It is not clear how these kinases regulate such different biological outcomes. Genetic studies suggest that guidance receptors signal through several downstream effectors to regulate axon guidance. Abl family kinases may represent only one component of a larger set of cytoskeletal regulators that mediate a specific response. In addition, Abl family kinases possess multiple distinct protein–protein interaction domains. These features may enable distinct receptors to engage Abl family kinases differently to achieve distinct biological outcomes. Finally, it is possible that Abl family kinases are required to maintain a specific cellular structure (e.g., filopodia) that are required to detect different classes of guidance cues. The identification of a growing list of downstream mediators of Abl family kinase signaling will facilitate the dissection of how these molecules contribute to these multifarious developmental processes.

longitudinal pathways (indicated with asterisks). Although homozygous *abl/abl* mutants have normal commissures, loss of D-Abl function enhances the commissural defects of *fra/fra;abl/abl* double mutants. (C) Abl and Arg are required for dendrite branch maintenance. The schematic depicts cortical dendrite arbor development in wild type and *abl*$^{-/-}$ *arg*$^{-/-}$ neurons. Cortical dendrites develop normally in *abl*$^{-/-}$*arg*$^{-/-}$ mice and appear normal at postnatal day 21 (P21). Dendrite branches are lost in *abl*$^{-/-}$*arg*$^{-/-}$ mice, leading to smaller dendrite arbors by adulthood.

8.4. Mechanisms by Which Abl Family Kinases Control Neuronal Morphogenesis

Studies of nonneuronal cultured cells indicate that Abl family kinases mediate signaling from a diverse set of adhesion, growth factor, and guidance receptors (Table 8.1). Studies of cultured neurons have begun to elucidate the cellular and molecular mechanisms by which Abl family kinases control neuronal morphogenesis.

8.4.1. Abl Family Kinases Regulate Neurite Outgrowth in Response to Integrin Engagement

In the developing brain, growing axons and dendrites migrate on and within the extracellular matrix (ECM), the semisolid meshwork that surrounds cells. Neurons use heterodimeric adhesion receptors called integrins to attach to specific ECM proteins (DeMali et al., 2003; Hynes, 2002; Jalali et al., 2001; Schwartz, 2001; Schwartz et al., 1995). Binding of integrin receptors to ECM proteins potently stimulates neurite outgrowth *in vitro* (Reichardt and Tomaselli, 1991), suggesting that integrin–ECM interactions may regulate neuronal morphogenesis *in vivo*. Disruption of integrins in *Drosophila* lead to gaps in longitudinal axon bundles and defects in peripheral axon targeting (Hoang and Chiba, 1998). Integrins are widely expressed in the mammalian brain (Clegg et al., 2003; Pinkstaff et al., 1999; Schmid and Anton, 2003), where they are likely to play important roles in axon and dendrite morphogenesis.

TABLE 8.1. Receptors that can bind to and/or interact with Abl

Receptor	*In vitro/In vivo*	Proposed functions of the receptor	References
robo	*In vitro/In vivo*	Receptor for Slit, repulsive axon guidance in response to secreted Slit proteins, dendrite branching and neuronal migration in vertebrates	Bashaw et al., 2000; Wills et al., 2002
Dlar	*In vitro/In vivo*	Motor axon guidance	Wills et al., 1999a,b
fas I and II	*In vivo*	CNS axon pathways	Elkins et al., 1990
fra	*In vitro/In vivo*	Growth cone guidance neurite outgrowth, cell spreading and filopodia extensions	Forsthoefel et al., 2005
TrkA	*In vitro/In vivo*	Receptor for nerve growth factor (NGF), Survival and neuronal differentiation	Koch et al., 2000; Yano et al., 2000
EphB2	*In vivo*	axon behavior actin cytoskeleton dynamics	Yu et al., 2001; Harbott et al., 2005
N-cadherins	*In vivo*	Proper axon outgrowth and fasciculation, development of neuronal connections	Rhee et al., 2002

The interactions depicted here are from the available biochemical or genetic data showing the interactions between Abl family kinases and a diverse set of cell adhesion, growth factors, and guidance receptors. Dlar = receptor protein-tyrosine phosphatases; fas I and II = fasciclin I and II; fra = netrin receptor frazzled; TrkA = tyrosine kinase receptor A; and EphB2 = ephrins.

Adhesion of embryonic mouse cortical neurons to the integrin ligand laminin promotes neurite outgrowth and branching beyond that observed in control neurons plated on poly-L-ornithine. Integrin-mediated neuronal adhesion to the guidance molecule Semaphorin-7A also promotes increased neurite branching. The responses to both of these adhesive cues are deficient in *arg* $^{-/-}$ cortical neurons or in wild-type neurons treated with the Abl/Arg kinase inhibitor STI571 (Moresco et al., 2005). Similar experiments show that *abl*$^{-/-}$ mouse cortical neurons plated on laminin have fewer neurite branches than wild-type neurons (Woodring et al., 2002). These observations demonstrate that Abl and Arg are required for adhesion-dependent increases in neurite branching.

Integrins activate the kinase activity of Abl and Arg to promote changes in neuronal structure. Integrin-mediated adhesion activates Abl and Arg kinase activity (Hernandez et al., 2004; Lewis et al., 1996; Woodring et al., 2001). Overexpression of Arg in cultured mouse cortical neurons leads to increased elongation and branching of both axons and dendrites (Moresco et al., 2005). This response is not observed in neurons overexpressing a kinase-inactive Arg point mutant. Overexpression of a mutationally hyperactived Abl kinase mutant in rat cortical neurons promotes significantly increased neurite branching when compared to neurons overexpressing Abl alone (Woodring et al., 2002). Together, these experiments support the notions that increased Abl and Arg kinase activities promote increased neurite branching and elongation following integrin-mediated adhesion.

8.4.2. Abl Family Kinases Regulate Neuronal Morphogenesis by Inhibiting Rho

The RhoA (Rho), Rac1 (Rac), and Cdc42 GTPases are master regulators of neuronal morphogenesis (Lee et al., 2000; Li et al., 2002; Luo et al., 1996; Nakayama et al., 2000; Tashiro et al., 2000). These proteins act as molecular switches that regulate actin cytoskeletal dynamics by cycling between an inactive GDP-bound form and an active GTP-bound form. Rho family GTPases are controlled by two classes of regulatory molecules: guanine nucleotide exchange factors (GEFs), activate Rho family GTPases by promoting exchange of GTP for GDP, and GTPase-activating proteins (GAPs), inhibit Rho family GTPases by stimulating Rho to hydrolyze its GTP (Kaibuchi et al., 1999). A third class of regulators, guanine nucleotide dissociation inhibitors (GDIs) inhibit Rho family GTPases by preventing exchange of GDP for GTP.

Inhibition of Abl family kinases by treatment with STI571 results in simplification of dendrite arbor structure in cultured neurons (Jones et al., 2004; Moresco et al., 2005; Woodring et al., 2002;). This phenotype is similar to that observed in neurons overexpressing active Rho (Jones et al., 2004). Exposure to STI571 leads to a modest elevation of Rho activity in cultured hippocampal neurons. Inhibition of Rho-kinase (ROCK), a major effector of Rho, blocks the inhibitory effects of STI571 on dendrite formation (Jones et al., 2004). v-Abl, an oncogenic form of Abl that is constitutively active for kinase activity, promotes increased dendrite complexity in cultured hippocampal neurons. Coexpression of an activated Rho mutant suppresses the effects of v-Abl on dendrite arbor formation. Together, these data suggest that Abl family kinases regulate neurite outgrowth and branching by inhibiting Rho activity (Figure 8.3).

Abl family kinases likely inhibit Rho activity by regulating Rho GEF and GAP regulatory proteins. The Rho inhibitor p190RhoGAP is a major Arg substrate in the postnatal mouse brain (Hernandez et al., 2004). Integrin-mediated adhesion stimulates p190RhoGAP phosphorylation in wild type but not *arg*$^{-/-}$ neurons. Arg-dependent phosphorylation of p190RhoGAP on tyrosine-1105 activates its RhoGAP activity. p190RhoGAP promotes neurite outgrowth when expressed in N2A neuroblastoma cells (Hernandez et al., 2004). Coexpression of Arg with p190RhoGAP leads to increased neuritogenesis. Treatment of N2A cells with STI571 also inhibits adhesion-dependent neuritogenesis, suggesting that Abl family kinases can regulate this process under normal physiological concentrations. Active Rho is a potent stimulator of actomyosin contractility in cells. p190RhoGAP is required for axon branch stabilization of mushroom body (MB) neurons in the *Drosophila* brain (Billuart et al., 2001). The accumulated evidence indicates that Abl family kinases inhibit Rho to stabilize nascent neurite branches (Figure 8.3).

8.5. Additional Mechanisms by Which Abl Family Kinases Might Regulate Axon/Dendrite Protrusion and Elongation

In addition to their effects on dendrite stability, Abl family kinases promote neurite branching and elongation. Although it is not completely clear how Abl family kinases regulate these processes in neurons, studies of nonneuronal cells have provided insights into how Abl family kinases regulate protrusions of the cell membrane.

8.5.1. Abl Promotes Filopodial Protrusion Through Phosphorylation of Dok Family Proteins

Expression of Abl in *abl*$^{-/-}$*arg*$^{-/-}$ fibroblasts leads to the appearance of numerous F-actin–rich microspikes (Woodring et al., 2002). The downstream of kinase (Dok) proteins serve as essential downstream effectors of Abl in this process. Coexpresssion of Abl with Dok-1 or Dok-2 leads to increased formation of actin microspikes (Master et al., 2003; Woodring et al., 2004). Dok proteins are modular adapter proteins that contain a pleckstrin homology (PH) domain, a phosphotyrosine binding (PTB) domain, and a C-terminal tail with multiple tyrosine phosphorylation sites. Abl and Dok-1 both localize to filopodial tips (Woodring et al., 2004). In response to integrin-mediated adhesion, Abl phosphorylates Dok1 on Y361. This phosphorylated residue acts as a binding site for Nck, which binds and activates the cytoskeletal regulatory proteins PAK (Zhao et al., 2000), N-WASp (Rohatgi et al., 2001), and WASp (Rivero-Lezcano et al., 1995). These data suggest that Abl promotes actin-based filopodial protrusions by assembling an Abl:Dok-1:Nck complex to promote filopodia formation and elongation.

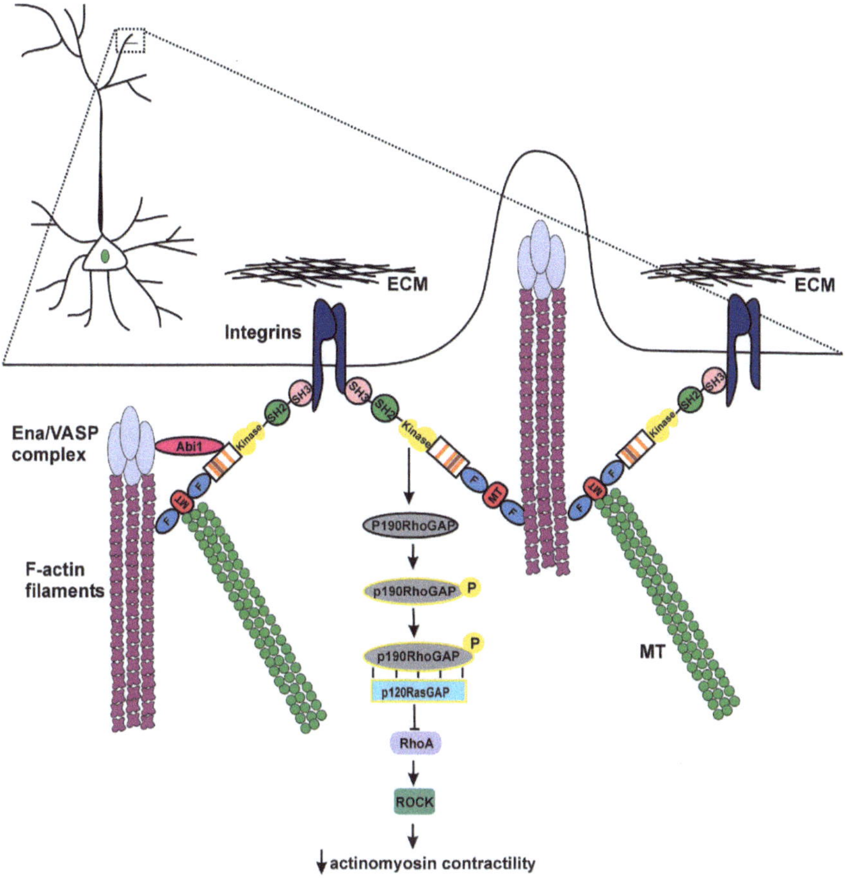

FIGURE 8.3. Mechanisms by which Abl family kinases regulate neuronal morphogenesis. The model presented here shows an enlarged region of neurite branch. Upon adhesion to the ECM, integrin receptors bind to and activate Abl family kinases. Abl and Arg may use three distinct mechanisms to regulate neurite branch formation and stabilization. First, Abl/Arg may interact with Abi family proteins to regulate the ability of Ena/VASP to promote filopodia formation. Second, Arg may phosphorylate and activate p190RhoGAP leading to inactivation of RhoA and relaxation of actomyosin contractility. Finally, Abl/Arg may directly regulate cytoskeletal structure by mediating interactions between F-actin and microtubules. These pathways may act together to regulate the formation or stability of new neurite branches.

8.5.2. Abl Family Kinases may Control Neurite Outgrowth by Regulating Ena/VASP Family Proteins

Ena/VASP proteins, which include the fly Ena protein, Unc-34 in *Caenorhabditis elegans*, and the mammalian Enabled (Mena), vasodilator-stimulated phosphoprotein (VASP), and Ena/VASP-like protein (EVL), promote actin filament elongation in a number of different cellular structures (Krause et al., 2003). As discussed previously, genetic studies indicate that Abl family kinases antagonize the function of Ena/VASP-family proteins. Ena/VASP family proteins regulate the formation of growth cone filopodia (Lebrand et al., 2004), which act as sensors of guidance cues. Abl family kinases and Ena/VASP family proteins may modulate axon and dendrite extension and pathfinding by regulating the structure and function of filopodia (Figure 8.3).

Although they are known to have antagonistic functions, the molecular mechanism(s) by which Abl inhibits Ena/VASP activity is (are) unclear. The Abl SH3 domain can bind to a central proline-rich domain (PRD) in Ena/VASP proteins (Gertler et al., 1996). Abl also phosphorylates Ena *in vitro* (Comer et al., 1998; Wills et al., 1999a). This phosphorylation prevents the Abl SH3 domain from binding to the PRD in Ena, suggesting that it could also restrict the binding of additional proteins *in vivo*. Abl-interactor (Abi) proteins stimulate phosphorylation of Ena/VASP family proteins by Abl family kinases *in vivo*. This finding suggests that Abi proteins could stabilize interactions between Ena/VASP and Abl family kinases *in vivo*.

8.5.3. Do Abl Family Kinases Mediate Interactions Between F-actin and Microtubules in Neurons?

Growth cone advance and turning requires dynamic interactions between the F-actin and microtubule cytoskeletons. In the growth cone, highly dynamic microtubules extend along F-actin bundles in the peripheral domain. Drug treatments that reduce microtubule dynamics result in growth cones that do not recognize substrate boundaries or steer correctly toward adhesive cues (Suter et al., 2004; Tanaka and Kirschner, 1995). These observations demonstrate that interactions between microtubules and F-actin are required for proper growth cone guidance and motility. Abl family kinases have emerged as potential mediators of F-actin–microtubule interactions and promotes lamellipodial protrusions.

The Orbit protein acts downstream of D-Abl in the regulation of motor axon guidance (Lee et al., 2004). CLASP, a vertebrate ortholog of Orbit binds to a subset of microtubules that extend into the periphery of *Xenopus* growth cones. Thus, a critical downstream mediator of Abl function can mediate interactions between F-actin and microtubules in the growth cone. Abl family kinases may also directly mediate interactions between F-actin and microtubules in the growth cone. The Arg C-terminal domain contains a microtubule-binding domain sandwiched between its two F-actin–binding domains (see Figure 8.1 for details). During fibroblast adhesion to fibronectin, Arg concentrates at sites in the cell periphery and promotes F-actin–rich lamellipodial protrusions (Miller et al., 2004). Microtubules concentrate and insert into these structures. Disruption of microtubules with nocodazole prevents Arg concentration at the cell periphery. An Arg mutant that cannot bind microtubules do not produce lamellipodial protrusions, although it is localized normally to the cell periphery. Together, these data suggest that Arg: microtubule interactions are required for Arg to promote lamellipodial protrusions. By binding to F-actin at the cell periphery Arg might serve as a target for growth cone protrusion. Abl and Arg both localize to growth cones, but this localization has not been performed with adequate resolution to determine whether either kinase might be bridging interactions between the two cytoskeletal systems.

8.6. Conclusions

Genetic studies have revealed numerous examples where Abl family kinases regulate neuronal morphogenesis. Work in the past few years has begun to shed light on the cellular and molecular mechanisms by which these kinases act in neurons to coordinate cytoskeletal changes in response to discrete cues. In particular, Abl family kinases have been shown to modulate several important cytoskeletal regulatory proteins. The future goal of this field is to determine how Abl family kinases interact with these proteins to properly regulate axon and dendrite morphogenesis and stability in the intact organism.

Acknowledgments. We thank our colleagues for many helpful discussions and Scott Boyle for comments on the chapter. Work in the Koleske lab is supported by a grant from the Public Health Service (NS39475), the Kavli Institute for Neuroscience at Yale, and a Scholar Award to A.J.K. from the Leukemia and Lymphoma Society.

References

Bashaw, G.J., Kidd, T., Murray, D., Pawson, T., and Goodman, C.S., 2000, Repulsive axon guidance: Abelson and Enabled play opposing roles downstream of the roundabout receptor, *Cell* **101:** 703–715.

Bennett, R.L., and Hoffmann, F.M., 1992, Increased levels of the Drosophila Abelson tyrosine kinase in nerves and muscles: Subcellular localization and mutant phenotypes imply a role in cell–cell interactions, *Development* **116:** 953–966.

Billuart, P., Winter, C.G., Maresh, A., Zhao, X., and Luo, L., 2001, Regulating axon branch stability: The role of p190 RhoGAP in repressing a retraction signaling pathway, *Cell* **107:** 195–207.

Clegg, D.O., Wingerd, K.L., Hikita, S.T., and Tolhurst, E.C., 2003, Integrins in the development, function and dysfunction of the nervous system, *Front. Biosci.* **8:** d723–d750.

Comer, A.R., Ahern-Djamali, S.M., Juang, J.L., Jackson, P.D., and Hoffmann, F.M., 1998, Phosphorylation of Enabled by the Drosophila Abelson tyrosine kinase regulates the in vivo function and protein–protein interactions of Enabled, *Mol. Cell. Biol.* **18:** 152–160.

DeMali, K.A., Wennerberg, K., and Burridge, K., 2003, Integrin signaling to the actin cytoskeleton, *Curr. Opin. Cell Biol.* **15:** 572–582.

Elkins, T., Zinn, K., McAllister, L., Hoffmann, F.M., and Goodman, C.S. (1990). Genetic analysis of a Drosophila neural cell adhesion molecule: interaction of fasciclin I and Abelson tyrosine kinase mutations. *Cell* **60:** 565–575.

Forsthoefel, D.J., Liebl, E.C., Kolodziej, P.A., and Seeger, M.A., 2005, The Abelson tyrosine kinase, the Trio GEF and Enabled interact with the Netrin receptor Frazzled in Drosophila, *Development* **132:** 1983–1994.

Gertler, F.B., Bennett, R.L., Clark, M.J., and Hoffmann, F.M., 1989, Drosophila abl tyrosine kinase in embryonic CNS axons: A role in axonogenesis is revealed through dosage-sensitive interactions with disabled, *Cell* **58:** 103–113.

Gertler, F.B., Comer, A.R., Juang, J.L., Ahern, S.M., Clark, M.J., Liebl, E.C., and Hoffmann, F.M. (1995). Enabled, a dosage-sensitive suppressor of mutations in the Drosophila Abl tyrosine kinase, encodes an Abl substrate with SH3 domain-binding properties. *Genes Dev* **9:** 521–533.

Gertler, F.B., Niebuhr, K., Reinhard, M., Wehland, J., and Soriano, P., 1996, Mena, a relative of VASP and Drosophila Enabled, is implicated in the control of microfilament dynamics, *Cell* **87:** 227–239.

Harbott, L.K., and Nobes, C.D. (2005). A key role for Abl family kinases in EphA receptor-mediated growth cone collapse. *Mol Cell Neurosci* **30:** 1–11.

Hernandez, S.E., Settleman, J., and Koleske, A.J., 2004, Adhesion-dependent regulation of p190RhoGAP in the developing brain by the Abl-related gene tyrosine kinase, *Curr. Biol.* **14:** 691–696.

Hoang, B., and Chiba, A., 1998, Genetic analysis on the role of integrin during axon guidance in Drosophila, *J. Neurosci.* **18:** 7847–7855.

Hsouna, A., Kim, Y.S., and VanBerkum, M.F., 2003, Abelson tyrosine kinase is required to transduce midline repulsive cues, *J. Neurobiol.* **57:** 15–30.

Hynes, R.O., 2002, Integrins: Bidirectional, allosteric signaling machines, *Cell* **110:** 673–687.

Jalali, S., del Pozo, M.A., Chen, K., Miao, H., Li, Y., Schwartz, M.A., Shyy, J.Y., and Chien, S., 2001, Integrin-mediated mechanotransduction requires its dynamic interaction with specific extracellular matrix (ECM) ligands, *Proc. Natl. Acad. Sci. USA* **98:** 1042–1046.

Jones, S.B., Lu, H.Y., and Lu, Q., 2004, Abl tyrosine kinase promotes dendrogenesis by inducing actin cytoskeletal rearrangements in cooperation with Rho family small GTPases in hippocampal neurons, *J. Neurosci.* **24:** 8510–8521.

Kaibuchi, K., Kuroda, S., and Amano, M., 1999, Regulation of the cytoskeleton and cell adhesion by the Rho family GTPases in mammalian cells, *Annu. Rev. Biochem.* **68:** 459–486.

Koch, A., Mancini, A., Stefan, M., Niedenthal, R., Niemann, H., and Tamura, T. (2000). Direct interaction of nerve growth factor receptor, TrkA, with non-receptor tyrosine kinase, c-Abl, through the activation loop. *FEBS Lett* **469:** 72–76.

Krause, M., Dent, E.W., Bear, J.E., Loureiro, J.J., and Gertler, F.B., 2003, Ena/VASP proteins: Regulators of the actin cytoskeleton and cell migration, *Annu. Rev. Cell Dev. Biol.* **19:** 541–564.

Lanier, L.M., and Gertler, F.B., 2000, From Abl to actin: Abl tyrosine kinase and associated proteins in growth cone motility, *Curr. Opin. Neurobiol.* **10:** 80–87.

Lebrand, C., Dent, E.W., Strasser, G.A., Lanier, L.M., Krause, M., Svitkina, T.M., et al., 2004, Critical role of Ena/VASP proteins for filopodia formation in neurons and in function downstream of netrin-1, *Neuron* **42:** 37–49.

Lee, T., Winter, C., Marticke, S.S., Lee, A., and Luo, L., 2000, Essential roles of Drosophila RhoA in the regulation of neuroblast proliferation and dendritic but not axonal morphogenesis, *Neuron* **25:** 307–316.

Lee, H., Engel, U., Rusch, J., Scherrer, S., Sheard, K., and Van Vactor, D., 2004, The microtubule plus end tracking protein Orbit/MAST/CLASP acts downstream of the tyrosine kinase Abl in mediating axon guidance, *Neuron* **42:** 913–926.

Lewis, J.M., Baskaran, R., Taagepera, S., Schwartz, M.A., and Wang, J.Y., 1996, Integrin regulation of c-Abl tyrosine kinase activity and cytoplasmic-nuclear transport, *Proc. Natl. Acad. Sci. USA* **93:** 15174–15179.

Li, W., Li, Y., and Gao, F.B., 2005, Abelson, enabled, and p120 catenin exert distinct effects on dendritic morphogenesis in Drosophila, *Dev. Dyn.* **234:** 512–522.

Li, Z., Aizenman, C.D., and Cline, H.T., 2002, Regulation of rho GTPases by crosstalk and neuronal activity in vivo, *Neuron* **33:** 741–750.

Liebl, E.C., Forsthoefel, D.J., Franco, L.S., Sample, S.H., Hess, J.E., Cowger, J.A., et al., 2000, Dosage-sensitive, reciprocal genetic interactions between the Abl tyrosine kinase and the putative GEF trio reveal trio's role in axon pathfinding, *Neuron* **26:** 107–118.

Lohmann, C., Myhr, K.L., and Wong, R.O., 2002, Transmitter-evoked local calcium release stabilizes developing dendrites, *Nature* **418:** 177–181.

Luo, L., Hensch, T.K., Ackerman, L., Barbel, S., Jan, L.Y., and Jan, Y.N., 1996, Differential effects of the Rac GTPase on Purkinje cell axons and dendritic trunks and spines, *Nature* **379:** 837–840.

Master, Z., Tran, J., Bishnoi, A., Chen, S.H., Ebos, J.M., Van Slyke, P., et al., 2003, Dok-R binds c-Abl and regulates Abl kinase activity and mediates cytoskeletal reorganization, *J. Biol. Chem.* **278:** 30170–30179.

McAllister, A.K., 2000, Cellular and molecular mechanisms of dendrite growth, *Cereb. Cortex* **10:** 963–973.

McWhirter, J.R., and Wang, J.Y., 1993, An actin-binding function contributes to transformation by the Bcr-Abl oncoprotein of Philadelphia chromosome-positive human leukemias, *EMBO J.* **12:** 1533–1546.

Miao, Y.J., and Wang, J.Y., 1996, Binding of A/T-rich DNA by three high mobility group-like domains in c-Abl tyrosine kinase, *J. Biol. Chem.* **271:** 22823–22830.

Miller, A.L., Wang, Y., Mooseker, M.S., and Koleske, A.J., 2004, The Abl-related gene (Arg) requires its F-actin-microtubule cross-linking activity to regulate lamellipodial dynamics during fibroblast adhesion, *J. Cell Biol.* **165:** 407–419.

Moresco, E.M., Donaldson, S., Williamson, A., and Koleske, A.J., 2005, Integrin-mediated dendrite branch maintenance requires Abelson (Abl) family kinases, *J. Neurosci.* **25:** 6105–6118.

Nakayama, A.Y., Harms, M.B., and Luo, L., 2000, Small GTPases Rac and Rho in the maintenance of dendritic spines and branches in hippocampal pyramidal neurons, *J. Neurosci.* **20:** 5329–5338.

Pendergast, A.M., 2002, The Abl family kinases: Mechanisms of regulation and signaling, *Adv. Cancer Res.* **85:** 51–100.

Pinkstaff, J.K., Detterich, J., Lynch, G., and Gall, C., 1999, Integrin subunit gene expression is regionally differentiated in adult brain, *J. Neurosci.* **19:** 1541–1556.

Redmond, L., Kashani, A.H., and Ghosh, A., 2002, Calcium regulation of dendritic growth via CaM kinase IV and CREB-mediated transcription, *Neuron* **34:** 999–1010.

Reichardt, L.F., and Tomaselli, K.J., 1991, Extracellular matrix molecules and their receptors: Functions in neural development, *Annu. Rev. Neurosci.* **14:** 531–570.

Rhee, J., Mahfooz, N.S., Arregui, C., Lilien, J., Balsamo, J., and VanBerkum, M.F. (2002). Activation of the repulsive receptor Roundabout inhibits N-cadherin-mediated cell adhesion. *Nat Cell Biol* **4:** 798–805.

Rivero-Lezcano, O.M., Marcilla, A., Sameshima, J.H., and Robbins, K., 1995, Wiskott-Aldrich syndrome protein physically associates with Nck through Src homology 3 domains, *Mol. Cell Biol.* **15:** 5725–5731.

Rohatgi, R., Nollau, P., Ho, H.Y., Kirschner, M.W., and Mayer, B.J., 2001, Nck and phosphatidylinositol 4,5-bisphosphate synergistically activate actin polymerization through the N-WASP-Arp2/3 pathway, *J. Biol. Chem.* **27:** 26448–26452.

Ruchhoeft, M.L., Ohnuma, S., McNeill, L., Holt, C.E., Harris, W.A., 1999, The neuronal architecture of Xenopus retinal ganglion cells is sculpted by rho-family GTPases in vivo, *J. Neurosci.* **19:** 8454–8463.

Schmid, R.S., and Anton, E.S., 2003, Role of integrins in the development of the cerebral cortex, *Cereb. Cortex* **13:** 219–224.

Schwartz, M.A., 2001, Integrin signaling revisited, *Trends Cell Biol.* **11:** 466–470.

Schwartz, M.A., Schaller, M.D., and Ginsberg, M.H., 1995, Integrins: Emerging paradigms of signal transduction, *Annu. Rev. Cell Dev. Biol.* **11:** 549–599.

Suter, D.M., Schaefer, A.W., and Forscher, P., 2004, Microtubule dynamics are necessary for SRC family kinase-dependent growth cone steering, *Curr. Biol.* **14:** 1194–1199.

Tanaka, E., and Kirschner, M.W., 1995, The role of microtubules in growth cone turning at substrate boundaries, *J. Cell Biol.* **128:** 127–137.

Tashiro, A., Minden, A., and Yuste, R., 2000, Regulation of dendritic spine morphology by the rho family of small GTPases: Antagonistic roles of Rac and Rho, *Cereb. Cortex* **10:** 927–938.

Vaillant, A.R., Zanassi, P., Walsh, G.S., Aumont, A., Alonso, A., and Miller, F.D., 2002, Signaling mechanisms underlying reversible, activity-dependent dendrite formation, *Neuron* **34:** 985–998.

Van Etten, R.A., Jackson, P.K., Baltimore, D., Sanders, M.C., Matsudaira, P.T., and Janmey, P.A., 1994, The COOH terminus of the c-Abl tyrosine kinase contains distinct F- and G-actin binding domains with bundling activity, *J. Cell Biol.* **124:** 325–340.

Wen, S.T., Jackson, P.K., and Van Etten, R.A., 1996, The cytostatic function of c-Abl is controlled by multiple nuclear localization signals and requires the p53 and Rb tumor suppressor gene products, *EMBO J.* **15:** 1583–1595.

Whitford, K.L., Dijkhuizen, P., Polleux, F., and Ghosh, A., 2002, Molecular control of cortical dendrite development, *Annu. Rev. Neurosci.* **25:** 127–149.

Wills, Z., Bateman, J., Korey, C.A., Comer, A., and Van Vactor, D., 1999a, The tyrosine kinase Abl and its substrate enabled collaborate with the receptor phosphatase Dlar to control motor axon guidance, *Neuron*, **22:** 301–312.

Wills, Z., Marr, L., Zinn, K., Goodman, C.S., and Van Vactor, D., 1999b, Profilin and the Abl tyrosine kinase are required for motor axon outgrowth in the Drosophila embryo, *Neuron* **22:** 291–299.

Wills, Z., Emerson, M., Rusch, J., Bikoff, J., Baum, B., Perrimon, N., et al., 2002, A Drosophila homolog of cyclase-associated proteins collaborates with the Abl tyrosine kinase to control midline axon pathfinding, *Neuron* **36:** 611–622.

Woodring, P.J., Hunter, T., and Wang, J.Y., 2001, Inhibition of c-Abl tyrosine kinase activity by filamentous actin, *J. Biol. Chem.* **276:** 27104–27110.

Woodring, P.J., Litwack, E.D., O'Leary, D.D., Lucero, G.R., Wang, J.Y., and Hunter, T., 2002, Modulation of the F-actin cytoskeleton by c-Abl tyrosine kinase in cell spreading and neurite extension, *J. Cell Biol.* **156:** 879–892.

Woodring, P.J., Meisenhelder, J., Johnson, S.A., Zhou, G.L., Field, J., Shah, K., et al., 2004, c-Abl phosphorylates Dok1 to promote filopodia during cell spreading, *J. Cell Biol.* **165:** 493–503.

Yano, H., Cong, F., Birge, R.B., Goff, S.P., and Chao, M.V. (2000). Association of the Abl tyrosine kinase with the Trk nerve growth factor receptor. *J Neurosci Res* **59:** 356–364.

Yu, H.H., Zisch, A.H., Dodelet, V.C., and Pasquale, E.B. (2001). Multiple signaling interactions of Abl and Arg kinases with the EphB2 receptor. *Oncogene* **20:** 3995–4006.

Zhao, Z.S., Manser, E., and Lim, L., 2000, Interaction between PAK and nck: A template for Nck targets and role of PAK autophosphorylation, *Mol. Cell Biol.* **20:** 3906–3917.

Zukerberg, L.R., Patrick, G.N., Nikolic, M., Humbert, S., Wu, C.L., Lanier, L.M., et al., 2000, Cables links Cdk5 and c-Abl and facilitates Cdk5 tyrosine phosphorylation, kinase upregulation, and neurite outgrowth, *Neuron* **26:** 633–646.

9 Membrane Glycolipids in Neurotrophin Receptor-Mediated Signaling

José Abad-Rodríguez

9.1. Glycosphingolipids: Structure and Metabolism

Glycolipids are oligosaccharide-containing lipids that show a consistent assymetric distribution in the plasma membrane of animal cells. They are found at the outer leaflet of the membrane bilayer exposing their sugar chains to the extracellular milieu. These molecules are present in the plasma membrane of all eukaryotic cells where they represent, in average, 5% of the total lipidic content of the external monolayer. Their expression patterns are specific for different cell types, species, and developmental stages. In the case of plants and bacteria, glycolipids are mostly glycerol-derived lipids while in animal cells they are synthesized on a ceramide basis constituting the family of the glycosphingolipids (GSL).

9.1.1. Molecular Structure of Glycosphingolipids

The structure of GSL consists of a hydrophobic moiety, the ceramide, and a hydrophilic head formed by a polysaccharide chain. The ceramides are characterized by the amidic binding of a fatty acid to the amino group of the long-chain amino alcohols called sphingosines (dashed rectangle in Figure 9.1A). Neutral GSL, one of the most widely distributed glycolipid types, carry a polar head composed by different combinations of neutral monosaccharides. A very simple member of this family, the galactosylceramide or galactocerebroside (Figure 9.1B) contains only a molecule of galactose in its polar head and is the main component of the myelin. It accounts for 40% of the lipids in the outer monolayer of myelinating cells while being practically absent from the membranes of other cell types. Glycolipid complexity reaches the highest level in the family of gangliosides, which are characterized by the presence in their structure of one or more residues of *N*-acetylneuraminic acid (also known as sialic acid or NANA, Figure 9.1C), conferring the whole molecule a net negative charge. These GSL derive from a glucosylceramide molecule (Figure 9.1D) to which galactose, *N*-acetylgalactosamine, and sialic acid residues are sequentially bound (Figure 9.2). Little quantities of gangliosides are found in most cell types, while they are the

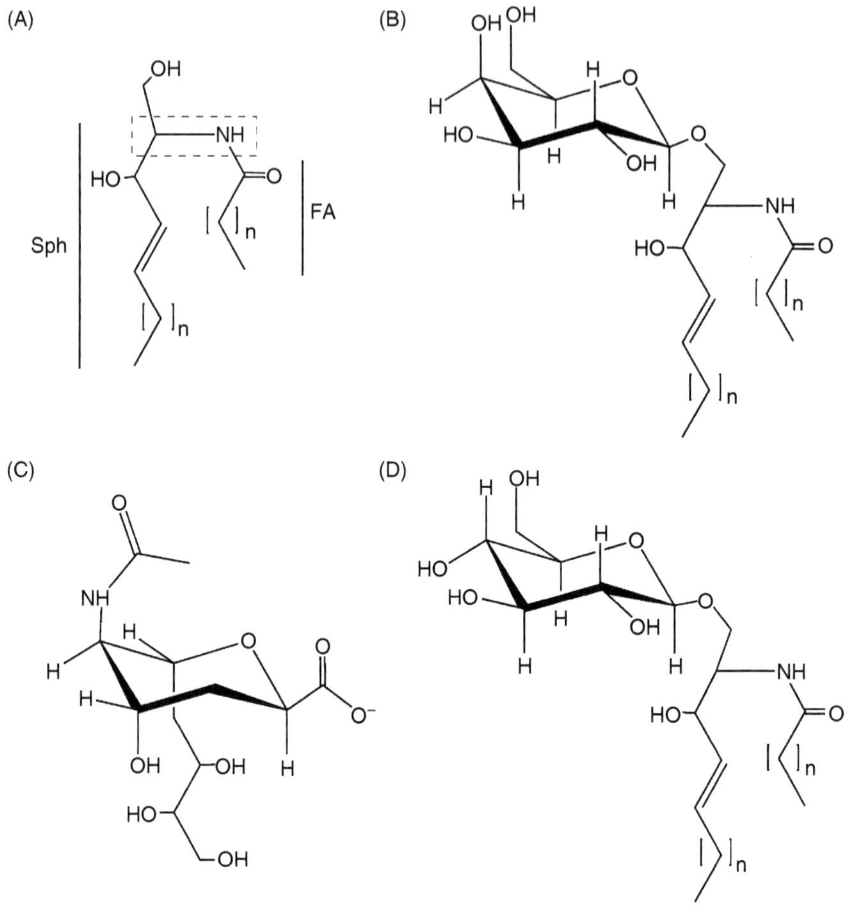

FIGURE 9.1. Basic molecular components in glycosphingolipid structure. (A) Ceramide structure (Sph = sphingosine; FA = fatty acid); (B) Galactosyl-ceramide; (C) *N*-acetyl-neuraminic acid (NANA or sialic acid); (D) Glucosyl-ceramide.

most abundant glycolipid in plasma membrane of nerve cells (5–10% of the total lipid mass). Although a huge amount of studies have shown fundamental ganglioside implications in most of the main physiological or pathological cellular and morphogenic events (Hakomori, 1981; Ledeen, 1989), ganglioside abundance in the nervous system emphasizes their special relevance in neuronal function.

9.1.2. Biosynthesis of Glycosphingolipids

In Figure 9.2, the scheme of sequential ganglioside biosynthesis from ceramide is shown and indicates the enzymes involved in each step. The names of the gangliosides follow the nomenclature of Svennerholm (Svennerholm, 1964) as recommended by the IUPAC (Chester, 1998).

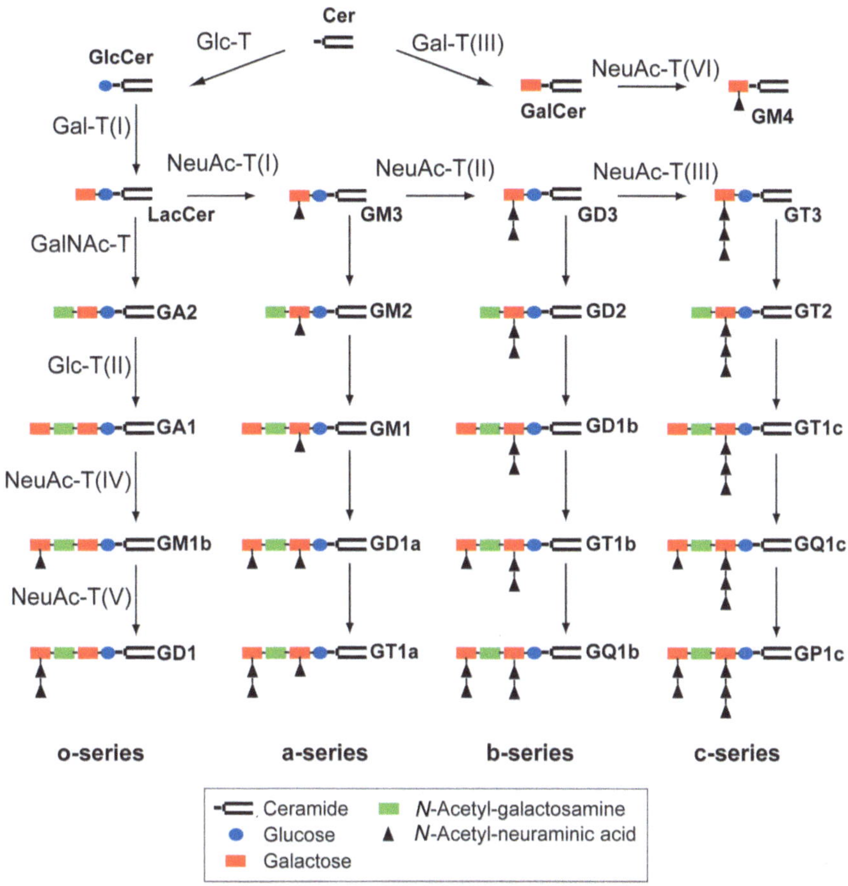

FIGURE 9.2. Biosynthesis pathways of gangliosides. The names of the gangliosides, (bold characters) follow the nomenclature of Svennerholm (Svennerholm, 1964) as recommended by the IUPAC (Chester, 1998). The enzymes involved are indicated above or on the left side of the reaction arrows.

In synthesis, the enzyme serine-palmitoyl transferase, located on the membranes of the endoplasmic reticulum, synthesizes ceramide that is transported to the Golgi apparatus. In its cytosolic surface a glucose residue is then transferred to the ceramide by the UDPglucose-ceramide glucosyltransferase (glucosylceramide synthase encoded by Glc-T) to produce glucosylceramide. The following step in the synthetic pathway takes place in the lumenal side of the Golgi and results in the addition of a galactose residue to render lactosylceramide, the precursor to most gangliosides. The simple and widely distributed ganglioside GM3 is then produced by ulterior binding of a sialic acid to lactosylceramide and functions as structural basis for the biosynthesis of the more complex gangliosides that occurs in the distal Golgi (Tettamanti, 2004).

It is important to mention that, after synthesis is completed in the Golgi apparatus, gangliosides are addressed to the plasma membrane, due to their biophysical characteristics (structural rigidity of the ceramide and "tight-packing" capacity with cholesterol), and are organized in special membrane micro domains known as detergent-resistant micro domains or rafts. As a consequence, the study of the role of gangliosides in the modulation of any cellular event should consider the raft context where gangliosides are immersed.

9.1.3. *Catabolism of Glycosphingolipids*

Like their biosynthesis, the degradation of gangliosides also occurs on the surface of internal cell membranes. After the gangliosides are removed from the plasma membrane by endocytosis, they arrive in the form of small vesicles (with diameters of 50–100 nm) to lysosomes for degradation. The degradation of GSLs with short oligosaccharide chains by water-soluble hydrolases requires small, non-enzymatic cofactors called sphingolipid activator proteins. Factors such as membrane curvature, membrane pressure, and lipid composition (in particular the presence of anionic lipids) enable the efficient degradation of GSLs in the small vesicles (Huwiler et al., 2000; Tettamanti, 2004). Genetic defects have been described for many of the steps of ganglioside and GSL degradation, resulting in a family of disorders related to sphingolipid storage (Jeyakumar et al., 2002; Walkley, 2003). These disorders generally involve severe neurodegeneration owing, in part, to the relatively high rate of ganglioside synthesis in the brain.

9.2. Effects of Glycosphingolipid Biosynthesis Deficiencies on Nervous System Development

The substantial changes in ganglioside brain content and composition during embryonic development (Yu et al., 2004a) would suggest that these GSL might be essential for the correct development of the nervous system. This idea is redimensioned by the fact that the silencing of single glycosyltransferase genes involved in their biosynthesis produces relatively subtle phenotypes. This reinforces the possibility that, in absence of a given ganglioside, the existing ones could act as functional substitutes. In contrast, double-null mice lacking entire ganglioside families, or those presenting the total elimination of glucosylceramide-derived GSL, display deleterious or even lethal phenotypes.

9.2.1. *Single Knockout Mice*

A paradigmatic mild phenotype is observed in *N*-acetylneuraminic acid transferase (NeuAc-T) (II) knockout mice (Figure 9.2; Table 9.1). These animals present a deficiency on the b- and c-ganglioseries but show an unchanged morphology of the brain and any other nervous tissue that is consistent with the absence of behavioral defects. These mutant mice, however, exhibit a reduced regeneration capacity of axotomized hypoglossal nerves, suggesting a role of b- and c-series gangliosides in nerve repair (Kawai et al., 2001; Okada et al., 2002). In the case of NeuAc-T (I) knockout mice

TABLE 9.1. Neurological effects of the suppression of GSL biosynthetic enzymes

Ko enzyme	Absent GSL	Expressed GSL	Phenotype severity	Neurol. Disorders	References
NeuAc-T(I) (EC 2.4.99.9)	Ganglio-series: a,b,c	Ganglio-series: o	Normal development. High muscle sensitivity to insulin	Undetected	Yamashita et al., 2003
NeuAc-T(II) (EC 2.4.99.8)	Ganglio-series: b,c	Ganglio-series: a, o	Normal development and CNS morphology.	Impaired regeneration of injured hypoglossal nerve	Kawai et al., 2001; Okada et al., 2002
GalNAc-T (EC 2.4.1.92)	Most complex ganglio-series	GM3, GD3	Near to normal life span	Optic and sciatic nerve dysmyelination, axon degeneration, slow neural conduction	Chiavegatto et al., 2000; Liu et al., 1999; Sheikh et al., 1999; Takamiya et al., 1996
GalNAc-T*/NeuAc-T(II) (EC 2.4.1.92)/ (EC 2.4.99.8)	Complex ganglio-series	GM3	Short life span, sudden death	Peripheral nerve degeneration, audiogenic lethal seizures	Inoue et al., 2002; Kawai et al., 2001
GalNAc-T/NeuAc-T(I) (EC 2.4.1.92)/ (EC 2.4.99.9)	Ganglio-series	Lactosyl-ceramide	Early postweaning death	Axon degeneration, perturbed white matter and axon-glia interaction, motor impairement	Yamashita et al., 2005
Glc-T (EC 2.4.1.80)	All	None	Embryonic lethal	Nonstratified tissues	Yamashita et al., 1999b

*GalNAc-T = *N*-acetylgalactosyl transferase; Glc-T = glucosyl transferase; NeuAc-T = *N*-acetylneuraminic acid transferase.

(Figure 9.2; Table 9.1), the animals lack a-, b-, and c-ganglioseries but still develop a quite normal nervous system. The major phenotypic defect is a high sensitivity to insulin due to an augmented phosphorylation state of the insulin receptor in muscle (Yamashita et al., 2003). In mice expressing only simple gangliosides (GM3, GD3) by elimination of *N*-acetylgalactosyl transferase (GalNAc-T) (Figure 9.2; Table 9.1), in spite of a near-normal life span, some degree of axonal degeneration and demyelization is detected in optic and sciatic nerves accompanied by a reduction in the speed of neural conduction. In old mice, these defects lead to motor defects (Chiavegatto et al., 2000; Liu et al., 1999; Sheikh et al., 1999; Takamiya et al., 1996).

9.2.2. Double Knockout Mice

Double mutants, as indicated earlier, produce much more deleterious neural phenotypes. GalNAc-T/NeuAc-T (II) double-null mice (Figure 9.2; Table 9.1) accumulate high quantities of GM3 because of the impossibility to step further in complex ganglioside biosynthesis. Their life span is shortened compared to wild-type animals, and they undergo sudden death by the induction of audiogenic lethal seizures. This phenotype is concomitant to a deep degeneration of peripheral nerves (Inoue et al., 2002; Kawai et al., 2001).

Even more drastic is the case of the GalNAc-T/NeuAc-T (I) double-null mice (Figure 9.2; Table 9.1). These mice are unable to produce any ganglioside of the ganglio-series, the major ganglioside class in the central nervous system (CNS), although they synthesize lactosylceramide. Their embryonic development is quite normal and they are born without evident problems. Soon after birth they begin to suffer tremors, hind limbs weakness, and deep motor problems. The neurological phenotype occurs in parallel to a catastrophic neurodegenerative process corroborated by shrinkage of the brain, apoptosis in the cerebral cortex, astrogliosis, severe axonal degeneration, white matter vacuolation, and a deep alteration of the axon-oligodendrocyte contact. In particular, the paranodal loops of myelin in the node of Ranvier appear loosely associated to the axonal membrane, a fact that may lead to axonal degeneration and oligodendrocyte dysfunction (Yamashita et al., 2005).

The extreme case of a complete depletion of ganglioside biosynthesis leads to an embryonic lethal phenotype, showing a general failure in the formation of the whole organism. In fact, glucosyl transferase (Glc-T) knockout mice (Figure 9.2; Table 9.1), lacking the major ganglioside synthesis pathway, die in gastrulation stage and are not able to form any properly differentiated and stratified tissue (Yamashita et al., 1999b).

9.3. Modifications of Membrane Glycosphingolipid Population Affect Neurotrophin Signaling

9.3.1 Effects of Exogenous Glycosphingolipids

Pioneering research on the effect of gangliosides in neuronal development was performed by the addition of crude preparations of these GSL, extracted from brain or spinal cord, to neuroblastoma or primary neuron cell cultures. From these works (Ledeen, 1984), it became clear that gangliosides play a role in neuritogenesis, neurite outgrowth, axonal development, and even in nervous system repair. Concomitantly to the development of improved purification methods, the availability of individual gangliosides allowed the specific evaluation of their effects.

9.3.1.1. Exogenous Gangliosides Stimulate Neuritogenesis and Axon Growth

In an exhaustive study, Byrne et al. (1983) reported that, from a highly purified bovine brain ganglioside preparation, 10 individual gangliosides out of 11 presented different levels of neurite-stimulating activity on N2A neuroblastoma cells. Only GM4 was not active. Further work (Leskawa and Hogan, 1985) confirmed that asialogangliosides were not active in N2A differentiation and described differential activities of GM1 and GT1b. GM1 stimulated the appearance of longer neurites while GT1b enhanced the number of neurites per cell and the arborization level. Parallel research further demonstrated the capacity of GM1 to promote differentiation and neurite outgrowth on neuroblastoma cell lines (Facci et al., 1984) and neurons in culture (Skaper et al., 1985). More complex gangliosides as GQ1b also showed neuritogenic effects on GOTO and NB-1 neuroblastoma cells (Tsuji et al., 1983).

9.3.1.2. Neurotrophin Signaling Is Modulated by Exogenous Gangliosides

As neuronal development is strongly influenced by neurotrophin activity, some authors raised the hypothesis that gangliosides could regulate neuronal differentiation through neurotrophin signaling modulation (Katoh-Semba et al., 1984; Schwartz and Spirman, 1982). GM1 potentiates nerve growth factor (NGF) effect *in vivo* (Cuello et al., 1989; Fong et al., 1995; Hadjiconstantinou and Neff, 1998; Panni et al., 1998), and *in vitro* it enhances the NGF-induced phosphorylation, dimerization, and activation of tyrosine kinase (Trk) receptors (Farooqui et al., 1997; Ferrari et al., 1995; Mutoh et al., 1993, 1995; Rabin and Mocchetti, 1995) and brain derived neurotrophic factor (BDNF)-induced TrkB activation (Pitto et al., 1998). The potentiation of NGF effect by GM1 *in vivo* is mediated by enhanced Trk phosphorylation as has been observed in rat striatum and hippocampus after intracerebroventricular administration of the ganglioside (Duchemin et al., 2002).

Trk family and p75^{NTR} neurotrophic factor receptors have been localized in membrane rafts (Kasahara and Sanai, 2000). As these micro domains are enriched in gangliosides, they provide an ideal microenviroment for close interactions between the receptors and the GSL.

The enhancement of NGF-induced TrkA autophosphorylation produced by GM1 is, at least in part, due to a detergent-resistant association of GM1 to TrkA protein (Mutoh et al., 1993, 1995). On the other hand, unglycosylated form of TrkA protein remains intracellular and is phosphorylated, even in the absence of the ligand, suggesting a constitutive kinase activity. In spite of its ability to interact with signaling molecules Shc and PLC-γ (Phospholipase C gamma), this form of Trk is unable to enhance neurite growth. A glycosylated form of Trk, in contrast, is targeted to the plasma membrane, which activates Ras/Raf/MAPK

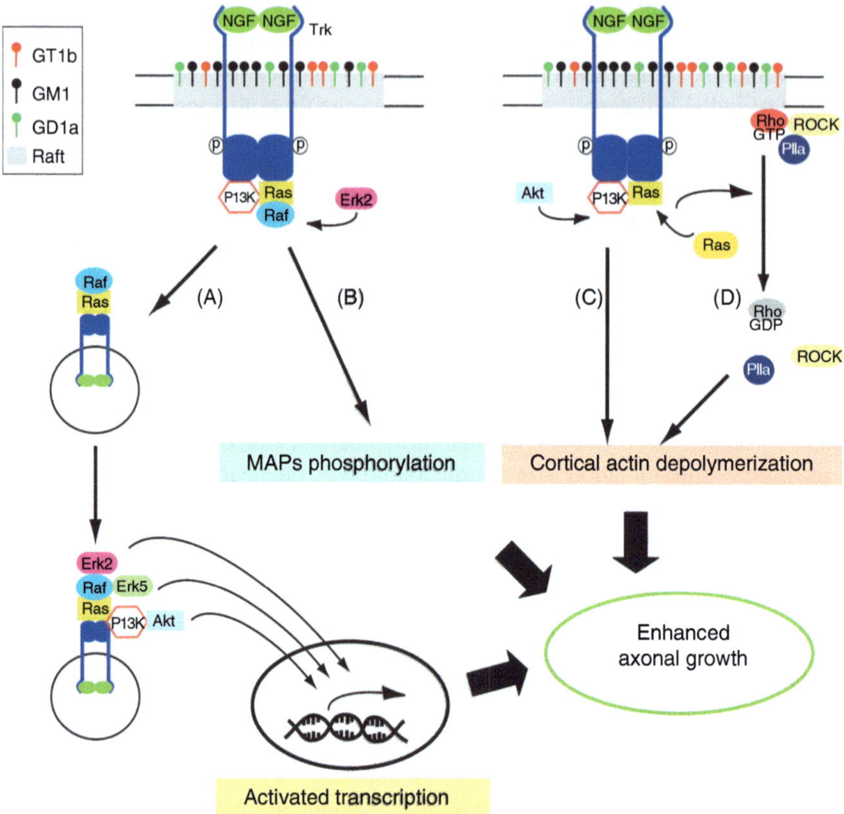

FIGURE 9.3. Trk neurotrophic signaling pathways leading to axonal growth. Upon neurotrophin binding, Trk receptors, interacting with GM1 in rafts, dimerize, undergo autophosphorylation, and recruit several downstream effectors to rafts. (A) Trk–Ras–Raf complex is internalized by clathrin-coated vesicles, retrogradely transported along the axon to the cell body where Erks and Akt are activated. These molecules enter the nucleus and activate transcription factors as CREB. (B) Local Erk activation can occur at the level of the plasma membrane leading to MAP phosphorylation and enhancement of microtubule dynamics. (C) Local Akt activation at the level of the plasma membrane leads to cortical actin depolymerization, the same final effect exerted by Rac activation and Rho–ROCK–Profilin IIa (PIIa) complex disaggregation showed in (D).

(mitogen-activated protein kinases) cascade (Figure 9.3) and enhances NGF-stimulated neuritogenic effect (Watson et al., 1999) by direct interaction with GM1 (Mutoh et al., 2000). Moreover, glycosylated TrkA, but not the nonglycosylated form, fractionates with GM1 in low density fractions corresponding to rafts (Mutoh et al., 2000).

Other groups have suggested that gangliosides activate transfected Trk receptors in 3T3 fibroblasts by inducing the synthesis and release of neurotrophins (Rabin et al., 2002). GM1 stimulates the release of NT-3 while LIGA20, a synthetic derivative of GM1, mediates the release of BDNF. These treatments lead to autophosphorylation of TrkC and TrkB, respectively, in the absence of direct ganglioside interaction with the neurotrophin binding site or with the tyrosine kinase domain (Rabin et al., 2002). In this system, GM1 does not stimulate NGF synthesis, thus TrkA cannot be activated by its ligand as occurs to TrkB and TrkC. This leaves the open possibility that TrkA could be activated by direct interaction with GM1, as indicated earlier. Still, it is likely that TrkA artificially expressed in nonneural cells could undergo specific modifications leading to glycosylation states that would not permit interaction with GM1. An accurate study of such modifications would help to clarify this question.

9.3.2. Studies on the Modification of Endogenous Glycosphingolipids

In spite of its usefulness, the use of exogenous gangliosides added in solution to cell cultures and tissue slices or injected *in vivo* has been controversial because it is not clear whether they behave in the same way than the endogenous ganglioside population. The same ganglioside may trigger different effects or remain inactive depending on how it is presented to the cells. As any amphypatic molecule in aqueous solution, gangliosides may stay as monomers or may form micelles depending on the concentration and the characteristics of the solvent. In any case, the molecular topology of the added gangliosides is difficult to be determined. To overcome these problems in the elucidation of the ganglioside role in neurotrophin signaling, a considerable effort has been carried out to study the effects of endogenous ganglioside changes. Different strategies have been utilized for this purpose based on genetic manipulation or pharmacological modulation of the enzymes involved in ganglioside metabolism or on the use of ganglioside-binding agents as toxins or antibodies in order to modify their interactions.

9.3.2.1. Trk Signaling Is Modulated by Endogenous GM1

The inhibition of Glc-T activity (Figure 9.2; Table 9.2) in PC12 cells using the ceramide analog D-threo-1-phenyl-2-decanoylamin-3-morpholino-propanol (D-PDMP) leads to extensive depletion of GSL derived from glucosyl ceramide. D-PDMP inhibits NGF-induced PC12 neurite outgrowth and the autophosphorylation of Trk. This prevents the activation of phosphoinositide-3-kinase (PI3K) and MAPK, downstream targets of Trk-initiated intracellular protein kinase cascades (Figure 9.3B and C). The specificity for Trk receptor is underlined by the fact that intracellular signaling pathway of epidermal growth factor is not affected by D-PDMP. All these effects are precluded by GM1 but not by the addition of other gangliosides, which supports the idea of a direct interaction between GM1 and TrkA underlying the regulation of the receptor (Mutoh et al., 1998).

NG-CR72 cells expressing a mutated version of galactosyl transferase (Gal-T) (II) (Figure 9.2; Table 9.2) lack GM1, among other gangliosides. In these cells, Trk protein is expressed but NGF does not elicit the autophosphorylation of the

TABLE 9.2. Effects of GSL synthesis modifications on neurotrophin signaling

Modification of enzyme activity	Cell type	Main effect on GSL expression	Cellular phenotype	Effect on neurotrophin receptor or signaling	References
Gal-T* (EC 2.4.1.62) null mutation	NG-CR72 hybrid Neuroblast/glioma	Lack of GM1	Nondescribed	No Trk in plasma membrane, no NGF-derived Trk phosphoryl	Mutoh et al., 2002
Gal-T (EC 2.4.1.62) transfection	NG-CR72	Recovered GM1	Nondescribed	Trk in plasma membrane, and NGF-Trk phosphoryl	Mutoh et al., 2002
Gal-T (EC 2.4.1.62) transfection.	PC12	Increased GM1	No NGF-stimulated neurite formation, reduced membrane fluidity	No NGF-derived Trk phosphoryl; no Erk activation. Trk; p75, Ras out of rafts	Nishio et al., 2004
Neu3 (EC 3.2.1.18) overexpression	Hippocampal neurons	Increased GM1 in plasma membrane	Enhanced axon growth, enhanced axon regeneration after section	Enhanced Trk phosphoryl. RhoA/ROCK/PIIa deactivation, local growth cone actin depolymerization	Da Silva et al., 2005; Rodríguez et al., 2001
GlucCer synthase (EC 2.4.1.80) inhibition by D-PMDP	PC12	Lack of GlucCer derived gangliosides	No NGF-stimulated neurite formation, rescue by GM1	No NGF-derived Trk phosphoryl, inhibition of MAPK	Mutoh et al., 1998
Neu3 (EC 3.2.1.18) inhibition by NeuAc2en or iRNA	Hippocampal neurons	Reduced GM1 in plasma membrane	Retarded axon growth	Low Trk phosphoryl, RhoA/ROCK/PIIa activation, no local growth cone actin depolymerization	Da Silva et al., 2005; Rodríguez et al., 2001

*Cer = ceramide; D-PMDP = D-threo-1-phenyl-2-decanoylamin-3-morpholino-propanol; Gal-T = galactosyl transferase; Gluc = glucose; NeuAc2en = 2,3,dehydro-2,deoxy-*N*-acetylneuraminic acid.

receptor that remains in intracellular compartments. The stable transfection of the wild-type enzyme restores GM1 expression and induces Trk sorting to the plasma membrane. Additionally, NGF returns to trigger its autophosphorylation and to activate the MAPK cascade (Figure 9.3) (Mutoh et al., 2002).

Considering the above, it is surprising that the observation that PC12-Gal-T (II) stable transfectants, showing increased expression of GM1, are not responsive to NGF stimulated neuritogenesis in contrast to wild-type cells. Dimerization and autophosphorylation of TrkA are fairly undetectable, and there is no activation of extracellular signal-regulated kinase (Erk1/2) after NGF treatment. The sucrose density gradient fractionation revealed that TrkA, $p75^{NTR}$, and Ras, localized to raft fractions in wild type cells, are found in nonraft fractions of transfectant cells, while GM1 or flotillin remain in rafts likely due to the reduced membrane fluidity observed in GM1 overexpressing cells (Nishio et al., 2004).

These results point to a precise mechanism of neurotrophin receptor regulation by gangliosides. GM1 would modulate Trk mediated signaling by direct interaction, as explained earlier, although its excess would lead to the modification of the intrinsic raft properties. These modifications would sort the receptors and the downstream signaling molecules to different membrane microenvironments to control their activity.

9.3.2.2. Local GM1 *Enhancement at a Growth Cone Determines the Axon*

Cytoskeletal components required for proper formation and growth of the axon and the dendrites are differentially regulated by members of Rho small GTPase family (Dickson, 2001; Luo, 2000). For instance, in hippocampal neurons, the local rearrangements of the actin cytoskeleton at the level of the plasma membrane, necessary for neuritogenesis and axonal determination/growth, are triggered by the spatially restricted inactivation of a specific Rho small-GTPase pathway. This pathway involves Ras homologous member A (RhoA), Rho-associated coiled-coil–containing protein kinase (ROCK), and profillin IIa (PIIa) (Da Silva et al., 2003, 2005). This last molecule is the one conferring neural specificity to the pathway as it is an actin-binding protein largely restricted to the brain (Witke et al., 2001).

Plasma membrane ganglioside sialidase (PMGS or Neu3) hydrolyzes sialic acid residues form polysialilated gangliosides to produce GM1 in the plasma membrane (Hasegawa et al., 2000; Miyagi et al., 1999; Wada et al., 1999). Its inhibition by sialic acid analogs [NeuAc2en (2,3,dehydro-2,deoxy-*N*-acetylneuraminic acid)] or by small RNA interference leads to the block of axon formation, while the overexpression accelerates axon growth (Da Silva et al., 2005; Rodriguez et al., 2001). This enzyme accumulates at the tip of a single neurite of the unpolarized rat neuron, coinciding with the highest level of actin instability, an early indication for the formation of the future axon (Da Silva et al., 2003). Neu3, enhances the local GM1 content within this particular neurite, induces the formation of Neu3:GM1:TrkA complex and the local phosphorylation of TrkA. This triggers PI3K/Rac1-dependent deactivation of RhoA. RhoA transformation to its GDP-bound form dismantles the GTP-RhoA/ROCK/PIIa complex and leads to the actin depolymerization in this neurite only (Figure 9.3D) (Da Silva et al., 2005). In summary, the early polarization of a ganglioside-transforming enzyme drives the locally restricted enhancement of its product GM1, induces spatial restricted modulation of a specific actin-regulatory molecular machinery before polarization and determines the axonal fate.

Three important conclusions arise from these studies: (i) A correct ganglioside synthesis is necessary for neurotrophin-stimulated Trk signaling, (ii) this signaling is only active if GM1 production remains within a certain range, and (iii) local variations of GM1 in axonal plasma membrane regulate Trk signaling, actin dynamics, and growth.

9.4. Glycosphingolipids as Rafts Components: Modulation of Neuronal Differentiation Signaling

As it was mentioned earlier, the study of GSL involvement in any cellular event should consider their presence in membrane rafts. These particular membrane microenvironments have been described as platforms with capacity to recruit different molecules involved in cell–cell interactions, transmembrane receptors, or signaling proteins (Galbiati et al., 2001; Golub et al., 2004; Simons and Toomre, 2000). Focusing on neurotrophin signaling proteins, many of them have been found associated to rafts in different circumstances (Table 9.3), although it is not known for most of them if there is a direct interaction of GSL underlying their regulation. Only for some transmembrane receptors, these interactions have been described. The last insights on this issue will be discussed within the next paragraphs.

9.4.1. Glial-Derived Neurotrophic Factor Signaling Depends on Rafts

The family of ligands related to glial-derived neurotrophic factor (GDNF)—GFL—sustains many neuronal populations in both, CNS and PNS. So far, four members of this family have been described, GDNF (Lin et al., 1993), neurturin (Kotzbauer et al., 1996), persephin (Milbrandt et al., 1998), and artemin (Baloh et al., 1998). GSLs use a complex receptor system formed by a common signaling component, the receptor tyrosine kinase RET, and the glial-derived factor receptor alpha (GFRα), a glycosylphosphatidylinositol (GPI)-linked coreceptor that confers specificity to the complex as there is one GFRα for each GFL (GFRα1–α4). On receptor tyrosine kinase (RET) activation, its interaction with Src kinases is enhanced leading to activation of Akt and MAP kinases.

The presence of GFLs induce the association of GFRα and RET in rafts. GFRα due to its GPI anchor is a component of rafts, while RET is a transmembrane protein localized to nonraft fractions in the absence of GFLs. On ligand binding, RET is translocated to rafts and forms the fully active receptor complex (Tansey et al., 2000).

When a nonraft transmembrane GFRα chimera form is over expressed, GFL-derived RET enrichment in rafts is impaired as well as Src kinases binding to

TABLE 9.3. Neurotrophic signaling molecules associated with lipid rafts

Molecule	Tissue/cell type	Raft association	References
Trk	Hippocampal neurons, PC12, brain	Transmembrane GM1 interaction	Mutoh et al., 1995; Wu et al., 1997
p75^NTR	PC12, cultured neurons	Transmembrane GT1b interaction	Fujitani et al., 2005a; Huang et al., 1999; Yamashita et al., 2002
GFRα	N2A, cerebellar granule cells	GPI anchor	Tansey et al., 2000
Ras	PC12	Acylation	Nishio et al., 2004
Rho	PC12	Prenylation	Fujitani et al., 2005a;
Rac	PC12	Prenylation	Michaelson et al., 2001
Raf	PC12	Ras interaction	Carey et al., 2003; Markus et al., 2002
RET	N2A, cerebellar granule cells	Transmembrane, GFRα interaction	Tansey et al., 2000
MAPK	PC12, cultured neurons	Raf interaction	Guirland et al., 2004; Perron and Bixby, 1999
PI3K	Cultured neurons	Trk adaptors interaction	Atwal et al., 2000
Fyn	Neuroblastoma	Acylation	Kalka et al., 2001;
Lyn	Neuroblastoma, cerebellar granule cells	Acylation	Kasahara et al., 1997;
Yes	Neuroblastoma	Acylation	Prinetti et al., 2000
N-Src	Brain	Acylation	Mukherjee et al., 2003

RET. In consequence Akt and p42/p43 MAPK are not activated resulting in reduction of GFL-induced neuroblastoma cell differentiation or GFL-enhanced survival of cerebellar granule cells (Tansey et al., 2000). RET still binds to non-raft GFRα form, is phosphorylated out of the raft, and the chimeric GFRα has intact Src-binding domains. All these data indicate the fundamental role of the presence of GFRα in rafts for GFL neurotrophic signaling.

GFRα1 is also required for the GDNF-induced p140^{NCAM} signaling through Fyn and focal-adhesion kinase (FAK) and the accumulation in rafts of the transmembrane isoform of this neuronal cell adhesion molecule. As Fyn and FAK are usual raft components, it is very likely that GFRα1 enhances GDNF-derived adhesion signaling by facilitating p140^{NCAM} contact with its effectors within the raft microenvironment (Paratcha et al., 2003).

The direct implication of GSLs in the GFRα association to rafts or in its interaction with RET or p140^{NCAM} is unknown. Cholesterol depletion in neuroblastoma cells precludes GFL/RET but not NGF/Trk signaling as measured by Akt and MAPK phosphorylation. This fact suggests a direct implication of cholesterol in GFL/RET signaling while Trk, directly interacting with GM1 (Mutoh et al., 1995, 2000), would not be cholesterol dependent. Anyway, this is not conclusive enough to completely rule out an eventual implication of GSLs in GFL/RET signaling.

9.4.2. *Special Case of p75*^{NTR}

p75^{NTR} has been historically referred to as the low-affinity neurotrophin receptor due to its capacity to bind neurotrophins (NGF, BDNF, NT3, NT4) with low affinity compared to their high-affinity receptors Trks (TrkA, TrkB, and TrkC). Currently, p75^{NTR} definition as low-affinity neurotrophin receptor should be avoided as neurotrophin precursors (proneurotrophins) bind to p75^{NTR} with high affinity (Lee et al., 2001). Although p75^{NTR} lacks a kinase domain, it cooperates with many protein partners to form multimeric receptor complexes and thus contribute to a plethora of signaling platforms with very different biological effects (Barker, 2004; Lu et al., 2005). The most classical p75^{NTR} contribution to neurotrophin signaling is mediated by its physical interaction with Trk receptors. This enhances the ability of each Trk to respond to and discriminate their preferred neurotrophin ligand, leading to the activation of survival pathways. In contrast, p75^{NTR} forming a complex with Sortilin triggers pro-apoptotic signals in response to pro-NGF (Nykjaer et al., 2004). Furthermore, p75^{NTR} on NGF-induced cAMP enhancement can be phosphorylated by protein kinase A (PKA) and promote axonal growth even in the absence of Trk proteins (Figure 9.4) (Nishio et al., 2004). On the other hand, p75^{NTR} is able to from a triple complex with Nogo receptor (NgR) and Lingo-1 to transduce axonal growth inhibitory signals of the myelin-associated inhibitors myelin-associated glycoprotein (MAG), Nogo, and oligodendrocyte-myelin glycoprotein (OmgP) (Figure 9.5) (Wang et al., 2002a; Wong et al., 2002; Yamashita et al., 2002).

The complexity of p75^{NTR} function requires special mechanisms in order to discriminate among many different extracellular ligands, and GSL play a major role on these mechanisms.

9.4.2.1. Gangliosides, p75^{NTR}, and Trks: Neuronal *Survival* and Differentiation by Neurotrophins

The enhancement of survival produced by mature neurotrophins is well established (Huang and Reichardt, 2001; Lewin and Barde, 1996). For instance, the administration of mature neurotrophins in certain brain zones rescues aging or injury-induced neuronal loss, and conversely, the elimination of particular neurotrophin genes leads to the loss of the neuronal population that depends on this neurotrophin for survival (Snider and Silos-Santiago, 1996). Survival effect of neurotrophins is mediated by Trks. Mice lacking individual Trks lose entire neurotrophin-dependent neuronal populations because of the absence of neurotrophin-derived survival signaling. Coexpression of p75^{NTR} and Trk is enough to enhance the avidity of each particular Trk for its ligand and the capacity of discrimination in favor of each specific neurotrophin. This "refinement" in neurotrophin reception can explain how Trks are able to maintain sufficient level of neuronal survival, for example, in peripheral tissues where neurotrophin concentrations are extremely low. As a matter of fact, p75^{NTR}-null mice undergo progressive peripheral neuropathy concomitant with the loss of sympathetic and sensory neurons (Lee et al., 1992; von Schack et al., 2001), very likely as a consequence of the failure on Trk activation in the absence of p75^{NTR}, the required coreceptor. Trk-p75^{NTR} interaction is regulated by

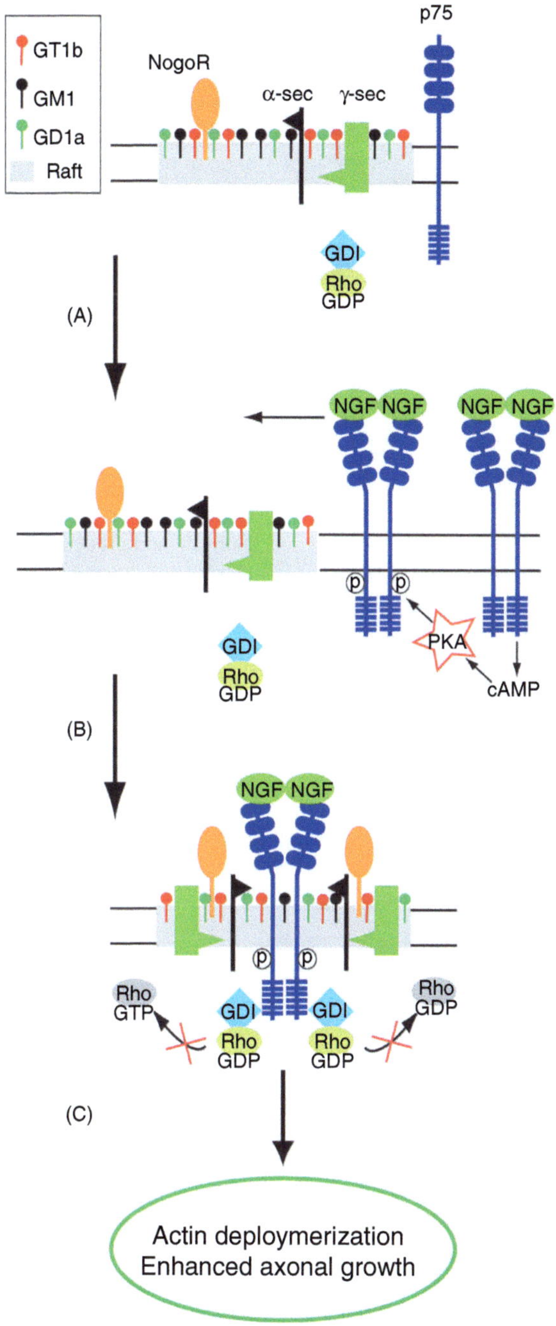

FIGURE 9.4. Neurotrophin-induced p75^{NTR} activation of axonal growth. (A) Neurotrophin binding to p75^{NTR} induces PKA-mediated receptor phosphorylation, dimerization, and translocation to rafts. (B) Within rafts, phosphorylated p75^{NTR} does not undergo α- and γ-cleavage, and stabilizes Rho-GDP–Rho-GDI (GDI) complex. (C) RhoA pathway remains deactivated with the concomitant enhancement of axonal growth.

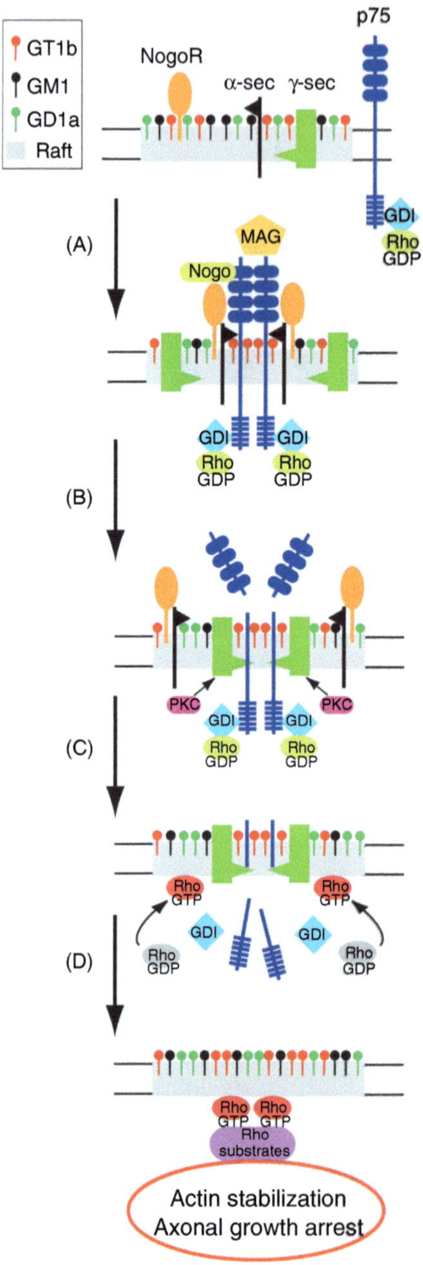

FIGURE 9.5. Inhibition of axonal growth by myelin-associated inhibitors through $p75^{NTR}$. (A) Exposition of the growing axon to myelin inhibitory components (Nogo, MAG) induces the recruitment of $p75^{NTR}$–Rho-GDI–Rho-GDP complex into rafts, with the participation of NgR. (B) Within the raft, $p75^{NTR}$ is first cleaved by α-secretase. (C) The resulting peptide after α-cleavage is further hydrolyzed by γ-secretase in a PKC-dependent manner. (D) Final $p75^{NTR}$ C-terminal peptide displaces Rho-GDI and Rho-GDP from the complex. The latter is converted to the GTP-bound form, triggering Rho pathway and producing the arrest of the axon.

their location within the plasma membrane. Their sorting to the same raft micro domain facilitates the interaction and enhances neurotrophin signaling. Even changes in the ganglioside population within the raft eventually regulate receptor interaction. Actually, $p75^{NTR}$ interacts with GT1b but not with GM1, although both gangliosides are neuronal raft components (Yamashita et al., 2002).

On neurotrophin binding, $p75^{NTR}$ undergoes specific phosphorylation in Ser304 by cAMP-PKA and moves into rafts. This is accompanied by an enhancement of neurite outgrowth and RhoA inactivation (Figure 9.3). As this occurs in cell lines lacking Trk proteins, meaning that $p75^{NTR}$ phosphorylated by PKA and included in rafts can transduce by itself the neurotrophic signal. Nevertheless, although no Trk enrichment in rafts is observed in Trk expressing cells after ligand binding, the eventual formation of $p75^{NTR}$-Trk complex with the Trk population normally resident in these micro domains cannot be ruled out (Nishio et al., 2004).

9.4.2.2. Gangliosides, p75NTR, and Nogo receptor: Axonal *Growth Inhibition by My*elin

Myelin-associated glycoproteins MAG, Nogo, and OMgp are the components of CNS myelin that prevent axonal regeneration in the adult vertebrae. Contact of these molecules with the growing axon induces the activation of RhoA pathway in a PKC-dependent manner (Sivasankaran et al., 2004). This leads to the local stabilization of the actin cytoskeleton at the tip of the neurite, which blocks the advance of the axon and elicits the final collapse of the growth cone (Figure 9.5) (Niederost et al., 2002; Schwab, 2004; Yiu and He, 2003).

Nogo binds to a membrane GPI-linked protein known as NgR (Fournier et al., 2001). Unexpectedly, also the other two myelin-associated inhibitors (MAG and OMgp) were subsequently identified as NgR ligands (Domeniconi et al., 2002; Liu et al., 2002; Wang et al., 2002b). NgR, due to its GPI-linked nature, is located in membrane raft fractions, although it does not have any cytoplasmic domain and needs a coreceptor spanning the membrane to mediate signal trasduction. NgR coreceptor is $p75^{NTR}$, thus all three myelin-associated axonal growth inhibitors depend on $p75^{NTR}$ to signal inside the cell (Wang et al., 2002a; Wong et al., 2002; Yamashita et al., 2002). This signaling involves as well LINGO-1, a CNS-specific transmembrane protein that is the third component of the receptor complex (Mi et al., 2004).

The ability of $p75^{NTR}$ to transduce inhibitory signals is regulated by its localization within the plasma membrane and by interactions with gangliosides and other components of rafts. So far, there is not a full picture about how myelin-$p75^{NTR}$ inhibitory signaling occurs, nevertheless some important insights that have recently came to light about the implication of GSLs in it will be discussed along the next paragraphs.

9.4.2.2.1. Myelin Inhibitors Induce p75 Recruitment to Rafts

Gangliosides are necessary for the effects of soluble MAG (MAG-Fc chimeric form) on Rho activity and neurite outgrowth. In contrast, the inhibitory effect of the Nogo peptide is independent of gangliosides. Although the addition of any of the two soluble molecules induces association of $p75^{NTR}$ to rafts (Fujitani et al., 2005b), most likely the mechanism of $p75^{NTR}$ recruitment and signal transduction are different in response to each ligand. Anyhow, the overall structure and composition of rafts are fundamental for this signaling. In fact, raft disruption by cholesterol depletion with β-methylcyclodextrin produces the displacement of NgR and $p75^{NTR}$ from the raft fractions, reduces MAG- and Nogo-induced RhoA activation and precludes axon growth inhibition (Yu et al., 2004b).

Illustrating the importance of gangliosides in myelin-p75^{NTR} signaling, p75^{NTR} recruitment to rafts can be achieved by direct MAG binding to complex gangliosides as GT1b or GD1a on the cell membrane (Fujitani et al., 2005b). Moreover, cross linking of gangliosides by antiganglioside antibodies inhibits neurite outgrowth (Vinson et al., 2001). In particular, anti-GD1a or GT1b IgM-type antibody induces p75^{NTR} raft recruitment and RhoA activation (Fujitani et al., 2005b). These data suggest that GD1a/GT1b enhance the clustering of the whole signaling complex for in MAG-Fc inhibition of axon growth.

How this ganglioside-regulated signaling complex activates RhoA pathway is one of the main subjects that remain obscure in myelin-derived inhibition. Two very recent works show apparently divergent ways to answer this question, although, as it is discussed within the next sections, they might be only different pieces fitting in the same puzzle.

9.4.2.2.2. p75 in Rafts Binds Directly to Rho-GDI but Not to Rho.

RhoA was thought to directly interact (only its inactive GDP-bound form) with p75^{NTR} as indicated by coimmunoprecipitation and two-hybrid experiments (Yamashita et al., 1999a). The same group of researchers, by p75^{NTR} immunoprecipitation of GST-RhoA expressing cells extracts, reinterpreted their results and concluded that p75^{NTR} interacts only indirectly with RhoA (Yamashita and Tohyama, 2003).

GDP-bound Rho proteins interact as well with the Rho-guanine–dissociation inhibitor (Rho-GDI), which prevents the conversion of Rho proteins to the active GTP-bound and its association to the membrane. Additionally, when Rho proteins are reconverted to the inactive form at the membrane, Rho-GDI can bind and carry them to the cytoplasm (Sasaki and Takai, 1998).

Precisely, Rho-GDI interacts directly with the fifth helical sequence of p75^{NTR} forming a complex upon ganglioside-p75^{NTR} clustering induced by MAG or Nogo (Figure 9.5A). According to the authors, this interaction would displace Rho-GDI from its binding to GDP-RhoA with the concomitant transformation to GTP-RhoA and activation of downstream RhoA pathway. Pep5, a 15-amino acid peptide corresponding to the RhoA-GDI binding site (framed by helices 5 and 6) of p75^{NTR} blocks the interaction between both molecules and precludes the neurite inhibitory effect of the myelin proteins (Yamashita and Tohyama, 2003).

There has been some controversy on this interpretation, mostly because some immunoprecipitation results in the same work clearly suggest that the binding of MAG or Nogo tends to stabilize a triple p75:Rho-GDI:RhoA complex. If so, RhoA would remain inactive after ligand binding.

9.4.2.2.3. p75^{NTR} Cleavage by Regulated Intramembrane Proteolysis Triggers Myelin Inhibitory Signaling

The key of this enigma, according to Marie T. Filbin's group (Domeniconi et al., 2005), can be hidden behind another result of Yamashita's work, which is the blocking activity of the p75^{NTR} cytoplasmic peptide Pep5. Again, gangliosides and rafts appear as fundamental players, this time for membrane proteolytic processes.

As I already mentioned, myelin inhibitors induce neurite inhibition by activating RhoA pathway in a PKC-dependent manner (Sivasankaran et al., 2004). On the other hand, in cell lines overexpressing $p75^{NTR}$, the receptor undergoes proteolytic cleavage after PKC activation with phorbol myristate acetate (PMA) (Jung et al., 2003; Kanning et al., 2003). The fragments resulting from this cleavage are, on one side, the extracellular domain and, on the other, a 30 kDa C-terminal piece. The latter undergoes subsequent cleavage to render a 25 kDa peptide, so called intracellular domain (ICD).

In fact, MAG induces regulated intramembrane proteolysis (RIP) of $p75^{NTR}$, first by α-secretase (ADAM metalloprotease) producing the 30 kDa fragment and then by γ-secretase complex in a PKC-dependent mode. γ-secretase cleaves this peptide inside the membrane and produces the 25 kDa ICD (Domeniconi et al., 2005). Secretases or PKC inhibition, as well as the expression of a noncleavable form of $p75^{NTR}$, block Rho activation and MAG-derived neurite inhibition. On the contrary, expression of the $p75^{NTR}$ ICD alone is enough to trigger inhibition in the absence of MAG.

According to these results, $p75^{NTR}$ inhibitory signaling may take place as follows: Ligand binding would stabilize the quaternary complex ligand: $p75^{NTR}$: Rho-GDI:Rho maintaining Rho in its GDP-bound inactive state (Figure 9.5A). Concomitantly, α- and γ-secretases are activated to process $p75^{NTR}$ and liberate the 25 kDa ICD to the cytoplasm (Figure 9.5B and C). This fragment contains the sequence of Pep5 that, as indicated earlier, can displace Rho-GDI from its interaction with GDP-RhoA and permit the conversion to GTP-Rho active form (Figure 9.5D) (Domeniconi et al., 2005).

9.4.2.2.4. Raft Glycosphingolipids Modulate RIP and p75 Inhibitory Signaling

The whole neuronal secretase machinery (α-, β-, and γ-secretases) participating in RIP can be localized in membrane rafts (Riddell et al., 2001; Vetrivel et al., 2004). This proteolytic processing is involved in the cleavage of several substrates, among others Notch (Selkoe and Kopan, 2003), APP (Amyloid precursor protein (Haass, 2004), or $p75^{NTR}$ (Jung et al., 2003; Kanning et al., 2003). It is logical to imagine that RIP activity is optimal when all the molecular players are clustered in discrete membrane zones. Of course, this should include the substrates, which habitual residence in rafts is still surprisingly controversial, at least in the three paradigmatic cases mentioned earlier. This different membrane compartmentalization of the enzymatic machinery and the substrates defines a new level of regulation by raft components of the signaling pathways downstream RIP.

The modification of lipidic raft composition can produce the exit of enzymatic complexes to nonraft membrane domains. For instance, moderate cholesterol loss in neuronal plasma membrane and concomitant subtle raft disruption, induce the displacement of the β-secretase to nonraft fractions, where most of the APP resides, enhancing the production of amyloidogenic Aβ peptide (Abad-Rodriguez et al., 2004). In other cases, the redistribution of lipids within the raft can permit the insertion of the substratum in these microdomains. This is the situation for RIP cleavage of $p75^{NTR}$ in myelin-associated neurite inhibition. A possible explanation could be that upon ligand binding, gangliosides GT1b and GD1a would get together and recruit $p75^{NTR}$ inside the micro domain, where proteolysis takes place. Afterward, the subsequent Rho activation drives to the final inhibitory effect (Domeniconi et al., 2005).

The exact molecular reasons for $p75^{NTR}$ raft recruitment are still unknown. The existence of a critical local concentration threshold of complex gangliosides, above which, $p75^{NTR}$ would get "trapped" is a good candidate to explain such an event. Additionally, the ligand could induce some change in the $p75^{NTR}$ molecule and increase its affinity for these gangliosides.

Anyhow, $p75^{NTR}$ is pushed into rafts either by neurite growth inhibitors (myelin) or enhancers (neurotrophins). In these microenvironments, GSL play a major role in the discrimination of the extracellular signal and in the triggering of the adequate intracellular pathway. It is tempting to speculate that different phosphorylation states of $p75^{NTR}$ would underlie this discrimination. For example, it is described earlier that neurotrophins induce $p75^{NTR}$ phosphorylation by PKA and its translocation to rafts (Higuchi et al., 2003). The position of such phosphorylation (Ser 304) at the intracellular juxtamembrane portion of the protein might impair intramembrane γ-cleavage, induce $p75^{NTR}$ dimerization, and avoid Rho activation by stabilizing the p75:Rho-GDI:Rho complex. Significantly, PKA activation in cerebellar and dorsal root ganglia neurons converts repulsion induced by MAG into attraction (Cai et al., 1999).

9.5. Concluding Remarks

There is already a huge amount of information to support the conclusion of a key role of GSLs and rafts in signal transduction and specially in neurotrophic signaling. Although some specific interactions of GSL with neurotrophin receptors have been described (GM1/Trk; GT1b/ $p75^{NTR}$), their exact nature is unknown. Differential effects due to interactions with the ceramides of GSLs, very similar in structure, would depend more on the presence of specific binding domains in the involved proteins than in ceramide differences. Contrarily, the elevated variability of carbohydrate moieties in GSLs and proteins could confer a high level of specificity for carbohydrate–carbohydrate and carbohydrate–protein differential interactions and would explain the extremely fine regulation occurring in rafts, where the same molecule ($p75^{NTR}$, for instance) can be included in different ways and exert completely opposite effects.

It is important to consider a different interpretation for the role of rafts as signaling platforms. Very recent works strongly point to the idea that, rather than the property of detergent insolubility, protein–protein interactions underlie the clustering of signaling molecules (Douglass and Vale, 2005; Lin and Shaw, 2005). Nevertheless, this does not exclude a more than probable role of GSLs in the stabilization, modulation or impairment of signaling effectors clustering and function. New powerful imaging technologies, such as Forsters resonance energy transfer (FRET) or total internal reflection fluorescence (TIRF), which allow to study the dynamics and interactions of single molecules at the plasma membrane level, will be of capital importance to solve this old controversy.

There is still a long way to go in order to unravel the full molecular mechanisms behind the GSL modulation of neurotrophic signals. Nevertheless, the important steps that have been taken during the last two decades permit us to envision further advances and the development of new GSL-based therapeutical approaches to fight nervous system degenerative disorders or to tackle nerve regeneration after any kind of injury.

References

Abad-Rodríguez, J., Ledesma, M.D., Craessaerts, K., Perga, S., Medina, M., Delacourte, A., et al., 2004, Neuronal membrane cholesterol loss enhances amyloid peptide generation. *J. Cell Biol.* **167:** 953–960.

Atwal, J.K., Massie, B., Miller, F.D., and Kaplan, D.R., 2000, The TrkB-Shc site signals neuronal survival and local axon growth via MEK and P13-kinase, *Neuron* **27:** 265–277.

Baloh, R.H., Tansey, M.G., Lampe, P.A., Fahrner, T.J., Enomoto, H., Simburger, K.S., et al., 1998, Artemin, a novel member of the GDNF ligand family, supports peripheral and central neurons and signals through the GFRalpha3-RET receptor complex, *Neuron* **21:** 1291–1302.

Barker, P.A., 2004, p75NTR is positively promiscuous: Novel partners and new insights, *Neuron* **42:** 529–533.

Byrne, M.C., Ledeen, R.W., Roisen, F.J., Yorke, G., and Sclafani, J.R., 1983, Ganglioside-induced neuritogenesis: Verification that gangliosides are the active agents, and comparison of molecular species, *J. Neurochem.* **41:** 1214–1222.

Cai, D., Shen, Y., De Bellard, M., Tang, S., and Filbin, M.T., 1999, Prior exposure to neurotrophins blocks inhibition of axonal regeneration by MAG and myelin via a cAMP-dependent mechanism, *Neuron* **22:** 89–101.

Carey, K.D., Watson, R.T., Pessin, J.E., and Stork, P.J., 2003, The requirement of specific membrane domains for Raf-1 phosphorylation and activation, *J. Biol. Chem.* **278:** 3185–3196.

Chester, M.A., 1998, IUPAC-IUB Joint Commission on Biochemical Nomenclature (JCBN, Nomenclature of glycolipids—recommendations 1997, *Eur. J. Biochem.* **257:** 293–298.

Chiavegatto, S., Sun, J., Nelson, R.J., and Schnaar, R.L., 2000, A functional role for complex gangliosides: Motor deficits in GM2/GD2 synthase knockout mice, *Exp. Neurol.* **166:** 227–234.

Cuello, A.C., Garofalo, L., Kenigsberg, R.L., and Maysinger, D., 1989, Gangliosides potentiate in vivo and in vitro effects of nerve growth factor on central cholinergic neurons, *Proc. Natl. Acad. Sci. USA* **86:** 2056–2060.

Da Silva, J.S., Medina, M., Zuliani, C., Di Nardo, A., Witke, W., and Dotti, C.G., 2003, RhoA/ROCK regulation of neuritogenesis via profilin IIa-mediated control of actin stability, *J. Cell Biol.* **162:** 1267–1279.

Da Silva, J.S., Hasegawa, T., Miyagi, T., Dotti, C.G., and Abad-Rodríguez, J., 2005, Asymmetric membrane ganglioside sialidase activity specifies axonal fate, *Nat. Neurosci.* **8:** 606–615.

Dickson, B.J., 2001, Rho GTPases in growth cone guidance, *Curr. Opin. Neurobiol.* **11:** 103–110.

Domeniconi, M., Cao, Z., Spencer, T., Sivasankaran, R., Wang, K., Nikulina, E., et al., 2002, Myelin-associated glycoprotein interacts with the Nogo66 receptor to inhibit neurite outgrowth, *Neuron* **35:** 283–290.

Domeniconi, M., Zampieri, N., Spencer, T., Hilaire, M., Mellado, W., Chao, M.V., et al., 2005, MAG induces regulated intramembrane proteolysis of the p75 neurotrophin receptor to inhibit neurite outgrowth, *Neuron* **46:** 849–855.

Douglass, A.D., and Vale, R.D., 2005, Single-molecule microscopy reveals plasma membrane microdomains created by protein-protein networks that exclude or trap signaling molecules in T cells, *Cell* **121:** 937–950.

Duchemin, A.M., Ren, Q., Mo, L., Neff, N.H., and Hadjiconstantinou, M., 2002, GM1 ganglioside induces phosphorylation and activation of Trk and Erk in brain, *J. Neurochem.* **81:** 696–707.

Facci, L., Leon, A., Toffano, G., Sonnino, S., Ghidoni, R., and Tettamanti, G., 1984, Promotion of neuritogenesis in mouse neuroblastoma cells by exogenous gangliosides. Relationship between the effect and the cell association of ganglioside GM1, *J. Neurochem.* **42:** 299–305.

Farooqui, T., Franklin, T., Pearl, D.K., and Yates, A.J., 1997, Ganglioside GM1 enhances induction by nerve growth factor of a putative dimer of TrkA, *J. Neurochem.* **68:** 2348–2355.

Ferrari, G., Anderson, B.L., Stephens, R.M., Kaplan, D.R., and Greene, L.A., 1995, Prevention of apoptotic neuronal death by GM1 ganglioside. Involvement of Trk neurotrophin receptors, *J. Biol. Chem.* **270:** 3074–3080.

Fong, T.G., Vogelsberg, V., Neff, N.H., and Hadjiconstantinou, M., 1995, GM1 and NGF synergism on choline acetyltransferase and choline uptake in aged brain, *Neurobiol. Aging* **16:** 917–923.

Fournier, A.E., GrandPre, T., and Strittmatter, S.M., 2001, Identification of a receptor mediating Nogo-66 inhibition of axonal regeneration, *Nature* **409:** 341–346.

Fujitani, M., Honda, A., Hata, K., Yamagishi, S., Tohyama, M., and Yamashita, T., 2005a, Biological activity of neurotrophins is dependent on recruitment of Rac1 to lipid rafts, *Biochem. Biophys. Res. Commun.* **327:** 150–154.

Fujitani, M., Kawai, H., Proia, R.L., Kashiwagi, A., Yasuda, H., and Yamashita, T., 2005b, Binding of soluble myelin-associated glycoprotein to specific gangliosides induces the association of p75NTR to lipid rafts and signal transduction, *J. Neurochem.* **94:** 15–21.

Galbiati, F., Razani, B., and Lisanti, M.P., 2001, Emerging themes in lipid rafts and caveolae, *Cell* **106:** 403–411.

Golub, T., Wacha, S., and Caroni, P., 2004, Spatial and temporal control of signaling through lipid rafts, *Curr. Opin. Neurobiol.* **14:** 542–550.

Guirland, C., Suzuki, S., Kojima, M., Lu, B., and Zheng, J.Q., 2004, Lipid rafts mediate chemotropic guidance of nerve growth cones, *Neuron* **42:** 51–62.

Haass, C., 2004, Take five–BACE and the gamma-secretase quartet conduct Alzheimer's amyloid beta-peptide generation, *EMBO J.* **23:** 483–488.

Hadjiconstantinou, M., and Neff, N.H., 1998, GM1 and the aged brain, *Ann. N. Y. Acad. Sci.* **845:** 225–231.

Hakomori, S., 1981, Glycosphingolipids in cellular interaction, differentiation, and oncogenesis, *Annu. Rev. Biochem.* **50:** 733–764.

Hasegawa, T., Yamaguchi, K., Wada, T., Takeda, A., Itoyama, Y., and Miyagi, T., 2000, Molecular cloning of mouse ganglioside sialidase and its increased expression in neuro2a cell differentiation, *J. Biol. Chem.* **275:** 14778.

Higuchi, H., Yamashita, T., Yoshikawa, H., and Tohyama, M., 2003, PKA phosphorylates the p75 receptor and regulates its localization to lipid rafts, *EMBO J.* **22:** 1790–1800.

Huang, C.S., Zhou, J., Feng, A.K., Lynch, C.C., Klumperman, J., DeArmond, S.J., et al., 1999, Nerve growth factor signaling in caveolae-like domains at the plasma membrane, *J. Biol. Chem.* **274:** 36707–36714.

Huang, E.J., and Reichardt, L.F., 2001, Neurotrophins: Roles in neuronal development and function, *Annu. Rev. Neurosci.* **24:** 677–736.

Huwiler, A., Kolter, T., Pfeilschifter, J., and Sandhoff, K., 2000, Physiology and pathophysiology of sphingolipid metabolism and signaling, *Biochim. Biophys. Acta* 1**485:** 63–99.

Inoue, M., Fujii, Y., Furukawa, K., Okada, M., Okumura, K., Hayakawa, T., et al., 2002, Refractory skin injury in complex knock-out mice expressing only the GM3 ganglioside, *J. Biol. Chem.* **277:** 29881–29888.

Jeyakumar, M., Butters, T.D., Dwek, R.A., and Platt, F.M., 2002, Glycosphingolipid lysosomal storage diseases: Therapy and pathogenesis, *Neuropathol. Appl. Neurobiol.* **28:** 343–357.

Jung, K.M., Tan, S., Landman, N., Petrova, K., Murray, S., Lewis, R., et al., 2003, Regulated intramembrane proteolysis of the p75 neurotrophin receptor modulates its association with the TrkA receptor, *J. Biol. Chem.* **278:** 42161–42169.

Kalka, D., von Reitzenstein, C., Kopitz, J., and Cantz, M., 2001, The plasma membrane ganglioside sialidase cofractionates with markers of lipid rafts, *Biochem. Biophys. Res. Commun.* **283:** 989–993.

Kanning, K.C., Hudson, M., Amieux, P.S., Wiley, J.C., Bothwell, M., and Schecterson, L.C., 2003, Proteolytic processing of the p75 neurotrophin receptor and two homologs generates C-terminal fragments with signaling capability, *J. Neurosci.* **23:** 5425–5436.

Kasahara, K., and Sanai, Y., 2000, Functional roles of glycosphingolipids in signal transduction via lipid rafts, *Glycoconj. J.* **17:** 153–162.

Kasahara, K., Watanabe, Y., Yamamoto, T., and Sanai, Y., 1997, Association of Src family tyrosine kinase Lyn with ganglioside GD3 in rat brain. Possible regulation of Lyn by glycosphingolipid in caveolae-like domains, *J. Biol. Chem.* **272:** 29947–29953.

Katoh-Semba, R., Skaper, S.D., and Varon, S., 1984, Interaction of GM1 ganglioside with PC12 pheochromocytoma cells: Serum- and NGF-dependent effects on neuritic growth (and proliferation), *J. Neurosci. Res.* **12:** 299–310.

Kawai, H., Allende, M.L., Wada, R., Kono, M., Sango, K., Deng, C., et al., 2001, Mice expressing only monosialoganglioside GM3 exhibit lethal audiogenic seizures, *J. Biol. Chem.* **276:** 6885–6888.

Kotzbauer, P.T., Lampe, P.A., Heuckeroth, R.O., Golden, J.P., Creedon, D.J., Johnson, E.M., Jr., et al., 1996, Neurturin, a relative of glial-cell-line-derived neurotrophic factor, *Nature* **384:** 467–470.

Ledeen, R.W., 1984, Biology of gangliosides: Neuritogenic and neuronotrophic properties, *J. Neurosci. Res.* **12:** 147–159.

Ledeen, R.W., 1989, Biosynthesis, metabolism and biological effect of gangliosides, in: *Neurobiology of Glycoconjugates*, R.U. Margolis, Margolis, R.K., eds., Plenum Press Corp., New York, pp. 43–83.

Lee, K.F., Li, E., Huber, L.J., Landis, S.C., Sharpe, A.H., Chao, M.V., et al., 1992, Targeted mutation of the gene encoding the low affinity NGF receptor p75 leads to deficits in the peripheral sensory nervous system, *Cell* **69:** 737–749.

Lee, R., Kermani, P., Teng, K.K., and Hempstead, B.L., 2001, Regulation of cell survival by secreted proneurotrophins, *Science* **294:** 1945–1948.

Leskawa, K.C., and Hogan, E.L., 1985, Quantitation of the in vitro neuroblastoma response to exogenous, purified gangliosides, *J. Neurosci. Res.* **13:** 539–550.

Lewin, G.R., and Barde, Y.A., 1996, Physiology of the neurotrophins, *Annu. Rev. Neurosci.* **19:** 289–317.

Lin, J., and Shaw, A.S., 2005, Getting downstream without a raft, *Cell* **121:** 815–816.

Lin, L.F., Doherty, D.H., Lile, J.D., Bektesh, S., and Collins, F., 1993, GDNF: A glial cell line-derived neurotrophic factor for midbrain dopaminergic neurons, *Science* **260:** 1130–1132.

Liu, B.P., Fournier, A., GrandPre, T., and Strittmatter, S.M., 2002, Myelin-associated glycoprotein as a functional ligand for the Nogo-66 receptor, *Science* **297:** 1190–1193.

Liu, H., Nakagawa, T., Kanematsu, T., Uchida, T., and Tsuji, S., 1999, Isolation of 10 differentially expressed cDNAs in differentiated Neuro2a cells induced through controlled expression of the GD3 synthase gene, *J. Neurochem.* **72:** 1781–1790.

Lu, B., Pang, P.T., and Woo, N.H., 2005, The yin and yang of neurotrophin action, *Nat. Rev. Neurosci.* **6:** 603–614.

Luo, L., 2000, Rho GTPases in neuronal morphogenesis, *Nat. Rev. Neurosci.* **1:** 173–180.

Markus, A., Patel, T.D., and Snider, W.D., 2002, Neurotrophic factors and axonal growth, *Curr. Opin. Neurobiol.* **12:** 523–531.

Mi, S., Lee, X., Shao, Z., Thill, G., Ji, B., Relton, J., et al., 2004, LINGO-1 is a component of the Nogo-66 receptor/p75 signaling complex, *Nat. Neurosci.* **7:** 221–228.

Michaelson, D., Silletti, J., Murphy, G., D'Eustachio, P., Rush, M., and Philips, M.R., 2001, Differential localization of Rho GTPases in live cells: Regulation by hypervariable regions and RhoGDI binding, *J. Cell Biol.* **152:** 111–126.

Milbrandt, J., de Sauvage, F.J., Fahrner, T.J., Baloh, R.H., Leitner, M.L., Tansey, M.G., et al., 1998, Persephin, a novel neurotrophic factor related to GDNF and neurturin, *Neuron* **20:** 245–253.

Miyagi, T., Wada, T., Iwamatsu, A., Hata, K., Yoshikawa, Y., Tokuyama, S., et al., 1999, Molecular cloning and characterization of a plasma membrane-associated sialidase specific for gangliosides, *J. Biol. Chem.* **274:** 5004–5011.

Mukherjee, A., Arnaud, L., and Cooper, J.A., 2003, Lipid-dependent recruitment of neuronal Src to lipid rafts in the brain, *J. Biol. Chem.* **278:** 40806–40814.

Mutoh, T., Tokuda, A., Guroff, G., and Fujiki, N., 1993, The effect of the B subunit of cholera toxin on the action of nerve growth factor on PC12 cells, *J. Neurochem.* **60:** 1540–1547.

Mutoh, T., Tokuda, A., Miyadai, T., Hamaguchi, M., and Fujiki, N., 1995, Ganglioside GM1 binds to the Trk protein and regulates receptor function, *Proc. Natl. Acad. Sci. USA* **92:** 5087–5091.

Mutoh, T., Tokuda, A., Inokuchi, J., and Kuriyama, M., 1998, Glucosylceramide synthase inhibitor inhibits the action of nerve growth factor in PC12 cells, *J. Biol. Chem.* **273:** 26001–26007.

Mutoh, T., Hamano, T., Tokuda, A., and Kuriyama, M., 2000, Unglycosylated Trk protein does not co-localize nor associate with ganglioside GM1 in stable clone of PC12 cells overexpressing Trk (PCtrk cells), *Glycoconj. J.* **17:** 233–237.

Mutoh, T., Hamano, T., Yano, S., Koga, H., Yamamoto, H., Furukawa, K., et al., 2002, Stable transfection of GM1 synthase gene into GM1-deficient NG108–15 cells, CR-72 cells, rescues the responsiveness of Trk-neurotrophin receptor to its ligand, NGF, *Neurochem. Res.* **27:** 801–806.

Niederost, B., Oertle, T., Fritsche, J., McKinney, R.A., and Bandtlow, C.E., 2002, Nogo-A and myelin-associated glycoprotein mediate neurite growth inhibition by antagonistic regulation of RhoA and Rac1, *J. Neurosci.* **22:** 10368–10376.

Nishio, M., Fukumoto, S., Furukawa, K., Ichimura, A., Miyazaki, H., Kusunoki, S., and Urano, T., 2004, Overexpressed GM1 suppresses nerve growth factor (NGF) signals by modulating the intracellular localization of NGF receptors and membrane fluidity in PC12 cells, *J. Biol. Chem.* **279:** 33368–33378.

Nykjaer, A., Lee, R., Teng, K.K., Jansen, P., Madsen, P., Nielsen, M.S., et al., 2004, Sortilin is essential for proNGF-induced neuronal cell death, *Nature* **427:** 843–848.

Okada, M., Itoh Mi, M., Haraguchi, M., Okajima, T., Inoue, M., Oishi, H., et al., 2002, b-series ganglioside deficiency exhibits no definite changes in the neurogenesis and the sensitivity to Fas-mediated apoptosis but impairs regeneration of the lesioned hypoglossal nerve, *J. Biol. Chem.* **277:** 1633–1636.

Panni, M.K., Cooper, J.D., and Sofroniew, M.V., 1998, Ganglioside GM1 potentiates NGF action on axotomised medial septal cholinergic neurons, *Brain Res.* **812:** 76–80.

Paratcha, G., Ledda, F., and Ibanez, C.F., 2003, The neural cell adhesion molecule NCAM is an alternative signaling receptor for GDNF family ligands, *Cell* **113:** 867–879.

Perron, J.C., and Bixby, J.L., 1999, Distinct neurite outgrowth signaling pathways converge on ERK activation, *Mol. Cell Neurosci.* **13:** 362–378.

Pitto, M., Mutoh, T., Kuriyama, M., Ferraretto, A., Palestini, P., and Masserini, M., 1998, Influence of endogenous GM1 ganglioside on TrkB activity, in cultured neurons, *FEBS Lett.* **439:** 93–96.

Prinetti, A., Marano, N., Prioni, S., Chigorno, V., Mauri, L., Casellato, R., et al., 2000, Association of Src-family protein tyrosine kinases with sphingolipids in rat cerebellar granule cells differentiated in culture, *Glycoconj. J.* **17:** 223–232.

Rabin, S.J., and Mocchetti, I., 1995, GM1 ganglioside activates the high-affinity nerve growth factor receptor trkA, *J. Neurochem.* **65:** 347–354.

Rabin, S.J., Bachis, A., and Mocchetti, I., 2002, Gangliosides activate Trk receptors by inducing the release of neurotrophins, *J. Biol. Chem.* **277:** 49466–49472.

Riddell, D.R., Christie, G., Hussain, I., and Dingwall, C., 2001, Compartmentalization of beta-secretase (Asp2) into low-buoyant density, noncaveolar lipid rafts, *Curr. Biol.* **11:** 1288–1293.

Rodríguez, J.A., Piddini, E., Hasegawa, T., Miyagi, T., and Dotti, C.G., 2001, Plasma membrane ganglioside sialidase regulates axonal growth and regeneration in hippocampal neurons in culture, *J. Neurosci.* **21:** 8387–8395.

Sasaki, T., and Takai, Y., 1998, The Rho small G protein family-Rho GDI system as a temporal and spatial determinant for cytoskeletal control, *Biochem. Biophys. Res. Commun.* **245:** 641–645.

Schwab, M.E., 2004, Nogo and axon regeneration, *Curr. Opin. Neurobiol.* **14:** 118–124.

Schwartz, M., and Spirman, N., 1982, Sprouting from chicken embryo dorsal root ganglia induced by nerve growth factor is specifically inhibited by affinity-purified antiganglioside antibodies, *Proc. Natl. Acad. Sci. USA* **79:** 6080–6083.

Selkoe, D., and Kopan, R., 2003, Notch and Presenilin: Regulated intramembrane proteolysis links development and degeneration, *Annu. Rev. Neurosci.* **26:** 565–597.

Sheikh, K.A., Sun, J., Liu, Y., Kawai, H., Crawford, T.O., Proia, R.L., et al., 1999, Mice lacking complex gangliosides develop Wallerian degeneration and myelination defects, *Proc. Natl. Acad. Sci. USA* **96:** 7532–7537.

Simons, K., and Toomre, D., 2000, Lipid rafts and signal transduction, *Nat. Rev. Mol. Cell. Biol.* **1:** 31–39.

Sivasankaran, R., Pei, J., Wang, K.C., Zhang, Y.P., Shields, C.B., Xu, X.M., et al., 2004, PKC mediates inhibitory effects of myelin and chondroitin sulfate proteoglycans on axonal regeneration, *Nat. Neurosci.* **7:** 261–268.

Skaper, S.D., Katoh-Semba, R., and Varon, S., 1985, GM1 ganglioside accelerates neurite outgrowth from primary peripheral and central neurons under selected culture conditions, *Brain Res.* **355:** 19–26.

Snider, W.D., and Silos-Santiago, I., 1996, Dorsal root ganglion neurons require functional neurotrophin receptors for survival during development, *Philos. Trans. R. Soc. Lond. B. Biol. Sci.* **351:** 395–403.

Svennerholm, L., 1964, The gangliosides, *J. Lipid Res.* **41:** 145–155.

Takamiya, K., Yamamoto, A., Furukawa, K., Yamashiro, S., Shin, M., Okada, M., et al., 1996, Mice with disrupted GM2/GD2 synthase gene lack complex gangliosides but exhibit only subtle defects in their nervous system, *Proc. Natl. Acad. Sci. USA* **93:** 10662–10667.

Tansey, M.G., Baloh, R.H., Milbrandt, J., and Johnson, E.M., Jr., 2000, GFRalpha-mediated localization of RET to lipid rafts is required for effective downstream signaling, differentiation, and neuronal survival, *Neuron* **25:** 611–623.

Tettamanti, G., 2004, Ganglioside/glycosphingolipid turnover: New concepts, *Glycoconj. J.* **20:** 301–317.

Tsuji, S., Arita, M., and Nagai, Y., 1983, GQ1b, a bioactive ganglioside that exhibits novel nerve growth factor (NGF)-like activities in the two neuroblastoma cell lines, *J. Biochem. Tokyo* **94:** 303–306.

Vetrivel, K.S., Cheng, H., Lin, W., Sakurai, T., Li, T., Nukina, N., et al., 2004, Association of gamma-secretase with lipid rafts in post-Golgi and endosome membranes, *J. Biol. Chem.* **279:** 44945–44954.

Vinson, M., Strijbos, P.J., Rowles, A., Facci, L., Moore, S.E., Simmons, D.L., et al., 2001, Myelin-associated glycoprotein interacts with ganglioside GT1b. A mechanism for neurite outgrowth inhibition, *J. Biol. Chem.* **276:** 20280–20285.

von Schack, D., Casademunt, E., Schweigreiter, R., Meyer, M., Bibel, M., and Dechant, G., 2001, Complete ablation of the neurotrophin receptor p75NTR causes defects both in the nervous and the vascular system, *Nat. Neurosci.* **4:** 977–978.

Wada, T., Yoshikawa, Y., Tokuyama, S., Kuwabara, M., Akita, H., and Miyagi, T., 1999, Cloning, expression, and chromosomal mapping of a human ganglioside sialidase, *Biochem. Biophys. Res. Commun.* **261:** 21–27.

Walkley, S.U., 2003, Neurobiology and cellular pathogenesis of glycolipid storage diseases, *Philos. Trans. R. Soc. Lond. B. Biol. Sci.* **358:** 893–904.

Wang, K.C., Kim, J.A., Sivasankaran, R., Segal, R., and He, Z., 2002a, P75 interacts with the Nogo receptor as a co-receptor for Nogo, MAG and OMgp, *Nature* **420:** 74–78.

Wang, K.C., Koprivica, V., Kim, J.A., Sivasankaran, R., Guo, Y., Neve, R.L., et al., 2002b, Oligodendrocyte-myelin glycoprotein is a Nogo receptor ligand that inhibits neurite outgrowth, *Nature* **417:** 941–944.

Watson, F.L., Porcionatto, M.A., Bhattacharyya, A., Stiles, C.D., and Segal, R.A., 1999, TrkA glycosylation regulates receptor localization and activity, *J. Neurobiol.* **39:** 323–336.

Witke, W., Sutherland, J.D., Sharpe, A., Arai, M., and Kwiatkowski, D.J., 2001, Profilin I is essential for cell survival and cell division in early mouse development, *Proc. Natl. Acad. Sci. USA* **98:** 3832–3836.

Wong, S.T., Henley, J.R., Kanning, K.C., Huang, K.H., Bothwell, M., and Poo, M.M., 2002, A p75(NTR) and Nogo receptor complex mediates repulsive signaling by myelin-associated glycoprotein, *Nat. Neurosci.* **5:** 1302–1308.

Wu, C., Butz, S., Ying, Y., and Anderson, R.G., 1997, Tyrosine kinase receptors concentrated in caveolae-like domains from neuronal plasma membrane, *J. Biol. Chem.* **272:** 3554–3559.

Yamashita, T., and Tohyama, M., 2003, The p75 receptor acts as a displacement factor that releases Rho from Rho-GDI, *Nat. Neurosci.* **6:** 461–467.

Yamashita, T., Tucker, K.L., and Barde, Y.A., 1999a, Neurotrophin binding to the p75 receptor modulates Rho activity and axonal outgrowth, *Neuron* **24:** 585–593.

Yamashita, T., Wada, R., Sasaki, T., Deng, C., Bierfreund, U., Sandhoff, K., et al., 1999b, A vital role for glycosphingolipid synthesis during development and differentiation, *Proc. Natl. Acad. Sci. USA* **96:** 9142–9147.

Yamashita, T., Higuchi, H., and Tohyama, M., 2002, The p75 receptor transduces the signal from myelin-associated glycoprotein to Rho, *J. Cell Biol.* **157:** 565–570.

Yamashita, T., Hashiramoto, A., Haluzik, M., Mizukami, H., Beck, S., Norton, A., et al., 2003, Enhanced insulin sensitivity in mice lacking ganglioside GM3, *Proc. Natl. Acad. Sci. USA* **100:** 3445–3449.

Yamashita, T., Wu, Y.P., Sandhoff, R., Werth, N., Mizukami, H., Ellis, J.M., et al., 2005, Interruption of ganglioside synthesis produces central nervous system degeneration and altered axon-glial interactions, *Proc. Natl. Acad. Sci. USA* **102:** 2725–2730.

Yiu, G., and He, Z., 2003, Signaling mechanisms of the myelin inhibitors of axon regeneration, *Curr. Opin. Neurobiol.* **13:** 545–551.

Yu, R.K., Bieberich, E., Xia, T., and Zeng, G., 2004a, Regulation of ganglioside biosynthesis in the nervous system, *J. Lipid Res.* **45:** 783–793.

Yu, W., Guo, W., and Feng, L., 2004b, Segregation of Nogo66 receptors into lipid rafts in rat brain and inhibition of Nogo66 signaling by cholesterol depletion, *FEBS Lett.* **577:** 87–92.

10
Wnt Signaling in Neurite Development

SILVANA B. ROSSO AND PATRICIA C. SALINAS

10.1. Summary

The highly complex organization and computational capacity of the nervous system is essential for its function. This complexity is achieved during early development by the intrinsic property of each neuronal cell type and the influence of extracellular cues. The polarized morphology of neurons with elaborate dendritic arbors and long thin axons is critical for brain function. Neurite development is established after neurons extend multiple processes that soon become polarized when one or two processes differentiate into axons and the rest into dendrites. Both axons and dendrites extend and branch in search for their appropriate synaptic targets. Members of the Wnt family of proteins are well-known embryonic morphogens that regulate cell fate decisions during early embryonic patterning. However, new studies show that Wnts also regulate different stages of neurite development from axon outgrowth, pathfinding to axon branching and remodeling. Wnts also modulate the formation of complex dendritic arborization thus increasing the ability of neurons to receive multiple inputs. Finally, Wnts can function as target release molecules that modulate presynaptic axon remodeling and synaptic assembly. Here we discuss advances that shed new light into the function of Wnt signaling in neurite development with particular emphasis on the formation of the vertebrate nervous system.

10.2. Introduction

The function of the nervous system depends on the morphological complexity of neurons and the establishment of proper neuronal connections. After migrating to their final destinations, neurons begin to form processes called neurites that later differentiate into functionally and structurally distinct axons and dendrites. While axons are specialized to propagate electrical signals and release neurotransmitters, dendrites are specialized in the reception and integration of these signals. These distinct specializations are acquired early in development during the establishment of cell polarity. Neurons are the most highly polarized cells, and their polarity is essential for informational processing and to ensure unidirectional signal flow required for proper neural connectivity. Subsequently neurite outgrowth and axon pathfinding ensure that neurites project and synapse to their appropriate targets. These processes from neuronal polarity to axon guidance are controlled by a combination of an intrinsic program of gene expression that is characteristic of each neuronal cell type and by extrinsic factors. This intrinsic program regulates the expression of structural proteins, membrane receptors, signaling molecules, and transport proteins that are distinctively localized to axons or dendrites. This specific distribution guaranties the acquisition of specific functional properties of axons and dendrites. Extrinsic factors also play an important role as regulators of neuronal outgrowth, axon guidance, and remodeling during the formation of synapses.

A combination of membrane and secreted molecules modulate the polarity of neurons and the growth, branching, and shape of neurites. Membrane bound proteins, such as cadherins, and Eph molecules as well as diffusible molecules, such as neurotrophins, netrin, semaphorin, and Slit, have been implicated in different aspects of neuronal development from axon guidance, branching, and synapse formation and growth (Dickson, 2002; Waites et al., 2005). In addition, growth factors, such as BMPs, FGFs, Shh, and Wnt proteins, previously known for their role as embryonic morphogens are now emerging as neuronal circuit modulators (Bovolenta, 2005; Salinas, 2005; Torroja et al., 2005). These extracellular signals bind to specific membrane receptors in developing neurites triggering a signal cascade that impinges into the neurite cytoskeleton but can also modulate gene expression.

Wnt signaling regulates different aspects of neuronal development from stem cell proliferation, neurogenesis, axon guidance, dendritic development to synapse formation (Ciani and Salinas, 2005). Wnts induce profound changes in neuronal morphology and behavior. In this chapter, we will discuss the function of Wnts and some key components of the pathway in neurite development with particular emphasis in cell polarity, axon behavior, and dendritic development.

10.3. Wnt Signaling Pathways

10.3.1. Wnt Receptors

Wnts are a large family of cysteine-rich secreted proteins that bind to cell surface receptors encoded by the frizzled (Fz) family (Logan and Nusse, 2004) and the atypical tyrosine kinase receptor Ryk (Inoue et al., 2004; Lu et al., 2004). In the mouse genome 19 Wnt genes, 9 Fz genes, and 1 Ryk receptor gene have been identified. Wnts and their downstream signaling components are highly conserved among animal species. Although very little is known about the structure of Wnts, these proteins are posttranslationally modified to carry a palmitate moiety (Willert et al., 2003). Wnts binds to the amino-terminal cysteine-rich domain

(CRD) of the seven transmembrane Fz receptors. A cytoplasmic carboxyl terminal tail of the receptor binds to dishevelled (Dvl) and may play a role in determining the specificity of responses (Boutros et al., 2000; Medina and Steinbeisser, 2000; Wong et al., 2003). Several studies have suggested that G-proteins could be involved in the Wnt signaling pathways (Liu et al., 2001; Slusarski et al., 1997). However, a loss of function study demonstrated the requirement of G-proteins for Wnt signaling (Katanaev et al., 2005).

Signaling through the Ryk receptor is less understood. Several Wnts have been shown to bind and signal through Ryk (Liu et al., 2005; Yoshikawa et al., 2003). The same Wnt protein can signal through Fz or Ryk, for example, Wnt1 has been shown to signal through both receptors (Holmen et al., 2002; Liu et al., 2005). Although Ryk can signal through an Fz-independent mechanism during axon guidance (Yoshikawa et al., 2003), Ryk has been shown to form a complex with Fz and Wnt and that this is essential for signaling during axon elongation (Lu et al., 2004). Moreover in *Caenorhabditis elegans*, Ryk and Fz seem to signal through parallel pathway during vulval development (Inoue et al., 2004). This different and apparently contradictory result could be explained by different requirements of these receptors depending on the cellular or developmental context.

10.3.2. Signal Transduction Pathways

Wnt molecules can signal through three different pathways (Figure 10.1): the canonical or Wnt/β-catenin, the planar cell polarity (PCP), and the Wnt/calcium pathways (Figure 10.1). Binding of Wnts to their receptors leads to activation of Dvl, a cytoplasmic scaffold protein required for signaling through the three pathways. In the *canonical or Wnt/β-catenin pathway*, the Wnt-Fz complex interacts with LRP5/6 coreceptors, members of a family of molecules that are related to low-density lipoprotein receptors (LDLRs). This ternary receptor complex leads to activation of Dvl, which induces the disassembly of a complex formed by axin, adenomatosis polys coli (APC), the serine/threonine kinase glycogen synthase kinase 3β (GSK-3β), and β-catenin. The activation of this pathway induces the inhibition of GSK-3β resulting in the stabilization and accumulation of β-catenin in the cytoplasm, its subsequent translocation to the nucleus. β-catenin forms a complex with TCF/LEF transcription factors that activate the transcription of target genes (Figure 10.1) (Logan and Nusse, 2004). The β-catenin pathway is involved in changes in cell fate decisions during development, and deregulation of this pathway results in tumor formation (Gregorieff and Clevers, 2005; Luu et al., 2004). The Wnt canonical pathway can also diverge downstream of GSK-3β to regulate the cytoskeleton in a transcriptional-independent manner (Figure 10.1) (Ciani et al., 2004). Inhibition of GSK-3β by Dvl leads to changes in the phosphorylation of its targets, such us microtubule associated protein 1B (MAP1B), resulting in changes of microtubule organization and increased microtubule stability (Ciani et al., 2004). In the *PCP pathway*, Dvl is required but not the LRP coreceptor, β-catenin, GSK-3β, and TCF. Instead, activation of Dvl triggers a cascade through Rho GTPases and *c*-Jun amino (N)-terminal kinase (JNK) (Boutros et al.,

1998; Ciani et al., 2004; Habas et al., 2003). This signaling pathway regulates tissue cell polarity, cell movement, and cell orientation (Heisenberg et al., 2000; Strutt et al., 1997). Finally, in the *Wnt/calcium pathway*, activation of Dvl triggered by Wnt/Fz interaction induces calcium influx and activation of two calcium sensitive enzymes—protein kinase C (PKC) and calcium/calmodulin-dependent protein kinase II (CaMKII) (Kuhl et al., 2000; Veeman et al., 2003). This pathway has been implicated in the regulation of cell movements during gastrulation and heart development (Sheldahl et al., 2003; Veeman et al., 2003).

Wnt signaling pathways are involved in multiple developmental processes from early embryonic patterning, cell proliferation, migration, cell differentiation to cell polarity (Cadigan and Nusse, 1997; Moon et al., 2002; Peifer and Polakis, 2000; Strutt, 2003). In the nervous system, Wnt proteins are required for early patterning by acting as posteriorising signals, for neural crest cell induction, neural precursor cell proliferation, and neurogenesis (Ciani and Salinas, 2005). Later, Wnt signaling plays a key role in neuronal circuit formation by regulating axon guidance (Lyuksyutova et al., 2003; Yoshimura et al., 2005; Zhou et al., 2004), dendrite morphogenesis (Rosso et al., 2005), and synapse formation (Hall et al., 2000; Packard et al., 2002).

10.4. Wnts and Neuronal Polarity

In the last few years, great progress has been made in the identification of the factors and the mechanisms that regulate the establishment and maintenance of neuronal polarity (Arimura et al., 2004; Bradke and Dotti, 2000; Luo, 2000). A general theme that emerges from studies in different developmental systems is that cell polarity is established by the local recruitment and local activation of signaling components to restricted cellular compartments or domains such as the leading edge of a migrating cell or the growth cone of a newly formed neurite (Rodriguez-Boulan and Powell, 1992; Wiggin et al., 2005). The restricted localization of molecules is achieved by the asymmetric segregation of determinants during cell division of neuronal progenitors and/or by the asymmetric or local exposure to extracellular factors. Although no direct evidence for Wnt proteins in neuronal polarity has been documented, some components of the Wnt pathway, such as APC and particularly GSK-3β, are emerging as key molecules in the regulation of neuronal polarity (Jiang et al., 2005; Yoshimura et al., 2005; Zhou et al., 2004).

10.4.1. GSK-3β

Modulation of the activity of GSK-3β has been implicated in the establishment and maintenance of the axon-dendrite polarity (Jiang et al., 2005; Yoshimura et al., 2005). GSK-3β can be inhibited by phosphorylation of the serine residue at position 9 (Sutherland et al., 1993). In developing hippocampal neurons, this phosphorylated form of GSK-3β is localized to axons but not to dendrites.

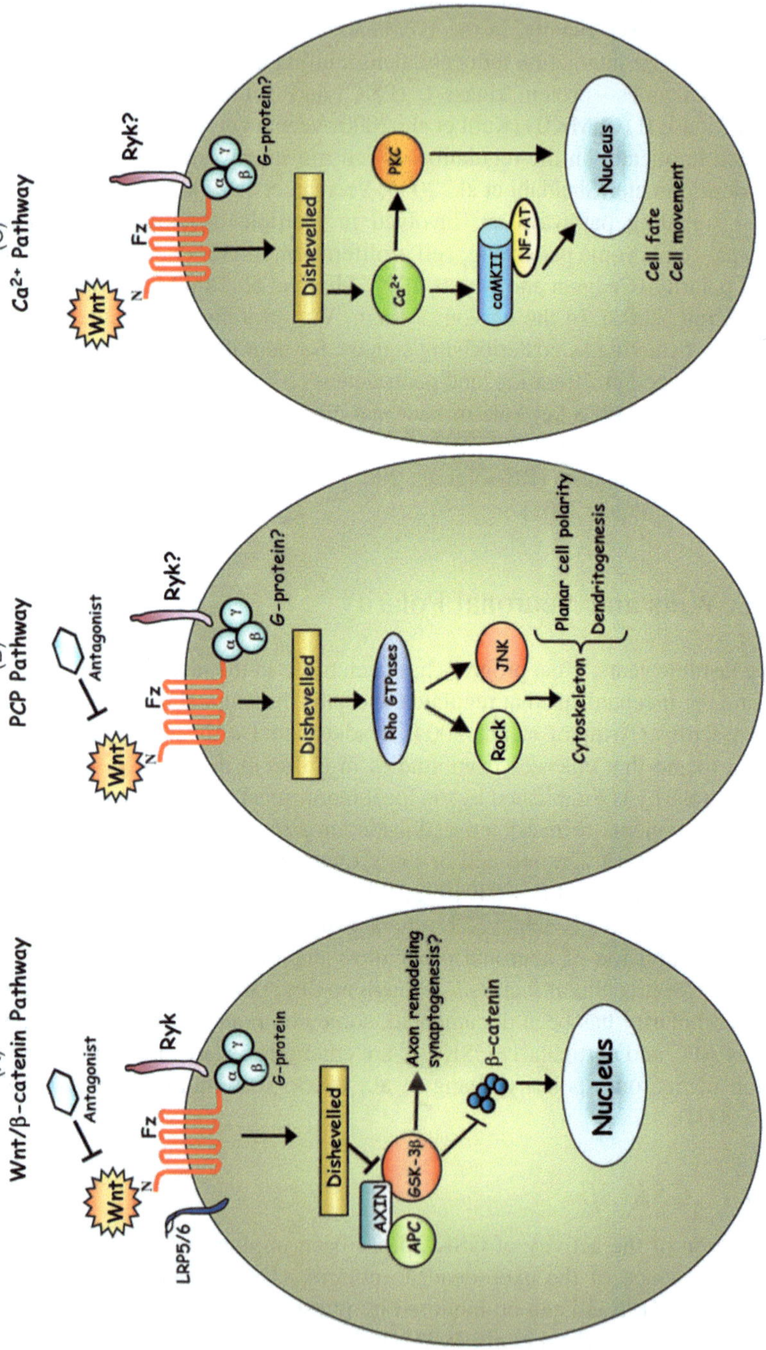

(A)
Wnt/β-catenin Pathway
Wnt
N
LRP5/6
Fz
Antagonist
Ryk
α
β
γ
G-protein
Dishevelled
AXIN
APC
GSK-3β
Axon remodeling
synaptogenesis?
β-catenin
Nucleus
(B)
PCP Pathway
Wnt
N
Fz
Antagonist
Ryk?
α
β
γ
G-protein?
Dishevelled
Rho GTPases
Rock
JNK
Cytoskeleton
Planar cell polarity
Dendritogenesis
(C)
Ca2+ Pathway
Wnt
N
Fz
Ryk?
α
β
γ
G-protein?
Dishevelled
Ca2+
PKC
caMKII
NF-AT
Nucleus
Cell fate
Cell movement

FIGURE 10.1. Wnt signaling pathways. (A) The canonical or Wnt/β-catenin pathway, Wnt binds to Fz receptor and low-density lipoprotein receptor-related protein 5 (LRP5) or LRP6 and activates Dvl. This pathway requires G-proteins, and Ryk can also function as a coreceptor of Fz for Wnts. Activation of Dvl leads to the inhibition of glycogen synthase kinase 3β (GSK-3β) and disassembly of a complex formed by APC, axin and GSK-3β that regulates the degradation of β-catenin. Inhibition of GSK-3β results in stabilization of β-catenin in the cytoplasm and its translocation to the nucleus where β-catenin activates transcription mediated by T-cell specific transcription factor (TCF)/lymphoid-enhancing factor (LEF). A pathway that diverges downstream of GSK-3β regulates microtubule stability by directly affecting the phosphorylation of microtubule associated proteins such as MAP1B. (B) The PCP pathway, Dvl is activated on binding of Wnt to the Fz receptor resulting in activation of small Rho GTPases. Rho and Rac activate the Rho kinase (Rock) and c-Jun amino (N)-terminal kinase (JNK), respectively, to regulate tissue polarization and dendritogenesis through changes in the cytoskeleton organization. **(C)** The Wnt/calcium (Ca^{2+}) pathway, activation of Dvl increases the intracellular levels of Ca^{2+} and activation of PKC and CaMKII. This pathway is involved in the regulation of cell fate decisions and cell movement through changes in gene transcription.

Expression of a constitutively active GSK-3β mutant eliminates axon formation whereas dendrite development is unaffected (Jiang et al., 2005). Conversely, inhibition of GSK-3β, by pharmacological or a peptide inhibitor or siRNA leads to the formation of multiple axons without affecting the total number of neuritis (Jiang et al., 2005). GSK-3β plays a crucial role in not only establishing but also maintaining neuronal polarity as inhibition of GSK-3β during very early neurite development (before stage 3) (Dotti et al., 1988) induces the formation of multiple axons at expenses of dendrites. In contrast, when neuronal polarity has been established, inhibition of GSK-3β results in the conversion of preexisting dendrites into axons (Jiang et al., 2005). These results demonstrate a role for GSK-3β in establishing axon-dendrite polarity and also in the maintenance of that polarity.

How can GSK-3β activity be modulated and transduced to maintain neuronal morphology? Due to its high level of basal activity, GSK-3β is regulated through inhibition (Doble and Woodgett, 2003). Although phosphorylation at the serine-9 residue is the main mechanism of inhibition, GSK-3β is also inhibited through a serine-9 phosphorylation independent mechanism (Sutherland et al., 1993). At present, the exact mechanism by which Wnts inhibit GSK-3β remains poorly understood. During the establishment of neuronal polarity, Akt and PTEN inactivate GSK-3β by serine-9 phosphorylation (Jiang et al., 2005). IP3K, a regulator of Akt, is localized to growth cones, and its activation results in localized inhibition of GSK-3β (Jiang et al., 2005). Although no Wnt proteins have been implicated in this process, these findings raise the interesting possibility that modulation of GSK-3β activity through upstream molecules, such as Wnts, could modulate axon-dendrite polarity.

10.4.2. GSK-3β and Its Cytoskeletal Targets

How does inhibition of GSK-3β stimulate axon formation at the expense of dendrites? For example, modulation of GSK-3β activity could regulate neuronal polarity by affecting the neurite cytoskeleton. Several studies have suggested that during early neurite outgrowth the fastest growing neurite becomes the axon. For example, actin-depolymerizing drugs induce the formation of multiple axons (Bradke and Dotti, 1999). Many proteins involved in the regulation of cytoskeletal dynamics and organization are direct targets of GSK-3β. For example, microtubule-associated proteins (MAPs), such us MAP1B, tau, and MAP2, can be phosphorylated by GSK-3β (Berling et al., 1994; Hanger et al., 1992; Lucas et al., 1998). Another protein that is directly phosphorylated by GSK-3β is APC, an important player in the Wnt signaling pathway and a microtubule plus-end binding protein that is accumulated at growth cones (Zhou et al., 2004; Zumbrunn et al., 2001). Phosphorylation of these proteins by GSK-3β changes their ability to bind microtubules and their function as modulators of microtubule dynamics (Baas and Qiang, 2005; Gonzalez-Billault et al., 2004; Zhou et al., 2004). Thus, changes in the phosphorylation of cytoskeletal proteins might contribute to axon determination and growth.

Consistent with the view that GSK-3β regulates neuronal polarity through changes in the cytoskeleton, a study has shown that GSK-3β phosphorylates CRMP-2 (collapsin response mediator protein-2) resulting in its inactivation (Yoshimura et al., 2005). CRMP-2 regulates neuronal polarity (Inagaki et al., 2001), possibly by promoting microtubule assembly (Fukata et al., 2002). In hippocampal neurons, CRMP-2 is enriched in the distal part of the growing axon, and overexpression of a nonphosphorylated form of CRMP-2 induces the formation of multiple axons. Conversely, expression of dominant negative mutants of CRMP-2 suppresses axonal growth (Arimura et al., 2004; Inagaki et al., 2001). The inhibitory effect of constitutively active GSK-3β is blocked by overexpression of nonphosphorylated form of CRMP-2 (Yoshimura et al., 2005), indicating that CRMP-2 is a downstream target of GSK-3β in the specification of axon-dendrite polarity.

GSK-3β has also been implicated in axon growth. Inhibition of GSK-3β by IP3K stimulates axon growth through the phosphorylation of APC. The localized GSK-3β inactivation at the growth cone promotes axon elongation by increasing the interaction of APC with microtubules (Zhou et al., 2004). Thus, inhibition of GSK-3β regulates two aspects of axon development: the establishment of axon identity and axon elongation.

10.5. Wnt Signaling and Neurite Develoment

10.5.1. Axon Behavior

A role for Wnt signaling in different aspects of neurite circuit formation is beginning to emerge. Wnts regulate axon guidance, axon branching, and the remodeling of terminal arbors during synapse formation. The first evidence for a role of Wnts in the control of axon behavior came from studies using cultured cerebellar neurons exposed to Wnt. Wnt7a induces profound changes in the axonal morphology and growth. Exposure to Wnt7a and Wnt3a induces the formation of axons with large growth cones and spread areas along the axon shaft and increased branching (Krylova et al., 2002; Lucas and Salinas, 1997) (Figure 10.2). Spreading along the axon shaft is particularly evident at the distal portion of the axon. These morphological changes are mimicked by pharmacological inhibitors of GSK-3β (Hall et al., 2000; Lucas and Salinas, 1997) suggesting that Wnt7a regulates axonal remodeling through inhibition of GSK-3β.

Studies in *Drosophila* and the vertebrate nervous system have begun to elucidate a role for Wnt signaling in axon pathfinding. In the fly, Wnt signaling through Derailed, the homolog of the Ryk receptor is required for the proper navigation of commissural axons at the midline (Yoshikawa et al., 2003). Derailed-expressing axons cross the midline through the anterior commissure but not through the posterior commissure where Dwnt5 repels these axons (Figure 10.3). In this system, Dwnt5 binds to Derailed but the Fz receptor appears not to be required (Yoshikawa et al., 2003). In mammals, Ryk is able to bind Wnt1 and

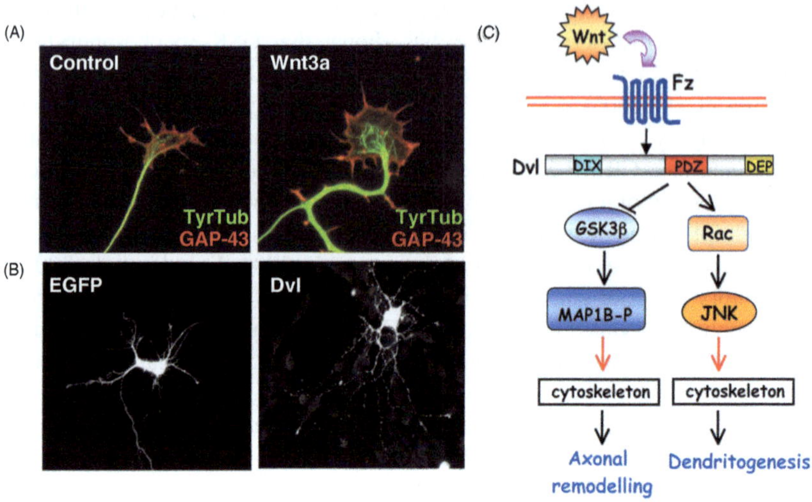

FIGURE 10.2. Wnt signaling regulates growth cone morphology and dendritic development through different pathways. (A) Dorsal root ganglia (DRG) neurons exposed to Wnt3a exhibit enlarged growth cones with the formation of looped microtubules. (B) Hippocampal neurons expressing Dvl exhibit longer and more complex dendrites when compared to EGFP controls. (C) Activation of Dvl leads to axonal remodeling and dendritogenesis through two different pathways. In axon, Wnts regulate remodeling through a divergent Wnt canonical pathway that results in the inhibition of GSK-3β and changes in microtubule stability through the phosphorylation of the MAP1B. In dendrites, Wnt signaling through Dvl leads to the activation of Rac and Jun N-terminal kinase (JNK) resulting in increased dendritic branching and growth however GSK-3β is not involved.

Wnt3a but in contrast to the fly the extracellular domain of Ryk forms a ternary complex with Fz and Wnt1 (Lu et al., 2004). The intracellular domain of Ryk can interact with Dvl and stimulates TCF transcription. In DRG explants Ryk is required for Wnt3a mediated stimulation of axon outgrowth. Moreover, expression of Ryk siRNA in transgenic mice results in axon pathfinding defects (Lu et al., 2004). These studies demonstrate that the Ryk receptor can function alone or as a coreceptor of Fz to regulate axon guidance.

Like in *Drosophila*, the Ryk receptor can also mediate repulsion in the vertebrate nervous system. Two Wnts, Wnt1 and Wnt5a, are expressed in a graded and descending fashion from cervical to thoracic regions of the spinal cord (Liu et al., 2005). Experiments show that this graded Wnt expression controls corticospinal tract (CST) axon projections. CST axons project posteriorly to the spinal cord and are repelled by the graded Wnts. This repulsion is mediated through Ryk receptor as injection of anti-Ryk antibodies block the repulsive activity of Wnts (Liu et al., 2005). Thus, Ryk can mediate repulsive and attractive axon guidance responses.

A study demonstrated the role for Wnt signaling retinotectal topographical mapping (Schmitt et al., 2005). Wnt3 is expressed in a graded fashion along the

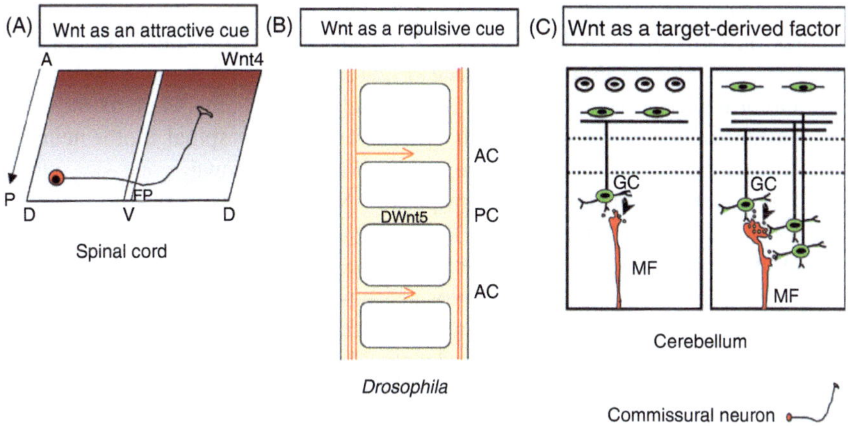

FIGURE 10.3. Wnts can function as axon guidance cues and as target-derived factors to control axonal behavior. (A) In spinal cord, Wnt4 functions as an attractive cue for spinal commissural axons. A = anterior; P = posterior; D = dorsal; V = ventral. (B) At the *Drosophila* midline, commissural axons that express the Derailed/Ryk receptor cross only through the anterior commissure (AC) as Wnt5 present in the posterior commissure (PC) repels these axons. (C) In the mouse cerebellum, Wnt7a is released by granule cells and acts retrogradely on mossy fiber axons to induce remodeling of axons, a process that precedes synaptic differentiation.

medial-lateral axis in the chick tectum. High levels of Wnt3 induce repulsion of retinal axons through Ryk. In contrast, low levels of Wnt3 induce attraction through Fz. This is the first demonstration that different concentrations of Wnts elicit different responses in axons (Schmitt et al., 2005).

Wnts do not only modulate the behavior of axons in transit to their targets but also function as synaptic target-derived signals to regulate the terminal arborization and remodeling of axons during the initial stages of synapse formation (Figure 10.3). In the cerebellum, Wnt7a is expressed in the cerebellar granule neurons at the time when mossy fibers, their presynaptic partners, make contacts with granule cell dendrites (Lucas and Salinas, 1997) and form the glomerular rosette (a multisynaptic structure). Initially mossy axons are relatively simple in morphology, but on contact with granule cells, they become extensively remodeled with increased diameter, spread areas along the axon shaft and growth cones with irregular shape (Hamori and Somogyi, 1983). *In vitro* exposure of mossy fiber axons to Wnt7a induces similar changes as those observed *in vivo* characterized by increased axon diameter, formation of spread regions along the axon shaft, and increased growth cone size and complexity (Hall et al., 2000). Electron-microscopy analyses of glomerular rosettes in the cerebellum of the Wnt7a mouse mutant revealed a significant defect in the remodeling of mossy fiber axons indicating the Wnt7a from granule cells is required for the morphological maturation of glomerular rosettes (Hall et al., 2000). Thus, Wnt7a functions as a retrograde signal for the remodeling of presynaptic mossy axons.

Further evidence for the role of Wnts in axonal morphology emerges from studies on the spinal cord, an excellent model system to identify factors and mechanisms involved in axon guidance (Figure 10.2). Wnt3 is expressed in lateral motor neurons, which innervate limb muscles, at the time when axons from proprioceptive sensory neurons form synapses with these spinal motor neurons (Krylova et al., 2002). Wnt3 increases branching and growth cone size while inhibiting axonal extension in neurotrophin-3 (NT3)-responsive sensory neurons but not in nerve growth factor (NGF)-responsive sensory neurons. This suggests that Wnt3 regulates axonal morphology of a specific sensory neuron population that project ventrally and contact motor neurons directly. An axonal remodeling activity for NT3 responsive neurons was identified in ventral spinal cord explants from limb levels but not from thoracic levels. This activity was blocked by Sfrp1, a secreted Wnt antagonist, suggesting that Wnt3 mediates axonal remodeling in NT3 responsive sensory neurons (Krylova et al., 2002). In this system inhibition of GSK-3β mimics the axonal remodeling effect of Wnt3 suggesting that Wnt signals through GSK-3β to modulate axonal remodeling. Taken together, these findings indicate that Wnt7a and Wnt3 act as retrograde signals to regulate axonal morphology through inhibition of the GSK-3β kinase possibly by directly affecting the organization and dynamics of the axonal cytoskeleton.

Wnt–Dvl signaling has profound effects on the axon cytoskeleton. Expression of Dvl induces changes in the organization and dynamics of microtubules. Dvl-expressing neurons exhibit large growth cones containing looped and unbundled microtubules and increased number of microtubules along the axon (Ciani et al., 2004; Krylova et al., 2000). Dvl increases microtubule stability as Dvl-expressing neurons are more resistant to microtubule depolymerizing drugs (Ciani et al., 2004; Krylova et al., 2000). Endogenous Dvl binds to axonal microtubules and regulates microtubule dynamics through inhibition of GSK-3β but through a transcriptional-independent mechanism (Ciani et al., 2004; Krylova et al., 2000). Thus, transcription through β-catenin and TCF is not required for Dvl function in microtubules. Instead, inhibition of GSK-3β activity by Dvl leads to changes in the phosphorylation of MAP1B increasing microtubule stability. Thus, Wnt signaling, through Dvl, directly modulates both the dynamics and organization of the neuronal cytoskeleton resulting in changes in cell and neurite morphology.

10.5.2. Dendritic Morphogenesis

The growth and morphological differentiation of dendrites are critical episodes in the formation of functional neuronal connections. Each neuron acquires its own dendritic morphology through the regulation of its cytoskeleton, which is modulated by a combination of intrinsic factors and environment cues (Cline, 2001; Jan and Jan, 2003; Scott and Luo, 2001; Whitford et al., 2002; Wong and Ghosh, 2002). Many extracellular factors have been identified as regulators of dendritic arborization. For example, semaphorin-3A acts as chemoattractant factor for growing apical dendrites (Polleux et al., 2000) whereas neurotrophins and Slit stimulate dendritic growth and branching (McAllister, 2000; Whitford et al.,

2002). Intracellular molecules, such as Rho GTPases, have been implicated in neuronal polarity and dendritic development (Luo, 2000; Van Aelst and Cline, 2004). However, little is known about how extracellular cues modulate these cellular switches and how these interactions translate into changes in the dendritic cytoskeleton and morphology.

A study has shown that Wnt signaling regulates dendrite morphogenesis. Activation of the Wnt signaling pathway through Dvl stimulates dendritic growth and branching in cultured hippocampal neurons (Rosso et al., 2005). Wnt7b, which is expressed in the hippocampus during the period of dendritogenesis, increases dendritic arborization by increasing dendritic length and the formation of primary, secondary, and tertiary branches. Sfrp1, a secreted Wnt antagonist, blocks these effects. Sfrp1 blocks endogenous Wnt activity present in hippocampal cultures that contributes to the normal dendritic development (Rosso et al., 2005). The Wnt activity is mimicked by expression of Dvl1 that is expressed in the hippocampus (Krylova et al., 2000). Dvl colocalizes with microtubules and is highly concentrated in the peripheral region of growth cones associated to the actin rich regions. Dvl1 mutant neurons exhibit shorter and less complex dendrite arbors when compared to neurons from wild-type mice (Rosso et al., 2005). These results demonstrate that Dvl1 is required for normal dendritic development in hippocampal neurons (Figure 10.2).

Epistatic and biochemical analyses revealed that Wnt7b/Dvl signaling regulates dendritic development through a noncanonical Wnt pathway. Activation of GSK-3β or inhibition of β-catenin mediated transcription does not block Dvl function in dendrites. In contrast, Wnt7b and Dvl modulate dendrite development through the regulation of Rho GTPases and JNK. Wnt7b and Dvl activate endogenous Rac, and this effect is blocked by Sfrp1. In addition, Rac dominant negative mutants block the robust Dvl effect in dendritic morphogenesis. Furthermore, JNK, a downstream effector of Rac, is activated by Wnt7b and Dvl. Moreover, inhibition of JNK blocks Dvl function in dendritic development (Rosso et al., 2005). These results demonstrate that the Wnt pathway regulates dendritic development through Rac and JNK. Thus, Dvl functions as a molecular link between Wnt factors and the very well-known cytoskeletal regulators Rho GTPases to control dendritic development.

β-catenin, a member of the Wnt signaling pathway, also regulates dendritic development. Overexpression of β-catenin increases dendritic arborization in hippocampal neurons (Yu and Malenka, 2003). However, β-catenin does not affect dendritogenesis through the β-catenin–transcription pathway. Instead, constitutively active β-catenin increases dendritic arborization through its interaction with N-cadherin and αN-catenin (neural-catenin). Conversely, sequestering endogenous β-catenin leads to a decrease in dendritic complexity caused by neural activity (high K^+ depolarization) suggesting that the level of endogenous β-catenin is important to the regulation of dendritic branching. In addition, dickkopf-1 (DKK-1), an extracellular Wnt antagonist (Glinka et al., 1998), blocks the dendritogenic effect of depolarization by high K^+ suggesting that neuronal activity regulates Wnt expression or release, which in turn modulates dendritic arborization.

Taken together, these studies demonstrate that Wnt signaling regulates dendritic development through a transcriptional independent pathway. It remains to be determined whether Wnt signaling and β-catenin–mediated adhesion regulate dendrite development through two independent pathways or whether these two pathways interact to modulate dendritic morphology.

10.6. Concluding Remarks

Here we have highlighted the most recent findings for the role of Wnt signaling in neuritogenesis, axon growth, and dendritic development. Some components of the Wnt pathway have been implicated in the establishment and maintenance of neuronal polarity. However, a clear demonstration for Wnts in this process needs to be established. Wnts through their Fz and Ryk receptors elicit different responses in neurites. The Wnt-GSK-3β and the Wnt-Dvl-Rac pathways in axon remodeling during synapse formation and in dendritic development, respectively, have been established. However, the mechanisms by which Wnts elicit attraction, repulsion, and axon branching remains to be elucidated. The intriguing observation that Wnts are expressed in the adult brain suggest that these signaling molecules could also regulate neurite remodeling in the adult, a process that has been implicated in synaptic plasticity. Future studies on Wnts will shed new light into the function of these molecules in different aspects of neuronal development and function. These studies will also unravel the strategies used by neurons to form highly complex neuronal circuits characteristic of the vertebrate nervous system.

References

Arimura, N., Menager, C., Fukata, Y., and Kaibuchi, K., 2004, Role of CRMP-2 in neuronal polarity, *J. Neurobiol.* **58:** 34–47.

Baas, P.W., and Qiang, L., 2005, Neuronal microtubules: When the MAP is the roadblock, *Trends Cell Biol.* **15:** 183–187.

Berling, B., Wille, H., Roll, B., Mandelkow, E.M., Garner, C., and Mandelkow, E., 1994, Phosphorylation of microtubule-associated proteins MAP2a,b and MAP2c at Ser136 by proline-directed kinases in vivo and in vitro, *Eur. J. Cell Biol.* **64:** 120–130.

Boutros, M., Paricio, N., Strutt, D.I., and Mlodzik, M., 1998, Dishevelled activates JNK and discriminates between JNK pathways in planar polarity and wingless signaling, *Cell* **94:** 109–118.

Boutros, M., Mihaly, J., Bouwmeester, T., and Mlodzik, M., 2000, Signaling specificity by Frizzled receptors in Drosophila, *Science* **288:** 1825–1828.

Bovolenta, P., 2005, Morphogen signaling at the vertebrate growth cone: A few cases or a general strategy? *J. Neurobiol.* **64:** 405–416.

Bradke, F., and Dotti, C.G., 1999, The role of local actin instability in axon formation, *Science* **283:** 1931–1934.

Bradke, F., and Dotti, C.G., 2000, Establishment of neuronal polarity: Lessons from cultured hippocampal neurons, *Curr. Opin. Neurobiol.* **10:** 574–581.

Cadigan, K.M., and Nusse, R., 1997, Wnt signaling: A common theme in animal development, *Genes Dev.* **11:** 3286–3305.

Ciani, L., and Salinas, P.C., 2005, WNTs in the vertebrate nervous system: From patterning to neuronal connectivity, *Nat. Rev. Neurosci.* **6:** 351–362.

Ciani, L., Krylova, O., Smalley, M.J., Dale, T.C., and Salinas, P.C., 2004, A divergent canonical WNT-signaling pathway regulates microtubule dynamics: Dishevelled signals locally to stabilize microtubules, *J. Cell Biol.* **164:** 243–253.

Cline, H.T., 2001, Dendritic arbor development and synaptogenesis, *Curr. Opin. Neurobiol.* **11:** 118–126.

Dickson, B.J., 2002, Molecular mechanisms of axon guidance, *Science* **298:** 1959–1964.

Doble, B.W., and Woodgett, J.R., 2003, GSK-3: Tricks of the trade for a multi-tasking kinase, *J. Cell Sci.* **116:** 1175–1186.

Dotti, C.G., Sullivan, C.A., and Banker, G.A., 1988, The establishment of polarity by hippocampal neurons in culture, *J. Neurosci.* **8:** 1454–1468.

Fukata, Y., Itoh, T.J., Kimura, T., Menager, C., Nishimura, T., Shiromizu, T., et al., 2002, CRMP-2 binds to tubulin heterodimers to promote microtubule assembly, *Nat. Cell Biol.* **4:** 583–591.

Glinka, A., Wu, W., Delius, H., Monaghan, A.P., Blumenstock, C., and Niehrs, C., 1998, Dickkopf-1 is a member of a new family of secreted proteins and functions in head induction, *Nature* **391:** 357–362.

Gonzalez-Billault, C., Jimenez-Mateos, E.M., Caceres, A., Diaz-Nido, J., Wandosell, F., and Avila, J., 2004, Microtubule-associated protein 1B function during normal development, regeneration, and pathological conditions in the nervous system, *J. Neurobiol.* **58:** 48–59.

Gregorieff, A., and Clevers, H., 2005, Wnt signaling in the intestinal epithelium: From endoderm to cancer, *Genes Dev.* **19:** 877–890.

Habas, R., Dawid, I.B., and He, X., 2003, Coactivation of Rac and Rho by Wnt/Frizzled signaling is required for vertebrate gastrulation, *Genes Dev.* **17:** 295–309.

Hall, A.C., Lucas, F.R., and Salinas, P.C., 2000, Axonal remodeling and synaptic differentiation in the cerebellum is regulated by WNT-7a signaling, *Cell* **100:** 525–535.

Hamori, J., and Somogyi, J., 1983, Differentiation of cerebellar mossy fiber synapses in the rat: A quantitative electron microscope study, *J. Comp. Neurol.* **220:** 365–377.

Hanger, D.P., Hughes, K., Woodgett, J.R., Brion, J.P., and Anderton, B.H., 1992, Glycogen synthase kinase-3 induces Alzheimer's disease-like phosphorylation of tau: Generation of paired helical filament epitopes and neuronal localisation of the kinase, *Neurosci. Lett.* **147:** 58–62.

Heisenberg, C.P., Tada, M., Rauch, G.J., Saude, L., Concha, M.L., Geisler, R., et al., 2000, Silberblick/Wnt11 mediates convergent extension movements during zebrafish gastrulation, *Nature* **405:** 76–81.

Holmen, S.L., Salic, A., Zylstra, C.R., Kirschner, M.W., and Williams, B.O., 2002, A novel set of Wnt-Frizzled fusion proteins identifies receptor components that activate beta -catenin-dependent signaling, *J. Biol. Chem.* **277:** 34727–34735.

Inagaki, N., Chihara, K., Arimura, N., Menager, C., Kawano, Y., Matsuo, N., et al., 2001, CRMP-2 induces axons in cultured hippocampal neurons, *Nat. Neurosci.* **4:** 781–782.

Inoue, T., Oz, H.S., Wiland, D., Gharib, S., Deshpande, R., Hill, R.J., et al., 2004, C. elegans LIN-18 is a Ryk ortholog and functions in parallel to LIN-17/Frizzled in Wnt signaling, *Cell* **118:** 795–806.

Jan, Y.N., and Jan, L.Y., 2003, The control of dendrite development, *Neuron* **40:** 229–242.

Jiang, H., Guo, W., Liang, X., and Rao, Y., 2005, Both the establishment and the maintenance of neuronal polarity require active mechanisms: Critical roles of GSK-3beta and its upstream regulators, *Cell* **120:** 123–135.

Katanaev, V.L., Ponzielli, R., Semeriva, M., and Tomlinson, A., 2005, Trimeric G protein-dependent frizzled signaling in Drosophila, *Cell* **120:** 111–122.

Krylova, O., Messenger, M.J., and Salinas, P.C., 2000, Dishevelled-1 regulates microtubule stability: A new function mediated by glycogen synthase kinase-3beta, *J. Cell Biol.* **151:** 83–94.

Krylova, O., Herreros, J., Cleverley, K.E., Ehler, E., Henriquez, J.P., Hughes, S.M., et al., 2002, WNT-3, expressed by motoneurons, regulates terminal arborization of neurotrophin-3-responsive spinal sensory neurons, *Neuron* **35:** 1043–1056.

Kuhl, M., Sheldahl, L.C., Malbon, C.C., and Moon, R.T., 2000, Ca(2+)/calmodulin-dependent protein kinase II is stimulated by Wnt and Frizzled homologs and promotes ventral cell fates in Xenopus, *J. Biol. Chem.* **275:** 12701–12711.

Liu, T., DeCostanzo, A.J., Liu, X., Wang, H., Hallagan, S., Moon, R.T., et al., 2001, G protein signaling from activated rat frizzled-1 to the beta-catenin-Lef-Tcf pathway, *Science* **292:** 1718–1722.

Liu, Y., Shi, J., Lu, C.C., Wang, Z.B., Lyuksyutova, A.I., Song, X., et al., 2005, Ryk-mediated Wnt repulsion regulates posterior-directed growth of corticospinal tract, *Nat. Neurosci.* **8:** 1151–1159.

Logan, C.Y., and Nusse, R., 2004, The Wnt signaling pathway in development and disease, *Annu. Rev. Cell Dev. Biol.* **20:** 781–810.

Lu, W., Yamamoto, V., Ortega, B., and Baltimore, D., 2004, Mammalian Ryk is a Wnt coreceptor required for stimulation of neurite outgrowth, *Cell* **119:** 97–108.

Lucas, F.R., and Salinas, P.C., 1997, WNT-7a induces axonal remodeling and increases synapsin I levels in cerebellar neurons, *Dev. Biol.* **192:** 31–44.

Lucas, F.R., Goold, R.G., Gordon-Weeks, P.R., and Salinas, P.C., 1998, Inhibition of GSK-3beta leading to the loss of phosphorylated MAP-1B is an early event in axonal remodelling induced by WNT-7a or lithium, *J. Cell Sci.* **111:** 1351–1361.

Luo, L., 2000, Rho GTPases in neuronal morphogenesis, *Nat. Rev. Neurosci.* **1:** 173–180.

Luu, H.H., Zhang, R., Haydon, R.C., Rayburn, E., Kang, Q., Si, W., et al., 2004, Wnt/beta-catenin signaling pathway as a novel cancer drug target, *Curr. Cancer Drug Targets* **4:** 653–671.

Lyuksyutova, A.I., Lu, C.C., Milanesio, N., King, L.A., Guo, N., Wang, Y., et al., 2003, Anterior-posterior guidance of commissural axons by Wnt-frizzled signaling, *Science* **302:** 1984–1988.

McAllister, A.K., 2000, Cellular and molecular mechanisms of dendrite growth, *Cereb. Cortex* **10:** 963–973.

Medina, A., and Steinbeisser, H., 2000, Interaction of Frizzled 7 and Dishevelled in Xenopus, *Dev. Dyn.* **218:** 671–680.

Moon, R.T., Bowerman, B., Boutros, M., and Perrimon, N., 2002, The promise and perils of Wnt signaling through beta-catenin, *Science* **296:** 1644–1646.

Packard, M., Koo, E.S., Gorczyca, M., Sharpe, J., Cumberledge, S., and Budnik, V., 2002, The Drosophila Wnt, wingless, provides an essential signal for pre- and postsynaptic differentiation, *Cell* **111:** 319–330.

Peifer, M., and Polakis, P., 2000, Wnt signaling in oncogenesis and embryogenesis—a look outside the nucleus, *Science* **287:** 1606–1609.

Polleux, F., Morrow, T., and Ghosh, A., 2000, Semaphorin 3A is a chemoattractant for cortical apical dendrites, *Nature* **404:** 567–573.

Rodriguez-Boulan, E., and Powell, S.K., 1992, Polarity of epithelial and neuronal cells, *Annu. Rev. Cell Biol.* **8:** 395–427.

Rosso, S.B., Sussman, D., Wynshaw-Boris, A., and Salinas, P.C., 2005, Wnt signaling through Dishevelled, Rac and JNK regulates dendritic development, *Nat. Neurosci.* **8:** 34–42.

Salinas, P.C., 2005, Signaling at the vertebrate synapse: New roles for embryonic morphogens? *J. Neurobiol.* **64:** 435–445.

Schmitt, A.M., Shi, J., Wolf, A.M., Lu, C.C., King, L.A., and Zou, Y., 2005, Wnt-Ryk signalling mediates medial-lateral retinotectal topographic mapping, *Nature* **439:** 31–37.

Scott, E.K., and Luo, L., 2001, How do dendrites take their shape? *Nat. Neurosci.* **4:** 359–365.

Sheldahl, L.C., Slusarski, D.C., Pandur, P., Miller, J.R., Kuhl, M., and Moon, R.T., 2003, Dishevelled activates Ca2+ flux, PKC, and CamKII in vertebrate embryos, *J. Cell Biol.* **161:** 769–777.

Slusarski, D.C., Corces, V.G., and Moon, R.T., 1997, Interaction of Wnt and a Frizzled homologue triggers G-protein-linked phosphatidylinositol signaling, *Nature* **390:** 410–413.

Strutt, D., 2003, Frizzled signalling and cell polarisation in Drosophila and vertebrates, *Development* **130:** 4501–4513.

Strutt, D.I., Weber, U., and Mlodzik, M., 1997, The role of RhoA in tissue polarity and Frizzled signaling, *Nature* **387:** 292–295.

Sutherland, C., Leighton, I.A., and Cohen, P., 1993, Inactivation of glycogen synthase kinase-3 beta by phosphorylation: New kinase connections in insulin and growth-factor signaling, *Biochem. J.* **296:** 15–19.

Torroja, C., Gorfinkiel, N., and Guerrero, I., 2005, Mechanisms of Hedgehog gradient formation and interpretation, *J. Neurobiol.* **64:** 334–356.

Van Aelst, L., and Cline, H.T., 2004, Rho GTPases and activity-dependent dendrite development, *Curr. Opin. Neurobiol.* **14:** 297–304.

Veeman, M.T., Axelrod, J.D., and Moon, R.T., 2003, A second canon. Functions and mechanisms of beta-catenin-independent Wnt signaling, *Dev. Cell* **5:** 367–377.

Waites, C.L., Craig, A.M., and Garner, C.C., 2005, Mechanisms of vertebrate synaptogenesis, *Annu. Rev. Neurosci.* **28:** 251–274.

Whitford, K.L., Marillat, V., Stein, E., Goodman, C.S., Tessier-Lavigne, M., Chedotal, A., et al., 2002, Regulation of cortical dendrite development by Slit-Robo interactions, *Neuron* **33:** 47–61.

Wiggin, G.R., Fawcett, J.P., and Pawson, T., 2005, Polarity proteins in axon specification and synaptogenesis, *Dev. Cell* **8:** 803–816.

Willert, K., Brown, J.D., Danenberg, E., Duncan, A.W., Weissman, I.L., Reya, T., et al., 2003, Wnt proteins are lipid-modified and can act as stem cell growth factors, *Nature* **423:** 448–452.

Wong, R.O., and Ghosh, A., 2002, Activity-dependent regulation of dendritic growth and patterning, *Nat. Rev. Neurosci.* **3:** 803–812.

Wong, H.C., Bourdelas, A., Krauss, A., Lee, H.J., Shao, Y., Wu, D., et al., 2003, Direct binding of the PDZ domain of Dishevelled to a conserved internal sequence in the C-terminal region of Frizzled, *Mol. Cell* **12:** 1251–1260.

Yoshikawa, S., McKinnon, R.D., Kokel, M., and Thomas, J.B., 2003, Wnt-mediated axon guidance via the Drosophila Derailed receptor, *Nature* **422:** 583–588.

Yoshimura, T., Kawano, Y., Arimura, N., Kawabata, S., Kikuchi, A., and Kaibuchi, K., 2005, GSK-3beta regulates phosphorylation of CRMP-2 and neuronal polarity, *Cell* **120:** 137–149.

Yu, X., and Malenka, R.C., 2003, Beta-catenin is critical for dendritic morphogenesis, *Nat. Neurosci.* **6:** 1169–1177.

Zhou, F.Q., Zhou, J., Dedhar, S., Wu, Y.H., and Snider, W.D., 2004, NGF-induced axon growth is mediated by localized inactivation of GSK-3beta and functions of the microtubule plus end binding protein APC, *Neuron* **42:** 897–912.

Zumbrunn, J., Kinoshita, K., Hyman, A.A., and Nathke, I.S., 2001, Binding of the adenomatous polyposis coli protein to microtubules increases microtubule stability and is regulated by GSK3 beta phosphorylation, *Curr. Biol.* **11:** 44–49.

11 Role of CRMP-2 in Neuronal Polarization

NARIKO ARIMURA, TAKESHI YOSHIMURA, AND KOZO KAIBUCHI

11.1. Introduction

The neuron is one of the most dramatic examples of a polarized cell forming two distinctive types of neurites, an axon and dendrites. Neuronal polarization begins in a symmetric cell with the elongation of one of the immature neurites as an axon. Polarization enables the two types of neurites to perform different functions: sending and receiving electrical signals in a vectorial fashion. Although researchers are accumulating a catalog of structural, molecular, and functional differences between axons and dendrites, the mechanisms involved in the establishment of neuronal polarity are not well understood.

In this review, we describe recent advances in the understanding of cellular events in the early development of an axon and dendrites. We also discuss the roles of the Rho family small GTPases (Rho GTPases), their upstream and downstream molecules, and collapsin response mediator protein-2 (CRMP-2) in neuronal polarity.

11.2. Morphological Change in Neuronal Polarization

A neuron typically develops a single axon and several dendrites. These two compartments acquire specific characteristics that enable them to transmit intercellular signals from several dendrites to an axon. These two types of neurites gradually establish their specific characteristics during the initial stage of neuronal development.

Hippocampal neurons represent the most widely used culture system for examining polarity. In culture, as *in situ*, hippocampal neurons develop one axon and several dendrites, which maintain their structural characteristics at the molecular level (Craig and Banker, 1994). During maturation, hippocampal neurons dramatically change their morphology. Banker and his colleagues divided these morphological events into five stages (Figure 11.1) (Dotti et al., 1988).

Shortly after plating, the neurons attached to the substratum form small protrusion veils and a few spikes (stage 1). Neurons that develop these truncated protrusions grow and establish several short neurites (stage 2). Because neurites

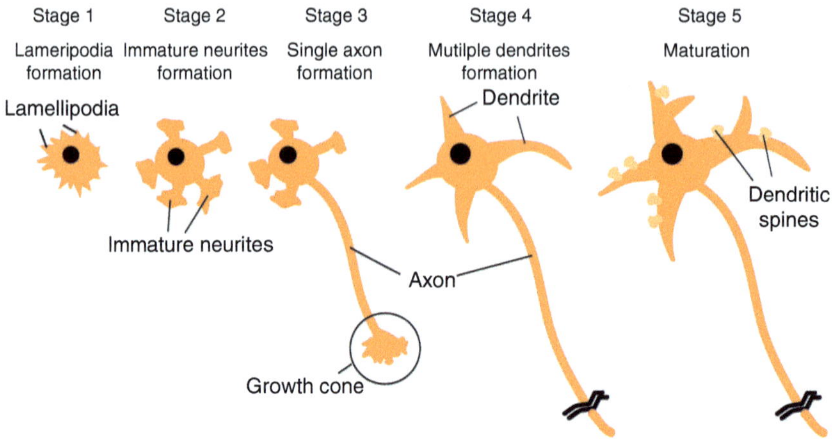

FIGURE 11.1. Establishment of neuronal polarity in cultured hippocampal neurons. Embryonic hippocampal neurons in culture acquire their characteristic polarized morphology in a sequence of well-defined stages. (Reproduced from Dotti et al., 1988, with permission.)

are roughly equal in length at this stage, it is difficult to identify special characteristics that could allow us to predict which one will become an axon. One neurite then starts to break the initial morphological symmetry, growing at a rapid rate (minutes or tens of minutes) (Esch et al., 1999), and neurons immediately establish the polarization (stage 3). A few days after the axon has begun its rapid growth, the remaining neurites elongate and acquire the characteristics of dendrites (stage 4). Typically, by 7 days after plating, cultured neurons form synaptic contacts and establish a neuronal network (stage 5). During development, neurons acquire their polarity from stage 2 to stage 3. However, we are only now beginning to understand the molecular mechanisms underlying neuronal polarization.

11.3. Extracellular Signals Regulating Neuronal Polarity

Because neurons develop polarity in culture without any directional gradients of extracellular cues, an internal polarization program appears to exist in neurons (Craig and Banker, 1994). Axons and dendrites face the proper direction under physiological conditions, however, suggesting that the polarization may be regulated by extracellular signals. In fact, studies have shown that extracellular substrate molecules can govern which neurite becomes an axon depending on the substrate preference of neurite elongation (Esch et al., 1999). When neurons are plated on alternating stripes of poly-D-lysine and either laminin or neuron-glia cell adhesion molecule (NgCAM), they usually form an axon on laminin or NgCAM. These observations suggest that the signals produced by the attachment of laminin/NgCAM with adhesion molecules, such as integrins, cause the rapid neurite growth and are sufficient to induce axon formation. This rapid axon formation is also observed when an immature neurite in stage 2 neurons contacts laminin-coated beads (Menager et al., 2004). Thus, signaling cascades accelerated by laminin through integrins may initiate neurite growth and the subsequent axon formation, and certain extracellular cues may determine the axon/dendrite fate during physiological development.

11.4. Downstream of the Extracellular Signals

11.4.1. PI 3-Kinase and PIP_3

Experiments have shown the importance of PI 3-kinase and its lipid product (phosphatidylinositol(3,4,5)triphosphate; PIP_3) in determining and maintaining internal polarity, especially in neurotrophils and dictyostelium (De Camilli et al., 1996; Iijima et al., 2002). Shi et al. (2003) reported that PI 3-kinase activity is localized at the tip of a newly specified axon in stage 3 neurons, and PI 3-kinase inhibitor prevents axon formation at stage 3. PI 3-kinase generates PIP_3, which is essential for the translocation of Akt (also known as protein kinase B) at the plasma membrane. Akt is phosphorylated there and is activated by phosphoinositide-dependent kinase (PDK) or other kinases, including integrin-linked kinase (ILK) (Hannigan et al., 2005). Phosphorylated Akt is enriched only at the tips of the growing axons, not at other neurites. These observations were further confirmed by other experimental methods. The pleckstrin homology domain of Akt tagged with green fluorescent protein (Akt-PH-GFP), which has been demonstrated to interact specifically with PIP_3 and PIP_2, is the useful tool for monitoring the activity of PI 3-kinase (Menager et al., 2004; Varnai and Balla, 1998). In the initial stage of neuronal polarization, the accumulation of Akt-PH-GFP is observed in the tip of one immature neurite (Figure 11.2). When this immature neurite contacts laminin-coated beads, it rapidly accumulates Akt-PH-GFP at its tip followed by dramatic growth of the neurite as an axon (Menager et al., 2004). These observations suggest that PI 3-kinase is specifically activated at the tips of future axons and that PIP_3 is essential for neuronal polarization. Thus, PI 3-kinase must be essential in internal programmed polarity, which appears to be regulated by extracellular signals.

11.4.2. Small GTPases

A highly dynamic area is located at the tips of axons, where drastic rearrangements of actin filaments and microtubules occur during neurite elongation (Baas and Buster, 2004; Bradke and Dotti, 2000). Cytochalasin D application to stage 2 neurons causes multiple axon-like neurites, implying that the reorganization of actin filaments is necessary for axon formation (Bradke and Dotti, 2000). Therefore, one of the major regulators of actin filaments and microtubules, Rho GTPases, may be involved in axon specification.

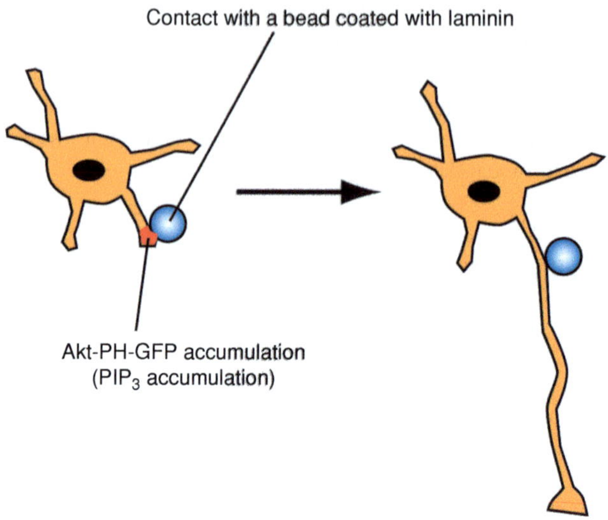

FIGURE 11.2. Schematic drawings of morphology of hippocampal neurons in response to the laminin-coated bead contact. One of the immature neurites in the neuron contacts a bead coated with laminin. This contact leads to the accumulation of Akt-PH-GFP (red color) at the contact site and a rapid elongation of this process.

Rho GTPases cycle between a GTP-bound active state and a GDP-bound inactive state, acting as molecular switches. Among the 10 known types of Rho GTPases, Cdc42, Rac1, and RhoA are the most intensively characterized. Cdc42 and Rac1 are responsible for the formation of membrane ruffling, lamellipodia, and RhoA is responsible for the formation of stress fibers and focal adhesions (Govek et al., 2005). In response to signals, several regulators control the nucleotide states of Rho GTPases, including a guanine nucleotide exchange factor (GEF), which promotes the exchange of GDP for GTP. The upstream signals mediate GEFs to activate Rho GTPases. Cdc42 and Rac1 are involved in neurite extension in N1E-115 neuroblastoma cells through the action of their specific effectors, including neural Wiskott-Aldrich syndrome protein (N-WASP) and WASP family verprolin-homologous protein-1 (WAVE1) (Abe et al., 2003). RhoA is implicated in neurite retraction through the action of Rho-kinase (Amano et al., 1998; Jalink et al., 1994). Cdc42 is thought to act upstream of Rac1 for neurite extension. T-cell lymphoma invasion and metastasis protein (Tiam1) serves as a GEF for Rac1 downstream of Cdc42 (see below) and promotes neurite extension.

The signaling cascades of Cdc42 and Rac1 are also implicated in neuronal polarization. Expression of the constitutively active mutant of Cdc42 or Tiam1 induces multiple axon-like neurites but impairs axonal maturation (Kunda et al., 2001; Nishimura et al., 2005; Schwamborn and Puschel, 2004), suggesting that cycling between GTP- and GDP-bound states of Cdc42 and Rac1 is needed for proper axonal maturation. In addition, Chuang et al. (2005) showed that the cytoplasmic dynein light chain Tctex-1 functions in initial neurite sprouting and axon outgrowth, acting through Rac1. Thus, it is likely that Cdc42/Rac1-mediated signal cascades regulate the reorganization of actin filaments and microtubules in the prospective axon, thereby specifying an axon. Given that Rac1 activates PI 3-kinase (Govek et al., 2005), the signal initially evoked by PI 3-kinase appears to terminate at PI 3-kinase itself. This positive feedback loop may be a driving force in axon specification and maturation.

11.4.3. Par Complex

A polarity complex of Par3, Par6, and atypical protein kinase C (aPKC) functions in various cell-polarization events (Ohno, 2001; Wiggin et al., 2005), including axon formation (Nishimura et al., 2004; Shi et al., 2003). This protein complex is critical for anterior/posterior polarity of the single-cell embryo of *Caenorhabditis elegans* (*C. elegans*) as well as epithelial cells and neuroblast cells of *Drosophila*.

In hippocampal neurons, Par3 and Par6 are localized at the tips of axons. Suppression of Par3 or Par6 function inhibits axon formation. Ectopic expression of Par3 forms the multiple axon-like neurites but impairs axonal maturation, indicating that proper localization and the activities of Par3 and Par6 are necessary for axon maturation (Nishimura et al., 2005; Shi et al., 2003). PI 3-kinase activity is required for the proper localization of the Par complex and Cdc42 at the tips of axons (Nishimura et al., 2005; Shi et al., 2003).

Cdc42-GTP binds to Par6 and determines its localization, but until recently little was known about the downstream signals from the Cdc42/Par complex. Par3 directly interacts with Still life (Sif)- and Tiam1-like exchange factor (STEF)/Tiam1, which are GEFs for Rac1 (Chen and Macara, 2005; Nishimura et al., 2005). In addition to the Par3/Par6 complex, STEF accumulates at the tips of the growing axons. The dominant negative or knockdown of Par3 inhibits the Cdc42-induced Rac1 activation in N1E-115 neuroblastoma cells (Nishimura et al., 2005). Thus, during axon formation the Par3/Par6 complex appears to mediate the signal from Cdc42 to Rac1 (Nishimura et al., 2005). It has been reported, however, that *Drosophila* Par6, aPKC, and Bazooka/Par3 are not involved in axon or dendrite specification *in vivo* (Rolls and Doe, 2004). In *Drosophila*, the effects of Par deletion might be compensated by the extracellular environment and/or other signaling cascades linking Cdc42 and Rac1. Further work is needed to elucidate the roles of Par proteins in polarization of mammalian neurons.

A member of the Ras subfamily of GTPases, Rap1B, functions as a positional signal and organizes cell architecture (Schwamborn and Puschel, 2004). Rap1 is a mammalian homolog of Bud1p/Rsr1p, which determines the position of incipient budding sites in *Saccharomyces cervisiae*. In hippocampal neurons, Rap1B is localized at the tips of axons preceding the accumulation of Cdc42 and the Par complex. Rap1B activation induces the formation of multiple axon-like neurites and the accumulation of the Par complex in each axon, whereas the inactivation of Rap1B leaves the neurons without an axon. This loss of axon formation is rescued by the coexpression of active forms of Cdc42 and Par6, suggesting that Rap1B acts upstream of Cdc42 and the Par complex. By collecting the essential regulators of axon formation, Rap1B localization at the single neurite appears to be a decisive step in determining which neurite becomes an axon. Because the PI 3-kinase inhibitor did not prevent axon formation induced by Rap1B expression, PI 3-kinase seems to function upstream of Rap1B (Schwamborn and Puschel, 2004). It would be better for our understanding to examine how Rap1B is excluded from the other immature neurites.

11.4.4. Akt and GSK-3β

The signaling cascade from PI 3-kinase to Akt/glycogen synthase kinase-3β (GSK-3β) is involved in cell polarity in various cell types, including neurons (Cole et al., 2004; Jiang et al., 2005; Yoshimura et al., 2005, 2006). PI 3-kinase activates Akt by the phosphorylation of Akt at Thr308 and Ser473 via PIP_3, PDK, and ILK. GSK-3β is constitutively active; activated Akt phosphorylates GSK-3β at Ser9 and inactivates its kinase activity (Eldar-Finkelman, 2002; Frame and Cohen, 2001).

The decreased activity of GSK-3β is required for neuronal polarization (Jiang et al., 2005; Yoshimura et al., 2005, 2006). Akt localizes at the tips of axons, and the form of GSK-3β phosphorylated at Ser9 (i.e., the inactive form of GSK-3β) is also found at the tips of growing axons. PI 3-kinase inhibitor prevents the phosphorylation of GSK-3β at Ser9, whereas the constitutively active form of Akt induces supernumerary axons. The phosphatase and tensin homolog deleted on chromosome 10 (PTEN) dephosphorylates the lipid products of PI 3-kinase, thereby regulating the contents of PIP_3 and the Akt activity. Depletion of PTEN induces the formation of multiple axons (Jiang et al., 2005). These results suggest that the signaling cascade from PI 3-kinase to GSK-3β is involved in the neuronal polarization. Furthermore, the overexpression of constitutively active GSK-3β inhibits axon formation, whereas the knockdown or specific inhibitors of GSK-3β induce the formation of multiple axons. Surprisingly, the inhibition of GSK-3β at later stages causes the dramatic change of preexisting dendrites to axons (Jiang et al., 2005). These results indicate the significance of PI 3-kinase/PIP_3/Akt/GSK-3β signaling in neuronal polarity.

11.5. Substrates of GSK-3β

Concerning the substrates of GSK-3β in neuronal polarity, studies have shown that GSK-3β phosphorylates CRMP-2 at Thr514 and inactivates it (Brown et al., 2004; Cole et al., 2004; Uchida et al., 2005; Yoshimura et al., 2005). Overexpression of CRMP-2 induces multiple axons, and the inhibition of CRMP-2 functions impairs axon formation (Inagaki et al., 2001). Thus, CRMP-2 is important in axon specification (see in a later section).

In cultured hippocampal neurons, about 30% of CRMP-2 is constitutively phosphorylated at Thr514, and this phosphorylation is decreased by GSK-3 inhibitors (Yoshimura et al., 2005). CRMP-2 phosphorylated at Thr514 is enriched in the distal part of the growing axons but clearly not at the axonal growth cones, suggesting that there is a pool of CRMP-2 not phosphorylated at Thr514 at the growth cones of growing axons. The phosphorylation of CRMP-2 lowers its axon-inducing activity (Yoshimura et al., 2005). Similar to the inhibition of GSK-3β, the expression of the nonphosphorylated form of CRMP-2 induces the formation of multiple axon-like neurites. The expression of constitutively active GSK-3β inhibits axon formation, whereas the nonphosphorylated form of CRMP-2 counteracts the inhibitory effects of GSK-3β, indicating that GSK-3β regulates neuronal polarity through the phosphorylation of CRMP-2. The neurotrophic factors, neurotrophin-3 (NT-3), and brain-derived neurotrophic factor (BDNF), induce the inactivation of GSK-3β and dephosphorylation of CRMP-2. The depletion of CRMP-2 prevents the NT-3-induced axon outgrowth. Thus, the regulation of GSK-3β and the CRMP-2 activity is of central importance in neuronal polarization.

In addition, GSK-3β phosphorylates MAP1B (Goold et al., 1999; Trivedi et al., 2005) and the adenomatous polyposis coil gene product (APC) (Zumbrunn et al., 2001). The phosphorylation of MAP1B by GSK-3β suppresses detyrosination of microtubules and decreases the number of stable microtubules. The binding of APC to microtubules increases microtubule stability, whereas the phosphorylation of APC decreases the interaction with microtubules. In addition, hippocampal neurons derived from double-knockout mice of tau and MAP1B, which have redundant functions, show a defect in axon formation at stage 3 (Takei et al., 2000). Given these results, the inactivation of GSK-3β may increase the stability of microtubules through microtubule-associating molecules such as MAP1B and APC.

11.6. Molecular Mechanisms Involving CRMP-2 in Neuronal Polarity

The CRMP/TOAD/Ulip/DRP family is highly expressed in the developing nervous system, and CRMP-2 homologues have been identified from various species (Byk et al., 1998; Gaetano et al., 1997; Goshima et al., 1995; Hamajima et al., 1996; Minturn et al., 1995; Wang and Strittmatter, 1996). CRMP-62, the chick CRMP-2 (98% identity), is required for the growth cone collapse of chick dorsal root ganglion (DRG) neurons induced by semaphorin-3A (Sema3A; also known as collapsin-1) (Goshima et al., 1995). UNC-33, the *C. elegans* homologue of CRMP-2, is identified by a mutation resulting in severely uncoordinated movement, abnormalities in axon guidance, and superabundance of microtubules in neurons (Hedgecock et al., 1985; Li et al., 1992). In hippocampal neurons, the overexpression of CRMP-2 induces the formation of multiple axons (Inagaki et al., 2001). Because the induced axon-like neurites possess synaptophysin-positive synaptic terminals, CRMP-2 appears to be a powerful regulator of neuronal polarization. Furthermore, the overexpression of CRMP-2 can cause a preexisting dendrite to become an axon in stage 4, suggesting that the overexpressed

CRMP-2 confers the axonal identity not only on immature neurites but also on established dendrites. These observations indicate that CRMP-2 plays a critical role in axon formation of hippocampal neurons by establishing and maintaining neuronal polarity.

We searched for the CRMP-2 binding proteins to unravel the molecular mechanisms of CRMP-2-induced axon formation, and we identified the interacting molecules, including tubulin heterodimer, Numb, and the Specifically Rac1-associated protein 1 (Sra-1) (Figure 11.3) (Fukata et al., 2002; Kawano et al., 2005; Nishimura et al., 2003). CRMP-2 regulates microtubule dynamics by binding to tubulin heterodimers. Because CRMP-2 copolymerizes with tubulin heterodimers into microtubules and promotes tubulin polymerization *in vitro*, CRMP-2 appears to function as a partner protein of tubulin heterodimers to assist in microtubule assembly. Given the enriched localization of CRMP-2 in the growing axons, it is likely that the CRMP-2/tubulin complexes concentrated in the distal parts of axons modulate microtubule dynamics by changing the rate of microtubule assembly (Fukata et al., 2002).

Numb colocalizes with CRMP-2 in the central region of axonal growth cones (Nishimura et al., 2003). Numb is associated with L1, a neuronal cell adhesion molecule that is recycled through endocytosis at the growth cone. Numb has been shown to interact with Eps15 and α-adaptin, a subunit of the AP-2 complex, and to be involved in the clathrin-dependent endocytosis at the plasma membrane (Berdnik et al., 2002; Salcini et al., 1997; Santolini et al., 2000). Depletion of CRMP-2 prevents the L1 endocytosis, indicating that CRMP-2 is involved in Numb-mediated endocytosis of L1.

We found that CRMP-2 interacts with Sra-1 and the WAVE1 complex (Kawano et al., 2005), which are regulators of the actin cytoskeleton (Kobayashi et al., 1998; Luo et al., 1994; Schenck et al., 2003; Zallen et al., 2002). CRMP-2 associates with actin filaments indirectly, presumably through Sra-1. The knockdown of Sra-1 and WAVE1 by RNA interference cancels CRMP-2–induced axon outgrowth and the formation of multiple axons in cultured hippocampal neurons. These results suggest that CRMP-2 forms a complex with Sra-1/WAVE1 and participates in the reorganization of the actin cytoskeleton in neurons.

We identified CRMP-2 as a substrate of Rho-kinase in the brain (Arimura et al., 2000). Rho-kinase phosphorylates CRMP-2 at Thr555 *in vitro*. The phosphorylation of CRMP-2 at Thr555 is observed in DRG neurons during lysophosphatidic acid– or ephrin-A5 (a repulsive guidance cue)- induced growth cone collapse, whereas the phosphorylation is not detected during Sema3A-induced growth cone collapse (Arimura et al., 2000, 2005). Sema3A stimulation enhances the phosphorylation of CRMP-2 at Ser522 or Thr509 by Cdk5/p35 and GSK-3β in DRG neurons (Brown et al., 2004; Uchida et al., 2005). Thus, phosphorylation by Rho-kinase or GSK-3β appears to be regulated separately. CRMP-2 phosphorylation by Rho-kinase or GSK-3β cancels the binding to tubulin heterodimers, microtubules, and Numb but not to actin (Arimura et al., 2005). Electron microscopy revealed that CRMP-2 localizes on microtubules, clathrin-coated pits, and actin filaments at growth cones of DRG neurons, whereas CRMP-2 phosphorylated at Thr555 localizes only on actin filaments. Similar to the phosphorylation by GSK-3β, the phosphorylation of CRMP-2 by Rho-kinase weakens its ability to enhance neurite elongation. Taken together, these results suggest that the phosphorylations at the C-terminus of CRMP-2 inactivate the molecule by preventing the association with interacting partners (Arimura et al., 2005).

11.7. Roles of CRMP-2 in Axonal Transport

Intracellular trafficking is fundamental for the establishment and maintenance of cell polarity. Many proteins and vesicles are selectively transported along microtubules to either axons or dendrites by members of the kinesin and dynein families (Hirokawa and Takemura, 2005). However, it is largely unknown how proteins, especially polarity-regulating proteins, are specifically transported to the premature axons.

Kinesins are a family of motor proteins that use the energy of adenosine triphosphate (ATP) hydrolysis to move cargo along microtubules (Kamal and Goldstein, 2002). Kinesin-1 is a tetramer of two kinesin heavy chains (KHCs; also known as KIF5) and two kinesin light chains (KLCs), which are associated in a 1:1 stoichiometric ratio (Brady, 1985; Johnson et al., 1990; Vale et al., 1985). Kinesin-1 associates with several cargo receptors (Hirokawa and Takemura, 2005) such as glutamate receptor–interacting protein-1 (GRIP1) (Setou et al., 2002), amyloid precursor protein (APP) (Kamal et al., 2000), and *c*-jun NH2-terminal kinase (JNK)–interacting protein (JIP/SYD) (Verhey et al., 2001).

We found that Kinesin-1 forms a complex with soluble tubulin in a CRMP-2–dependent manner (Kimura et al., 2005), and CRMP-2 binds KLC and tubulin heterodimer simultaneously. The movement of GFP-tubulin in a photobleached area was significantly weakened by knockdown of KLC or CRMP-2 (Kimura et al., 2005). These results indicate that the CRMP-2/Kinesin-1 complex regulates the transport of soluble tubulin to the distal part of the growing axons. Furthermore, CRMP-2/Kinesin-1 transports the Sra-1/WAVE1 complex to the distal part of axons (Kawano et al., 2005). Together with previous observations, our results suggest that CRMP-2 serves as a cargo receptor between Kinesin-1 and CRMP-2–interacting molecules, which are involved in axon specification and elongation. In addition, Par3 and APC are transported to the distal part of the growing axon by Kinesin-2 (Jimbo et al., 2002; Nishimura et al., 2004). The CRMP family consists of five members, each having a unique C-terminus. This variety may enable the CRMP family to associate with different motor proteins and/or cargos. Further studies are necessary to address this issue.

11.8. Remaining Questions

Significant progress has been made toward understanding the intracellular events of neuronal polarization. However, more questions must be answered before we can elucidate the whole picture. First, what extracellular cues govern neuronal polarity *in situ*? The extracellular matrix (ECM) is a good candidate, but its role in axon formation *in situ* must be examined. Second, if certain receptors and adhesion molecules, such as integrins, participate in axon specification, how are the signals conveyed to PI 3-kinase? The Ras family of GTPases, including H-Ras and R-Ras, may be activated at the tips of axons and stimulate PI 3-kinase (Dickson et al., 1992; Hancock, 2003; Rodriguez-Viciana et al., 1994; Vojtek et al., 1993). Ras is important for axon specification through PI 3-kinase in hippocampal neurons (Yoshimura et al., 2006). Ras regulates the phosphorylation of CRMP-2 via GSK-3β. These findings suggest that Ras plays critical roles in neuronal polarization via the PI 3-kinase/Akt/GSK-3β/CRMP-2 pathway. Third, how does PIP_3 activate the downstream signal cascades? Although several signaling molecules, such as Akt and Cdc42/Rac1, are known to be activated via PIP_3, the full details regarding PIP_3 in neuronal polarity

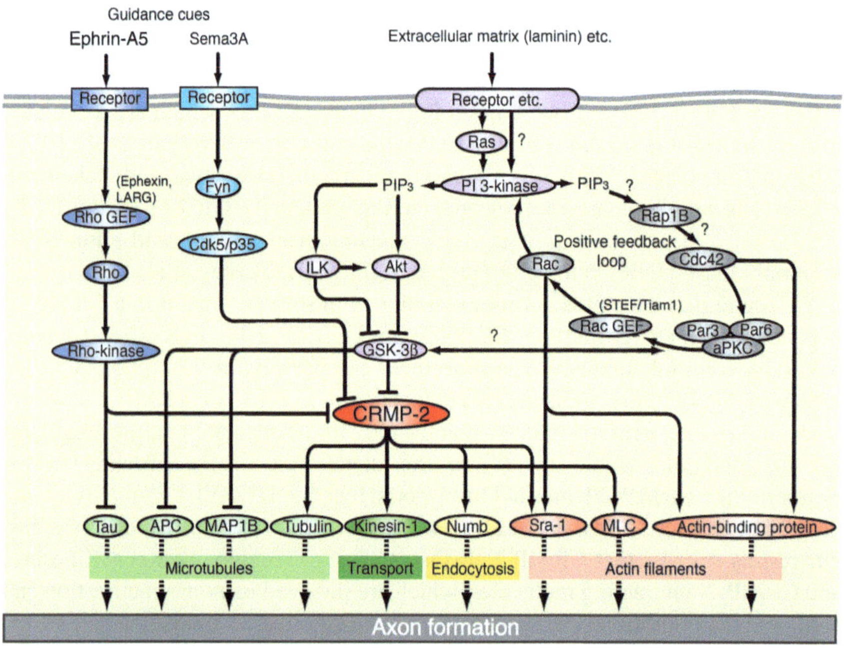

FIGURE 11.3. Signals and polarity-regulating proteins in neuronal polarization. The ECM activates PI 3-kinase through interaction with adhesion molecules or receptors (right), thereby producing PIP_3. PIP_3 activates Akt, which inactivates GSK-3β, resulting in an increase of nonphosphorylated CRMP-2 at axonal growth cones. PIP_3 activates Cdc42, which, in turn, recruits the Par6/Par3/aPKC complex. The Par complex stimulates STEF/Tiam1 and consequently activates Rac1, thereby enhancing the axon outgrowth. In contrast, Rho and Rho-kinase are activated in response to a repulsive guidance cue, ephrin-A5 (left). Rho-kinase phosphorylates CRMP-2 and myosin light chain (MLC). Sema3A, in contrast, activates Cdk5/p35 and GSK-3β (middle). Both repulsive guidance cues increase the phosphorylated CRMP-2 and support growth cone collapse by decreasing the ability of CRMP-2 to associate with binding partners. The relationship between the Akt/GSK-3β pathway and the Par3/Par6 pathway in neuronal polarity is not yet clear.

are still unknown. Finally, how are the signals and polarity-regulating proteins spatially and temporally regulated, and how are they related to each other in axon specification? Analyses of the spatial-temporal dynamics of signals and polarity-regulating proteins are required for a complete picture of neuronal polarization.

References

Abe, T., Kato, M., Miki, H., Takenawa, T., and Endo, T., 2003, Small GTPase Tc10 and its homologue RhoT induce N-WASP-mediated long process formation and neurite outgrowth, *J. Cell Sci.* **116:** 155–168.

Amano, M., Chihara, K., Nakamura, N., Fukata, Y., Yano, T., Shibata, M., et al., 1998, Myosin II activation promotes neurite retraction during the action of Rho and Rho-kinase, *Genes Cells* **3:** 177–188.

Arimura, N., Inagaki, N., Chihara, K., Menager, C., Nakamura, N., Amano, M., et al., 2000, Phosphorylation of collapsin response mediator protein-2 by Rho-kinase. Evidence for two separate signaling pathways for growth cone collapse, *J. Biol. Chem.* **275:** 23973–23980.

Arimura, N., Menager, C., Kawano, Y., Yoshimura, T., Kawabata, S., Hattori, A., et al., 2005, Phosphorylation by Rho kinase regulates CRMP-2 activity in growth cones, *Mol. Cell Biol.* **25:** 9973–9984.

Baas, P.W., and Buster, D.W., 2004, Slow axonal transport and the genesis of neuronal morphology, *J. Neurobiol.* **58:** 3–17.

Berdnik, D., Torok, T., Gonzalez-Gaitan, M., and Knoblich, J.A., 2002, The endocytic protein α-Adaptin is required for numb-mediated asymmetric cell division in Drosophila, *Dev. Cell* **3:** 221–231.

Bradke, F., and Dotti, C.G., 2000, Establishment of neuronal polarity: Lessons from cultured hippocampal neurons, *Curr. Opin. Neurobiol.* **10:** 574–581.

Brady, S.T., 1985, A novel brain ATPase with properties expected for the fast axonal transport motor, *Nature* **317:** 73–75.

Brown, M., Jacobs, T., Eickholt, B., Ferrari, G., Teo, M., Monfries, C., et al., 2004, a2-chimaerin, cyclin-dependent Kinase 5/p35, and its target collapsin response mediator protein-2 are essential components in semaphorin 3A-induced growth-cone collapse, *J. Neurosci.* **24:** 8994–9004.

Byk, T., Ozon, S., and Sobel, A., 1998, The Ulip family phosphoproteins: Common and specific properties, *Eur. J. Biochem.* **254:** 14–24.

Chen, X., and Macara, I.G., 2005, Par-3 controls tight junction assembly through the Rac exchange factor Tiam1, *Nat. Cell Biol.* **7:** 262–269.

Chuang, J.Z., Yeh, T.Y., Bollati, F., Conde, C., Canavosio, F., Caceres, A., et al., 2005, The dynein light chain Tctex-1 has a dynein-independent role in actin remodeling during neurite outgrowth, *Dev. Cell* **9:** 75–86.

Cole, A.R., Knebel, A., Morrice, N.A., Robertson, L.A., Irving, A.J., Connolly, C.N., et al., 2004, GSK-3 phosphorylation of the Alzheimer epitope within collapsin response mediator proteins regulates axon elongation in primary neurons, *J. Biol. Chem.* **279:** 50176–50180.

Craig, A.M., and Banker, G., 1994, Neuronal polarity, *Annu. Rev. Neurosci.* **17:** 267–310.

De Camilli, P., Emr, S.D., McPherson, P.S., and Novick, P., 1996, Phosphoinositides as regulators in membrane traffic, *Science* **271:** 1533–1539.

Dickson, B., Sprenger, F., Morrison, D., and Hafen, E., 1992, Raf functions downstream of Ras1 in the Sevenless signal transduction pathway, *Nature* **360:** 600–603.

Dotti, C.G., Sullivan, C.A., and Banker, G.A., 1988, The establishment of polarity by hippocampal neurons in culture, *J. Neurosci.* **8:** 1454–1468.

Eldar-Finkelman, H., 2002, Glycogen synthase kinase 3: An emerging therapeutic target, *Trends Mol. Med.* **8:** 126–132.

Esch, T., Lemmon, V., and Banker, G., 1999, Local presentation of substrate molecules directs axon specification by cultured hippocampal neurons, *J. Neurosci.* **19:** 6417–6426.

Frame, S., and Cohen, P., 2001, GSK3 takes centre stage more than 20 years after its discovery, *Biochem. J.* **359:** 1–16.

Fukata, Y., Itoh, T.J., Kimura, T., Menager, C., Nishimura, T., Shiromizu, T., et al., 2002, CRMP-2 binds to tubulin heterodimers to promote microtubule assembly, *Nat. Cell Biol.* **4:** 583–591.

Gaetano, C., Matsuo, T., and Thiele, C.J., 1997, Identification and characterization of a retinoic acid-regulated human homologue of the unc-33-like phosphoprotein gene (hUlip) from neuroblastoma cells, *J. Biol. Chem.* **272:** 12195–12201.

Goold, R.G., Owen, R., and Gordon-Weeks, P.R., 1999, Glycogen synthase kinase 3b phosphorylation of microtubule-associated protein 1B regulates the stability of microtubules in growth cones, *J. Cell Sci.* **112:** 3373–3384.

Goshima, Y., Nakamura, F., Strittmatter, P., and Strittmatter, S.M., 1995, Collapsin-induced growth cone collapse mediated by an intracellular protein related to UNC-33, *Nature* **376:** 509–514.

Govek, E.E., Newey, S.E., and Van Aelst, L., 2005, The role of the Rho GTPases in neuronal development, *Genes Dev.* **19:** 1–49.

Hamajima, N., Matsuda, K., Sakata, S., Tamaki, N., Sasaki, M., and Nonaka, M., 1996, A novel gene family defined by human dihydropyrimidinase and three related proteins with differential tissue distribution, *Gene* **180:** 157–163.

Hancock, J.F., 2003, Ras proteins: Different signals from different locations, *Nat. Rev. Mol. Cell Biol.* **4:** 373–384.

Hannigan, G., Troussard, A.A., and Dedhar, S., 2005, Integrin-linked kinase: A cancer therapeutic target unique among its ILK, *Nat. Rev. Cancer* **5:** 51–63.

Hedgecock, E.M., Culotti, J.G., Thomson, J.N., and Perkins, L.A., 1985, Axonal guidance mutants of *Caenorhabditis elegans* identified by filling sensory neurons with fluorescein dyes, *Dev. Biol.* **111:** 158–170.

Hirokawa, N., and Takemura, R., 2005, Molecular motors and mechanisms of directional transport in neurons, *Nat. Rev. Neurosci.* **6:** 201–214.

Iijima, M., Huang, Y.E., and Devreotes, P., 2002, Temporal and spatial regulation of chemotaxis, *Dev. Cell* **3:** 469–478.

Inagaki, N., Chihara, K., Arimura, N., Menager, C., Kawano, Y., Matsuo, N., et al., 2001, CRMP-2 induces axons in cultured hippocampal neurons, *Nat. Neurosci.* **4:** 781–782.

Jalink, K., van Corven, E.J., Hengeveld, T., Morii, N., Narumiya, S., and Moolenaar, W.H., 1994, Inhibition of lysophosphatidate- and thrombin-induced neurite retraction and neuronal cell rounding by ADP ribosylation of the small GTP-binding protein Rho, *J. Cell Biol.* **126:** 801–810.

Jiang, H., Guo, W., Liang, X., and Rao, Y., 2005, Both the establishment and the maintenance of neuronal polarity require active mechanisms: Critical roles of GSK-3β and its upstream regulators, *Cell* **120:** 123–135.

Jimbo, T., Kawasaki, Y., Koyama, R., Sato, R., Takada, S., Haraguchi, K., et al., 2002, Identification of a link between the tumour suppressor APC and the kinesin superfamily, *Nat. Cell Biol.* **4:** 323–327.

Johnson, C.S., Buster, D., and Scholey, J.M., 1990, Light chains of sea urchin kinesin identified by immunoadsorption, *Cell Motil. Cytoskeleton* **16:** 204–213.

Kamal, A., and Goldstein, L.S., 2002, Principles of cargo attachment to cytoplasmic motor proteins, *Curr. Opin. Cell Biol.* **14:** 63–68.

Kamal, A., Stokin, G.B., Yang, Z., Xia, C.H., and Goldstein, L.S., 2000, Axonal transport of amyloid precursor protein is mediated by direct binding to the kinesin light chain subunit of kinesin-I, *Neuron* **28:** 449–459.

Kawano, Y., Yoshimura, T., Tsuboi, D., Kawabata, S., Kaneko-Kawano, T., Shirataki, H., et al., 2005, CRMP-2 Is involved in kinesin-1-dependent transport of the Sra-1/WAVE1 complex and axon formation, *Mol. Cell Biol.* **25:** 9920–9935.

Kimura, T., Arimura, N., Fukata, Y., Watanabe, H., Iwamatsu, A., and Kaibuchi, K., 2005, Tubulin and CRMP-2 complex is transported via Kinesin-1, *J. Neurochem.* **93:** 1371–1382.

Kobayashi, K., Kuroda, S., Fukata, M., Nakamura, N., Nagase, T., Nomura, N., et al., 1998, p140Sra-1 (specifically Rac1-associated protein) is a novel specific target for Rac1 small GTPase, *J. Biol. Chem.* **273:** 291–295.

Kunda, P., Paglini, G., Quiroga, S., Kosik, K., and Caceres, A., 2001, Evidence for the involvement of Tiam1 in axon formation, *J. Neurosci.* **21:** 2361–2372.

Li, W., Herman, R.K., and Shaw, J.E., 1992, Analysis of the *Caenorhabditis elegans* axonal guidance and outgrowth gene unc-33, *Genetics* **132:** 675–689.

Luo, L., Liao, Y.J., Jan, L.Y., and Jan, Y.N., 1994, Distinct morphogenetic functions of similar small GTPases: Drosophila Drac1 is involved in axonal outgrowth and myoblast fusion, *Genes Dev.* **8:** 1787–1802.

Menager, C., Arimura, N., Fukata, Y., and Kaibuchi, K., 2004, PIP_3 is involved in neuronal polarization and axon formation, *J. Neurochem.* **89:** 109–118.

Minturn, J.E., Fryer, H.J., Geschwind, D.H., and Hockfield, S., 1995, TOAD-64, a gene expressed early in neuronal differentiation in the rat, is related to unc-33, a *C. elegans* gene involved in axon outgrowth, *J. Neurosci.* **15:** 6757–6766.

Nishimura, T., Fukata, Y., Kato, K., Yamaguchi, T., Matsuura, Y., Kamiguchi, H., et al., 2003, CRMP-2 regulates polarized Numb-mediated endocytosis for axon growth, *Nat. Cell Biol.* **5:** 819–826.

Nishimura, T., Kato, K., Yamaguchi, T., Fukata, Y., Ohno, S., and Kaibuchi, K., 2004, Role of the PAR-3-KIF3 complex in the establishment of neuronal polarity, *Nat. Cell Biol.* **6:** 328–334.

Nishimura, T., Yamaguchi, T., Kato, K., Yoshizawa, M., Nabeshima, Y., Ohno, S., et al., 2005, PAR-6-PAR-3 mediates Cdc42-induced Rac activation through the Rac GEFs STEF/Tiam1, *Nat. Cell Biol.* **7:** 270–277.

Ohno, S., 2001, Intercellular junctions and cellular polarity: The PAR-aPKC complex, a conserved core cassette playing fundamental roles in cell polarity, *Curr. Opin. Cell Biol.* **13:** 641–648.

Rodriguez-Viciana, P., Warne, P.H., Dhand, R., Vanhaesebroeck, B., Gout, I., Fry, M.J., et al., 1994, Phosphatidylinositol-3-OH kinase as a direct target of Ras, *Nature* **370:** 527–532.

Rolls, M.M., and Doe, C.Q., 2004, Baz, Par-6 and aPKC are not required for axon or dendrite specification in Drosophila, *Nat. Neurosci.* **7:** 1293–1295.

Salcini, A.E., Confalonieri, S., Doria, M., Santolini, E., Tassi, E., Minenkova, O., et al., 1997, Binding specificity and in vivo targets of the EH domain, a novel protein-protein interaction module, *Genes Dev.* **11:** 2239–2249.

Santolini, E., Puri, C., Salcini, A.E., Gagliani, M.C., Pelicci, P.G., Tacchetti, C., et al., 2000, Numb is an endocytic protein, *J. Cell Biol.* **151:** 1345–1352.

Schenck, A., Bardoni, B., Langmann, C., Harden, N., Mandel, J.L., and Giangrande, A., 2003, CYFIP/Sra-1 controls neuronal connectivity in Drosophila and links the Rac1 GTPase pathway to the fragile X protein, *Neuron* **38:** 887–898.

Schwamborn, J.C., and Puschel, A.W., 2004, The sequential activity of the GTPases Rap1B and Cdc42 determines neuronal polarity, *Nat. Neurosci.* **7:** 923–929.

Setou, M., Seog, D.H., Tanaka, Y., Kanai, Y., Takei, Y., Kawagishi, M., et al., 2002, Glutamate-receptor-interacting protein GRIP1 directly steers kinesin to dendrites, *Nature* **417:** 83–87.

Shi, S.H., Jan, L.Y., and Jan, Y.N., 2003, Hippocampal neuronal polarity specified by spatially localized mPar3/mPar6 and PI 3-kinase activity, *Cell* **112:** 63–75.

Takei, Y., Teng, J., Harada, A., and Hirokawa, N., 2000, Defects in axonal elongation and neuronal migration in mice with disrupted tau and map1b genes, *J. Cell Biol.* **150:** 989–1000.

Trivedi, N., Marsh, P., Goold, R.G., Wood-Kaczmar, A., and Gordon-Weeks, P.R., 2005, Glycogen synthase kinase-3b phosphorylation of MAP1B at Ser1260 and Thr1265 is spatially restricted to growing axons, *J. Cell Sci.* **118:** 993–1005.

Uchida, Y., Ohshima, T., Sasaki, Y., Suzuki, H., Yanai, S., Yamashita, N., et al., 2005, Semaphorin3A signalling is mediated via sequential Cdk5 and GSK3β phosphorylation of CRMP2: Implication of common phosphorylating mechanism underlying axon guidance and Alzheimer's disease, *Genes Cells* **10:** 165–179.

Vale, R.D., Reese, T.S., and Sheetz, M.P., 1985, Identification of a novel force-generating protein, kinesin, involved in microtubule-based motility, *Cell* **42:** 39–50.

Varnai, P., and Balla, T., 1998, Visualization of phosphoinositides that bind pleckstrin homology domains: Calcium- and agonist-induced dynamic changes and relationship to myo-[3H]inositol-labeled phosphoinositide pools, *J. Cell Biol.* **143:** 501–510.

Verhey, K.J., Meyer, D., Deehan, R., Blenis, J., Schnapp, B.J., Rapoport, T.A., et al., 2001, Cargo of kinesin identified as JIP scaffolding proteins and associated signaling molecules, *J. Cell Biol.* **152:** 959–970.

Vojtek, A.B., Hollenberg, S.M., and Cooper, J.A., 1993, Mammalian Ras interacts directly with the serine/threonine kinase Raf, *Cell* **74:** 205–214.

Wang, L.H., and Strittmatter, S.M., 1996, A family of rat CRMP genes is differentially expressed in the nervous system, *J. Neurosci.* **16:** 6197–6207.

Wiggin, G.R., Fawcett, J.P., and Pawson, T., 2005, Polarity proteins in axon specification and synaptogenesis, *Dev. Cell* **8:** 803–816.

Yoshimura, T., Kawano, Y., Arimura, N., Kawabata, S., Kikuchi, A., and Kaibuchi, K., 2005a, GSK-3β regulates phosphorylation of CRMP-2 and neuronal polarity, *Cell* **120:** 137–149.

Yoshimura, T., Arimura, N., Kawano, Y., Kawabata, S., Wang, S., and Kaibuchi, K., 2006, Ras regulates neuronal polarity via the PI 3-kinase/Akt/GSK-3β/CRMP-2 pathway, *Biochem. Biophys. Res. Comm.* **340:** 62–68.

Zallen, J.A., Cohen, Y., Hudson, A.M., Cooley, L., Wieschaus, E., and Schejter, E.D., 2002, SCAR is a primary regulator of Arp2/3-dependent morphological events in Drosophila, *J. Cell Biol.* **156:** 689–701.

Zumbrunn, J., Kinoshita, K., Hyman, A.A., and Nathke, I.S., 2001, Binding of the adenomatous polyposis coli protein to microtubules increase microtubule stability and is regulated by GSK3 phosphorylation, *Curr. Biol.* **11:** 44–49.

12
Regulation of Axon Branching

Katherine Kalil, Erik W. Dent, and Fangjun Tang

12.1. Introduction

During development growing axons navigate over long distances to reach often distant targets. They are guided by highly motile growth cones at their tips that respond to attractive and inhibitory guidance cues in the environment (Dickson, 2002; Tessier-Lavigne and Goodman, 1996). However, in many CNS pathways the growth cone of the primary axon does not extend directly into targets. Instead, after the primary growth cone has grown past the target, branches develop interstitially (O'Leary et al., 1990) from the axon shaft and extend into targets (Figure 12.1). Collateral branching, like growth cone turning, changes the direction of axon outgrowth and is thus an important form of axon guidance (Kennedy and Tessier-Lavigne, 1995) particularly in the mammalian CNS. Axon branching enables a single neuron to connect with multiple targets and is therefore essential for the assembly of complex neural circuits (Kornack and Giger, 2005). Interstitial branching has been well documented in pathways such as the corpus callosum and corticospinal tract (Halloran and Kalil, 1994; Kuang and Kalil, 1994; O'Leary et al., 1990) that arise from the cerebral cortex and innervate the contralateral cortex and spinal cord respectively.

Given the fundamental importance of axon branching for establishing synaptic connectivity, surprisingly little is known about how branching is regulated. However, evidence suggests that axon guidance at the growth cone and branch points share common mechanisms (Dent et al., 2003; Kalil et al., 2000). Thus guidance factors, acting through receptors in the cell membrane, are attractive or repellent to growth cones during pathfinding and have been shown to promote or inhibit axon branching. For example, semaphorin-3A (Sema3A), a member of a large family of generally inhibitory guidance cues (Pasterkamp et al., 2003), repels cortical axons *in situ* and *in vitro* (Bagnard et al., 1998; Polleux et al., 1998, 2000) and reduces cortical axon branching (Bagnard et al., 1998; Bagri et al., 2003; Dent et al., 2004). Netrins are generally positive guidance cues. They promote outgrowth of a wide variety of axons (Manitt and Kennedy, 2002) and in explant cocultures have been shown to attract growing cortical axons (Metin

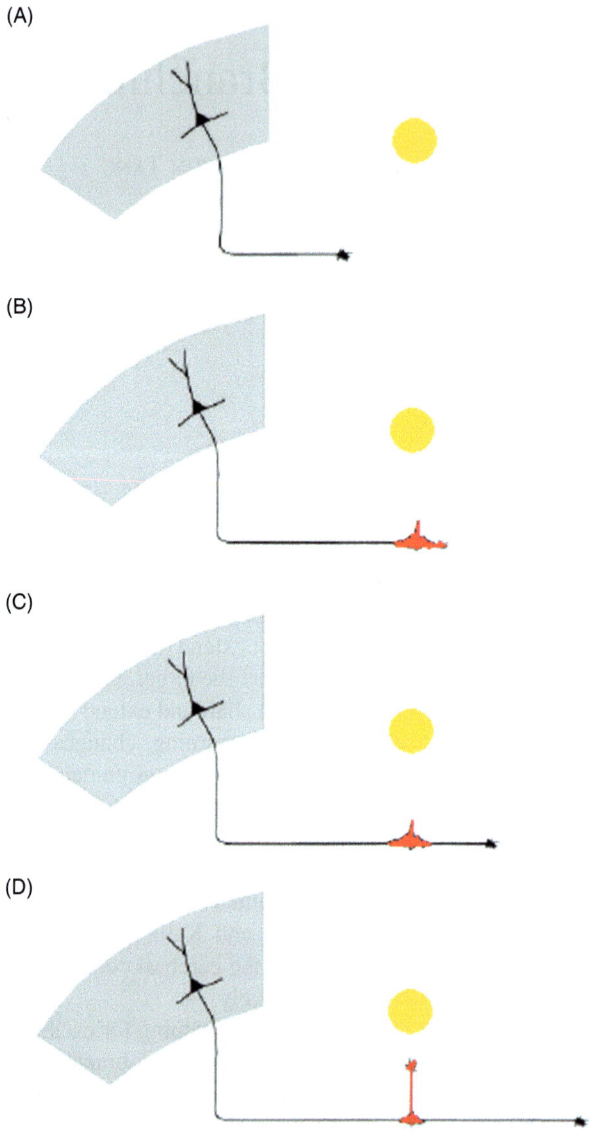

FIGURE 12.1. A model for interstitial branching by cortical neurons. Schematics represent different stages in axon branching *in vivo.* (A) Growth cone of an efferent cortical axon is advancing along a pathway toward its target, indicated by the yellow circle. (B) In response to target-derived signals, the primary growth cone pauses in the vicinity of the target for extended time periods and enlarges. (C) After the primary growth cone has resumed forward advance, remnants of the reorganized growth cone appear as filopodial or lamellar activity along the axon shaft. (D) A branch form these active membrane regions extends toward the target after various time delays. Typically, branches extend interstitially some distance behind the primary growth cone. (With permission from Szebenyi et al., 1998. Copyright 1998 by the Society for Neuroscience.)

et al., 1997; Richards et al., 1997). Netrin-1 was found to promote branching of cortical axons (Dent et al., 2004).

Observations of developing neurons in culture have also shown that axon branching is mechanistically linked to behaviors of the primary growth cone (Dent et al., 2003; Kalil et al., 2000). Axons of retinal ganglion neurons extend filopodia and lamellipodia following growth cone collapse (Davenport et al., 1999) and the primary growth cones of cortical axons enlarge during prolonged pausing behaviors to leave remnants behind on the axon shaft after the axon resumes forward extension (Szebenyi et al., 1998). These remnants subsequently give rise to interstitial axon branches. Increases and decreases in cortical axon branching by growth and guidance factors can occur through changes in behaviors and morphologies of the primary growth cone (Dent et al., 2003; Szebenyi et al., 2001).

Directed growth from the growth cone and branching from the axon shaft are also mechanistically linked by similar changes in cytoskeletal dynamics and their associated intracellular signaling pathways. In growth cones and developing collateral branches a similar dynamic reorganization of the cytoskeleton occurs in which actin–microtubule interactions are an essential element (Dent and Kalil, 2001; Dent et al., 1999; Gallo and Letourneau, 1998, 1999; Schaefer et al., 2002; Zhou and Cohen, 2004). Guidance cues promote or inhibit branching by reorganization of the cytoskeleton at both the growth cone and branch points on the axon shaft (Dent et al., 2004). Moreover, similar signaling pathways regulate growth cone behaviors and axon branching. For example, Rac GTPases, which are members of the Rho family GTPases that transduce extracellular signals to regulate the actin cytoskeleton, play a central role in axon growth, guidance, and branching. Findings in *Drosophila* (Ng et al., 2002) suggest that independent regulation of axon outgrowth and branching may be a developmental mechanism common to many species. In mushroom body neurons it was found that progressive loss of three Rac GTPases leads first to defects in axon branching, then guidance and finally growth, suggesting that growth, guidance, and branching may be mechanistically linked but require different levels of Rac GTPase activity which is involved in regulation of the actin cytoskeleton.

Calcium signaling pathways in growth cones have been well established as a mechanism for regulating axon outgrowth and guidance (Gomez and Spitzer, 2000; Henley and Poo, 2004). Measurements of calcium activity in cortical neurons (Tang and Kalil, 2005; Tang et al., 2003) demonstrated that calcium transients regulate axon outgrowth and branching in a frequency-dependent manner suggesting the involvement of activity-dependent mechanisms. Consistent with this view local depolarization enhances local axonal growth, providing a mechanism for the selection of one collateral over another (Singh and Miller, 2005).

Studies *in vivo* and *in vitro* have shown that axon branching often occurs after the primary axon has ceased extension and independent axon outgrowth and branching has been well documented in efferent cortical pathways (Bastmeyer and O'Leary, 1996; O'Leary and Terashima, 1988; O'Leary et al., 1990; Yamamoto et al., 1997). Direct visualization of fluorescently labeled axons extending in living brain slices (Halloran and Kalil, 1994) revealed that in the corpus callosum the primary axonal growth cone enlarges and pauses beneath the contralateral cortex while axon branches extend dorsally toward cortical targets. Although similar cytoskeletal and signaling mechanisms appear to regulate growth cone behaviors and axon branching, the mechanisms that slow or stop extension of the primary axon along its pathway while permitting growth of the axon's collaterals into targets are not well understood. In this chapter we will discuss common mechanisms that regulate guidance of CNS axons by the primary growth cone and branching of collaterals from the axon shaft with respect to independent regulation of axon outgrowth and branching. Establishment of neural connectivity during development involves not only neurite outgrowth but pruning of excess or inappropriate axon branches by retraction or degeneration (Luo and O'Leary, 2005) but will not be discussed here. Some axons in adult animals are capable of robust sprouting after injury (Schwab, 2002), but the focus of this review will be the mechanisms that regulate CNS axon branching during development.

12.2. Effects of Growth and Guidance Molecules on Axon Branching

12.2.1. Attractive and Inhibitory Guidance Molecules

Families of evolutionarily conserved guidance cues have been discovered that either attract or repel axonal growth cones. Repulsive guidance molecules collapse growth cones and inhibit their advance. For example, Slit proteins repel axons at the midline in *Drosophila* (Brose et al., 1999) and regulate midline crossing in regions of the mammalian CNS such as the optic chiasm (Plump et al., 2002) and corpus callosum (Bagri et al., 2003; Shu and Richards, 2001). Surprisingly, while Slit proteins collapse and repel growth cones they are capable of inducing branching on dorsal root ganglion (DRG) sensory axons (Brose and Tessier Lavigne, 2000; Wang et al., 1999). Sema3A has also been shown to promote branches from retinal axons after growth cone collapse (Campbell et al., 2001). Both the Slit proteins and the semaphorins, particularly Sema3A, are expressed at high levels in the developing mammalian cortex and act as chemorepellants for cortical axons *in vitro* (Bagnard et al., 1998; Polleux et al., 1998). Application of Sema3A to cultures of early postnatal sensorimotor cortical neurons not only collapsed more than 60% of growth cones within about 5 min but also reduced the development of axon branches by about 50% over several days (Dent et al., 2004). While Sema3A reduced branch length by over 60%, it had no effect on the length of the primary axon (Dent et al., 2004). Thus Sema3A specifically reduces branching but has no effect on axon outgrowth per se. Studies have shown that Sema3A and the Slit proteins repel cortical axons but respectively attract or promote the growth and branching of their dendrites (Polleux et al., 2000; Whitford et al., 2002). Thus, in many situations the growth of neuronal processes can be differentially regulated by a single guidance factor.

Other guidance molecules such as the ephrins can also induce growth cone repulsion (Huber et al., 2003) and have been shown to regulate axon branching. For example, members of the Eph family of receptor tyrosine kinases and their ephrin ligands play important roles in the development of the retinotectal projection. In the mammalian retinotectal system, retinal axons initially overshoot their future targets during development. Topographic specificity is achieved not by guidance of terminal growth cones to tectal targets but by interstitial axon branching of retinal axons at appropriate locations (Simon and O'Leary, 1992) in the tectal target. Gradients of ephrin-As were shown to regulate development of this topographic map (Yates et al., 2001), and map formation was thought to depend on a combination of branch promoting and suppressing activities. Studies using anatomical mapping of retinotectal connections in EphA7 mutant mice in concert with membrane stripe assays with chick retinal axons revealed that opposing gradients of ephrin-As and EphA7 are required for development of appropriate retinotectal topographic mapping (Rashid et al., 2005). These mechanisms of map formation do not involve the primary growth cone. Rather, formation of appropriate termination zones in the tectum depends upon suppression of retinal axon branching by EphA7 anterior to future termination zones.

The netrin family of guidance molecules can promote axon outgrowth and guide growth cones in many regions of the nervous system (Manitt and Kennedy, 2002). Studies of cortical axon outgrowth in explant cultures (Metin et al., 1997; Richards et al., 1997), for example, have shown a potent effect of netrin-1 in attracting growing axons, but effects of netrins on axon branching were unknown. Several studies have now shown that netrin-1 can elicit a large and rapid increase in axon branching. In cultured cortical neurons spontaneous development of axon branches typically requires several days (Szebenyi et al., 1998) and involves pausing and enlargement of the primary growth cone at future branch points. In contrast, netrin-1 applied to hippocampal (Lebrand et al., 2004) or to cortical neurons (Dent et al., 2004; Tang and Kalil, 2005) promotes formation of filopodia directly from the axon shaft within minutes. Time lapse imaging of cortical neurons (Dent et al., 2004) revealed that over several hours many of these filopodia develop into branches tipped by growth cones (Figure 12.2).

12.2.2. Growth Factors

Growth factors, such as nerve growth factor (NGF), fibroblast growth factor (FGF-2), brain derived neurotrophic factor (BDNF), and neurotrophin-3, which promote axon outgrowth are also potent stimulators of axon branching. Direct visualization of axon arborization revealed that BDNF, a target derived neurotrophin, plays an important role in axon branching and remodeling (Cohen-Cory and Fraser, 1995). Injection of BDNF into the optic tectum of living tadpoles increased the branching and complexity of optic axon arbors whereas neutralizing antibodies to BDNF reduced their arborization and complexity. These effects occurred rapidly and were specific to BDNF. BDNF had opposite effects on dendritic arbors of retinal ganglion cells (Lom and Cohen-Cory, 1999).

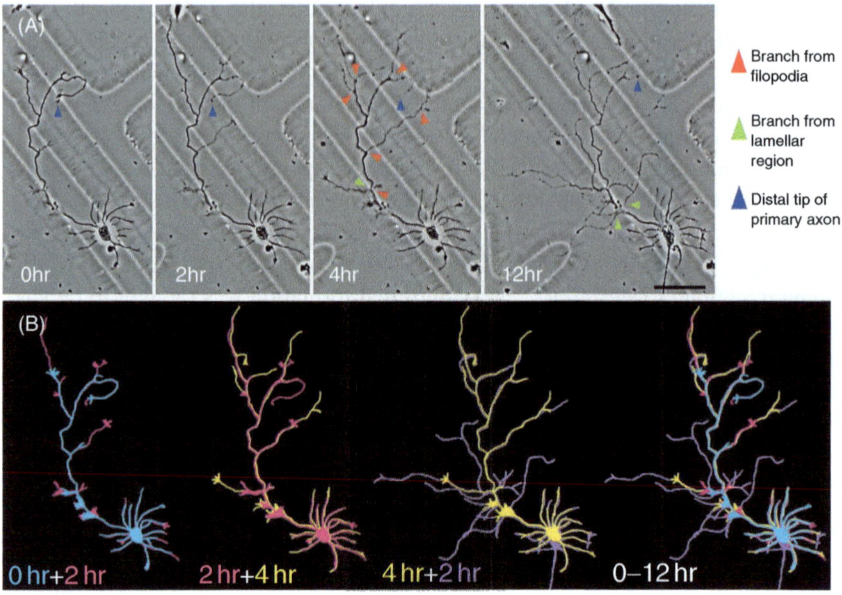

FIGURE 12.2. Netrin-1 induces axon branching from filopodial protrusions. (A) Phase-contrast image of a single postnatal day 1 cortical neuron observed over 12 h in time lapse. Branches emanating from filopodial protrusions are marked with a red arrowhead; branches forming from spread lamellar regions of the axon shaft are marked with a green arrowhead. The distal tip of the primary axon is marked with a blue arrowhead in all frames. (B) Composite tracings of the neuron in A overlaid on the previous time point showing the progression of branch formation and extension. The time points are color coded for clarity and correspond to the color of the tracing of the neuron. Scale bar, 50 μm. (With permission from Dent et al., 2004. Copyright 2004 by the Society for Neuroscience.)

Injection of BDNF into the tadpole retina reduced dendritic arborization, whereas function blocking antibodies increased arbor complexity. Target derived NGF is also a potent regulator of axonal sprouting. Localized sprouting of axon collateral branches, which arose as filopodia, was induced on cultured DRG neurons when NGF covalently conjugated to polystyrene beads was applied directly to the axons (Gallo and Letourneau, 1998). FGF-2 locally applied in the form of coated beads (Szebenyi et al., 2001) was also shown to induce cortical axon branching in regions where the beads contacted the axon shaft. However, in contrast to the rapid effects of locally applied netrin-1, development of axon branches in close proximity to the FGF-2 coated beads requires several days in culture.

While growth and guidance factors can evoke filopodial protrusions and branches directly from the axon, observations in living cortical slices (Halloran and Kalil, 1994) and in dissociated cell cultures (reviewed in Kalil et al., 2000) have shown that pausing behaviors and enlargement of the primary growth cone demarcate branch points along cortical axons from which branches can develop interstitially. Moreover effects on branching by global application of FGF-2 and

Sema3A to cortical neurons (Dent et al., 2004; Szebenyi et al., 2001) are mediated by effects on the primary growth cone. Thus FGF-2 increases axon branching by inducing growth cones to pause and enlarge whereas Sema3A collapses cortical growth cones and the resulting absence of large pausing growth cones reduces branching by about 50% (Dent et al., 2004). Bath application of netrin-1 actually decreases the numbers of large paused growth cones, and direct time-lapse observations showed that many branches resulting from netrin-1 treatment extend directly from the axon shaft rather than from expanded branch points. The rapid growth of branches on axons of netrin-1 treated neurons, in contrast to the growth of branches over several days in control and FGF-2 treated cultures, suggests that different intracellular signaling mechanisms may be involved. Time-lapse observations in living brain slices have demonstrated that branching of cortical axons in response to cues from pontine targets can occur by extension of filopodial processes directly from the axon shaft (Bastmeyer and O'Leary, 1996). Evidence suggests that axon pruning also involves diverse axon remodeling processes (Kantor and Kolodkin, 2003) including selective retraction (Bagri et al., 2003) and local degeneration (Watts et al., 2003) of axon branches. It seems likely that in different contexts axon branches may also develop by different cellular mechanisms.

Many of the factors that influence axon branching have no effect on axon length. Thus, netrin-1 increases while Sema3A decreases branch numbers as well as length, neither treatment affects axon length. Although it is generally the case that cues attractive or repellent to cortical axons increase or decrease their branching respectively, these results suggest that axon outgrowth and axon branching are independently regulated. This is consistent with the situation *in vivo* where axons stop elongating while collaterals are extending toward their targets (Bagri et al., 2003).

12.3. Cytoskeletal Reorganization in Growth Cones and Branches

As motile growth cones extend toward their targets, they advance, retract, turn, and branch. Changes in the direction of axon growth by growth cone guidance behaviors and axon branching require reorganization of the microtubule and actin cytoskeleton (Dent and Gertler, 2003; Dent et al., 2003; Dickson, 2002; Kalil and Dent, 2005; Kalil et al., 2000; Luo, 2002). Microtubules are long hollow polymers of tubulin that form dense bundles in axons and dendrites. They not only lend structural support to neuronal processes but serve as railways along which materials are transported from the cell body into neurites. The dynamic properties of microtubules, i.e., their ability to grow and shrink in dynamic instability, enable them to explore growth cone lamellipodia and filopodia. In growth cones, the fast growing (plus) ends of microtubules face outward toward the periphery. Actin filaments play an essential role in driving motility and guidance of the growth cone (Luo, 2002). In veil-like lamellipodia, actin filaments form a meshwork but within filopodia, the spiky fingerlike protrusions from the growth cone periphery that are important for exploring the environment, actin filaments form straight bundles (Figure 12.3). Actin filament dynamics involve cycles of net assembly of actin filaments at the leading edge of the growth cone and disassembly of filaments proximally (Forscher and Smith, 1988).

12.3.1. Microtubule Dynamics

Visualizing fluorescently labeled microtubules in living neurons with high resolution time-lapse microscopy has revealed the dynamic cytoskeletal changes that underlie growth cone motility and guidance behaviors (Dent and Kalil, 2001; Schaefer et al., 2002) as well as axon branching. Microtubules are bundled together in the axon shaft and also extend into the central region of the growth cone as well as the periphery. They are splayed apart in growth cones that are extending, but when growth cones stall microtubules form prominent loops in the central region (Figure 12.3) (Tanaka and Kirschner, 1991). Similarly, at axon branch points microtubules undergo localized debundling (Gallo and Letourneau,

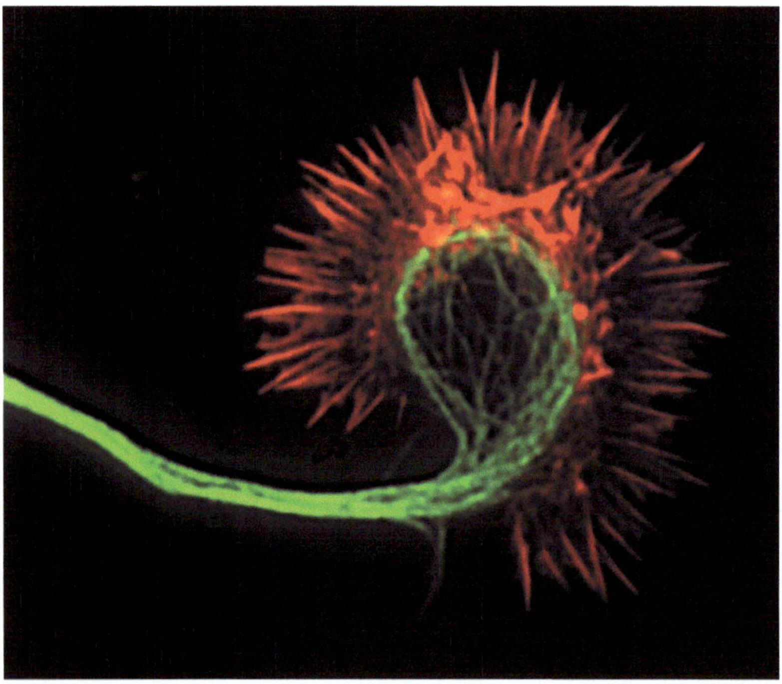

FIGURE 12.3. A large pausing cortical axonal growth cone *in vitro*. Microtubules, pseudocolored green, are stained with an anti-tubulin fluorescent antibody. Actin filaments, pseudocolored red, are stained with rhodamine phalloidin. Microtubules are bundled together in the axon, splay apart in the central region of the growth cone, and form a prominent loop surrounding the central region. Actin filaments extend into the lamellipodia and form straight bundles in spiky filopodia. In the transition region between the growth cone center and periphery microtubules overlap (yellow) with actin filaments.

1998). Live cell imaging of cortical neurons revealed movements of individual microtubules and dynamic reorganization of microtubule arrays in axons and their growth cones (Dent et al., 1999). Microtubule reorganization at the growth cone and at axon branch points are strikingly similar (Kalil et al., 2000). Along the axon shaft, microtubules splay apart and locally fragment prior to development of branches. During prolonged growth cone pausing behaviors microtubules are maintained in a tightly bundled loop. During transitions of the growth cone from pausing to growth states, microtubules reorganize from looped to splayed configurations followed by fragmentation (Dent and Kalil, 2001; Dent et al., 1999). Reorganization and fragmentation permits short microtubules to explore new directions of growth with forward and backward movements within the growth cone periphery and in branches extending from the axon shaft (Figure 12.4). Observations of long-term branching events showed that microtubules invade and

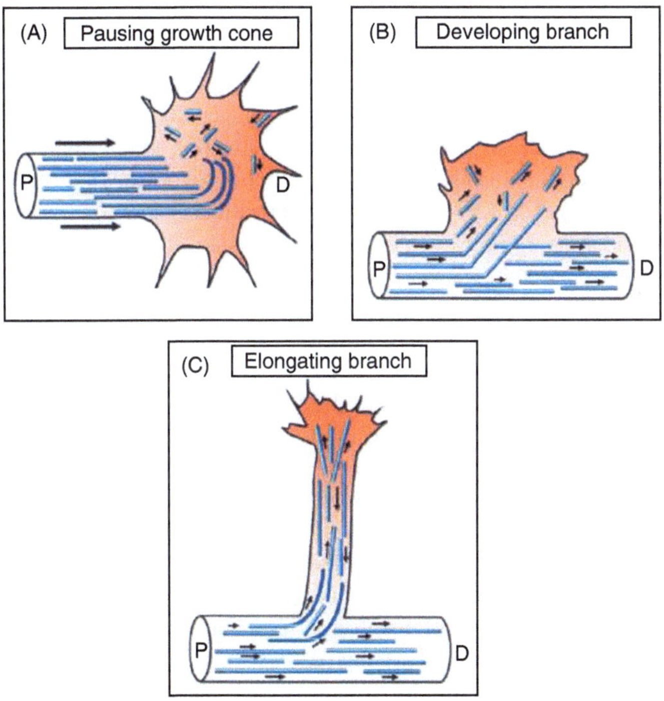

FIGURE 12.4. Schematic model for microtubule reorganization during axon outgrowth and branching. (A) Microtubules (blue) in the pausing growth cone form a loop in the central region. As microtubules splay apart in one region of the loop, short microtubules fragment from longer microtubules and explore the lamellipodium. (B) A branch is emerging form the axon shaft in a region where local splaying and fragmentation of microtubules occurs. Short microtubule fragments invade the developing branch. (C) The axon branch is extending as microtubules elongate within it. Arrows indicate directions of microtubule movement. P and D refer to proximal and distal segments of the axon. (With permission from Dent et al., 2003.)

remain in branches favored for further growth but withdraw from branches that regress, suggesting that microtubules can stabilize developing branches.

These findings suggest that the cytoskeletal mechanisms underlying axon branching involve the reorganization of the microtubule array into a more labile form whereby microtubules can become debundled and fragmented. This plasticity would permit microtubules to explore the growth cone periphery and newly forming branches. The phenomenon of retrograde microtubule movements suggests a mechanism whereby some axonal processes can regress while others grow. At present the role of microtuble loops in regulating growth cone advance and the mechanisms governing their formation and disassembly are unknown. Does the formation of loops retard growth cone extension? Or is this rearrangement a concomitant of growth cone arrest that causes growing microtubules to bend into loops? How do microtubules become stabilized? Evidence suggests that plus-end tracking proteins (+TIPS) regulate the dynamics of microtubules by stabilizing them. Tip proteins bind selectively to the plus ends of microtubules and contribute to their stabilization by reducing the frequency of shortening or promoting their growth. Evidence in *Xenopus* neurons showed that overexpression of a +TIP protein CLASP causes microtubules to form loops in the central region of the growth cone thereby slowing growth cone advance (Lee et al., 2004). In stable synapses at the *Drosophila* neuromuscular junction microtubules also form loops that disassemble during growth and sprouting (Roos et al., 2000) suggesting important parallels for microtubule reorganization as a regulatory mechanism for growth and stability in growth cones as well as developing synapses.

12.3.2. Actin–Microtubule Interactions

Microtubules in growth cones predominate in the central region and aotin filaments predominate in the growth cone periphery (Figure 12.3), but there is also extensive overlap between these two cytoskeletal elements in the growth cone transition region between the central domain and the periphery. Live cell imaging studies have shown that microtubules do not remain confined to the central region but penetrate into lamellipodia in the growth cone periphery and even invade filopodia where they align with actin filament bundles (Dent and Kalil, 2001; Schaefer et al., 2002). During changes in the direction of axon growth, microtubules reorient toward sites of focal F-actin accumulation (Lin and Forscher, 1993; O'Connor and Bentley, 1993). Direction of microtubules toward regions of attenuated F-actin flow during growth cone-target interactions (Lin et al., 1994) or capture of dynamic microtubule ends by actin filaments during growth cone turning (Bentley and O'Connor, 1994; Tanaka and Sabry, 1995) suggest the importance of F-actin–microtubule interactions in growth cone behaviors.

Microtubules and actin filaments are highly dynamic, undergoing continual cycles of growth and shortening during polymerization and depolymerization. At the same time actin filaments and microtubules can also move and undergo dramatic changes in their organization and location within the growth cone. In cortical neurons (Dent and Kalil, 2001) branching from the growth cone and the axon shaft is always preceded by splaying apart of looped or bundled microtubules accompanied by localized accumulation of F-actin. Dynamic microtubules colocalize with F-actin in transition regions of growth cones and axon branch points. Interactions between microtubules and actin filaments involve their coordinated polymerization and depolymerization (Dent and Kalil, 2001). Drugs that attenuate either microtubule or actin dynamics, such as nocodazole and latrunculin respectively, abolish F-actin–microtubule interactions at the growth cone and axon branch points. Consequently, these drug treatments inhibit axon branching and cause axons to adopt tortuous curved morphologies in contrast to the straight axon trajectories of untreated neurons. However, these drug treatments do not affect axon length. These findings show that actin–microtubule interactions are essential for initiating axon growth in new directions. Subsequent studies of invertebrate neuronal growth cones also showed that actin filaments are important for the guidance of microtubules into the periphery (Zhou et al., 2002) particularly in filopodia where microtubules were shown to extend along actin filament bundles that serve to guide the anterograde and retrograde transport of microtubules (Schaefer et al., 2002). Similarly, dynamic interactions between actin filaments and microtubules have been visualized in migrating epithelial cells (Salmon et al., 2002) suggesting that coupling of F-actin and microtubule movements is a common feature of migrating cells (Rodriguez et al., 2003). Taken together, these results imply that factors, such as extracellular cues, that influence the dynamics of either F-actin or microtubules may affect both cytoskeletal elements through bidirectional signaling pathways.

12.3.3 Effects of Growth and Guidance Molecules on the Axonal Cytoskeleton

12.3.3.1. Cytoskeletal Dynamics

To understand how extracellular cues regulate growth cone guidance behaviors and axon branching, it is necessary to understand their effects on the organization of the actin and microtubule cytoskeleton. Sema3A induces collapse of axonal growth cones which is known to involve actin depolymerization at the leading edge of the growth cone (Fan et al., 1993; Fritsche et al., 1999; Fournier et al., 2000) mediated by the Rho family of GTPases (Liu and Strittmatter, 2001). Sema3A (Bagnard et al., 1998; Polleux et al., 1998) and several of the Slit proteins repel cortical axons and can reduce their branching (Bagnard et al., 1998). How do growth and guidance factors affect axon branching through changes in cytoskeletal dynamics? As mentioned previously, exposure of cortical neurons to Sema3A reduces branching by more than 50% without affecting axon length in contrast to FGF-2 and netrin-1 which increase axon branching by 2–3 times. Within an hour of exposure to Sema3A, approximately 60% of cortical axonal growth cones collapse. However, large paused growth cones do not completely collapse but show some retraction of actin based filopodia and lamellipodia. Rapid sequential imaging of microtubules and actin filaments in growth cones after application of Sema3A (Dent et al., 2004) showed that within a few minutes Sema3A depolymerizes bundled actin in filopodia and attenuates motility and protrusion of lamellipodia (Figure 12.5) (Dent et al., 2004). Microtubules decrease their exploratory behaviors in the transition region and collapse into the microtubule loop. These changes resemble the attenuation of cytoskeletal

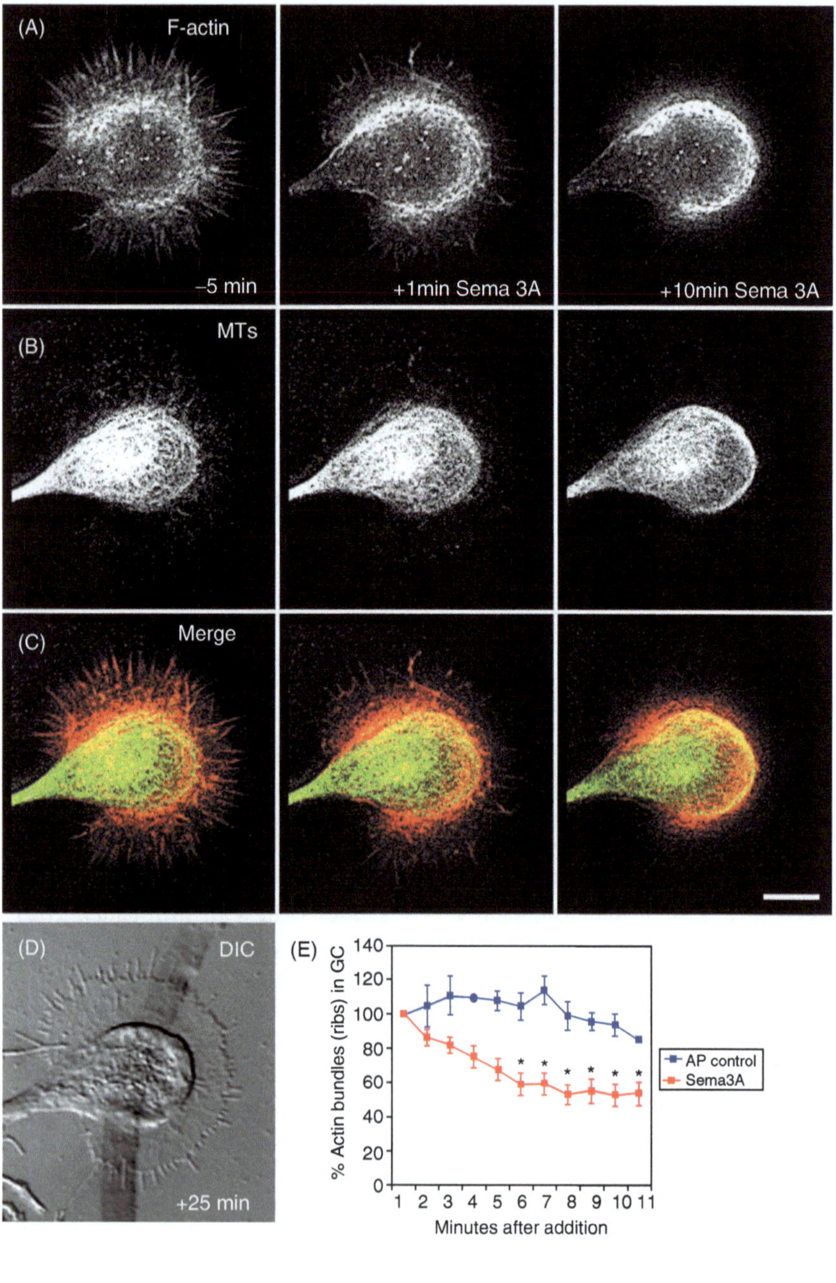

dynamics after application of the depolymerizing drugs nocodazole and latrunculin (Dent and Kalil, 2001) but are less severe. These results are consistent with the finding that cultures treated with Sema3A have almost no large paused growth cones and suggest that Sema3A reduces actin dynamics which in turn attenuates microtubule dynamics. Thus the reduction in axon branching may result from the failure of microtubules to invade the growth cone periphery and delineate new directions of growth at axon branch points. FGF-2 which increases cortical axon branching and enlarges growth cones over time (Szebenyi et al., 2001) promotes actin polymerization and the formation of microtubule loops. Application of netrin-1 also promotes actin polymerization resulting in increases in filopodia on the growth cones and along the axon shafts of hippocampal (Lebrand et al., 2004) and cortical (Dent et al., 2004) neurons, leading to increased axon branching. Cortical axon branching (Dent et al., 2004) occurs through increases in actin filaments which cause microtubules to splay apart in the axon shaft and the growth cone thereby permitting dynamic microtubules to interact with the newly polymerized actin bundles and initiate outgrowth in new directions.

These results suggest the importance of cytoskeletal dynamics in axon branching. In further support of this view a study of KIF2A, a kinesin family member that interacts directly with microtubules by inhibiting their ability to elongate, showed that when KIF2A is knocked out hippocampal axons have much longer collateral branches than in wild-type neurons. This result suggests that KIF2A is required for suppressing supernumerary branches on hippocampal axons (Homma et al., 2003). In mice lacking KIF2A microtubule depolymerizing activity is lower suggesting that KIF2A normally depolymerizes microtubules and may suppress outgrowth of collaterals by regulating microtubule dynamics. Thus, the absence of KIF2A would result in overextension of axon branches. A study of embryonic *Xenopus* spinal neurons also demonstrates an important role for microtubule dynamics in growth cone guidance behaviors (Buck and Zheng, 2002). In this study attraction and repulsion of growth cones by gradients of glutamate and netrin-1 respectively

FIGURE 12.5. Sema3A rapidly decreases the number of F-actin bundles and dynamic microtubules in the peripheral region of the growth cone without collapse of the growth cone. (A and B) Images of a living large paused growth cone from a postnatal day 3 cortical neuron coinjected with Alexa488 phalloidin to label F-actin and with rhodamine tubulin to label microtubules (MTs). Within 1 min after application of Sema3A, many of the actin bundles in the growth cone disappear, concomitant with the loss of dynamic microtubules extending from the central microtubule loop. By 10 min after Sema3A application, the intensity of an F-actin ring surrounding the peripheral microtubule loop increases. (C) Merged images of the same growth cone in (A) and (B), where F-actin is pseudocolored red and microtubules are green. (D) A DIC image of the same growth cone in (A–C) showing that the filopodia and broad lamellar region have not retracted. (E) Graph plotting the number of actin bundles over time. Semaphorin addition resulted in a significant decrease in actin bundles over time compared with supernatant collected form AP vector-infected cells (*$p < 0.05$; paired t test). All error bars are ±SEM. Scale bar, 10 μm. (With permission from Dent et al., 2004. Copyright 2004 by the Society for Neuroscience.)

were completely blocked when microtubule dynamics were inhibited by the drugs nocodazole or taxol. Local stabilization of microtubules in one side of the growth cone induced attractive steering, whereas local microtubule depolymerization caused growth cones to steer away, demonstrating an instructive role for microtubules in directional steering of the growth cone. Although actin dynamics were not visualized directly, inhibition of actin polymerization with cytochalasin and Rho GTPase signaling by toxin B both blocked taxol attraction of the growth cones, supporting the view that steering of the growth cone in response to guidance cues requires coordinated actin–microtubule interactions.

12.3.3.2. Actin-Associated Proteins

Actin-associated proteins have been identified that regulate actin filament assembly and some of these have been implicated in inducing protrusions at the leading edge of motile cells (Bear et al., 2001; Hu and Reichardt, 1999; Machesky and Insall, 1999; Pantaloni et al., 2001; Small et al., 2002). The Ena/VASP family of actin regulatory proteins, consisting of *Drosophila* Enabled (Ena), the mammalian homolog of Ena (Mena), vasodilator-stimulated phosphoprotein (VASP), and Ena/VASP-like (EVL), is associated with the Abelson (Abl) tyrosine kinase and is involved in cytoskeletal dynamics (Krause et al., 2003). In *Drosophila*, overexpression of Abl and deletion of Ena induces a bypass phenotype in which the motor axon extends beyond the normal choice point and fails to stop and branch (Wills et al., 1999). In mice, deletion of Mena causes defects in axon guidance in the corpus callosum and hippocampal commissure (Lanier et al., 1999). Mena localizes to the tips of filopodia in hippocampal growth cones (Lanier et al., 1999) consistent with the role of Ena/VASP proteins in regulating actin dynamics in the growth cone. Abl and its associated Ena/VASP proteins are hypothesized to link extracellular signaling pathways to actin cytoskeletal dynamics that modulate growth cone motility (Korey and Van Vactor, 2000; Lanier and Gertler, 2000). An important role for Ena/VASP proteins was demonstrated in netrin-1 induced filopodia formation on hippocampal axons (Lebrand et al., 2004), which requires phosphorylation of Ena/VASP proteins through activation of protein kinase A (PKA). Ena/VASP proteins promote elongation at the growing (barbed) end of actin filaments by antagonizing capping proteins that normally terminate actin filament elongation (Bear et al., 2002). Inactivation of Ena/VASP activity reduces the length and number of filopodia through a reduction of bundled actin filaments, whereas elevation of Ena/VASP increases filopodia formation by increasing actin filament bundles. Thus, the actin remodeling effects of Ena/VASP proteins play a central role in netrin-induced filopodia which can progress to axon branching. The effects of neurotrophins, such as BDNF, on increasing filopodial length and number also occur through the activation of an actin-associated protein, ADF/cofilin (actin depolymerizilng factor), that enhances actin dynamics (Gehler et al., 2004).

Thus in neuronal growth cones and at axon branch points the actin and microtubule cytoskeleton is the ultimate target of signaling pathways from extracellular guidance cues. Although these signaling pathways have been the subject of intense study (Huber et al., 2003), it is still not well understood how guidance cues modulate the dynamics of actin filaments and microtubules and regulate their interactions. However, it is known that, through the actions of cytoskeletal-associated regulatory proteins, reorganization of the cytoskeleton and changes in cytoskeletal dynamics underlie the ability of growth cones to turn toward gradients of attractive guidance cues (such as netrin-1) or away from repulsive cues (such as Sema3A). This model of growth cone guidance (Figure 12.6) could also apply to the promotion and inhibition of axon branching by growth and guidance cues.

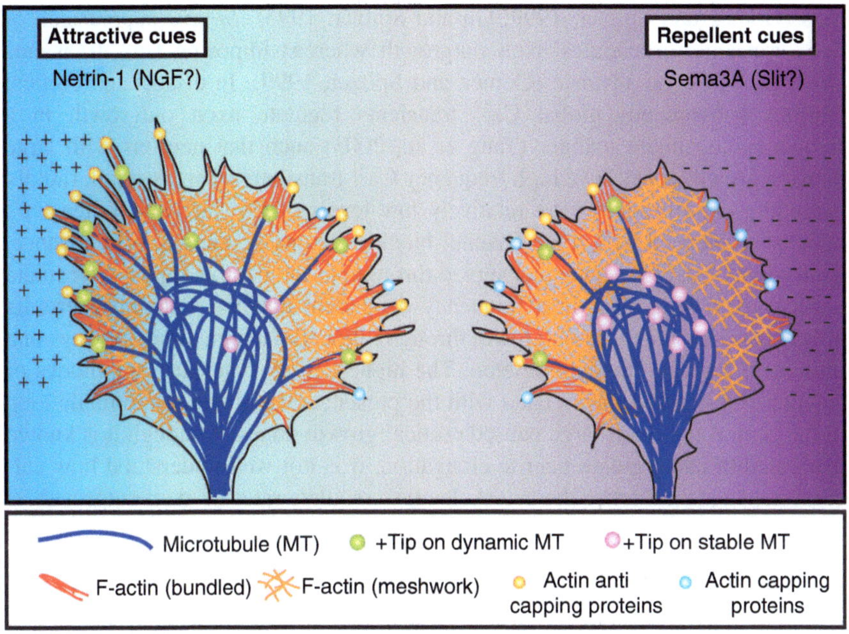

FIGURE 12.6. Model for cytoskeletal reorganization and dynamics in growth cones responding to attractive and repellent cues. The side of the growth cone facing away from guidance cues is the unstimulated side. The growth cone contains bundled stable microtubules in the center, a meshwork of actin filaments in the lamellipodium, actin filament bundles that penetrate filopodia and dynamic microtubules that extend into the periphery to interact with actin filaments. The attractive cue netrin-1 increases the number of filopodia and promotes actin filament elongation by actions of anticapping proteins such as Ena/VASP. (Capping proteins normally terminate actin filament elongation.) In the growth cone periphery an increased number of actin filaments bundles promotes interaction with dynamic microtubules and their associated +TIP proteins. These interactions could lead to attractive turning behavior. The repellent cue Sema3A causes a dissolution of actin filament bundles and the subsequent loss of dynamic microtubules, which leads to growth cone collapse and repulsive turning. Slit might induce similar microtubule looping through the +TIP protein CLASP leading to growth cone pausing and/or repulsion. (Reprinted from Kalil and Dent (2005). Touch and go: guidance cues signal to the growth cone cytoskeleton, *Curr. Opin. Neurobiol.* **15:** 521–526, with permission from Elsevier.)

12.4. Regulation of Axon Outgrowth and Branching by Calcium Signaling

12.4.1. Axon Outgrowth

Studies of neuronal development have revealed diverse roles for intracellular calcium signaling in the growth and guidance of axons (Berridge et al., 2003; Bolsover, 2005; Gomez and Spitzer, 2000; Henley and Poo, 2004; Kater and Mills, 1991; Letourneau et al., 1994). In the *Xenopus* spinal cord, axon outgrowth is regulated by frequencies of Ca^{2+} transients in the growth cones of spinal neurons (Gomez and Spitzer, 1999; Gu and Spitzer, 1995). *In vivo* suppression of Ca^{2+} transients accelerates axon outgrowth whereas imposing Ca^{2+} transients slows growth cone advance (Gomez and Spitzer, 1999). In dissociated cortical neurons spontaneous global Ca^{2+} transients regulate axon outgrowth in a frequency-dependent manner (Tang et al., 2003) such that neurons with large pausing growth cones have high frequency Ca^{2+} transients whereas small, rapidly extending growth cones have relatively low levels of Ca^{2+} activity. When Ca^{2+} activity is silenced by L-type channel blockers axons accelerate their advance. Thus global Ca^{2+} transients that spread throughout the neuron appear to regulate cortical axon outgrowth in a frequency dependent manner. Calcium exerts its effects on rates of neurite outgrowth through intracellular signaling pathways that ultimately regulate the cytoskeleton. The highest frequencies and amplitudes of calcium transients are associated with the presence of a stable microtubule loop in the central region of large paused cortical growth cones. Although it is known that calcium can regulate neurite elongation, it is not well understood how calcium transients influence the cytoskeleton to regulate rates of axon outgrowth.

12.4.2 Axon Guidance

Calcium signaling is also an important regulator of axon guidance (Henley and Poo, 2004). During guidance behaviors localized changes in the organization and dynamics of the cytoskeleton are likely to be regulated by changes in intracellular signaling events in restricted regions of the growth cone. Spatially restricted elevation of intracellular Ca^{2+} concentration in the growth cone induced by guidance cues has been shown to initiate turning behaviors in the direction of higher calcium (Hong et al, 2000; Zheng, 2000; Zheng et al., 1994). Induction of localized elevations in Ca^{2+} concentration also promote protrusion of filopodia from growth cones (Davenport and Kater, 1992; Gomez et al., 2001; Silver et al., 1990) and from axons (Lau et al., 1999). In response to gradients of netrin-1, intracellular Ca^{2+} concentrations are elevated in the growth cone to regulate changes in guidance behaviors (Hong et al., 2000; Ming et al., 2002; Nishiyama et al., 2003). Attractive turning is preceded by protrusion of lamellipodia and filopodia (Zheng, 2000). These effects involve relative Ca^{2+} concentrations that form a gradient across the growth cone (Henley and Poo, 2004; Hong et al., 2000; Zheng et al., 1994).

12.4.3. Axon Branching

Could netrin-1 also promote axon branching through localized Ca^{2+} activity in the axon? Netrin-1 was found to promote rapid and extensive axon branching by evoking repetitive Ca^{2+} transients in restricted regions of cortical axons (Tang and Kalil, 2005). Netrin-1 was also shown to modulate endogenous Ca^{2+} activity by increasing the frequency of Ca^{2+} transients which could be evoked in the entire axon or could begin in localized axon regions and spread throughout the axon. Experimental reduction of Ca^{2+} levels revealed that Ca^{2+} signaling is essential for netrin-1 induced axon branching and the use of specific receptor blockers showed that release from intracellular stores is a major source of this Ca^{2+} activity. Fluorescence imaging of netrin-1–induced Ca^{2+} transients showed that within minutes, new branches protruded from those regions of the axon exhibiting high frequency Ca^{2+} transients (Figure 12.7) demonstrating that localized calcium transients can promote rapid development of branches.

When netrin-1 was applied locally to an axon, Ca^{2+} transients often occurred in patterns, frequencies and amplitudes that were different and distinct in the axon and its branches (Figure 12.8). This raises the possibility that Ca^{2+} transients of different frequencies and amplitudes could differentially regulate the growth of different branches on the same axon. This is consistent with the finding that higher levels of electrical activity confer a competitive advantage on the growth of one axon branch over another (Singh and Miller, 2005) (Section 12.5). Studies of axon guidance have emphasized the role of local elevations in intracellular Ca^{2+} during growth cone turning (Henley and Poo, 2004; Wen et al., 2004), whereas axon branching appears to be dependent on increases in the frequency of localized Ca^{2+} transients. Netrin-1 attracts cortical axons *in vivo* (Serafini et al., 1996) and directed growth and branching of cortical axons by netrin-1 has also been demonstrated in explant cocultures (Richards et al., 1997). Although the role of netrin-1 in axon branching during development *in vivo* is not known, it is possible that similar calcium signaling mechanisms elicited by target-derived cues such as netrin-1 may promote branching to appropriate targets at specific locations along the axon.

12.4.4 Downstream Targets of Calcium Signaling During Axon Branching

What features of Ca^{2+} signaling activate specific downstream targets? Changes in the frequency of Ca^{2+} transients are known to activate genes for different neurotransmitters (Borodinsky et al., 2004; Gomez and Spitzer, 2000). Different frequencies as well as patterns of Ca^{2+} transients may also target different kinases and phosphatases in the cytoplasm to elicit different cellular responses (Tomida et al., 2003) and specific cellular events may be regulated by calcium transients at optimal frequencies (Eshete and Fields, 2001). For example, calcium/calmodulin-depedent protein kinase II (CaMKII) is known to function as a Ca^{2+} spike frequency detector (Hudmon and Schulman, 2002) and mitogen-activated protein kinases (MAPKs) are also sensitive to intracellular Ca^{2+} changes (Cullen and

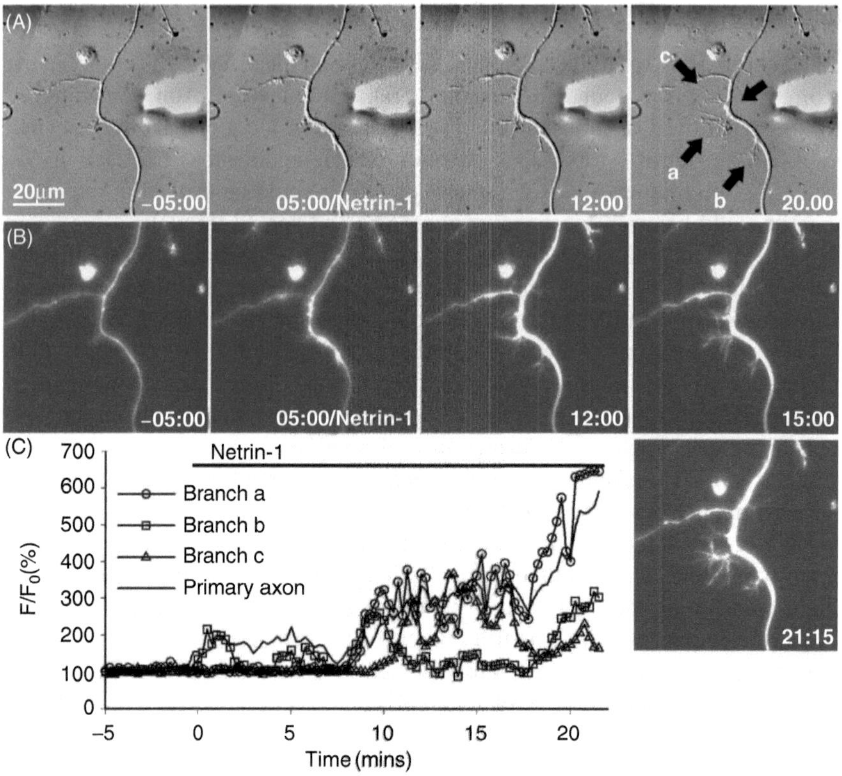

FIGURE 12.7. Localized netrin-1 application evokes repetitive calcium transients that are simultaneous with development of axon branches in the same region of the axon. (A) DIC time-lapse images showing protrusion of filopodia in response to local application of netrin-1. Arrows indicate three different branches. The branch at the arrow begins as a short filopodium but by 20 min has elongated and developed a complex growth cone. (B) The corresponding time-lapse fluorescence images show high levels of repetitive calcium transients that coincide with newly developing branches. Calcium activity begins prior to filopodial protrusion and continues during development of branches over 20 min. (C) Measurements of calcium transients in the primary axon and the three branches corresponding to lines of the arrows. The growth cone developing from the region of the arrow **a** shows the largest changes in calcium activity. (With permission from Tang and Kalil, 2005. Copyright 2005 by the Society for Neuroscience.)

Lockyer, 2002). Ca^{2+} oscillations are self-renewing, repetitive, and persistent. Moreover, the mechanism of frequency modulation to decode information from calcium signals has advantages over mechanisms based on changes in amplitude of Ca^{2+} concentration alone (Cullen and Lockyer, 2002) because discrete Ca^{2+} oscillations can be more easily distinguished by calcium decoders such as CaMKII (De Koninck and Schulman, 1998) and PKC (Oancea and Meyer, 1998).

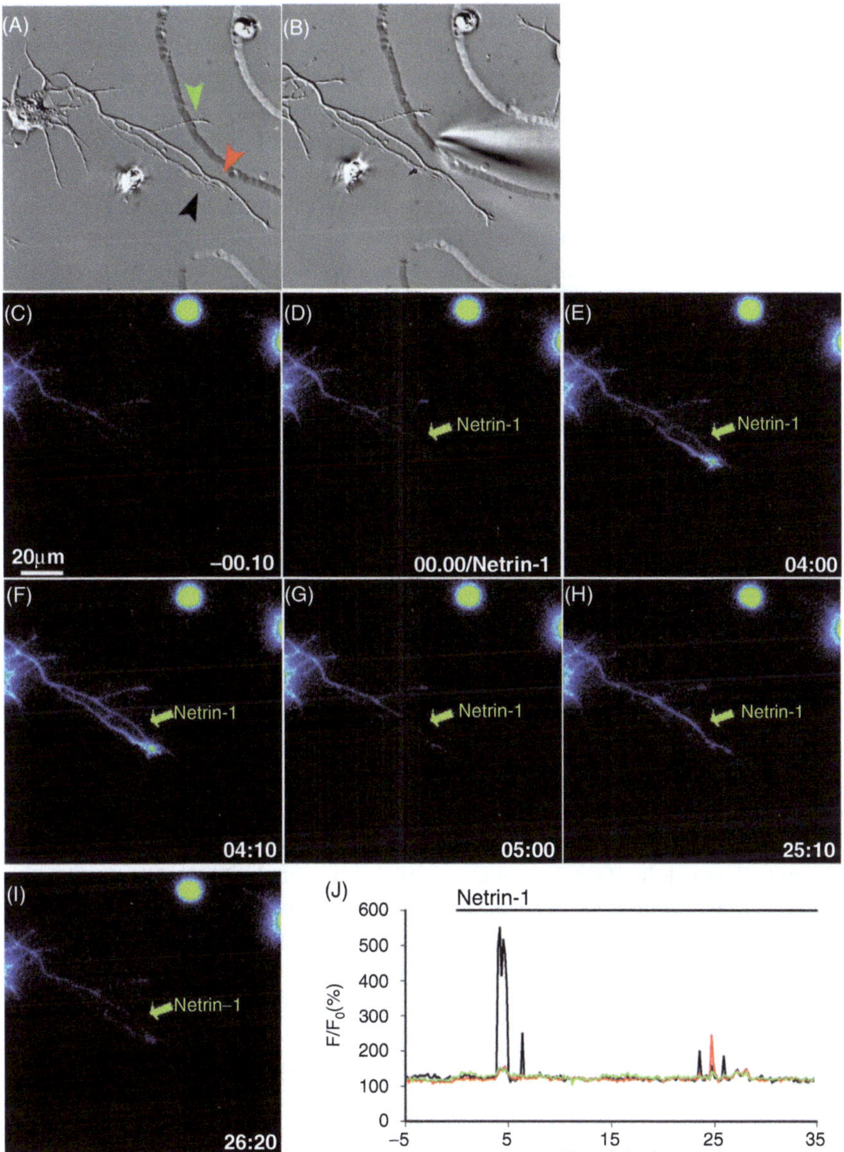

FIGURE 12.8. Repetitive calcium transients in cortical axons are evoked by netrin-1 application within minutes. (A and B) DIC images and (C–I) time-lapse images in pseudocolor showing changes in calcium activity in an axon (red arrowhead) and its two branches (green and black arrowheads) after local application of netrin-1 from the pipette shown in (B) and indicated by the green arrows in the fluorescence images. Localized calcium transients occur in one branch at 4 min, (E), in another smaller branch at 4:10, (F), and in the primary axon at 25 min, (H). (J) Measurements of calcium transients in the axon (red), a large branch (black) and a small branch (green) corresponding to the arrows in the DIC image in (A). Calcium transients induced by netrin-1 can occur at different frequencies and different times in an axon and its branches.

The downstream targets of Ca^{2+} signaling involved in axon branching were studied with methods involving pharmacological inhibition, overexpression, transfection with mutagenized constructs, and RNAi knockdown. Results of these experiments in cortical neurons (Tang and Kalil, 2005) showed that CaMKII and MAPK, both of which are known to be influenced by frequencies of Ca^{2+} transients in functions related to memory and synaptic plasticity, are major components of the netrin-1 calcium signaling pathway essential for axon branching. CaMKII promotes axon outgrowth as well as axon branching whereas MAPK affects only axon branching but not axon outgrowth. Activation of specific kinases could be a mechanism whereby netrin-1 Ca^{2+} signaling promotes axon branching independent of axon outgrowth. Calcium may activate different targets to elicit opposing effects on the axon such as inhibiting axon outgrowth but promoting the development of new branches.

CaMKII is known to play an important role in neurite outgrowth, growth cone motility (Goshima et al., 1993), and growth cone guidance (Kuhn et al., 1998; Zheng et al., 1994). Wen et al. (2004) showed that pharmacological inhibition of CaMKII prevented netrin-1 induced repulsive growth cone turning and that local high increases in the amplitude of Ca^{2+} signals induced turning behaviors. CaMKII and calcineurin phosphatase was shown to act as a switch to control the direction of Ca^{2+} dependent growth cone turning. Relatively large localized Ca^{2+} elevations activate CaMKII to cause attraction whereas lower Ca^{2+} signals induce repulsion through calcineurin and phosphatase-1. Others have found that CaMKI but not CaMKII promotes axon extension and growth cone motility (Wayman et al., 2004). Similarly, both positive (Fink et al., 2003; Gaudilliere et al., 2004; Jourdain et al., 2003) and inhibitory (Redmond et al., 2002; Wu and Cline, 1998) effects for CaMKII have been described for dendritic growth and branching. MAPKs have also been shown to influence netrin-1 induced axon guidance (Campbell and Holt, 2003; Forcet et al., 2002; Ming et al., 2002). Consistent with these studies, results on cortical neurons (Tang and Kalil, 2005) show that whereas knockdown of CaMKII inhibits axon outgrowth as well as axon branching, MAPK specifically affects axon branching but not axon outgrowth. Overexpression of CaMKII in the presence of MAPK inhibitors greatly reduces axon branching suggesting that CaMKII can activate MAPKs (Chen et al., 1998), although interactions between these signaling pathways are not well understood.

12.4.5. Effects of Calcium Signaling Pathways on the Cytoskeleton

Ultimately signaling pathways activated by guidance cues must converge on the actin–microtubule cytoskeleton to induce growth cone turning behaviors and axon branching (Dent and Gertler, 2003). As we have discussed, axon branching requires dynamic actin–microtubule interactions (Dent and Kalil, 2001) and netrin-1 elicits filopodial protrusion through actin polymerization (Dent et al., 2004). Netrin-1 is known to activate Ena/VASP proteins (Lebrand et al., 2004) and Rho GTPases which regulate actin polymerization (Li et al., 2002; Shekarabi and Kennedy, 2002) as well as activating microtubule associated protein (MAP1B) which is thought to play a role in actin–microtubule interactions (Del Rio et al., 2004). Thus netrin-1 induced calcium signaling could influence regulators of cytoskeletal dynamics required for axon branching. CaMKII is a major target of netrin-1 induced calcium signaling (Tang and Kalil, 2005). Studies of regulation of dendritic morphology and plasticity suggest that CaMKII may be an important link to the cytoskeleton. In dendritic filopodia, resembling spines, αCaMKII was shown to interact with βCaMKII and translocate to dendritic spines thereby targeting CaMKII to the actin cytoskeleton (Shen et al., 1998). This is consistent with the role of CaMKII in stabilizing dendritic arbors (Wu and Cline, 1998). Although these studies do not identify an exact role for CaMKII in regulating actin polymerization, there is evidence (Chen et al., 2003) that CaMKII can activate Cdc42, which promotes polymerization of actin during formation of filopodia. Thus repetitive calcium transients induced by netrin-1 could provide a mechanism for translocating CaMKII to actin filaments in nascent axonal filopodia. An important unanswered question is exactly how major targets of calcium signaling, such as CaMKII and MAPK, regulate cytoskeletal dynamics required for development of new branches.

12.5. Activity-Dependent Regulation of Axon Branching

The involvement of Ca^{2+} signaling, a measure of electrical activity, suggests a role for activity dependent mechanisms in the regulation of axon branching (Tang and Kalil, 2005). During development, electrical activity shapes the growth and branching of axon arbors and thus the formation of neural circuits (Katz and Shatz, 1996; Zhang and Poo, 2001). Several studies have investigated some of the mechanisms whereby activity shapes neural connectivity in visual system targets. For example, *in vivo* (Chandrasekaran et al., 2005) the importance of electrical activity in concert with molecular cues was demonstrated in refining the retinoptopic map. In the absence of spontaneous retinal waves the refinement of retinal ganglion axon arbors into a precise map in the superior colliculus is degraded. In newborn rat visual cortical slices, imaging of axons extending horizontally in layers 2/3 of the cortex combined with electrophysiological recording (Uesaka et al., 2005) showed that the complexity of axon branching gradually increased over time with a concomitant dramatic increase in frequency of spontaneous firing. When spontaneous action potential firing was silenced with TTX or synaptic transmission was blocked with pharmacological manipulation of glutaminergic transmission, the numbers of axon branches fell to about half that of control slices. These results suggest that neural activity regulates the early development of branching in cortical neurons.

Activity can shape axon arbors by regulating retraction or elimination of axon branches (Luo and O'Leary, 2005) but evidence has shown that activity also controls the extent and rate of branch formation. In an *in vivo* study imaging of retinal axon arbors in the developing optic tectum of zebrafish revealed how suppression of activity influences axon branching (Hua et al., 2005). By individually transfecting retinal ganglion cells with a specific type of potassium (K^+) channel it was possible to electrically silence the transfected cells by hyperpolarization. Overexpression of the inward rectifier K^+ channel was found to suppress calcium spiking associated with spontaneous electrical activity. In these neurons arbor length and complexity are reduced in comparison with adjacent untransfected cells with normal electrical activity. These results show that competitive mechanisms based on relative activity influence the development of functional networks through growth of axon arbors. Importantly the competition effect is expressed in the formation of new axon branches rather than on the stability of branches. Studies of axon branching *in vitro* also support a model of competitive growth of branches based on electrical activity. In cultures of sympathetic neurons (Singh and Miller, 2005) in the presence of NGF local depolarization of axon collaterals enhances their growth whereas unstimulated inactive axon collaterals from the same or neighboring neurons are at a competitive disadvantage for growth. Depolarization by patterned electrical stimulation, resulting in pulses of increased intracellular calcium, mediates this effect through the activation of a CaMKII-MEK signaling pathway. These findings provide a mechanism for selection of one axon branch over another through electrical activity that promotes the growth of one set of axon branches while inhibiting unstimulated axon collaterals. The CaMKII and MAPK signaling pathways are also essential for netrin-1 induced axon branching of cortical neurons (Tang and Kalil, 2005). Similarly, the ability of cortical axon collaterals to exhibit calcium transients of different amplitudes and frequencies could also provide an activity dependent mechanism for the competitive growth of one axon branch over another.

12.6. Conclusions

Axon branching is essential for establishing neural connectivity. Branches can arise along the length of the axon as interstitial collaterals as well as forming terminal arbors. Although pathfinding by axonal growth cones and branching from the axon shaft are different processes, both serve to guide axons in new directions toward their targets. Thus it is not surprising that growth cone guidance and axon branching involve similar extracellular guidance cues, cytoskeletal dynamics and intracellular calcium signaling pathways. The terminal growth cone and developing axon branches are often at long distances from each other and from the cell body. Therefore, as we have discussed, many of the cytoskeletal and signaling events that cause the growth cone to turn or a branch to form are highly localized. Thus different regions of the axon, in response to local target-derived attractive cues, can respond by extending a branch in exactly the right place to establish appropriate connectivity. In other regions of the axon, inhibitory cues could function to prevent branching to inappropriate targets. Localized calcium signals, which may be activated by growth and guidance molecules, also play an important role in activity dependent regulation of axon branching. Although it is not completely understood how these complex branching events are orchestrated during development or how extension of the primary axon is regulated independent of axon branching, it seems likely that these localized signaling mechanisms can elicit local changes along the axon to promote the growth of axon collaterals. At present it is not well understood how extracellular molecules, acting through their membrane receptors, influence intracellular signaling pathways that ultimately regulate cytoskeletal dynamics. These events rapidly transform the relatively static axon shaft to motile protrusions leading to branch formation. Similar mechanisms might promote axon sprouting after injury in the mature nervous system.

Acknowledgments. This work was supported by National Institutes of Health (NIH) grant NS14428 and a grant from the Whitehall Foundation to K. Kalil and NIH F32-NS045366 to E.W. Dent.

References

Bagnard, D., Lohrum, M., Uziel, D., Puschel, A.W., and Bolz, J., 1998, Semaphorins act as attractive and repulsive guidance signals during the development of cortical projections, *Development* **125:** 5043–5053.

Bagri, A., Cheng, H.J., Yaron, A., Pleasure, S.J., and Tessier-Lavigne, M., 2003, Stereotyped pruning of long hippocampal axon branches triggered by retraction inducers of the semaphorin family, *Cell* **113:** 285–299.

Bastmeyer, M., and O'Leary, D.D., 1996, Dynamics of target recognition by interstitial axon branching along developing cortical axons, *J. Neurosci.* **16:** 1450–1459.

Bear, J.E., Krause, M., and Gertler, F.B., 2001, Regulating cellular actin assembly, *Curr. Opin. Cell Biol.* **13:** 158–166.

Bear, J.E., Svitkina, T.M., Krause, M., Schafer, D.A., Loureiro, J.J., Strasser, G.A., et al., 2002, Antagonism between Ena/VASP proteins and actin filament capping regulates fibroblast motility, *Cell* **109:** 509–521.

Bentley, D., and O'Connor, T.P., 1994, Cytoskeletal events in growth cone steering, *Curr. Opin. Neurobiol.* **4:** 43–48.

Berridge, M.J., Bootman, M.D., and Roderick, H.L., 2003, Calcium signalling: Dynamics, homeostasis and remodeling, *Nat. Rev. Mol. Cell Biol.* **4:** 517–529.

Bolsover, S.R., 2005, Calcium signalling in growth cone migration, *Cell Calcium* **37:** 395–402.

Borodinsky, L.N., Root, C.M., Cronin, J.A., Sann, S.B., Gu, X., and Spitzer, N.C., 2004, Activity-dependent homeostatic specification of transmitter expression in embryonic neurons, *Nature* **429:** 523–530.

Brose, K., and Tessier-Lavigne, M., 2000, Slit proteins: Key regulators of axon guidance, axonal branching, and cell migration, *Curr. Opin. Neurobiol.* **10:** 95–102.

Brose, K., Bland, K.S., Wang, K.H., Arnott, D., Henzel, W., Goodman, C.S., et al., 1999, Slit proteins bind Robo receptors and have an evolutionarily conserved role in repulsive axon guidance, *Cell* **96:** 795–806.

Buck, K.B., and Zheng, J.Q., 2002, Growth cone turning induced by direct local modification of microtubule dynamics, *J. Neurosci.* **22:** 9358–9367.

Campbell, D.S., and Holt, C.E., 2003, Apoptotic pathway and MAPKs differentially regulate chemotropic responses of retinal growth cones, *Neuron* **37:** 939–952.

Campbell, D.S., Regan, A.G., Lopez, J.S., Tannahill, D., Harris, W.A., and Holt, C.E., 2001, Semaphorin 3A elicits stage-dependent collapse, turning, and branching in Xenopus retinal growth cones, *J. Neurosci.* **21:** 8538–8547.

Chandrasekaran, A.R., Plas, D.T., Gonzalez, E., and Crair, M.C., 2005, Evidence for an instructive role of retinal activity in retinotopic map refinement in the superior colliculus of the mouse, *J. Neurosci.* **25:** 6929–6938.

Chen, H.J., Rojas-Soto, M., Oguni, A., and Kennedy, M.B., 1998, A synaptic Ras-GTPase activating protein (p135 SynGAP) inhibited by CaM kinase II, *Neuron* **20:** 895–904.

Chen, N., Furuya, S., Doi, H., Hashimoto, Y., Kudo, Y., and Higashi, H., 2003, Ganglioside/calmodulin kinase II signal inducing cdc42-mediated neuronal actin reorganization, *Neuroscience* **120:** 163–176.

Cohen-Cory, S., and Fraser, S.E., 1995, Effects of brain-derived neurotrophic factor on optic axon branching and remodelling in vivo, *Nature* **378:** 192–196.

Cullen, P.J., and Lockyer, P.J., 2002, Integration of calcium and Ras signaling, *Nat. Rev. Mol. Cell Biol.* **3:** 339–348.

Davenport, R.W., and Kater, S.B., 1992, Local increases in intracellular calcium elicit local filopodial responses in Helisoma neuronal growth cones, *Neuron* **9:** 405–416.

Davenport, R.W., Thies, E., and Cohen, M.L., 1999, Neuronal growth cone collapse triggers lateral extensions along trailing axons, *Nat. Neurosci.* **2:** 254–259.

De Koninck, P., and Schulman, H., 1998, Sensitivity of CaM kinase II to the frequency of Ca2+ oscillations, *Science* **279:** 227–230.

Del Rio, J.A., Gonzalez-Billault, C., Urena, J.M., Jimenez, E.M., Barallobre, M.J., Pascual, M., et al., 2004, MAP1B is required for Netrin 1 signaling in neuronal migration and axonal guidance, *Curr. Biol.* **14:** 840–850.

Dent, E.W., and Gertler, F.B., 2003, Cytoskeletal dynamics and transport in growth cone motility and axon guidance, *Neuron* **40:** 209–227.

Dent, E.W., and Kalil, K., 2001, Axon branching requires interactions between dynamic microtubules and actin filaments, *J. Neurosci.* **21:** 9757–9769.

Dent, E.W., Callaway, J.L., Szebenyi, G., Baas, P.W., and Kalil, K., 1999, Reorganization and movement of microtubules in axonal growth cones and developing interstitial branches, *J. Neurosci.* **19:** 8894–8908.

Dent, E.W., Tang, F., and Kalil, K., 2003, Axon guidance by growth cones and branches: Common cytoskeletal and signaling mechanisms, *Neuroscientist* **9:** 343–353.

Dent, E.W., Barnes, A.M., Tang, F., and Kalil, K., 2004, Netrin-1 and semaphorin 3A promote or inhibit cortical axon branching, respectively, by reorganization of the cytoskeleton, *J. Neurosci.* **24:** 3002–3012.

Dickson, B.J., 2002, Molecular mechanisms of axon guidance, *Science* **298:** 1959–1964.

Eshete, F., and Fields, R.D., 2001, Spike frequency decoding and autonomous activation of Ca2+-calmodulin-dependent protein kinase II in dorsal root ganglion neurons, *J. Neurosci.* **21:** 6694–6705.

Fan, J., Mansfield, S.G., Redmond, T., Gordon-Weeks, P.R., and Raper, J.A., 1993, The organization of F-actin and microtubules in growth cones exposed to a brain-derived collapsing factor, *J. Cell Biol.* **121:** 867–878.

Fink, C.C., Bayer, K.U., Myers, J.W., Ferrell, J.E., Jr., Schulman, H., and Meyer, T., 2003, Selective regulation of neurite extension and synapse formation by the beta but not the alpha isoform of CaMKII, *Neuron* **39:** 283–297.

Forcet, C., Stein, E., Pays, L., Corset, V., Llambi, F., Tessier-Lavigne, M., et al., 2002, Netrin-1-mediated axon outgrowth requires deleted in colorectal cancer-dependent MAPK activation, *Nature* **417:** 443–447.

Forscher, P., and Smith, S.J., 1988, Actions of cytochalasins on the organization of actin filaments and microtubules in a neuronal growth cone, *J. Cell Biol.* **107:** 1505–1516.

Fournier, A.E., Nakamura, F., Kawamoto, S., Goshima, Y., Kalb, R.G., and Strittmatter, S.M., 2000, Semaphorin3A enhances endocytosis at sites of receptor-F-actin colocalization during growth cone collapse, *J. Cell Biol.* **149:** 411–422.

Fritsche, J., Reber, B.F., Schindelholz, B., and Bandtlow, C.E., 1999, Differential cytoskeletal changes during growth cone collapse in response to hSema III and thrombin, *Mol. Cell Neurosci.* **14:** 398–418.

Gallo, G., and Letourneau, P.C., 1998, Localized sources of neurotrophins initiate axon collateral sprouting, *J. Neurosci.* **18:** 5403–5414.

Gallo, G., and Letourneau, P.C., 1999, Different contributions of microtubule dynamics and transport to the growth of axons and collateral sprouts, *J. Neurosci.* **19:** 3860–3873.

Gaudilliere, B., Konishi, Y., de la Iglesia, N., Yao, G., and Bonni, A., 2004, A CaMKII-NeuroD signaling pathway specifies dendritic morphogenesis, *Neuron* **41:** 229–241.

Gehler, S., Shaw, A.E., Sarmiere, P.D., Bamburg, J.R., and Letourneau, P.C., 2004, Brain-derived neurotrophic factor regulation of retinal growth cone filopodial dynamics is mediated through actin depolymerizing factor/cofilin, *J. Neurosci.* **24:** 10741–10749.

Gomez, T.M., and Spitzer, N.C., 1999, In vivo regulation of axon extension and pathfinding by growth-cone calcium transients, *Nature* **397:** 350–355.

Gomez, T.M., and Spitzer, N.C., 2000, Regulation of growth cone behavior by calcium: New dynamics to earlier perspectives, *J. Neurobiol.* **44:** 174–183.

Gomez, T.M., Robles, E., Poo, M., and Spitzer, N.C., 2001, Filopodial calcium transients promote substrate-dependent growth cone turning, *Science* **291:** 1983–1987.

Goshima, Y., Ohsako, S., and Yamauchi, T., 1993, Overexpression of Ca2+/calmodulin-dependent protein kinase II in Neuro2a and NG108–15 neuroblastoma cell lines promotes neurite outgrowth and growth cone motility, *J. Neurosci.* **13:** 559–567.

Gu, X., and Spitzer, N.C., 1995, Distinct aspects of neuronal differentiation encoded by frequency of spontaneous Ca2+ transients, *Nature* **375:** 784–787.

Halloran, M.C., and Kalil, K., 1994, Dynamic behaviors of growth cones extending in the corpus callosum of living cortical brain slices observed with video microscopy, *J. Neurosci.* **14:** 2161–2177.

Henley, J., and Poo, M.M., 2004, Guiding neuronal growth cones using Ca2+ signals, *Trends Cell Biol.* **14:** 320–330.

Homma, N., Takei, Y., Tanaka, Y., Nakata, T., Terada, S., Kikkawa, M., et al., 2003, Kinesin superfamily protein 2A (KIF2A) functions in suppression of collateral branch extension, *Cell* **114:** 229–239.

Hong, K., Nishiyama, M., Henley, J., Tessier-Lavigne, M., and Poo, M., 2000, Calcium signalling in the guidance of nerve growth by netrin-1, *Nature* **403:** 93–98.

Hu, S., and Reichardt, L.F., 1999, From membrane to cytoskeleton: Enabling a connection, *Neuron* **22:** 419–422.

Hua, J.Y., Smear, M.C., Baier, H., and Smith, S.J., 2005, Regulation of axon growth in vivo by activity-based competition, *Nature* **434:** 1022–1026.

Huber, A.B., Kolodkin, A.L., Ginty, D.D., and Cloutier, J.F., 2003, Signaling at the growth cone: Ligand-receptor complexes and the control of axon growth and guidance, *Annu. Rev. Neurosci.* **26:** 509–563.

Hudmon, A., and Schulman, H., 2002, Neuronal CA2+/calmodulin-dependent protein kinase II: The role of structure and autoregulation in cellular function, *Annu. Rev. Biochem.* **71:** 473–510.

Jourdain, P., Fukunaga, K., and Muller, D., 2003, Calcium/calmodulin-dependent protein kinase II contributes to activity-dependent filopodia growth and spine formation, *J. Neurosci.* **23:** 10645–10649.

Kalil, K., and Dent, E.W., 2005, Touch and go: Guidance cues signal to the growth cone cytoskeleton, *Curr. Opin. Neurobiol.* **15:** 521–526.

Kalil, K., Szebenyi, G., and Dent, E.W., 2000, Common mechanisms underlying growth cone guidance and axon branching, *J. Neurobiol.* **44:** 145–158.

Kantor, D.B., and Kolodkin, A.L., 2003, Curbing the excesses of youth: Molecular insights into axonal pruning, *Neuron* **38:** 849–852.

Kater, S.B., and Mills, L.R., 1991, Regulation of growth cone behavior by calcium, *J. Neurosci.* **11:** 891–899.

Katz, L.C., and Shatz, C.J., 1996, Synaptic activity and the construction of cortical circuits, *Science* **274:** 1133–1138.

Kennedy, T.E., and Tessier-Lavigne, M., 1995, Guidance and induction of branch formation in developing axons by target-derived diffusible factors, *Curr. Opin. Neurobiol.* **5:** 83–90.

Korey, C.A., and Van Vactor, D., 2000, From the growth cone surface to the cytoskeleton: One journey, many paths, *J. Neurobiol.* **44:** 184–193.

Kornack, D.R., and Giger, R.J., 2005, Probing microtubule +TIPs: Regulation of axon branching, *Curr. Opin. Neurobiol.* **15:** 58–66.

Kuang, R.Z., and Kalil, K., 1994, Development of specificity in corticospinal connections by axon collaterals branching selectively into appropriate spinal targets, *J. Comp. Neurol.* **344:** 270–282.

Kuhn, T.B., Williams, C.V., Dou, P., and Kater, S.B., 1998, Laminin directs growth cone navigation via two temporally and functionally distinct calcium signals, *J. Neurosci.* **18:** 184–194.

Lanier, L.M., and Gertler, F.B., 2000, Actin cytoskeleton: Thinking globally, actin' locally, *Curr. Biol.* **10:** R655–657.

Lanier, L.M., Gates, M.A., Witke, W., Menzies, A.S., Wehman, A.M., Macklis, J.D., et al., 1999, Mena is required for neurulation and commissure formation, *Neuron* **22:** 313–325.

Lau, P.M., Zucker, R.S., and Bentley, D., 1999, Induction of filopodia by direct local elevation of intracellular calcium ion concentration, *J. Cell Biol.* **145:** 1265–1275.

Lebrand, C., Dent, E.W., Strasser, G.A., Lanier, L.M., Krause, M., Svitkina, T.M., et al., 2004, Critical role of Ena/VASP proteins for filopodia formation in neurons and in function downstream of netrin-1, *Neuron* **42:** 37–49.

Lee, H., Engel, U., Rusch, J., Scherrer, S., Sheard, K., and Van Vactor, D., 2004, The microtubule plus end tracking protein Orbit/MAST/CLASP acts downstream of the tyrosine kinase Abl in mediating axon guidance, *Neuron* **42:** 913–926.

Letourneau, P.C., Snow, D.M., and Gomez, T.M., 1994, Regulation of growth cone motility by substratum bound molecules and cytoplasmic [Ca^{2+}], *Prog. Brain Res.* **103:** 85–98.

Li, X., Saint-Cyr-Proulx, E., Aktories, K., and Lamarche-Vane, N., 2002, Rac1 and Cdc42 but not RhoA or Rho kinase activities are required for neurite outgrowth induced by the Netrin-1 receptor DCC (deleted in colorectal cancer) in N1E-115 neuroblastoma cells, *J. Biol. Chem.* **277:** 15207–15214.

Lin, C.H., and Forscher, P., 1993, Cytoskeletal remodeling during growth cone-target interactions, *J. Cell Biol.* **121:** 1369–1383.

Lin, C.H., Thompson, C.A., and Forscher, P., 1994, Cytoskeletal reorganization underlying growth cone motility, *Curr. Opin. Neurobiol.* **4:** 640–647.

Liu, B.P., and Strittmatter, S.M., 2001, Semaphorin-mediated axonal guidance via Rho-related G proteins, *Curr. Opin. Cell Biol.* **13:** 619–626.

Lom, B., and Cohen-Cory, S., 1999, Brain-derived neurotrophic factor differentially regulates retinal ganglion cell dendritic and axonal arborization in vivo, *J. Neurosci.* **19:** 9928–9938.

Luo, L., 2002, Actin cytoskeleton regulation in neuronal morphogenesis and structural plasticity, *Annu. Rev. Cell Dev. Biol.* **18:** 601–635.

Luo, L., and O'Leary, D.D., 2005, Axon retraction and degeneration in development and disease, *Annu. Rev. Neurosci.* **28:** 127–156.

Machesky, L.M., and Insall, R.H., 1999, Signaling to actin dynamics, *J. Cell Biol.* **146:** 267–272.

Manitt, C., and Kennedy, T.E., 2002, Where the rubber meets the road: Netrin expression and function in developing and adult nervous systems, *Prog. Brain Res.* **137:** 425–442.

Metin, C., Deleglise, D., Serafini, T., Kennedy, T.E., and Tessier-Lavigne, M., 1997, A role for netrin-1 in the guidance of cortical efferents, *Development* **124:** 5063–5074.

Ming, G.L., Wong, S.T., Henley, J., Yuan, X.B., Song, H.J., Spitzer, N.C., et al., 2002, Adaptation in the chemotactic guidance of nerve growth cones, *Nature* **417:** 411–418.

Ng, J., Nardine, T., Harms, M., Tzu, J., Goldstein, A., Sun, Y., et al., 2002, Rac GTPases control axon growth, guidance and branching, *Nature* **416:** 442–447.

Nishiyama, M., Hoshino, A., Tsai, L., Henley, J.R., Goshima, Y., Tessier-Lavigne, M., et al., 2003, Cyclic AMP/GMP-dependent modulation of Ca2+ channels sets the polarity of nerve growth-cone turning, *Nature* **423:** 990–995.

O'Connor, T.P., and Bentley, D., 1993, Accumulation of actin in subsets of pioneer growth cone filopodia in response to neural and epithelial guidance cues in situ, *J. Cell Biol.* **123:** 935–948.

O'Leary, D.D., and Terashima, T., 1988, Cortical axons branch to multiple subcortical targets by interstitial axon budding: Implications for target recognition and "waiting periods," *Neuron* **1:** 901–910.

O'Leary, D.D., Bicknese, A.R., De Carlos, J.A., Heffner, C.D., Koester, S.E., Kutka, L.J., et al., 1990, Target selection by cortical axons: Alternative mechanisms to establish axonal connections in the developing brain, *Cold Spring Harb. Symp. Quant. Biol.* **55:** 453–468.

Oancea, E., and Meyer, T., 1998, Protein kinase C as a molecular machine for decoding calcium and diacylglycerol signals, *Cell* **95:** 307–318.

Pantaloni, D., Le Clainche, C., and Carlier, M.F., 2001, Mechanism of actin-based motility, *Science* **292:** 1502–1506.

Pasterkamp, R.J., Peschon, J.J., Spriggs, M.K., and Kolodkin, A.L., 2003, Semaphorin 7A promotes axon outgrowth through integrins and MAPKs, *Nature* **424:** 398–405.

Plump, A.S., Erskine, L., Sabatier, C., Brose, K., Epstein, C.J., Goodman, C.S., et al., 2002, Slit1 and Slit2 cooperate to prevent premature midline crossing of retinal axons in the mouse visual system, *Neuron* **33:** 219–232.

Polleux, F., Giger, R.J., Ginty, D.D., Kolodkin, A.L., and Ghosh, A., 1998, Patterning of cortical efferent projections by semaphorin-neuropilin interactions, *Science* **282:** 1904–1906.

Polleux, F., Morrow, T., and Ghosh, A., 2000, Semaphorin 3A is a chemoattractant for cortical apical dendrites, *Nature* **404:** 567–573.

Rashid, T., Upton, A.L., Blentic, A., Ciossek, T., Knoll, B., Thompson, I.D., et al., 2005, Opposing gradients of ephrin-As and EphA7 in the superior colliculus are essential for topographic mapping in the mammalian visual system, *Neuron* **47:** 57–69.

Redmond, L., Kashani, A.H., and Ghosh, A., 2002, Calcium regulation of dendritic growth via CaM kinase IV and CREB-mediated transcription, *Neuron* **34:** 999–1010.

Richards, L.J., Koester, S.E., Tuttle, R., and O'Leary, D.D., 1997, Directed growth of early cortical axons is influenced by a chemoattractant released from an intermediate target, *J. Neurosci.* **17:** 2445–2458.

Rodriguez, O.C., Schaefer, A.W., Mandato, C.A., Forscher, P., Bement, W.M., and Waterman-Storer, C.M., 2003, Conserved microtubule-actin interactions in cell movement and morphogenesis, *Nat. Cell Biol.* **5:** 599–609.

Roos, J., Hummel, T., Ng, N., Klambt, C., and Davis, G.W., 2000, Drosophila Futsch regulates synaptic microtubule organization and is necessary for synaptic growth, *Neuron* **26:** 371–382.

Salmon, W.C., Adams, M.C., and Waterman-Storer, C.M., 2002, Dual-wavelength fluorescent speckle microscopy reveals coupling of microtubule and actin movements in migrating cells, *J. Cell Biol.* **158:** 31–37.

Schaefer, A.W., Kabir, N., and Forscher, P., 2002, Filopodia and actin arcs guide the assembly and transport of two populations of microtubules with unique dynamic parameters in neuronal growth cones, *J. Cell Biol.* **158:** 139–152.

Schwab, M.E., 2002, Increasing plasticity and functional recovery of the lesioned spinal cord, *Prog. Brain Res.* **137:** 351–359.

Serafini, T., Colamarino, S.A., Leonardo, E.D., Wang, H., Beddington, R., Skarnes, W.C., et al., 1996, Netrin-1 is required for commissural axon guidance in the developing vertebrate nervous system, *Cell* **87:** 1001–1014.

Shekarabi, M., and Kennedy, T.E., 2002, The netrin-1 receptor DCC promotes filopodia formation and cell spreading by activating Cdc42 and Rac1, *Mol. Cell Neurosci.* **19:** 1–17.

Shen, K., Teruel, M.N., Subramanian, K., and Meyer, T., 1998, CaMKIIbeta functions as an F-actin targeting module that localizes CaMKIIalpha/beta heterooligomers to dendritic spines, *Neuron* **21:** 593–606.

Shu, T., and Richards, L.J., 2001, Cortical axon guidance by the glial wedge during the development of the corpus callosum, *J. Neurosci.* **21:** 2749–2758.

Silver, R.A., Lamb, A.G., and Bolsover, S.R., 1990, Calcium hotspots caused by L-channel clustering promote morphological changes in neuronal growth cones, *Nature* **343:** 751–754.

Simon, D.K., and O'Leary, D.D., 1992, Responses of retinal axons in vivo and in vitro to position-encoding molecules in the embryonic superior colliculus, *Neuron* **9:** 977–989.

Singh, K.K., and Miller, F.D., 2005, Activity regulates positive and negative neurotrophin-derived signals to determine axon competition, *Neuron* **45:** 837–845.

Small, J.V., Stradal, T., Vignal, E., and Rottner, K., 2002, The lamellipodium: Where motility begins, *Trends Cell Biol.* **12:** 112–120.

Szebenyi, G., Callaway, J.L., Dent, E.W., and Kalil, K., 1998, Interstitial branches develop from active regions of the axon demarcated by the primary growth cone during pausing behaviors, *J. Neurosci.* **18:** 7930–7940.

Szebenyi, G., Dent, E.W., Callaway, J.L., Seys, C., Lueth, H., and Kalil, K., 2001, Fibroblast growth factor-2 promotes axon branching of cortical neurons by influencing morphology and behavior of the primary growth cone, *J. Neurosci.* **21:** 3932–3941.

Tanaka, E.M., and Kirschner, M.W., 1991, Microtubule behavior in the growth cones of living neurons during axon elongation, *J. Cell Biol.* **115:** 345–363.

Tanaka, E., and Sabry, J., 1995, Making the connection: Cytoskeletal rearrangements during growth cone guidance, *Cell* **83:** 171–176.

Tang, F., and Kalil, K., 2005, Netrin-1 induces axon branching in developing cortical neurons by frequency-dependent calcium signaling pathways, *J. Neurosci.* **25:** 6702–6715.

Tang, F., Dent, E.W., and Kalil, K., 2003, Spontaneous calcium transients in developing cortical neurons regulate axon outgrowth, *J. Neurosci.* **23:** 927–936.

Tessier-Lavigne, M., and Goodman, C.S., 1996, The molecular biology of axon guidance, *Science* **274:** 1123–1133.

Tomida, T., Hirose, K., Takizawa, A., Shibasaki, F., and Iino, M., 2003, NFAT functions as a working memory of Ca2+ signals in decoding Ca2+ oscillation, *EMBO J.* **22:** 3825–3832.

Uesaka, N., Hirai, S., Maruyama, T., Ruthazer, E.S., and Yamamoto, N., 2005, Activity dependence of cortical axon branch formation: A morphological and electrophysiological study using organotypic slice cultures, *J. Neurosci.* **25:** 1–9.

Wang, K.H., Brose, K., Arnott, D., Kidd, T., Goodman, C.S., Henzel, W., et al., 1999, Biochemical purification of a mammalian slit protein as a positive regulator of sensory axon elongation and branching, *Cell* **96:** 771–784.

Watts, R.J., Hoopfer, E.D., and Luo, L., 2003, Axon pruning during Drosophila metamorphosis: Evidence for local degeneration and requirement of the ubiquitin-proteasome system, *Neuron* **38:** 871–885.

Wayman, G.A., Kaech, S., Grant, W.F., Davare, M., Impey, S., Tokumitsu, H., et al., 2004, Regulation of axonal extension and growth cone motility by calmodulin-dependent protein kinase I, *J. Neurosci.* **24:** 3786–3794.

Wen, Z., Guirland, C., Ming, G.L., and Zheng, J.Q., 2004, A CaMKII/calcineurin switch controls the direction of Ca(2+)-dependent growth cone guidance, *Neuron* **43:** 835–846.

Whitford, K.L., Dijkhuizen, P., Polleux, F., and Ghosh, A., 2002, Molecular control of cortical dendrite development, *Annu. Rev. Neurosci.* **25:** 127–149.

Wills, Z., Bateman, J., Korey, C.A., Comer, A., and Van Vactor, D., 1999, The tyrosine kinase Abl and its substrate enabled collaborate with the receptor phosphatase Dlar to control motor axon guidance, *Neuron* **22:** 301–312.

Wu, G.Y., and Cline, H.T., 1998, Stabilization of dendritic arbor structure in vivo by CaMKII, *Science* **279:** 222–226.

Yamamoto, N., Higashi, S., and Toyama, K., 1997, stop and branch behaviors of geniculocortical axons: a time-lapse study in organotypic cultures, *J. Neurosc:* **17:** 3653–3663.

Yates, P.A., Roskies, A.L., McLaughlin, T., and O'Leary, D.D., 2001, Topographic-specific axon branching controlled by ephrin-As is the critical event in retinotectal map development, *J. Neurosci.* **21:** 8548–8563.

Zhang, L.I., and Poo, M.M., 2001, Electrical activity and development of neural circuits, *Nat. Neurosci.* **4(Suppl.):** 1207–1214.

Zheng, J.Q., 2000, Turning of nerve growth cones induced by localized increases in intracellular calcium ions, *Nature* **403:** 89–93.

Zheng, J.Q., Felder, M., Connor, J.A., and Poo, M.M., 1994, Turning of nerve growth cones induced by neurotransmitters, *Nature* **368:** 140–144.

Zhou, F.Q., and Cohan, C.S., 2004, How actin filaments and microtubules steer growth cones to their targets, *J. Neurobiol.* **58:** 84–91.

Zhou, F.Q., Waterman-Storer, C.M., and Cohan, C.S., 2002, Focal loss of actin bundles causes microtubule redistribution and growth cone turning, *J. Cell Biol.* **157:** 839–849.

13 Comparative Analysis of Neural Crest Cell and Axonal Growth Cone Dynamics and Behavior

FRANCES LEFCORT, TIM O'CONNOR, AND PAUL M. KULESA

13.1. Introduction

A fundamental problem that must be solved in the developing nervous system is for neurons, which are typically born at a considerable distance from their target destination, to navigate accurately toward and form connections with their appropriate targets. This same problem must be overcome for neural crest cells (NCCs), which are a multipotent, migratory population of cells that emerge along the vertebrate body axis and invade the embryonic periphery in a stereotypical pattern and must cease migration at specific target sites. Proper neural crest migration is essential for normal craniofacial, cardiovascular, and trunk patterning and is highly conserved among vertebrates. In the trunk, the neural crest emigrate from the dorsal neural tube and migrate over a lengthy and complicated terrain to reach their appropriate targets where they will stop and differentiate into the dorsal root ganglia, sympathetic and parasympathetic ganglia, and the enteric nervous system. The neuron solves this pathfinding problem by extending an axon that terminates in the highly motile growth cone. Depending on the neuron type and the substrate on which it is growing, growth cones can range from having a simple bulbous structure to a broad expansive lamellipodia with few or numerous long and/or short filopodia. Growth cones have a dual function in that they must sense their surrounding environment and transduce this information into an appropriate motile response. This allows growth cones to direct the extension of an axonal process to their appropriate target and therefore establish the connections necessary for nervous system function. Although NCCs do not extend a growth cone, they do extend lamellipodia that are tipped by multiple filopodia that interact with environmental cues. A major advance in studying both neural crest migration and growth cone dynamics is the use of live confocal or multiphoton imaging of cells moving in their normal environments *in vivo*, in conjunction with the revolution in fluorescent cell tracers (Lichtman and Fraser, 2001; Yuste and Konnerth, 2005). Thus, the future is bound to be rich in insights into the mechanisms mediating both neural crest migration and growth cone behavior. The goal of this review is to compare the molecular and cellular mechanisms by which axonal growth cones and

NCCs successfully pathfind to their respective targets to identify common principles that mediate formation of the nervous system.

13.2. Mechanics of Movement: Growth Cone

There is a wealth of literature describing the mechanisms of growth cone movement including a detailed analysis of the role of the cytoskeleton in growth cone steering. The classic description of a growth cone is that its peripheral edges are occupied by the actin cytoskeleton, while the central part of the growth cone is the domain of microtubules (Figure 13.1) (Bridgman and Dailey, 1989; Forscher and Smith, 1988; Gallo and Letourneau, 2004; Zhou and Cohan, 2004). While this description is generally true for most growth cones, there are numerous descriptions of microtubules extending to the actin-dominated periphery of growth cones *in vitro* and *in vivo* (Sabry et al., 1991; Tanaka et al., 1995; Zhou et al., 2002). Both actin and microtubules are loosely arranged with their growing or "plus" ends oriented away from the cell body (Baas et al., 1988; Heidemann et al., 1981; Lewis and Bridgman, 1992). This orientation could provide a simple mechanism of protrusion as both actin and microtubules could potentially stimulate extension by the coordinated assembly of their structures. However, while it is clear that assembly plays an important role in growth and guidance, the mechanism of extension appears more complicated than simple filament assembly. A major question at present then is how these two cytoskeletal systems coordinate their activity resulting in neurite extension.

Actin filaments have long been known to be important in directing correct neurite outgrowth during development, although they are not necessarily required for axon extension (Bentley and Toroian-Raymond, 1986; Chien et al., 1993; Marsh and Letourneau, 1984). Actin filaments and bundles are assembled at the periphery of the growth cone, while at the same time myosin motors move the filament to the center of the growth cone where the filaments are disassembled (Forscher and Smith, 1988). This treadmilling has been observed in all growth cones examined. Earlier observations suggested that actin filaments may act as a physical barrier for microtubule penetration (Forscher and Smith, 1988), however this may actually be due to the continuous retrograde flow of actin toward the center of the growth cone rather than the presence of actin per se (Bentley and O'Connor, 1994; Tanaka and Sabry, 1995; Zhou et al., 2002). In addition to the actin bundles in the peripheral domain, in some growth cones there is a meshwork of parallel actin filaments that are arranged in a loosely orthogonal pattern near the center of the growth cone that may also impede the protrusion of microtubules (Schaefer et al., 2002). Removal of the actin network results in enhanced extension of microtubules into the peripheral domain of the growth cone, suggesting a possible scenario of selective microtubule assembly into regions of low-actin concentration (Bradke and Dotti, 1999; Forscher and Smith, 1988).

Previously it has been proposed that the microtubule cytoskeleton merely provides the structural support or consolidation that is necessary for growth cone extension.

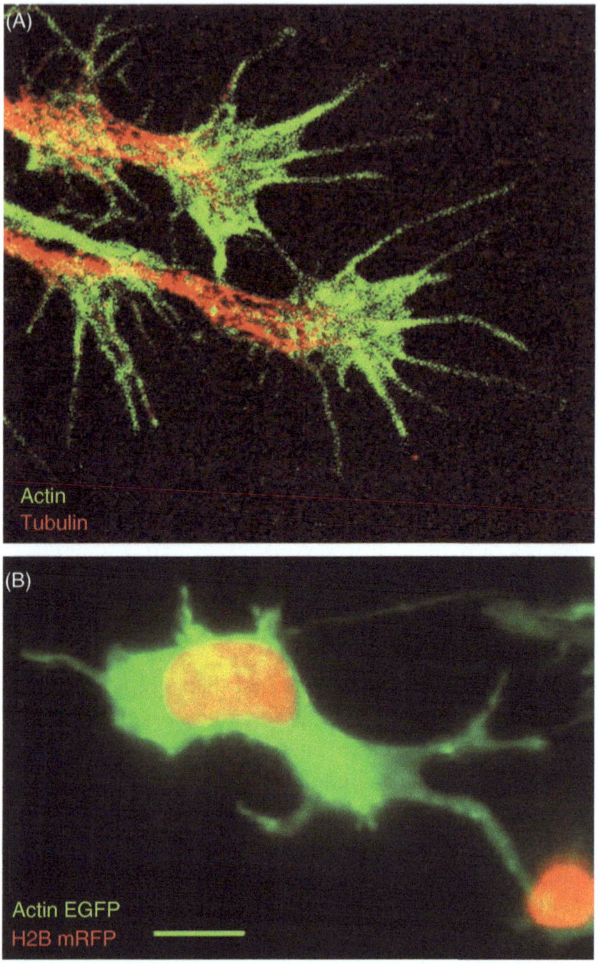

FIGURE 13.1. (A) Chick dorsal root ganglia neurons extending growth cones *in vitro* stained with antibodies to actin (green) and tubulin (red). Scale bar = 5 μm. (B) A typical migrating NCC was coinjected with actin-EGFP (green) and H2B-mRFP (red). The GFP signal shows the presence of actin throughout the cell and filopodial extensions, while the H2B-mRFP is localized to the nucleus of cells. The NCCs display filopodial extensions in many directions. Scalebar = 15 μm.

Microtubules are generally found bundled in the axon and then splay out as they enter into the center of the growth cone (Bridgman and Dailey, 1989; Gordon-Weeks, 2004). They are continually dynamic, growing, and shrinking as they explore the periphery of the growth cone (Dent and Kalil, 2001; Schaefer et al., 2002; Tanaka et al., 1995; Zhou et al., 2002). Microtubules that extend against a filamentous actin network either extend along bundles of actin or are swept back by the retrograde flow (Gordon-Weeks, 1991; Schaefer et al., 2002; Zhou et al., 2002). In numerous cases, microtubules are observed to interact and possibly stabilize when they contact actin filament bundles. In fact, rather than extending into regions of low F-actin distribution, microtubules appear to prefer to extend into regions of actin accumulation or increased density (Bentley and O'Connor, 1994; Buck and Zheng, 2002; Lin and Forscher, 1993; O'Connor and Bentley, 1993; Suter and Forscher, 2000). Similarly, when retrograde movement of actin is reduced, microtubules will selectively extend into the region of reduced flow (Lin and Forscher, 1995). Therefore, the impediment to microtubule protrusion appears to be due to retrograde flow rather than the presence of actin filaments, as slowing of the retrograde movement results in penetration of microtubules. Similarly, the localized loss of actin filament bundles within a growth cone causes microtubules to redistribute toward the region of the growth cone containing the largest amount of F-actin bundles (Zhou et al., 2002). Collectively these observations suggest that assembly, extension, and stabilization of microtubules in the periphery of the growth cone may result from an association with an accumulation of F-actin rather than a reduction in its concentration. The key questions therefore are: What is the nature of the microtubule-actin interaction? How is it regulated and how are the dynamic properties of the cytoskeleton regulated?

Growth cones without actin do not turn appropriately in turning assays, although microtubules are active and explore the periphery of the growth cone (Challacombe et al., 1996). In addition, taxol-induced growth cone turning is only successful when actin dynamics are intact (Buck and Zheng, 2002). These observations suggest a regulatory effect of actin filament dynamics on microtubule behavior during growth cone extension. Then how do actin filament dynamics regulate microtubule extension? As reported previously, intracellular examination has shown an association of microtubules with actin filament bundles. This association may be due to the capture of dynamic microtubules into a complex associated with stable actin filaments or alternatively, microtubule and F-actin assembly may be directed to the same location due to enhanced conditions for cytoskeletal growth. These scenarios are not mutually exclusive as microtubule extension and stabilization into the periphery of the growth cone could be due to an association with stable, F-actin bundles that can accumulate due to reduced retrograde flow as well as enhanced assembly in the same region. With reduced retrograde flow, F-actin filaments may provide direct protein links to encourage assembly or "pull" on the microtubules directly. Although there is little evidence for a direct pulling effect between microtubules and actin, there is an abundance of force generating myosin II localized in growth cones (Bridgman, 2002; Bridgman et al., 2001; Brown and Bridgman, 2003). Thus a simple protein interaction and mechanical transduction of force may be sufficient for selective microtubule advance.

13.3. Mechanics of Movement: Neural Crest Cells

Following their induction and specification while still in the neuroepithelium, NCCs then undergo an epithelial-to-mesenchymal transition (EMT) thereby facilitating the onset of their migration to their target destinations. This EMT requires reorganization of the cytoskeleton that is triggered by changes in cell adhesion and is not the subject of this review (Halloran and Berndt, 2003; Hay, 1995). The cytoskeletal dynamics that mediate neural crest migration have not been analyzed to the degree that they have been for growth cones. There are a number of studies examining movement *in vitro*, but it is only recently that investigators working on NCCs have been able to take advantage of confocal time-lapse imaging combined with targeted cell labeling to investigate cell dynamics in migrating NCC *in vivo*. The *in vitro* literature does allow some comparisons with the growth cone literature. For example, disruption of actin filaments with cytochalasin causes the immediate arrest of NCC movement, in contrast, cytochalasin treatment of growth cones *in vivo* does not cause growth arrest, although filopodia are disrupted and hence normal pathfinding goes awry (Bentley and Torian-Raymond, 1986; Haendel et al., 1996). A critical role for the F-actin depolymerizing factor, n-cofilin, in NCC migration has been demonstrated as targeted deletion of this gene in mice causes impaired delamination and migration of NCC (Gurniak et al., 2005). Fusion protein constructs targeted to specific cytoskeletal elements now allow the visualization and analysis of NCC filopodial and lamellipodial dynamics (Figure 13.1). The NCC interactions involve beautifully coordinated lamellipodial and filopodial interactions between cells in motion. The filopodial extensions may be long and stretch over several cell diameters away from the cell body to contact a nonneighboring cell or nonlocal microenvironment. The filopodial dynamics between neighboring cells appears to correlate with a cell's choice of direction, raising further speculation on the role of the cellular extensions in NCC guidance. The NCC morphologies are very dynamic but show a correlation between the number of filopodial extensions displayed by an individual cell and its position at the front versus the middle of a migratory stream. Neural crest cells display a wide variety of filopodial extensions that may be short connections (10–20 μm in length, 1–2 μm in width) between neighboring cells or long connections (>70 μm) to nonlocal neighbors. Each individual NCC may have several short filopodial connections with a number of local neighboring NCCs. The connections are dynamic, extending, and retracting as the NCC moves downstream.

Time-lapse analyses have revealed three distinctive NCC migratory behaviors that involve filopodial connections and potential directional cues from cell–cell interactions (Teddy and Kulesa, 2004). First, dividing NCCs often maintain a physical connection between progeny in the form of a thin (1–2 μm wide) filopodium. As the cells move apart, the length of the filopodium increases until it breaks at an arbitrary point between the two cells. If one of the NCC progeny migrates downstream, the trailing cell will follow the general trajectory of the lead cell, migrating at least to the position at which the filopodial connection breaks. Second, an NCC may extend a filopodial extension and contact the back end of a lead cell. In one scenario, the front end of the filopodium may continually extend to the position of the lead cell. The soma of the trailing cell moves forward to trace the position of the extending filopodium. In a third scenario, after extending a filopodium to contact a lead cell, the filopodium retracts and the trailing cell moves forward toward the position of the contact. Further investigations will help yield clues as to what type of cell–cell signals may be involved in the NCC connections.

Over the last 30 years, there has been a steady picture emerging of NCC migratory patterning (Halloran and Berndt, 2003; Helms and Schneider, 2003; Santagati and Rijli, 2003). Data from static end-point analyses (Sechrist et al., 1993; Serbedzija et al., 1992) and beautiful, innovative video time-lapse imaging have provided a glimpse of NCC migratory pathways (Erickson et al., 1980; Newgreen et al., 1979). Recent advances in live embryo imaging have begun to reveal the complex migratory behaviors of individual NCC (Krull et al., 1995; Schilling and Kimmel, 1994). In the chick, NCC cells have been observed to move rapidly downstream in a directed motion or meander within a stream and even turn in the reverse direction and move back toward the neural tube (Kulesa and Fraser, 1998). There is also evidence that some NCC migrate collectively, forming into chain-like arrays of cells that extend from the neural tube toward the target site (Kasemeier-Kulesa et al., 2005; Kulesa and Fraser, 1998). Cell-tracking analysis of fluorescently labeled NCCs shows that there are differences in average cell speeds and directionalities depending on the cell's position within a stream and whether the cell is within a stream or chain.

13.4. Individual Neural Crest Cell Migratory Behaviors

The formation of an NCC migratory stream emerges from a complex set of individual cell migratory behaviors. Within a typical stream, individual cell movements range from directed motion toward the lateral periphery to low directional wandering within a stream (Kulesa and Fraser, 1998, 2000). Some NCCs even turn back toward the neural tube and migrate in the reverse direction. Although NCCs emerge all along the vertebrate body axis, cell movements near the dorsal midline vary depending on the axial level of origin. In the head, NCC streams correlate with specific segments of the hindbrain, called rhombomeres (r) (Lumsden and Keynes, 1989; Lumsden and Krumlauf, 1996). Cranial NCC streams exit laterally from rhombomere 1(r1) + r2, r4, and r6. The NCC stream that emerges lateral to r4 and extends into the second branchial arch is narrower and denser than the other two cranial streams. Using the r4 NCC stream as a typical migratory stream, the leading front of the r4 NCC stream resembles a fan-shape that tapers in width back toward the neural tube. At the neural tube, the stream expands toward the middle of r3 and r5. Here, the trajectories of r3 and r5 NCCs give the appearance of merging freeway on-ramps to the main r4 super highway. Some cells that migrate from r3 and r5 contribute to neighboring streams (Birgbauer et al., 1995; Sechrist et al., 1993). Other NCCs that migrate into the regions lateral to r3 and r5 stop and collapse filopodia or change direction and migrate to join a neighboring stream (Kulesa and Fraser, 1998). In the chick, when r4 NCCs are transplanted lateral to r3, the cells disperse and join a neighboring migratory stream, suggesting local inhibitory signals prevent NCC streams from mixing (Farlie et al., 1999).

In the trunk, NCC streams emerge from the neural tube and are sculpted through the segmental structures adjacent to the vertebrate neural tube called somites

(Bronner-Fraser, 1986; Rickmann et al., 1985). Trunk NCCs follow one of two migratory routes, either migrating ventrally or dorsolaterally (Loring and Erickson, 1987). The NCCs migrate through the rostral portion of somites and avoid the caudal regions (Krull, 2001). The trunk NCC streams have a similar shape to the r4 migratory stream in that cells from adjacent regions emerge from the neural tube and are sculpted through the rostral portion of a somite (Krull et al., 1995). Recent advances exploiting a sagittal slice explant technique that allows for confocal imaging of ventral moving trunk NCCs shows surprising cell migratory behaviors (Kasemeier-Kulesa et al., 2005). Many trunk NCCs migrate in chain-like arrays through the rostral somite from the region of the forming dorsal root ganglia to the presumptive sympathetic ganglia. When the cells reach the dorsal aorta, instead of coalescing to form a discrete ganglion, the cells spread in the anteroposterior direction and mix with neighboring NCC subpopulations. Then, as if on cue, the cells begin to sort out and resegregate at discrete sites along the dorsal aorta. Thus, there appears to be several complex trunk NCC migratory behaviors that are well coordinated to produce a metameric pattern of sympathetic ganglia.

13.5. Collective Neural Crest Cell Migratory Behaviors

Neural crest cell migration in chain-like arrays has been observed both in the head and trunk in chick embryos (Kasemeier-Kulesa et al., 2005; Kulesa and Fraser, 1998). Chains form when cells align in a collective unit and migrate with a similar average speed. Typical NCC chains consist of five to seven cells and may stretch over 100 μm in length. Time-lapse confocal analysis reveals that NCCs in chains migrate slower than individual cells but with a higher directionality. Neural crest cell chains form in distinct spatial locations along the neural tube, including adjacent to r1, r7, and in the trunk, along the ventral migratory pathway. The chains structures persist until an arbitrary time, when somewhere along the length of the chain, the cell connections break. The chain disassembles and the cells migrate as individuals. This is somewhat analogous to axonal fasciculation—axons contact each other via filopodia which if attractive or permissive, then promote the fasciculation of axons that can travel together as an ensemble. Furthermore, NCCs that break free from a chain often migrate in aberrant directions and do not end up in their normal target location (Kasemeier-Kulesa et al., 2005). Thus traveling as a collective, either as an NCC chain or as an axon fascicle, maximizes the likelihood that a given cell or axon will pathfind correctly to its appropriate target.

13.6. Neural Crest Cell Morphologies

Growth cone morphologies are often correlated with their specific location: pioneer or lead neurons tend to have a complex growth cone with numerous filopodia, while follower axons that fasciculate with the pioneer axons tend to have more simple growth cones with few filopodia. At choice points however, growth cones become more elaborate as they sense and must integrate a myriad of cues in the local environment (Bonner et al., 2003; Bovolenta and Mason, 1987; Tosney and Landmesser, 1985). Similarly, NCCs display dynamic cell shape changes depending on their position within cell migratory streams. At the stream fronts, NCCs have many lamellipodial and filopodial extensions in nearly all directions. Neural crest cells at the stream fronts tend to have a higher directionality and a slower average cell speed. This may be due in part to the cells at the front having less interaction with neighboring cells and thus a better ability to sense the environment and any potential attraction cues toward the periphery. Midstream, NCCs display a bipolar shape. The direction of the cellular extension tends to align with the direction of the location of a particular peripheral target. Some of the filopodial extensions may be quite long (>70 μm) and extend in between local neighbors to reach six or more cell diameters away from the cell body.

13.7. Extrinsic Guidance Cues

The common visual aspects of cell migratory behaviors that appear to be shared between NCCs and axon growth cones begs the question whether there is a common set of molecular mechanisms? Many of the major guidance cues for both growth cones and NCC migration are common: e.g., Slit/robo, ephrin/Eph receptors, semaphorins, integrins/extracellular matrix, proteoglycans, and cadherins (Huber et al., 2003). A major question in the development of NCC patterning is to understand the mechanisms that ensure the stereotypical formation of NCC migratory streams. The local environmental cues appear to sculpt NCC into discrete streams and direct NCCs to specific migratory routes (Santiago and Erickson, 2002). This interplay between attraction and local inhibitory cues that guide lead NCCs, but set up discrete cell migratory streams is actively being investigated. Currently, it is thought that a combination of intrinsic cues, endowed by the neural tube on premigratory neural crest, and extrinsic environmental cues adjacent to the neural tube sculpt NCCs along stereotypical pathways (Graham et al., 2004; Trainor and Krumlauf, 2000). However, although several molecular families have been implicated in NCC guidance, including Slit/robo and Eph/ephrins, the coordination of these potential mechanisms is still unclear.

Clues from investigation of trunk NCC migration may help to provide some insights. Trunk NCCs migrate in dense streams of cells that are restricted to the rostral portion of the somites. In the trunk, NCCs travel along one of two distinct migratory routes, just adjacent to the dorsal neural tube. In the chick, the first wave of NCCs to emerge from the neural tube travel ventrally adjacent to the neural tube and/or just medial to the dermamyotome and through the sclerotome portions of the somite (Bronner-Fraser, 1986; Rickmann et al., 1985). These cells eventually give rise to most of the peripheral nervous system including the sympathetic ganglia and dorsal root ganglia. Later emerging NCCs take a dorsal route just beneath the ectoderm. These cells give rise to pigment cells in the skin

(Bronner-Fraser, 1994; Erickson and Goins, 1995; Le Douarin, 1982). The mechanisms by which NCCs are restricted to the rostral portions of somites and how early and late NCCs distinguish between the two distinct migratory routes are not fully understood. However, it is thought that environmental clues play a large role in a NCC's decision.

Slits belong to a large family of secreted proteins and have been shown to function as an extracellular cue for cells ranging from neurons to leukocytes (Wong et al., 2001). The robo family of receptors mediates the effects of secreted Slit molecules. Slit–robo signaling has been widely studied in axon pathfinding (Brose and Tessier-Lavigne, 2000), and its role in NCC guidance is relatively new. In the NCC system, both robo1 and 2 are expressed in migrating trunk NCCs, and one model proposes that that Slits expressed in ventral mesenchymal tissues help to prevent trunk NCCs from invading nontarget peripheral locations and thereby confining early migrating cells to the ventro-medial pathway (Jia et al., 2005). When overexpressing a robo1 dominant-negative receptor, NCCs migrate onto the dorso-lateral pathway at a time when their migration would normally be restricted to only the ventro-medial pathway (Jia et al., 2005). There may, however, be multiple guidance mechanisms that coordinate the sorting of the trunk NCC into distinct migratory pathways. Another functioning repellant mechanism is via Eph/ephrin signaling. Ephrins and their Eph receptors have been demonstrated to regulate many critical events in development, including neurual crest migration and axon guidance (Davy and Soriano, 2005; Poliakov et al., 2004). Ephrin-B and EphB2/B3, by repelling crest cells from the caudal half-somite, drive crest cells to migrate through the rostral half (Krull et al., 1997; McLennan and Krull, 2002). However, later emerging crest cells that will migrate into the ectoderm to become melanocytes are not repelled but in fact are attracted to dorsally expressed ephrins (Santiago and Erickson, 2002). When plated on ephrin-B1 *in vitro*, NCCs rounded up and their actin cytoskeleton was disrupted; in contrast, late emigrating NCCs (presumptive melanocytes) extended actin-filled microspikes on ephrin-B1. Thus for NCC, the same ligand, ephrin-B1 can act both as a repulsive and attractive cue. In fact, members of the ephrin-A family have also been shown to promote NCC migration (McLennan and Krull, 2002). It has been demonstrated that Eph receptors exert their effects by remodeling the actin cytoskeleton through the regulation of Rho GTPases in axons, it will be of interest to determine whether the same mechanisms are at work in NCCs (Murai and Pasquale, 2005; Sahin et al., 2005).

Another family of guidance molecules that regulate growth cone pathfinding, the semaphorins, are expressed and secreted in the hindbrain and trunk and significantly influence the migration patterns of NCCs and when misexpressed, induce aberrant NCC migration (Bron et al., 2004; Kawasaki et al, 2002; Osborne et al., 2005; Yu and Moens, 2005). Similarly, both NCCs and growth cones extend rapidly on many extracellular matrix constituents including fibronectin, laminin, collagen, PG-M/versican, and thombospondin (Perris and Perissinotto, 2000) and express the major class of ECM receptors, integrins (Strachan and Condic, 2004; Testaz et al., 1999). Furthermore, disruptions in integrin function *in vivo* can impair both neural crest migration and axonal pathfinding (Kil et al., 1998; review see Clegg et al., 2003). Both NCCs and several axonal populations respond to many classes of proteoglycans in the extracellular environment; chondroitin sulfate proteoglycans distributed in the perinotochordal region inhibit both motor axons outgrowth and neural crest migration (Krull, 2001; Oakley et al., 1994; Tosney and Oakley, 1990;) and of course inhibit axon growth in the regenerating CNS (Silver, 1994). Thus, both growth cones and migrating NCCs share expression of many of the same guidance receptors and respond similarly to the same sets of environmental attractive and repulsive cues.

13.8. Response to Diffusible Cues Including Classical Morphogens

One of the most striking similarities between growth cone navigation and neural crest migration is the number of shared extrinsic diffusible cues to which they both respond. Morphogens classically implicated in patterning the embryo, such as Wnts, BMPs, and Sonic hedgehog (Shh), all exert profound effects on both growth cones and migrating NCCs (Bovolenta, 2005; Charon and Tessier-Lavigne, 2005; Salie et al., 2005). Wnts influence multiple events during the lifetime of an NCC including its induction, delamination, migration, and in the case of melanoblasts and sensory neurons, their differentiation (De Calisto et al., 2005; Garcia-Castro et al., 2002; Lewis et al., 2004). Wnts also exert multiple effects on developing neurons including organization of cytoskeleton and axon guidance (Arevalo and Chao, 2005; Patapoutian and Reichardt, 2000; Zou, 2004). Wnt activities on NCCs are transduced via the canonical, β-catenin–mediated and the noncanonical, PCP or Wnt-Ca^{2+}, pathways (DeCalisto et al., 2005; Hari et al., 2002; Lee et al., 2004). Abrogation of the noncanonical Wnt signaling pathway inhibits neural crest migration in *Xenopus* embryos and alters the morphology of NCCs. In lieu of a broad lamellipodia, cells in which the noncanonical Wnt pathway was blocked extended numerous long filopodia suggesting the model that the Wnt noncanonical pathway is required for the stabilization of lamellipodia which is thus necessary for migration (De Castilllo et al., 2005). Disruption in Shh signaling abrogates several steps in neural crest development in addition to causing aberrations in axonal guidance (Ahlgren and Bronner-Fraser, 1999; Bourikas et al., 2005; Charron et al., 2003; Fedtsova et al., 2003; Fu et al., 2004). Another classical family of morphogens, the BMPs, although critically important for dorsalizing cell fate in the spinal cord including influencing axonal outgrowth in the cord, also exert a major effect on NCCs: they induce the differentiation of sympathetic neurons from NCCs and are required for the formation of the enteric nervous system (Butler and Dodd, 2003; White et al., 2001; Schneider et al., 1999; Goldstein et al., 2005). Lastly, chemokines have been shown to exert important functions on both neurons and NCCs through the receptor, CXCR4. Disruptions in this receptor interferes with both axonal pathfinding (Li et al., 2005; Lieberam et al., 2005; Pujol et al., 2005) and the cessation of NCC migration so that they neither stop properly to give rise to the dorsal root ganglion nor assemble a normal trigeminal ganglion (Belmadani et al, 2005; Knaut et al., 2005). Given that axonal outgrowth in the spinal cord overlaps temporally with neural crest migration, it is perhaps not so surprising that both neurons and NCCs express many of the same morphogen receptors and that these shared molecules induce albeit distinct, yet cell-context specific, responses.

13.9. Transduction Mechanisms: Growth Cones

How growth cones and NCC filopodia sense their environment and convert this information into the appropriate growth response is an area of extensive investigation. It is generally accepted that the stimulation of guidance receptors leads to intracellular signaling cascades whose final common path of action is principally on the growth cone cytoskeleton. Filopodia on growth cones generate a localized transient elevation in Ca^{2+} that propagates back to the growth cone, thereby promoting growth cone turning (Gomez et al., 2001). It will be of considerable interest to determine whether filopodia mediate the same effects on NCCs and whether they do so via local changes in Ca^{2+} signaling. There has been an emphasis on directly analyzing the cytoskeleton in neurons as they extend *in vitro* and *in vivo*. Given the importance and interrelated function of the actin and microtubule cytoskeletal network, it is not surprising that there are at least two major signaling pathways in growth cones that may be responsible for their function. The Rho family of GTPases has been well described in modulating the assembly of actin filaments while the Wnt signaling pathway, through GSK-3β, has been shown to regulate microtubule dynamics.

Actin filament assembly is regulated by a balance of activity between Rac1, Cdc42, and RhoA (Dickson, 2001; Gallo and Letourneau, 2004). In general, Rac and Cdc42 stimulate process outgrowth by enhancing the assembly of actin filaments and the actin meshwork underlying lamellipodial and filopodial extension (Brown et al., 2000; Jurney et al., 2002; Wahl, 2000). In contrast, the Rho pathway generally reduces neurite outgrowth, in part due to collapsing the F-actin network (Huber et al., 2003; Liu and Stritmmater, 2001). These GTPases appear to interact at the level of several important proteins including myosin light chain kinase and actin depolymerizing factor/cofilin. Rho kinase, a downstream effector of RhoA, stimulates the phosphorylation of ADF (through LIM kinase) resulting in the inactivation of ADF (Aizawa et al., 2001). In addition, RhoA activation results in the activation of myosin II through myosin light chain phosphorylation and myosin light chain phosphatase inhibition (Bito et al., 2000; Brown and Bridgman, 2004). The combined effect is an increase in myosin activity and reduced actin filaments available for assembly, resulting in a reduction of neurites and/or growth cone collapse. In contrast, Cdc42 and Rac regulated p21-activated kinase (PAK) inhibits myosin II activity, although ADF activity also seems to be reduced (Lundquist, 2003; Sarmiere and Bamburg, 2004). This pathway may also regulate actin bundle formation, possibly through the Arp2/3 complex, which has been shown to activate bundle formation in cells (Vadlamudi et al., 2004). How the reduction in myosin activity and increased F-actin bundle formation leads to enhanced outgrowth still remains unclear, however it may suggest a potential signaling cascade that underlies lamellipodial protrusion and reduced retrograde flow respectively.

Actin filament distribution and their dynamics appear to be key regulators of microtubule extension and process outgrowth. However, it has become evident that microtubules themselves can also be the targets for signaling pathways in growth cones and that they in turn may regulate the assembly and dynamics of the actin network. For example, the reorganization of microtubules in response to asymmetric F-actin bundle formation has a reciprocal effect as the microtubules may regulate lammellipodial formation and motility through GTPase Rac1 activity (Buck and Zheng, 2002; Rochlin et al., 1999). Neurotrophin mediated signaling has shown that microtubule assembly can also be directly regulated in growth cones (Zhou et al., 2004). Experimental evidence has implicated PI3-kinase (PI3K) as an important signaling molecule downstream of neurotrophin-mediated neurite outgrowth (Segal, 2003). Downstream effectors of PI3K include the known actin regulators, such as Rac, in addition to other signaling molecules such as Akt, integrin linked kinase (ILK), and GSK-3β (Segal, 2003). GSK-3β has been studied extensively due to its key role in Wnt signaling, where it forms a complex with axin and adenomatous polyposis coli (APC). In addition to their role in Wnt signaling, GSK-3β and APC are found in growth cones where APC has been shown to bind to the plus ends of microtubules through another microtubule plus-end–binding protein (EB-1) (Askham et al., 2000; Morrison et al., 2002; Zhou et al., 2004). The binding of the APC/EB-1 complex reduces shrinking of microtubules and promotes their stabilization (Zumbrunn et al., 2001). This activity is abolished by GSK-3β activation and APC phosphorylation (Zhou et al., 2004). In the presence of neurotrophins then, PI3K activity inhibits the activity of GSK3β, allowing for the activation of the APC/EB-1 complex and the subsequent stabilization and extension of microtubules.

How then are microtubule and actin filament dynamics coordinated for net axon extension? Actin filaments appear to act as both a barrier and a guide for microtubules. In addition, actin and microtubules may stimulate reciprocal signaling pathways to regulate assembly and stabilization. A key factor appears to be the coordinated extension of microtubules along specific actin bundles or meshwork. In nonneuronal cells, APC has been shown to localize with actin and the plasma membrane, possibly providing a link to the microtubule network (Bienz, 2002; Nathke et al., 1996). The downstream Rho effector protein, mDia, is also a candidate for linking the cytoskeletal networks. Overexpression of mDia results in the alignment of microtubules parallel to actin bundles (Ishizaki et al., 2001). This activity appears to be mediated by the positioning of microtubule ends to the sites of mDia accumulation along actin filaments. It is possible then that signaling mechanisms to microtubules promote their stabilization in part to allow for their capture by linking proteins such as mDia. This further stabilizes the microtubule network, allowing for increased extension. This may have a positive feedback on the actin cytoskeleton in such a manner as to increase actin polymerization and accumulation, resulting in further microtubule capture and assembly.

While there are undoubtedly additional signaling molecules that impact on the growth cone cytoskeleton, their ultimate outcome will be to regulate the extension and stabilization of the microtubule network. In order to have a complete understanding of this process it will be important to identify the molecules that regulate the interactions between the cytoskeletal networks. Ultimately, it will be necessary to understand how these interactions lead to the necessary forces responsible for protrusion and extension in order to appreciate how guidance cues steer growth cones to their targets.

13.10. Transduction Mechanisms: Neural Crest Cells

As described previously for growth cones, an emerging consensus in the field is that a cell's response to attractive and repulsive cues involves signaling to the actin cytoskeleton, in particular to small Rho GTPases. The small rho GTPases have been shown to play a role in axon guidance (Guan and Rao, 2003) and are also involved in epithelial sheet movements. During the sealing of the *Drosophila* dorsal epidermis, epithelial sheets fuse together to close the hole. Clever confocal time-lapse imaging studies have revealed that long filopodia extended from leading edge cells contact the opposite sides of the dorsal opening. The filopodial contact appears to ensure a proper alignment of one side to the other. When the assembly of these actin-based protrusions is perturbed, by using a dominant-negative Cdc42, the adhesion and fusion of opposite sides is prevented (Jacinto et al., 2000).

In relation to the NCCs, short and long filopdia have been revealed in high resolution imaging in chick (Teddy and Kulesa, 2004). The filopodial extensions appear to play a role in an individual NCC's choice of direction. Trailing NCCs that extend and contact lead cells with a filopodial extension often migrate in the direction of the contact, with or without retracting the filopodium (Teddy and Kulesa, 2004). In preliminary experiments, when Cdc42 is underexpressed, NCCs extend fewer filopodia and tend not to follow lead cell trajectories (Rupp and Kulesa, unpublished). Perturbation of Cdc42 and Rho have shown that there is a functional involvement of Rho GTPases in the Slit-robo pathway (Wong et al., 2001). Introduction of a constitutively active form of Cdc42 into migratory mouse cells from the anterior subventricular zone inhibited the Slit-mediated repulsion of the cells *in vitro* (Wong et al., 2001). Whether similar signaling interactions occur in NCCs is not known. In the dorsal neural tube, it has been shown that BMPs induce the expression of RhoB in nascent NCCs, and inhibition of its activity prevents the delamination of NCCs from the neural tube (Liu and Jessell, 1998). Further investigation into the role of the small Rho GTPases in NCC guidance may shed light on the function of cell–cell interactions and whether filopodia mimic the role of axon growth cones in sensing environmental cues and establishing directionality.

To summarize, the culmination of several years of study involving creative tissue and embryo culture designs and advances in *in situ* imaging, merged with video and confocal microscopy now offer a unique, subcellular framework to analyze NCC migratory behaviors. Data that include cell trajectories and speeds, filopodial and lamellipodial dynamics, and cell–cell interactions are revealing a level of sophistication of NCC migration that parallels characteristics of axon guidance and growth cone dynamics studies. These investigations will allow us to build a picture of NCC migratory behaviors at both the cellular and molecular level as a means of comparison to axon guidance and growth cone dynamics. Ultimately our goal is that these data will foster discussions and exchange of information between neurobiologists and developmental biologists and identify the common and/or distinct mechanisms mediating growth cone and NCC behavior.

References

Ahlgren, S.C., and Bronner-Fraser, M., 1999, Inhibition of sonic hedgehog signaling in vivo results in craniofacial neural crest cell death, *Curr. Biol.* **9:** 1304–1314.

Aizawa, H., Wakatsuki, S., Ishii, A., Moriyama, K., Sasaki, Y., Ohashi, K., et al., 2001, Phosphorylation of cofilin by LIM-kinase is necessary for semaphorin 3A-induced growth cone collapse, *Nat. Neurosci.* **4:** 367–373.

Arevalo, J.C., Chao, M.V., 2005, Axonal growth: where neurotrophins meet Wnts. *Curr. Opin. Cell. Biol.* **17(2):** 112–5.

Askham, J.M., Moncur, P., Markham, A.F., and Morrison, E.E., 2000, Regulation and function of the interaction between the APC tumour suppressor protein and EB1, *Oncogene* **19:** 1950–1958.

Baas, P.W., Deitch, J.S., Black, M.M., and Banker, G.A., 1988, Polarity orientation of microtubules in hippocampal neurons: Uniformity in the axon and nonuniformity in the dendrite, *Proc. Natl. Acad. Sci. USA* **85:** 8335–8339.

Belmadani, A., Tran, P.B., Ren, D., Assimacopoulos, S., Grove, E.A., and Miller, R.J., 2005, The chemokine stromal cell-derived factor-1 regulates the migration of sensory neuron progenitors, *J. Neurosci.* **25:** 3995–4003.

Bentley, D., and O'Connor, T.P., 1994, Cytoskeletal events in growth cone steering, *Curr. Opin. Neurobiol.* **4:** 43–48.

Bentley, D., and Toroian-Raymond, A., 1986, Disoriented pathfinding by pioneer neurone growth cones deprived of filopodia by cytochalasin treatment, *Nature* **323:** 712–715.

Bienz, M., 2002, The subcellular destinations of APC proteins, *Nat. Rev. Mol. Cell Biol.* **5:** 328–338.

Bito, H., Furuyashiki, T., Ishihara, H., Shibasaki, Y., Ohashi, K., Mizuno, K., et al., 2000, A critical role for a Rho-associated kinase, p160ROCK, in determining axon outgrowth in mammalian CNS neurons, *Neuron* **26:** 431–441.

Birgbauer, E., Sechrist, J., Bronner-Fraser, M., and Fraser, S., 1995, Rhombomeric origin and rostrocaudal reassortment of neural crest cells revealed by intravital microscopy, *Development* **121:** 935–945.

Bonner, J., Gerrow, K.A., and O'Connor, T.P., 2003, The tibial-1 pioneer pathways: An in vivo model for neuronal outgrowth and guidance, *Methods Cell Biol.* **71:** 171–193.

Bovolenta, P., and Mason, C., 1987, Growth cone morphology varies with position in the developing mouse visual pathway from retina to first targets, *J. Neurosci.* **7:** 1447–1460.

Bourikas, D., Pekarik, V., Baeriswyl, T., Grunditz, A., Sadhu, R., Nardo, M., et al., 2005, Sonic hedgehog guides commissural axons along the longitudinal axis of the spinal cord, *Nat. Neurosci.* **8:** 297–304.

Bovolenta, P., 2005, Morphogen signaling at the vertebrate growth cone: a few cases or a general strategy? *J. Nerobiol.* **64(4):** 405–16.

Bradke, F., and Dotti, C.G., 1999, The role of local actin instability in axon formation, *Science* **283:** 1931–1934.

Bridgman, P.C., 2002, Growth cones contain myosin II bipolar filament arrays, *Cell Motil. Cytoskeleton* **52:** 91–96.

Bridgman, P.C., and Dailey, M.E., 1989, The organization of myosin and actin in rapid frozen nerve growth cones, *J. Cell Biol.* **108:** 95–109.

Bridgman, P.C., Dave, S., Asnes, C.F., Tullio, A.N., and Adelstein, R.S., 2001, Myosin IIB is required for growth cone motility, *J. Neurosci.* **21:** 6159–6169.

Bron, R., Eickholt, B.J., Vermeren, M., Fragale, N., and Cohen, J., 2004, Functional knockdown of neuropilin-1 in the developing chick nervous system by siRNA hairpins phenocopies genetic ablation in the mouse. *Dev. Dyn.* **230(2):** 299–308.

Bronner-Fraser, M., 1986, An antibody to a receptor for fibronectin and laminin perturbs cranial neural crest development in vivo, *Dev. Biol.* **117:** 528–536.

Bronner-Fraser, M., 1994, Neural crest cell formation and migration in the developing embryo, *FASEB J.* **8:** 699–706.

Brose, K., and Tessier-Lavigne, M., 2000, Slit proteins: Key regulators of axon guidance, axonal branching, and cell migration, *Curr. Opin. Neurobiol.* **10:** 95–102.

Brown, M.D., Cornejo, B.J., Kuhn, T.B., and Bamburg, J.R., 2000, Cdc42 stimulates neurite outgrowth and formation of growth cone filopodia and lamellipodia, *J. Neurobiol.* **43:** 352–364.

Brown, M.E., and Bridgman, P.C., 2003, Retrograde flow rate is increased in growth cones from myosin IIB knockout mice, *J. Cell Sci.* **116:** 1087–1094.

Brown, M.E., and Bridgman, P.C., 2004, Myosin function in nervous and sensory systems, *J. Neurobiol.* **58:** 118–130.

Buck, K.B., and Zheng, J.Q., 2002, Growth cone turning induced by direct local modification of microtubule dynamics, *J. Neurosci.* **22:** 9358–9367.

Butler, S.J., and Dodd, J., 2003, A role for BMP heterodimers in roof plate-mediated repulsion of commissural axons, *Neuron* **38:** 389–401.

Challacombe, J.F., Snow, D.M., and Letourneau, P.C., 1996, Actin filament bundles are required for microtubule reorientation during growth cone turning to avoid an inhibitory guidance cue, *J. Cell Sci.* **109:** 2031–2040.

Charron, F., Stein, E., Jeong, J., McMahon, A.P., and Tessier-Lavigne, M., 2003, The morphogen sonic hedgehog is an axonal chemoattractant that collaborates with netrin-1 in midline axon guidance, *Cell* **113:** 11–23.

Charron, F., Tessier-Lavigne, M., 2005, Novel brain wiring functions for classical morphogens: a role as graded positional cues in axon guidance, *Development* **132(10):** 2251–62.

Chien, C.B., Rosenthal, D.E., Harris, W.A., and Holt, C.E., 1993, Navigational errors made by growth cones without filopodia in the embryonic Xenopus brain, *Neuron* **11:** 237–251.

Clegg, D.O., Wingerd, K.L., Hikita, S.T., and Tolhurst, E.C., 2003, Integrins in the development, function and dysfunction of the nervous system, *Front. Biosci.* **8:** d723–d750.

Davy, A., and Soriano, P., 2005, Ephrin signaling in vivo: Look both ways, *Dev. Dyn.* **232:** 1–10.

DeCalisto, J., Araya, C., Marchant, L., Riaz, C.F., and Mayor, R., 2005, Essential role of non-canonical Wnt signaling in neural crest migration, *Development* **132:** 2587–2597.

Dent, E.W., and Kalil, K., 2001, Axon branching requires interactions between dynamic microtubules and actin filaments, *J. Neurosci.* **21:** 9757–9769.

Dickson, B.J., 2001, Rho GTPases in growth cone guidance, *Curr. Opin. Neurobiol.* **11:** 103–110.

Erickson, C.A., and Goins, T.L., 1995, Avian neural crest cells can migrate in the dorsolateral path only if they are specified as melanocytes, *Development* **121:** 915–924.

Erickson, C.A., Tosney, K.W., and Weston, J.A., 1980, Analysis of migratory behavior of neural crest and fibroblastic cells in embryonic tissues, *Dev. Biol.* **77:** 142–156.

Farlie, P.G., Kerr, R., Thomas, P., Symes, T., Minichiello, J., Hearn, C.J., et al., 1999, A paraxial exclusion zone creates patterned cranial neural crest cell outgrowth adjacent to rhombomeres 3 and 5, *Dev. Biol.* **213:** 70–84.

Fedtsova, N., Perris, R., and Turner, E.E., 2003, Sonic hedgehog regulates the position of the trigeminal ganglia, *Dev. Biol.* **261:** 456–469.

Forscher, P., and Smith, S.J., 1988, Actions of cytochalasins on the organization of actin filaments and microtubules in a neuronal growth cone, *J. Cell Biol.* **107:** 1505–1516.

Fu, M., Lui, V.C., Sham, M.H., Pachnis, V., and Tam, P.K., 2004, Sonic hedgehog regulates the probiferetion, differentiation, and migration of enteric neural crest cells in gut. *J. Cell. Biol.* **166(5):** 673–84.

Gallo, G., and Letourneau, P.C., 2004, Regulation of growth cone actin filaments by guidance cues, *J. Neurobiol.* **58:** 92–102.

Garcia-Castro, M.I., Marcelle, C., Bronner-Fraser, M., 2002, Ectodermal Wnt function as a neural crest inducer. *Science.* **(5582):** 848–51.

Goldstein, A.M., Brewer, K.C., Doyle, A.M., Nagy, N., and Roberts, D.J., 2005, BMP signaling is necessary for neural crest cell migration and ganglion formation in the enteric nervous system, *Mech. Dev.* **122:** 821–833.

Gomez, T.M., Robles, E., Poo, M., and Spitzer, N.C., 2001, Filopodial calcium transients promote substrate dependent growth cone turning, *Science* **291:** 1983–1987.

Gordon-Weeks, P.R., 1991, Evidence for microtubule capture by filopodial actin filaments in growth cones, *Neuroreport* **2:** 573–576.

Gordon-Weeks, P.R., 2004, Microtubules and growth cone function, *J. Neurobiol.* **58:** 70–83.

Graham, A., Begbie, J., and McGonnell, I., 2004, Significance of the cranial neural crest, *Dev. Dyn.* **229:** 5–13.

Guan, K.L., and Rao, Y., 2003, Signalling mechanisms mediating neuronal responses to guidance cues. *Nat. Rev. Neurosci.* **4(12):** 941–56.

Gurniak, C.B., Perlas, E., and Witke, W., 2005, The actin depolymerizing factor n-cofilin is essential for neural tube morphogenesis and neural crest cell migration, *Dev. Biol.* **278:** 231–241.

Haendel, M.A., Bollinger, K.E., and Baas, P.W., 1996, Cytoskeletal changes during neurogenesis in cultures of avian neural crest cells, *J. Neurocytol.* **25:** 289–301.

Halloran, M.C., and Berndt, J.D., 2003, Current progress in neural crest cell motility and migration and future prospects for the zebrafish model system, *Dev. Dyn.* **228:** 497–513.

Hari, L., Brault, V., Kleber, M., Lee, H.Y., Ille, F., Leimeroth, R., et al., 2002, Lineage-specific requirements of beta-catenin in neural crest development, *J. Cell Biol.* **159:** 867–880.

Hay, E.D., 1995, An overview of epithelio-mesenchymal transformation, *Acta Anat.* **154:** 8–20.

Heidemann, S.R., Landers, J.M., and Hamborg, M.A., 1981, Polarity orientation of axonal microtubules, *J. Cell Biol.* **91:** 661–665.

Helms, J., and Schneider, R.A., 2003, Cranial skeletal biology, *Nature* **423:** 326–331.

Huber, A.B., Kolodkin, A.L., Ginty, D.D., and Cloutier, J.F., 2003, Signaling at the growth cone: Ligand-receptor complexes and the control of axon growth and guidance, *Annu. Rev. Neurosci.* **26:** 509–563.

Ishizaki, T., Morishima, Y., Okamoto, M., Furuyashiki, T., Kato, T., and Narumiya, S., 2001, Coordination of microtubules and the actin cytoskeleton by the Rho effector mDia1, *Nat. Cell Biol.* **3:** 8–14.

Jacinto, A., Wood, W., Balayo, T., Turmaine, M., Martinez-Aria, A. and Martin, P., 2000, Dynamic actin-based epithelial adhesion and cell matching during Drosophila dorsal closure. *Curr. Biol.* **10(22):** 1420–6.

Jia, L., Cheng, L., and Raper, J., 2005, Slit/Robo signaling is necessary to confine early neural crest cells to the ventral migratory pathway in the trunk. *Dev. Biol.* **282(2):** 411–21.

Jurney, W.M., Gallo, G., Letourneau, P.C., and McLoon, S.C., 2002, Rac1-mediated endocytosis during ephrin-A2- and semaphorin 3A-induced growth cone collapse, *J. Neurosci.* **22:** 6019–6028.

Kasemeier-Kulesa, J.C., Kulesa, P.M., and Lefcort, F., 2005, Imaging neural crest cell dynamics during formation of dorsal root ganglia and sympathetic ganglia, *Development* **132:** 235–245.

Kawasaki, T., Bekku, Y., Suto, F., Kitsukawa, T., Taniguchi, M., Nagatsu, I., et al., 2002, Requirement of neuropilin 1-mediated Sema3A signals in patterning of the sympathetic nervous system, *Development* **129:** 671–680.

Kil, S.H., Krull, C.E., Cann, G., Clegg, D., and Bronner-Fraser, M., 1998, The alpha4 subunit of integrin is important for neural crest cell migration, *Dev. Biol.* **202:** 29–42.

Knaut, H., Blader, P., Strahle, U., and Schier, A.F., 2005, Assembly of trigeminal sensory ganglia by chemokine signaling, *Neuron* **47:** 653–666.

Krull, C.E., 2001, Segmental organization of neural crest migration, *Mech. Dev.* **105:** 37–45.

Krull, C.E., Collazo, A., Fraser, S.E., and Bronner-Fraser, M., 1995, Segmental migration of trunk neural crest: Time-lapse analysis reveals a role for PNA-binding molecules, *Development.* **121:** 3733–43.

Krull, C.E., Lansford, R., Gale, N.W., Collazo, A., Marcelle, C., Yancopoulos, G.D., et al., 1997, Interactions of Eph-related receptors and ligands confer rostrocaudal pattern to trunk neural crest Migration, *Curr. Biol.* **7:** 571–580.

Kulesa, P.M., and Fraser, S.E., 1998, Neural crest cell dynamics revealed by time-lapse video microscopy of whole chick explant cultures, *Dev. Biol.* **204:** 327–344.

Kulesa, P.M., and Fraser, S.E., 2000, In ovo time-lapse analysis of chick hindbrain neural crest cell migration shows cell interactions during migration to the branchial arches, *Development* **127:** 1161–1172.

Le Douarin, N., 1982, *The Neural Crest*, Cambridge Universty Press, Cambridge.

Lichtman, J.W., and Fraser, S.E., 2001, The neuronal naturalist: Watching neurons in their native habitat, *Nat. Neurosci.* **4(Suppl.):** 1215–1220.

Lumsden, A., and Keynes, R., 1989, Segmental patterns of neuronal development in the chick hindbrain, *Nature* **337:** 424–428.

Lumsden, A., and Krumlauf, R., 1996, Patterning the vertebrate neuraxis, *Science* **274:** 1109–1115.

Lee, H.Y., Kleber, M., Hari, L., Brault, V., Suter, U., Taketo, M.M., et al., 2004, Instructive role of Wnt/beta-catenin in sensory fate specification in neural crest stem cells, *Science.* **303:** 1020–1023.

Lewis, A.K., and Bridgman, P.C., 1992, Nerve growth cone lamellipodia contain two populations of actin filaments that differ in organization and polarity, *J. Cell Biol.* **119:** 1219–1243.

Lewis, J.L., Bonner, J., Modrell, M., Ragland, J.W., Moon, R.T., Dorsky, R.I., et al., 2004, Reiterated Wnt signaling during zebrafish neural crest development, *Development* **131:** 1299–1308.

Li, Q., Shirabe, K., Thisse, C., Thisse, B., Okamoto, H., Masai, I., et al., 2005, Chemokine signalingguides axons within the retina in zebrafish, *J. Neurosci.* **25:** 1711–1717.

Lieberam, I., Agalliu, D., Nagasawa, T., Ericson, J., and Jessell, T.M., 2005, A Cxcl12-CXCR4 chemokine signaling pathway defines the initial trajectory of mammalian motor axons, *Neuron* **47:** 667–679.

Lin, C.H., and Forscher, P., 1993, Cytoskeletal remodeling during growth cone-target interactions, *J. Cell Biol.* **121:** 1369–1383.

Lin, C.H., and Forscher, P., 1995, Growth cone advance is inversely proportional to retrograde F-actin flow, *Neuron* **14:** 763–771.

Liu, B.P., and Strittmatter, S.M., 2001, Semaphorin-mediated axonal guidance via Rho-related G proteins, *Curr. Opin. Cell Biol.* **13:** 619–626.

Liu, J.P., and Jessell, T.M., 1998, A role for rhoB in the delamination of neural crest cells from the dorsal neural tube, *Development* **125:** 5055–5067.

Loring, J.F., and Erickson, C.A., 1987, Neural crest cell migratory pathways in the trunk of the chick embryo, *Dev. Biol.* **121:** 220–236.

Lundquist, E.A., 2003, Rac proteins and the control of axon development, *Curr. Opin. Neurobiol.* **13:** 384–390.

McLennan, R., and Krull, C.E., 2002, Ephrin-as cooperate with EphA4 to promote trunk neural crest migration, *Gene Expr.* **10:** 295–305.

Murai, K.K., and Pasquale, E.B., 2005, New exchanges in eph-dependent growth cone dynamics, *Neuron* **46:** 161–163.

Marsh, L., and Letourneau, P.C., 1984, Growth of neurites without filopodial or lamellipodial activity in the presence of cytochalasin B, *J. Cell Biol.* **99:** 2041–2047.

Morrison, E.E., Moncur, P.M., and Askham, J.M., 2002, EB1 identifies sites of microtubule polymerisation during neurite development, *Brain Res. Mol. Brain Res.* **98:** 145–152.

Nathke, I.S., Adams, C.L., Polakis, P., Sellin, J.H., and Nelson, W.J., 1996, The adenomatous polyposis coli tumor suppressor protein localizes to plasma membrane sites involved in active cell migration, *J. Cell Biol.* **134:** 165–179.

Newgreen, D.F., Ritterman, M., and Peters, E.A., 1979, Morphology and behaviour of neural crest cells of chick embryo in vitro, *Cell Tissue Res.* **203:** 115–140.

Oakley, R.A., Lasky, C.J., Erickson, C.A., and Tosney, K.W., 1994, Glycoconjugates mark a transient barrier to neural crest migration int eh chicken embryo, *Development* **120:** 103–114.

O'Connor, T.P., and Bentley, D., 1993, Accumulation of actin in subsets of pioneer growth cone filopodia in response to neural and epithelial guidance cues in situ, *J. Cell Biol.* **123:** 935–948.

Osborne, N.J., Begbie, J., Chilton, J.K. Schmidt, H., and Eickholt, B.J., 2005, Semaphorin/neuropilin signaling influences the positioning of migratory neural crest cells within the hindbrain region of the chick, *Dev. Dyn.* **232:** 939–949.

Patapoutian, A., and Reichardt, L.F., 2000, Roles of Wnt proteins in neural development and maintenance, *Curr. Opin. Neurobiol.* **10:** 392–399.
Perris, R., and Perissinotto, D., 2000, Role of the extracellular matrix during neural crest cell migration, *Mech. Dev.* **95:** 3–21.
Poliakov, A., Cotrina, M., and Wilkinson, D.G., 2004, Diverse roles of eph receptors and ephrins in the regulation of cell migration and tissue assembly, *Dev. Cell* **7:** 465–480.
Pujol, F., Kitabgi, P., and Boudin, H., 2005, the chemokine SDF-1 differentially regulates axonal elongation and branching in hippocampel neurons. *J. Cell. Sci.* **118(pt 5):** 1071–80.
Rickmann, M., Fawcett, J.W., and Keynes, R.J., 1985, The migration of neural crest cells and the growth of motor axons through the rostral half of the chick somite, *J. Embryol. Exp. Morphol.* **90:** 437–455.
Rochlin, M.W., Dailey, M.E., and Bridgman, P.C., 1999, Polymerizing microtubules activate site-directed F-actin assembly in nerve growth cones, *Mol. Biol. Cell* **10:** 2309–2327.
Sabry, J.H., O'Connor, T.P., Evans, L., Toroian-Raymond, A., Kirschner, M., and Bentley, D., 1991, Microtubule behavior during guidance of pioneer neuron growth cones in situ, *J. Cell Biol.* **115:** 381–395.
Santagati, F., and Rijli, F., 2003, Cranial neural crest and the building of the vertebrate head, *Nature Rev. Neuro.* **4:** 806–818.
Santiago, A., and Erickson, C.A., 2002, Ephrin-B ligands play a dual role in the control of neural crest cell migration, *Development* **129:** 3621–3632.
Sahin, M., Greer, P.L., Lin, M.Z., Poucher, H., Eberhart, J., Schmidt, S., et al., 2005, Eph Dependent tyrosine phosphorylation of ephexin1 moduulates growth cone collapse, *Neuron* **46:** 191–204.
Salie, R., Niederkofler, V., and Arber, S., 2005, Patterning molecules: Multitasking in the nervous system, *Neuron* **45(2):** 189–192.
Sarmiere, P.D., and Bamburg, J.R., 2004, Regulation of the neuronal actin cytoskeleton by ADF/cofilin. *J. Neuroboil.* **58(1):** 103–17.
Schaefer, A.W., Kabir, N., and Forscher, P., 2002, Filopodia and actin arcs guide the assembly and transport of two populations of microtubules with unique dynamic parameters in neuronal growth cones. *J. Cell. Biol.* **158(1):** 139–52.
Schilling, T.F., and Kimmel, C.B., 1994, Segment and cell type lineage restrictions during pharyngeal arch development in the zebrafish embryo, *Development* **120:** 483–494.
Schneider, C., Wicht, H., Enderich, J., Wegner, M., and Rohrer, H., 1999, Bone morphogenetic proteins are required in vivo for the generation of sympathetic neurons. *Neuron.* **24(4):** 861–70.
Sechrist, J., Serbedzija, G.N., Scherson, T., Graser, S.E., and Bronner-Fraser, M., 1993, Segmental migration of the hindbrain neural crest does not arise from its segmental generation, *Development* **118:** 691–703.
Segal, R.A., 2003, Selectivity in neurotrophin signaling: Theme and variations, *Annu. Rev. Neurosci.* **26:** 299–330.
Serbedzija, G.N., Bronner-Fraser, M., Fraser, S.E., 1992, Vital dye analysis of cranial neural crest cell migration in the mouse embryo. *Development.* **116(2):** 297–307.
Silver, J., 1994, Inhibitory molecules in development and regeneration, *J. Neurol.* **242:** S22–S24.
Strachan, L.R., and Condic, M.L., 2004, Cranial neural crest recycle surface integrins in a substratum-dependent manner to promote rapid motility, *J. Cell Biol.* **167:** 545–554.

Suter, D.M., and Forscher, P., 2000, Substrate-cytoskeletal coupling as a mechanism for the regulation of growth cone motility and guidance, *J. Neurobiol.* **44:** 97–113.

Tanaka, E., and Sabry, J., 1995, Making the connection: Cytoskeletal rearrangements during growth cone guidance, *Cell* **83:** 171–176.

Tanaka, E., Ho, T., and Kirschner, M.W., 1995, The role of microtubule dynamics in growth cone motility and axonal growth, *J. Cell Biol.* **128:** 139–155.

Teddy, J.M., and Kulesa, P.M., 2004, In vivo evidence for short- and long-range cell communication in cranial neural crest cells, *Development* **131:** 6141–6151.

Testaz, S., Delannet, M., and Duband, J., 1999, Adhesion and migration of avian neural crest cells on fibronectin require the cooperating activities of multiple integrins of the (beta)1 and (beta)3 families, *J. Cell Sci.* **112:** 4715–4728.

Tosney, K.W., and Landmesser, L.T., 1985, Growth cone morphology and trajectory in the lumbosacral region of the chick embryo, *J. Neurosci.* **5:** 2345–2358.

Tosney, K.W., and Oakley, R.A., 1990, The perinotochordal mesenchyme acts as a barrier to axon advance in the chick embryo: Implications for a general mechanism of axonal guidance, *Exp. Neurol.* **109:** 75–89.

Trainor, P., and Krumlauf, R., 2000, Plasticity in mouse neural crest cells reveals a new patterning role for cranial mesoderm, *Nat. Cell Biol.* **2:** 96–102.

Vadlamudi, R.K., Li, F., Barnes, C.J., Bagheri-Yarmand, R., and Kumar, R., 2004, p41-Arc subunit of human Arp2/3 complex is a p21-activated kinase-1-interacting substrate, *EMBO Rep.* **5:** 154–160.

Wahl, S., Barth, H., Ciossek, T., Aktories, K., and Mueller, B.K., 2000, Ephrin-A5 induces collapse of growth cones by activating Rho and Rho kinase, *J. Cell Biol.* **149:** 263–270.

White, P.M., Morrisou, S.J., Orimoto, K., Kubu, C.J., Verdi, J.M., and Anderson, D.J., 2001, Neural crest stem cells undergo cell-intrinsic differentiation signals. *Neuron.* **29(1):** 57–71.

Wong, K., Ren, X.R., Huang, Y.Z., Xie, Y., Liu, G., Saito, H., et al., 2001, Signal transduction in neuronal migration: Roles of GTPase activating proteins and the small GTPase Cdc42 in the Slit-Robo pathway, *Cell* **107:** 209–221.

Yu, H.H., and Moens, C.B., 2005, Semaphorin signaling guides cranial neural crest cell migration in zebrafish, *Dev. Biol.* **280:** 373–385.

Yuste, R., and Konnerth, A., 2005, *Imaging in Neuroscience and Development*, Cold Spring Harbor Laboratory Press, Cold Spring Harbor, NY.

Zhou, F.Q., and Cohan, C.S., 2004, How actin filaments and microtubules steer growth cones to their targets? *J. Neurobiol.* **58:** 84–91.

Zhou, F.Q., Waterman-Storer, C.M., and Cohan, C.S., 2002, Focal loss of actin bundles causes microtubule redistribution and growth cone turning, *J. Cell Biol.* **157:** 839–849.

Zhou, F.Q., Zhou, J., Dedhar, S., Wu, Y.H., and Snider, W.D., 2004, NGF-induced axon growth is mediated by localized inactivation of GSK-3beta and functions of the microtubule plus end binding protein APC, *Neuron* **42:** 897–912.

Zou, Y., 2004, Wnt signaling in axon guidance, *Trends Neurosci.* **27:** 528–532.

Zumbrunn, J., Kinoshita, K., Hyman, A.A., and Nathke, I.S., 2001, Binding of the adenomatous polyposis coli protein to microtubules increases microtubule stability and is regulated by GSK3 beta phosphorylation, *Curr. Biol.* **11:** 44–49.

14 Mechanisms of Axon Regeneration

Jan M. Schwab and Zhigang He

14.1. Introduction

The inability of injured central neurons to regenerate their axons and rebuild their functional connections leads consequently to permanent loss of function following injury of central nervous system (CNS). The consequences of injury are not just a break in communication between neurons but also a cascade of events leading to neuronal degeneration and cell death. Thus, regeneration in the adult CNS requires multistep processes. First, the injured neurons must survive and then the damaged axons must extend to innervate their original targets. Once the contact is made, the axons need to be remyelinated and functional synapses need to form on the surface of the targeted neurons. Several strategies could be considered to achieve these objectives such as cell replacement, neurotrophic factor delivery, removal of the inhibitory molecules at the site of injury, bridging the lesion gap with artificial substrates, and modulation of the immune responses. In this chapter, we will focus our discussion on the issues relevant to extracellular impediments and intracellular mechanisms for neuritogenesis, which results in axon regeneration, particularly in the CNS.

14.2. Differences Between CNS and PNS Axonal Regeneration

As neuronal connections are made in the developing nervous system of mammals, axons progressively cease growing. In the CNS, lesions that occur at or around the perinatal period can trigger some degree of regeneration. However, the majority of lesioned axons in a postnatal organism are not repaired, often resulting in devastating and permanent functional deficits. This is in contrast to axons of the peripheral nervous system (PNS), where regeneration often occurs even in the adult. Thus, both extrinsic and intrinsic differences between CNS and PNS may provide reasonable clues for studying the regeneration failure in the adult CNS. As early as in 1911, Tello and Cajal showed that adult CNS neurons could regrow if they were provided access to the permissive environment of a sciatic nerve (Ramon y Cajal, 1928). Seventy years passed before Aguayo and colleagues replicated their results with new methods that definitively confirmed that adult CNS axons have regenerative ability. These observations clearly suggest that the environmental differences in the PNS and CNS might be primary determinants for the regenerative fate of injured axons.

In the past two decades, progress has been made in defining the environmental determinants of axonal regeneration in cellular and molecular terms. In contrast to Schwann cells, a major glial cell type of the PNS that produce components of PNS myelin, CNS has unique sets of glial cells including astrocytes, oligodendrocytes, and other cells such as microglia. Analyzing the products of these PNS or CNS distinct glial cells represents an important approach to study mechanisms of axon regeneration. In addition, it is also noted that the damaged CNS microenvironment is not easily accessible to the macrophage scavengers whereas most of the debris in the PNS are rapidly removed by reactive macrophages and Schwann cells. Thus, presumably, even the presence of some inhibitory molecules in the PNS may not stay long enough to inhibit axons from regeneration. In addition, macrophages have been identified to modify directly the nonpermissive property of the CNS (David et al., 1990).

Although injured axons in general are able to regenerate beyond the lesion site in the adult PNS, it is not always true that these regenerated fibers are able to achieve full functional recovery. The degree of functional recovery attained after a peripheral nerve injury is, to a large extent, dependent on the number of lesioned axons that have reinnervated their appropriate postsynaptic target(s) (Fu and Gordon, 1997). However, very little is known about the molecular mechanisms regulating selective targeting of regenerating peripheral axons. One explanation for this apparent oversight is the general conception that regenerating axons do not reinnervate their original targets selectively after peripheral nerve injury. However, there are several examples in which selective axon regeneration likely occurs. For example, physiologically distinct classes of fast and slow motor neurons have a propensity to reinnervate their appropriate class of muscle fibers preferentially (Soileau et al., 1987). Similarly, transected motor neurons reinnervate the diaphragm in the same topographically correct manner as occurs during development (Laskowski and Sanes, 1988), suggesting that many of the same guidance cues expressed during embryogenesis are reexpressed after peripheral nerve injury. A simple system in further studying the molecular mechanisms of this selective reinnervation issue is known as preferential motor reinnervation (PMR) of injured femoral motor axons (Brushart, 1988). Thus, it is conceivable that the regenerating axons in the PNS may provide a useful system to study the guidance, targeting selection and synapse connection in the adult nervous system. Transgenic models have been developed, which provide a powerful tool to study CNS axonal regeneration pattern in more detail and enable to detect regenerative phenomena on the level of monosynaptic synapse formation from sprouted motor axons on motor neurons (Bareyre et al., 2005; Kerschensteiner et al., 2005).

14.3. Mechanisms of Regeneration Failure in the Adult CNS

To further dissect whether the difference between the CNS and PNS environments is due to a lack of trophic support or the presence of inhibitory molecules in the CNS, Schwab and Thoenen (1985) compared neurite growth from dissociated perinatal CNS neurons in contact with adult PNS (sciatic nerve) or CNS (optic nerve) tissue explants in the presence of nerve growth factor (NGF). After 2 weeks in culture, they found massive fiber growth into the sciatic nerve, whereas few to no axons grew in the optic nerve. These findings further suggested the presence of inhibitory influences in the adult CNS environment. As the most apparent difference between the CNS and PNS is the presence of CNS-specific glial cell types, major advances in recent years have been made in our understanding of inhibitory properties of CNS glial cells, the molecules associated with their axonal growth inhibition, and the underlying mechanisms in the context of CNS injury and axonal regeneration. The situs of such a situation is depicted in Figure 14.1 illustrating injured axons forming axonal stumps (bulbs) and surrounding growth inhibitory molecules located in the myelin and the glial scar following spinal cord injury.

14.3.1. Inhibitory Environment

14.3.1.1. Regeneration Inhibitors Associated with CNS Myelin

Inhibitory factors in oligodendrocyte-derived CNS myelin have also proposed to play a crucial role in limiting axon regeneration. Direct experimental evidence for the ability of CNS myelin to inhibit neurite outgrowth was later provided by Martin Schwab and other labs, which performed elegant cell and tissue culture experiments to show that CNS myelin, but not PNS myelin, possesses potent neurite growth inhibitory activity (Carbonetto et al., 1987; Crutcher, 1989; Sagot et al., 1991; Savio and Schwab, 1989). They found that neurite outgrowth on frozen sections from developing and adult peripheral and central tissues was clearly worse on white matter than on gray matter. Likewise, in cocultures of dissociated glial cells and neurons, mature oligodendrocytes were strictly avoided by neurons. Subsequent studies from several laboratories have resulted in the identification of a number of putative inhibitors associated with CNS myelin. In culture, differentiated oligodendrocytes and CNS myelin exert a strong inhibitory effect on adhesion and outgrowth of primary neurons, neuroblastoma cells, and even on the spreading of nonneuronal cells (Schwab and Caroni, 1988; Savio and Schwab, 1989). Consistent with this, Keirstead and colleagues observed that temporary suppression of myelination in the developing chick spinal cord extends the permissive period for axonal regeneration after injury (Keirstead et al., 1992). Similarly, axonal regeneration over considerable distances was also achieved in myelin-free spinal cord (Keirstead et al., 1992; Vanek et al., 1998), while purified CNS myelin immobilized as a substrate was shown to potently inhibit neurite outgrowth *in vitro* (Savio and Schwab, 1989; Schwab and Caroni, 1989).

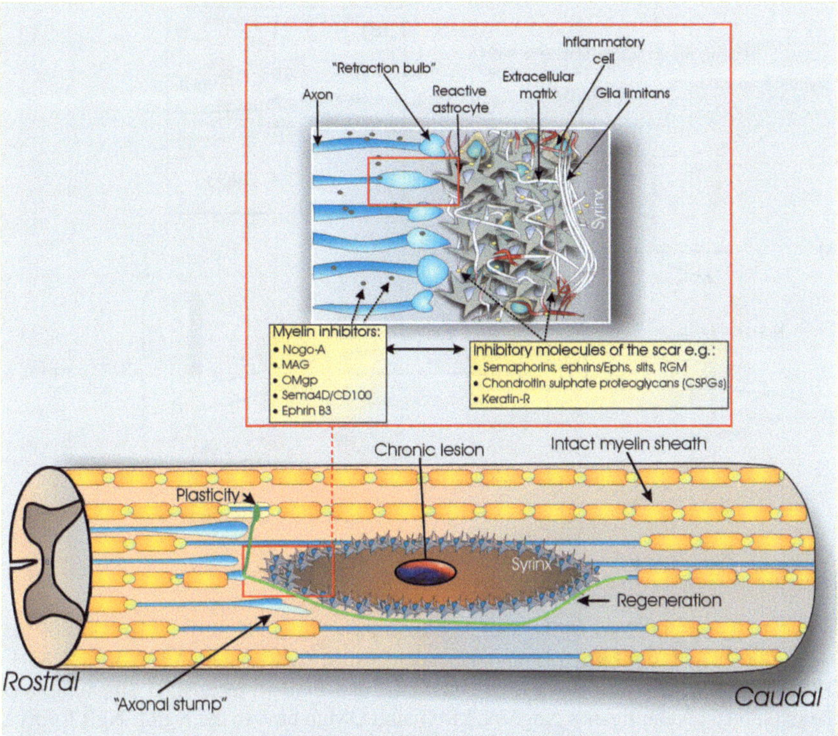

FIGURE 14.1. Simplified presentation of a chronic spinal cord lesion. Localization of the injured axon and its inhibitory environment (myelin, scar) and identification of growth inhibitors of axonal sprouting (see insert). Sprouting axons lead to regeneration (axonal growth and connection with the distal functionally corresponding axon end) and plasticity (axonal growth and connection with an intact axon). These phenomena are mostly located within the residual surviving spinal cord sections and, in general, do not pass through the primary damaged regions or through cavities.

Furthermore, anti-myelin antibodies have been used to neutralize the inhibitory effects of myelin (Caroni and Schwab, 1988a,b) and, more importantly, stimulate some sprouting/regeneration of the corticospinal tract (CST) *in vivo* (Bregman et al., 1995; Huang et al., 1999; Schnell and Schwab 1990; Schnell et al., 1994). Thus, it appears that something specific to, or greatly enriched in, CNS myelin actively inhibits regeneration. However, it is unclear whether all mature axons respond equally to myelin inhibitors.

14.3.1.2. Nogo-A, OMgp, and MAG Account for the Majority of the Inhibitory Activity of CNS Myelin

In the past years, three major myelin components have been identified as putative inhibitors of regeneration through biochemical approaches (Figure 14.2A). One

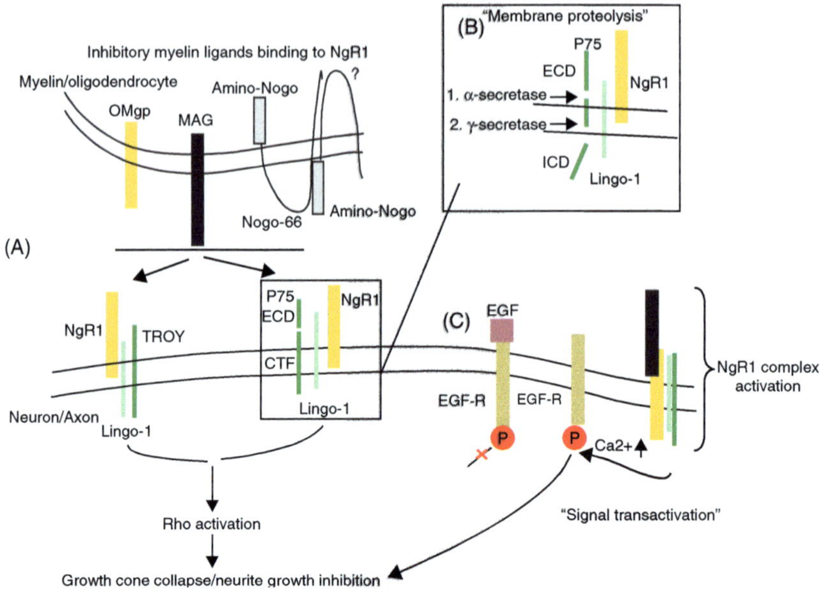

FIGURE 14.2. Insights into NgR1 operated axon growth inhibition. Incorporation of the recently identified molecular players and mechanisms, such as amino-Nogo, LINGO-1, TROY/TAJ, α- and γ-secretase $p75^{NTR}$ shedding, and EGF-R, into NgR1 signaling. (A) The inhibitory myelin ligands Nogo-A, MAG, and OMgp bind to the NgR1. NgR forms a trimeric complex with its coreceptors LINGO-1 and $p75^{NTR}$. Activation of the NgR complex triggers a rapid rise of intracellular Ca^{2+}, which results in the activation of the converging Rho pathway leading to axonal growth arrest and growth cone collapse. In the absence of $p75^{NTR}$, TROY acts as a functional homologue. (B) When $p75^{NTR}$ is part of the NgR complex, its sequential cleavage by α- and γ-secretases gives rise to an intracellular domain (ICD), which induce Rho activation. In more detail, the α-secretase cleaves the extracellular domain (ECD) of $p75^{NTR}$. The remaining membrane-bound carboxy-terminal fragment of $p75^{NTR}$ (CTF) is further cleaved by intramembrane proteolysis (RIP) through γ-secretase resulting in the cytoplasmic release of ICD, which "transmits" the inhibitory signal. (C) A link showing how the rise in intracellular Ca^{2+} induces the activation of Rho is provided by the EGF-R. Ca^{2+} rise triggered by the NgR1complex leads to EGF-R phosphorylation, which in turn induces Rho activation. However, extracellular binding of EGF-R to its ligand EGF does not. This phenomenon of inter-receptor crosstalk is also referred to as "*trans*-activation" and operated in consequence by the intracellular Ca^{2+} rise.

component is the myelin-associated glycoprotein (MAG), a transmembrane protein with five immunoglobulin-like domains in the extracellular region (Arquint et al., 1987; Salzer et al., 1987). Although MAG is capable of inhibiting axon outgrowth from different types of cultured neurons (Li et al., 1996; McKerracher et al., 1994; Mukhopadhyay et al., 1994; Tang et al., 1997), knockout animals have yielded conflicting data on the effects of removing MAG on axon regeneration *in vivo* (Bartsch et al., 1995; Schafer et al., 1996). In addition to MAG, another putative inhibitor, Nogo/NI-250 (Caroni and Schwab, 1988a,b), has attracted much attention because an anti-NI-250 monoclonal antibody, IN-1, has been shown to neutralize the growth-inhibitory effects of myelin-associated inhibitors both *in vitro* and *in vivo*. IN-1 treatment resulted in long-distance fiber growth and increased axonal sprouting within the adult CNS (Bregman et al., 1995; Schnell and Schwab, 1990; Schnell et al., 1994; Thallmair et al., 1998). The partial-peptide sequence of biochemically purified NI-250 (Spillmann et al., 1998) allowed several groups to clone the corresponding cDNA that encodes three Nogo splicing isoforms, designated Nogo-A, B, and C (Chen et al., 2000; Grandpre et al., 2000; Prinjha et al., 2000). The exact topology of the Nogo-A protein remains controversial, but it is believed to contain at least two transmembrane domains. Both the amino-terminal cytoplasmic (amino-Nogo) and extracellular (Nogo-66) domains of Nogo are able to inhibit axon growth *in vitro* (Chen et al., 2000; Fournier et al., 2001; Prinjha et al., 2000). The third major inhibitor in CNS myelin, oligodendrocyte myelin glycoprotein (OMgp), was identified as a potent inhibitor of neurite outgrowth *in vitro* (Wang et al., 2002a). Although it is possible that other myelin-associated inhibitors exist, our fractionation studies suggest that Nogo-A, OMgp, and MAG collectively account for the majority of the inhibitory activity associated with CNS myelin (Wang et al., 2002a).

14.3.1.3. Other Possible Inhibitors in CNS Myelin

Although Nogo-A, MAG, and OMgp may account for a majority of the *in vitro* inhibitory activity in CNS myelin, many lines of evidence suggest that other myelin components may also play a role in regeneration inhibition. For example, outgrowth inhibiting chondroitin sulfate proteoglycans (CSPGs) have been shown to be a minor component of CNS myelin (Niederost et al., 1999). A study suggests that the transmembrane semaphorin Sema4D/CD100 is expressed on oligodendrocytes and can be detected in CNS myelin (Moreau-Fauvarque et al., 2003). Recombinant proteins containing the ECD of Sema4D/CD100 are able to induce growth cone collapse and repel growing axons in a stripe assay. In addition, following spinal cord injury, Sema4D/CD100 is strongly upregulated in oligodendrocytes at the periphery of the lesion. Thus, even though Sema4D/CD100 may not be an abundant CNS myelin component, its injury-elicited upregulation may make it more relevant to axon regeneration. It will be interesting to examine whether other axonal repulsive molecules expressed in oligodendrocytes may also be injury inducible and whether they participate in restricting axon regeneration. The latest myelin inhibitory component discovered is ephrin-B3 (Benson et al., 2005). Also the repulsive guidance molecule (RGM) (Monnier et al., 2002) is expressed by oligodendrocytes upon CNS injury (Schwab et al., 2005a). Further studies will reveal whether RGM is present in CNS myelin. As has been suggested, the relative contributions of different types of myelin-associated inhibitors have not been fully characterized. Overall, these studies suggest that the effects of multiple inhibitors may be necessary to elicit complete blockage of axon regeneration upon injury in the adult CNS.

14.3.1.4. Inhibitors Associated with the Glial Scar

In many instances where CNS regeneration has been described following injury, regenerating nerve fibers either bypass or stop at the neuroglial scar that arises from the lesioned tissue. This glial scar is composed primarily of glial cells and connective tissue elements (Fawcett and Asher, 1999; Silver and Miller, 2004; Stichel and Muller, 1998), and therefore has been thought to act as a mechanical barrier that is impenetrable by regenerating axons (Figure 14.1). However, many lines of study have demonstrated that the glial scar contains components that are able to actively inhibit axon growth, including the glycoprotein tenascin-C and CSPGs (Davies et al., 1997, 1999; Grimpe and Silver, 2002; McKeon et al., 1991). The expression of several different CSPGs in the lesion site is increased following CNS injury (Grimpe and Silver, 2002), and enzymatic removal of these molecules using chondroitinase ABC (ChABC) has been reported to improve axon regeneration and functional recovery in different lesion models (Bradbury et al., 2002; Moon et al., 2001; Morgenstern et al., 2002). On formation, the molecular composition and inhibitory nature of the glial is not static. Long-term observations reveal that parts of the glial scar progressively become permissive whereas the areas of ECM deposition comprise the persistent inhibitory part of the scar (Camand et al., 2004). It appears that the expression of an extensive number of guidance molecules (netrins, semaphorins, ephrins, EphA receptors, Slit/robo, RGMs) are also dynamically regulated and expressed in the developing scar and/or by inflammatory cells following CNS injury constituting a map of overwhelming information for the mature axon to process. The putative repulsive axon guidance molecules of the class 3 semaphorins (De Winter et al., 2002) and members of the ephrin family (ephrin-B3) (Grimpe and Silver, 2002) are upregulated. Furthermore, EphA4 is also expressed in astrocytic scar tissue. Absence of the ephrin-B3 interaction partner EphA4 resulted in enhanced axonal regeneration and diminished scar formation *in vivo* (Goldshmit et al., 2004). Also the EphA3–A4, A6, A7, and A8 receptor (Irizarry-Ramirez et al., 2005; Willson et al., 2002, 2003) and EphB3 were upregulated upon CNS injury (Miranda et al., 1999). Furthermore, the netrins, comprising another family of guidance molecules, which were anchored in both the cellular membrane and the extracellular matrix, are regulated following CNS injury (Manitt and Kennedy, 2002). Their receptors (Unc- and DCC-[deleted in colorectal cancer] family, including neogenin) are also expressed by axons in the adult CNS (Dickson, 2002; Guan and Rao, 2003; Manitt and Kennedy, 2002). Netrin-1 is lesionally expressed by macrophages (Wehrle et al., 2005) following Spinal Cord Injury (SCI). The Slits, Slit-1, Slit-3, and their receptors robo-1, robo-2, and robo-3 (rig-1) were also detected at the lesion site following CNS injury (Wehrle et al., 2005). Last, the RGM (Monnier et al., 2002; Stahl et al., 1990) was identified in the scar (Schwab et al., 2005a,b), which operates both axon repulsion and survival via binding to its ligand neogenin (Matsunaga et al., 2004; Rajagopalan et al., 2004).

14.3.1.5. Other Inhibitory Cues Acting at the Lesion Site

Moreover, repulsive molecules, such as the coagulation protease thrombin, are found in the acute lesion which induces growth inhibition of the axonal fibers. Lesional thrombocytes additionally secrete the strongly repulsive lysophosphatidic acid (LPA) (Jalink and Moolenaar, 1992; Jalink et al., 1994; Moolenaar, 1995). Inflammatory messengers, such as the tumor necrosis factor-α (TNF-α), are secreted at the lesion site by microglia and astrocytes in reponse to injury. TNF-α reduces neurite outgrowth and branching *in vitro* by activating RhoA (Neumann et al., 2002). Not localized to cellular entities yet are the putative inhibitory candidates Bamacan/CSPG-6, semaphorin-6B, and the robo-related protein CDO (Kury et al., 2004).

14.3.1.6. The Lack of Neurotrophic Support in the Adult CNS

A well-documented repair strategy for stimulating axon regeneration is transplantation. For example, providing a PNS environment, such as peripheral nerve grafts, pure populations of Schwann cells, and fetal spinal cord tissue, have all been shown to stimulate axon regeneration. Some of the beneficial effects of these manipulations may be due to the synthesis of neurotrophic factors by the transplanted cells. Thus, there have been increased interests in the possibility of stimulating growth of injured spinal cord neurons by the exogenous supply of neurotrophic factors. For example, Schnell et al. (1994) found that despite an increase in sprouting of CST axons observed after NT3 treatment, elongation of fibers remains restricted to about 1 mm. However, if they combined NT3 treatment with an IN-1 antibody infusion, a small population of fibers was observed to elongate over a much longer distance. Similarly, NT3, but not GDNF or BDNF, has been shown to promote growth of lesioned adult rat sensory axons ascending in the dorsal columns of the spinal cord (Bregman et al., 1997). Thus, it is possible that supplying exogenous neurotrophic factors may be used as a complementary approach to encourage axon regeneration. However, several important questions remain unclear. For example, as the receptors of neurotrophins are present on both the surfaces of the axon and the soma, it remains unclear whether delivering neurotrophins to the axonal terminals (the lesion site) or their cell bodies will be more effective in promoting axon regeneration. This is important not only for designing therapeutic strategies but also for understanding the underlying mechanisms of neurotrophins in affecting axon regeneration.

14.3.2. Impaired Ability of the Axon to Regrowth

Besides the inhibitory environment, part of the limited regeneration capacity is attributed to a limited intrinsic ability of the axon itself to regrowth. Being the real substrate for regeneration, influence of the injury (axotomy) on the ability of the growth cone to sense inhibitors is now emerging.

14.3.2.1. Decreased Intrinsic Regenerative Ability in Mature Neurons

It is known that embryonic and in some cases perinatal CNS neurons will spontaneously regenerate both *in vitro* and *in vivo*, while their postnatal counterparts cannot. Even when a sciatic nerve graft is transplanted into the lesioned CNS, only small amounts of axons are able to regenerate and extend at a very slow rate. Thus, the failure in regeneration may also be attributable to the reduced growth capacity of mature neurons. A possible underlying mechanism for such a developmental reduction in axonal growth ability is a decrease of cyclo-Adenosine Monophosphate (cAMP) levels. At least in dorsal root ganglion (DRG) neurons, endogenous cAMP levels are greatly elevated in young neurons and decrease precipitously after birth. In addition, the superior regenerative growth of embryonic neurons *in vitro* can be abrogated by blocking the cAMP/PKA (protein kinase A) pathway. Injecting db-cAMP into DRGs can increase the regeneration of ascending axons from DRG neurons in a dorsal column hemisection model. Similarly, by culturing the retinal ganglion neurons *in vitro*, Goldberg et al. (2002) found that a dramatic decline of axonal growth ability occurs around the stage of birth. Axonal growth rate of retinal ganglion neurons from E20 is about 10 times faster than those from P8. They also found that such a developmental switch is the result of a contact-dependent signal from amacrine cells. The identification of such molecules will be important for understanding how the axonal growth ability is regulated.

14.3.2.2. The Injured Axon

For a long time, dystrophic-like axotomized axons elongating from mature postmitotic neurons were thought to be nonmotile neuritic residuum incapable of robust regeneration. Time-lapse observations of induced dystrophic-like axons by Jerry Silver's group challenged this idea (Tom et al., 2004). They showed that the so-called sterile endings (plump, foamy, sensing growth cone) are dynamic and constantly moving. This cycle repeats itself continuously and, although struggling growth cones often manage to move short distances, they always round up in a more compact ball and retract. *In-vitro* labeling with dextran showed that dystrophic endings differ substantially in the rate and trafficking of endocytosis, putatively affecting membrane remodeling at the growth cone and the expression of surface molecules or receptors (Tom et al., 2004). Accordingly, the high density of mitochondria in axotomized neurites (Sotelo and Palay, 1971) probably reflects the important energy-demand required for the permanent and fast oscillations of the dystrophic axon tip. The response to axotomy varies and depends on factors such as age, species, nature of the injury (compression vs. transaction), and distance of the injury from the parent soma (Tom et al., 2004).

14.3.3. Dynamic and Injury-Induced Growth Inhibition

By contrast with the glial scar, myelin was considered a constitutive static inhibitory barrier not reactive to lesion in the CNS. However, like in the scar, results suggest considerable add-on inhibition of myelin as a result of CNS injury (Schwab et al., 2005c). Furthermore, catastrophic events cause morphological and biochemical changes in the axon itself. This results in the accumulation of cytoskeleton components and intracellular transported proteins at the axon tip, which might modify the axon response to its inhibitory environment. Evidence for injury reactivity challenges our understanding of axon growth-inhibition in the injured CNS. These might be due to (i) qualitative and quantitative enrichment of the periaxonal environment by myelin/oligodendrocytes after CNS injury, (ii) increased axonal sensitivity to its inhibitory environment, and (iii) lesion-induced axonal signaling. Postlesional reactive myelin/oligodendrocyte-inhibition necessitates the development of novel screening approaches and therapeutic reagents to promote axonal regeneration. Moreover, we need to improve our understanding of the pathophysiology of the lesion to find more efficient experimental strategies to restore neurological function (Mandolesi et al., 2004).

14.4. Integration and Intra-Axonal Processing of the Inhibitory Signals

As Nogo-66 receptor (NgR1) is the pan-receptor for the inhibitory myelin ligands comprising most of the myelin derived inhibitory potency, its signaling became an area of intensive investigation. NgR1 elicited intra-axonal signaling might serve as a matrix for other repulsive cues derived from the glial scar.

14.4.1. NgR1 Complex

14.4.1.1. NgR1 Is a Common Receptor for the Three Myelin Inhibitors

Further insight into the signaling mechanisms mediating the inhibitory activity of Nogo-A came with the identification of a functional receptor for Nogo-66, NgR1, by expression cloning (Fournier et al., 2001). NgR1 can bind Nogo-66 with high affinity. It is widely expressed in the adult nervous system, but the expression levels vary in different populations and subpopulations of mature neurons (Hunt et al., 2002a,b; Wang et al., 2002c). Furthermore, early embryonic chick retinal ganglion cells that are normally insensitive to Nogo-66 become responsive upon expression of NgR1, suggesting that NgR1 is able to mediate the activity of Nogo-66. Strikingly, it was shown that both MAG and OMgp also exert their inhibitory activities through the same NgR1 (Figure 14.2A) whereas MAG binds in addition to NgR2. The neurons treated with phosphatidylinositol-specific phospholipase C, which can cleave NgR1 and other Glycosyl Phosphatidyl Inositol (GPI)-linked proteins from the axon surface, lost their responses to these myelin inhibitors. Although the receptor of the amino-terminus of Nogo-A remains elusive, it is likely that the majority of the inhibitory activity associated with CNS myelin is mediated by NgR1. Consistent with this notion, a soluble truncated version of NgR1 has been shown to antagonize neurite outgrowth inhibition elicited by CNS myelin (Fournier et al., 2002). However, it is still unknown whether NgR1 is the sole receptor for mediating these inhibitory activities.

14.4.1.2. P75/TROY and Lingo-1 Are Two Coreceptors of NgR1

Because NgR1 is a GPI-anchored axon surface molecule, it is unlikely to be sufficient for transducing the inhibitory signals across the plasma membrane. Previous studies have implicated p75, initially identified as a low-affinity receptor for neurotrophins (Hempstead et al., 2002; Roux and Barker, 2002), as a signaling coreceptor of NgR1 (Wang et al., 2002b; Wong et al., 2002). At least DRG sensory neurons and cerebellar granule neurons from p75 knockout mice have significantly reduced responses to myelin inhibition (Wang et al., 2002b; Yamashita et al., 2002). Along the same line, *in vivo*, axons of adult sympathetic neurons from p75 knockout mice overexpressing NGF can grow extensively in myelinated portions of the cerebellum and myelinated optic nerve tracts (Hannila and Kawaja, 1999; Walsh et al., 1999). However, no improvement in axon regeneration could be observed in p75 knockout mice after SCI (Song et al., 2004). In fact, p75 is only expressed in subpopulations of mature neurons, suggesting the existence of other p75 functional homologues as well. Results have implicated TROY, another TNF receptor family member, as a p75 functional homologue in mediating the activity of myelin inhibitors (Figure 14.2A). In addition to p75/TROY, another transmembrane protein Lingo-1 has been identified as the other signaling coreceptor in the complex (Mi et al., 2004). Nonneuronal cells transfected with NgR1, Lingo-1, and p75 or TROY could respond to myelin inhibitors by Rho activation, suggesting that these proteins are sufficient for reconstituting functional receptor complexes for myelin inhibitors.

14.4.2. Membranous Mechanism Interfering with Inhibitory Signal Integration

A new mechanism of NgR1 mediated intraneuronal signal integration was unraveled modifying one of the transducing components, p75. Upon stimulation by MAG, the NgR1 complex triggers intramembrane proteolysis (Figure 14.2B) of the coreceptor p75. p75 is sequentially cleaved by α- and γ-secretase in a PKC-dependent manner. The cleaved cytoplasmic product of p75, the ICD, is necessary for MAG induced inhibition of neurite outgrowth. This might also be a general mechanism for the p75/NgR1 mediated axon growth inhibition triggering the intra-axonal activation of RhoA (Domeniconi et al., 2005). Sequential ectodomain cleavage of a membrane bound molecule is also referred to as "shedding." "Shedding" acts as a "late" regulatory mechanism subsequently to gene transcription and posttranslational modifications. One of the proteases involved, the γ-secretase (presenelin assembled with three other membrane proteins), has also been implicated in the development of other neuropathological events such as APP cleavage giving rise to the amyloidogenic Aβ. Aβ is the principal neurotoxic component located in cerebral plaques found in the brain of patients with Alzheimer disease.

14.4.3. Intracellular Signaling of Axon Regeneration

Our knowledge of the signaling mechanisms underlying regeneration inhibition has largely been derived from two different approaches. First, the identification of the individual inhibitors and their receptor components has provided the molecular tools for identifying the intracellular signaling components. Second, because myelin-associated inhibitors and other types of axonal repellents, like semaphorins and slits, could elicit such similar axonal responses as growth cone collapse, the information from axon guidance studies can provide valuable insights into understanding the mechanisms of axon regeneration. Although we are still at the early stages of determining the detailed signaling mechanisms of myelin inhibition, several important intracellular molecules appear to be directly involved in myelin-elicited pathways. Among these are small GTPases, cyclic nucleotides, and intracellular calcium. While some molecules like the small GTPases are likely to be direct signal mediators, other molecules like cyclic nucleotides may act more indirectly by modulating the inhibitory signaling pathways.

14.4.3.1. Small GTPases as Critical Signaling Molecules

Evidence suggests a similar involvement of small GTPases in the signaling pathways downstream of myelin-associated inhibitors and CSPGs. The requirement of RhoA activation for the inhibitory activity of MAG and other myelin components was initially suggested by the observation that inhibiting RhoA activity by either C3 transferase or dominant-negative Rho in neurons allows neurite growth on inhibitory substrates (Lehmann et al., 1999). Subsequent studies provided direct biochemical evidence for RhoA activation by MAG and other myelin inhibitors (Niederost et al., 2002). In addition to RhoA, one of its major effectors termed RhoA-associated kinase (ROCK) has also been implicated in myelin-mediated reorganization of cytoskeleton structures. Like C3, a synthetic inhibitor of ROCK called Y27632 was also able to block the inhibitory activity of myelin inhibitors, adding ROCK to the list of major downstream components of the NgR-mediated inhibitory signal. *In vivo* experiments in rats and mice have even indicated that inactivation of RhoA or ROCK promotes axon regeneration and functional recovery after SCI (Chan et al., 2005).

14.4.3.2. Modulation by cAMP

In understanding the signaling mechanisms of guidance effects, it is important to consider other modulatory pathways that may participate indirectly in determining the specificity of response as suggested for axon guidance responses (Song et al., 1997). A well-documented example of this is the effect of intracellular cAMP levels on growth cones. By using a *Xenopus* spinal neuron-based growth cone turning assay, Poo and his colleagues found that depending on the level of cyclic nucleotides within a neuron, the response of the growth cone to many guidance cues can either be attractive or repulsive, with high levels favoring attraction and low levels favoring repulsion (Song et al., 1997). Filbin and her colleagues found that pretreatment (priming) of responding neurons with neurotrophins can block the inhibitory effects of MAG and perhaps other components of CNS myelin. In addition, they showed that this priming procedure elevates cAMP levels and activates PKA, thus providing a mechanism for overcoming neurite outgrowth inhibition (Cai et al., 1999). However, how cAMP/PKA impinges on the inhibitory pathway of myelin components remains unclear. Although studies from nonneuronal cells suggest a possibility that PKA may phosphorylate RhoA and regulate its activity, there is no definitive evidence for RhoA phosphorylation by PKA in the myelin inhibitory pathways. As mentioned earlier, cAMP-dependent PKA can phosphorylate p75 and affect its translocation to lipid rafts. A more recent study showed that elevated cAMP could also upregulate Arginase I and thus enhance the synthesis of polyamines (Cai et al., 2002). Increased cAMP levels resulted in regeneration of lesioned spinal axons (Qiu et al., 2002).

14.4.3.3. Calcium

The critical involvement of Ca^{2+} in mediating the axonal responses to environmental cues has been well documented in the past decade (Kater and Mills, 1991). In cultured *Xenopus* spinal neurons, a variety of guidance cues, such as netrin-1 and MAG, can induce a rise of intracellular calcium level in the growth cone. Also, preventing an elevation in cytoplasmic Ca^{2+} levels abolishes the growth cone turning response in gradients of individual cues. Moreover, photolytic release of caged Ca^{2+} or induction of Ca^{2+} release from internal stores with an extracellular gradient of ryanodine (in the absence of guidance cues) is sufficient to induce growth cone turning. Purified myelin fractions enriched in NI-35 also trigger a large and rapid increase of Ca^{2+} levels in the responding growth cone of rat DRG neurons. Depletion of the caffeine-sensitive intracellular calcium stores prevents the increase in $[Ca^{2+}]_i$ evoked by NI-35, thus demonstrating the involvement of calcium release from intracellular stores in the signaling pathways mediating NI-35-induced growth cone collapse. Studies also demonstrated that recombinant MAG triggers a Ca^{2+} influx. Even in nonneuronal cells expressing both NgR1 and p75, MAG-Fc can trigger a robust Ca^{2+} influx, suggesting the possibility that p75 may be sufficient to trigger the signal for calcium influx as elicited by myelin inhibitors.

However, it remains to be determined how the signal is transmitted from p75 to Ca^{2+} influx and how Ca^{2+} influx affects growth cone behavior. Studies implicated two known calcium effectors in affecting the axonal growth behaviors. By analyzing the phenotypes of mice lacking the expression of three isoforms of the Ca^{2+}-dependent transcription factor NFAT, Graef et al. (2003) demonstrated that calcineurin/NFAT signaling pathways are required for the neurite outgrowth promoting activity of neurotrophins and the chemoattractant netrin-1. At the same time, a class of Ca^{2+}-activated proteases calpains has also been implicated in responding to filopodial calcium transients to regulate growth cone motility. It will be interesting to examine whether these or other calcium downstream effectors act in the signaling pathways of myelin inhibition.

Unexpectedly, the epidermal growth factor receptor (EGF-R) was identified as calcium dependent effector candidate. Binding of the myelin inhibitors to the NgR1 complex triggers phosphorylation of the EGF-R in a calcium-dependent fashion (Koprivica et al., 2005). Functional relevance was confirmed by EGF-R kinase blockers which resulted in enhanced axon outgrowth of cerebellar granule neurons or DRG neurons when plated on a myelin substrate. Moreover, *in vivo* blockade of EGF-R kinase promoted nerve regeneration following optic nerve crush injury. Binding experiments reveal that EGF-R does not act as a receptor for myelin derived NgR1 ligands or is a part of the NgR1 receptor complex. It is noteworthy that the extent of axon regeneration was comparable to that induced by C3 transferase, a Rho activation inhibitor which is the downstream effector mechanism of axon growth inhibition. Large-scale screening also revealed that the inhibitory activity of the CSPGs located in the glial scar (the other axon regeneration impediment) was neutralized by EGF-R inhibitors.

Phosphorylation of EGF-R by binding to its ligand alone (not triggered by NgR1 receptor complex) did not block neurite outgrowth. The authors indicate that EGF-R activation alone is a "required but not sufficient signaling step" in the response to axon growth inhibitors (Koprivica et al., 2005). In more detail, EGF-R activation might interact with other cascades (or binding to molecules), which are switched on in response to the pleiotropic Ca^{2+} signal induced by the NgR1 complex. Or more directly, intracellular Ca^{2+} rise may lead to a different phoshorylation pattern of EGF-R compared to that induced by its own ligand EGF. This phenomenon of interreceptor "cross-talk" is also referred to as "*trans*-activation" (Figure 14.2C). Straightforward *in vivo* testing of this concept is facilitated by available EGF-R inhibitors, such as Erlotinib, which is approved for the treatment of cancer. Moreover, in addition to these guidance molecules, it is known that many pathways, particularly those involved in neuronal activity, are able to affect the intracellular calcium level.

In addition, also as a consequence of CNS injury, neurons depolarize and release K^* and glutamate, which lead to excitatory burst and further Ca^{2+} influx. In the core region of CNS injury where mechanical destruction is worsened by ischemic conditions, neurons can undergo anoxic depolarization and never repolarize. Thus, it is conceivable that calcium may function, at least in some circumstances, as a coincidence detector for different signaling pathways to affect the final axonal responses to myelin inhibitors and other cues.

14.4.3.4. Modulation by PKC

Protein kinases play a plethora of roles in intracellular signaling. PKC promotes neuronal cell death and can be inhibited by Gö6976 (Ghoumari et al., 2002). In the spinal cord the intrathecal delivery of Gö6976 induced anatomical regeneration of sensory fibers of the dorsal columns (Sivasankaran et al., 2004). In addition the PKC inhibitor was able to inhibit the Nogo-66 and MAG and CSPG-induced Rho activation (Figure 14.3). However, since Gö6976 also inhibits the neurotrophin receptors (Trk) other modes of action are likely.

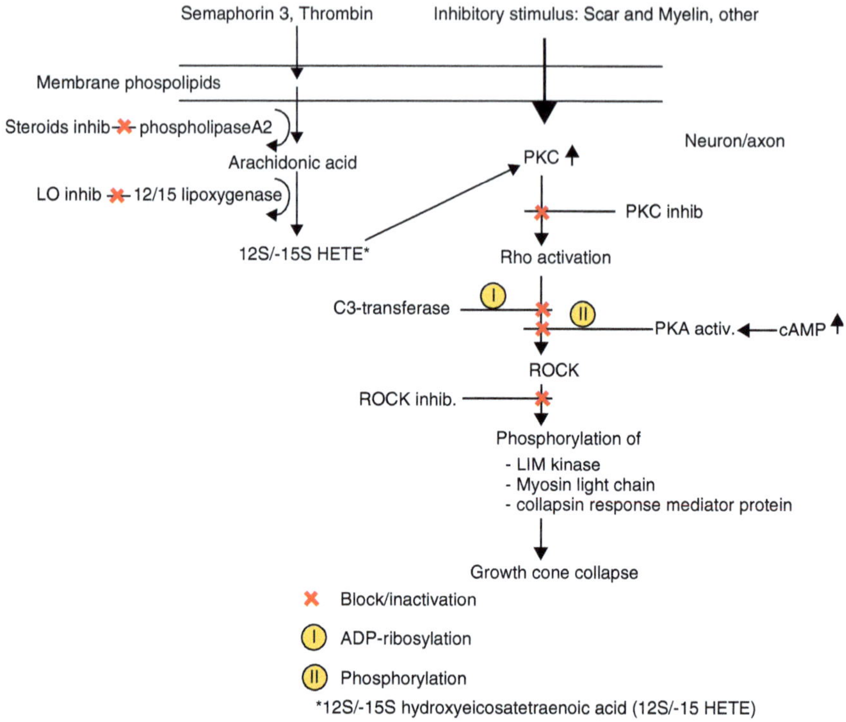

FIGURE 14.3. Lipid mediators in repulsive signaling: increase in PKC. Eicosanoids are lipid mediators, which are characterized by 20-carbon chain (C:20, eicosa = 20). Eicosanoids have been identified to be involved in semaphorin-3 and thrombin-induced growth cone collapse. Arachidonic acid (AA, C20:4 an omega-6 fatty acid member), which is well known as a substrate for the mostly pro-inflammatory prostaglandins is being generated after phospolipase-A2 (PLA2) catalyzation. Further AA catalyzation by the 12/15 lipoxygenase results in the product 12S/-15S HETE. Blocking of either PLA2 or 12/15 lipoxygenase protected growth cones from thrombin or semaphorin-3A–induced collapse. Incorporation and linkage with the signaling cascades known to date was provided recently. Both thrombin (the growth cone collapse inducer) and 12-HETE activate PKC (ε-Isoform) in the growth cone.

14.4.3.5. Lipids: Eicosanoids Acting Upstream of PKC

Eicosanoids are lipid mediators which are characterized by 20-carbon chain and constitute a widely unrecognized and fast reacting pool of fatty acids, which can be recruited and modified either from the phospholipid cell membrane or from intracellular stores. Eicosanoids have been identified to be involved in semaphorin-3 and thrombin induced growth cone collapse. Arachidonic acid (AA), which is well known as a substrate for the mostly pro-inflammatory prostaglandins is being generated after PLA2 catalyzation. Further AA catalyzation by the 12/15 lipoxygenase (LOX) results in the product 12S/-15S HETE (Figure 14.3). Blocking of either PLA2 or mouse 12/15 lipoxygenase protected growth cones from thrombin or semaphorin-3A induced collapse (De la Houssaye et al., 1999;

Mikule et al., 2002). Noteworthy, mouse 12/15 LOX corresponds to at least 3 LOX in humans: 12-LOX, 15-LOX type I and II (Funk et al., 2002). Incorporation and linkage with the inhibitory, PKC-dependent signaling cascade was provided recently (Mikule et al., 2003). Both thrombin (the growth cone collapse inducer) and 12-HETE activate PKC (ε-Isoform) in the growth cone. Furthermore, both 12-HETE and lipoxin-A4 (the lipoxygenase interaction product) activates the γ-isoform of PKC in the presence of Ca^{2+} *in vitro* (Hansson et al., 1986; Shearman et al., 1989). Nevertheless, a role of LOX and LOX product, such as LXA4, in growth cone collapse and underlying signaling in human neurons is only emerging.

14.4.3.6. Hierarchy of Signals to Reconnect with Deafferated Targets

The lesional reexpression of axon guidance molecules as "target specific information" (Schwab et al., 2005a,b; Wizenmann et al., 1993) in the form of guidance molecules (e.g., netrins, slits, robos, ephrins, semaphorins, RGM-A) and guidance-active cell adhesion molecules [e.g., L1 (Mohajeri et al., 1996), specific matrix proteins] emphasizes the multitude of influences that a regenerating axon is exposed to. Until now it is unclear which signals within the orchestra of inhibitors, guidance molecules, and neurotrophins dominate upon SCI, especially since the involved intracellular signal transduction pathways (Rho) overlap to a large extent and why. Thus, since neurotrophins, guidance molecules, and nerve growth inhibitors share the same RhoA second messenger system, the hierarchy for the navigating axon in the CNS lesions remains unclear. In addition, long-term interference with signaling might not be advantageous. Exemplified by the converging downstream mechanism of growth arrest, the activation of the Rho small GTPases, it appears that activated Rho is also essential for a number of physiologic processes such as structural refinement of neuronal circuits in the cortex and for learning and memory (Ramakers, 2002). During CNS development, the presence of guidance molecules helps the extending axon to navigate through the CNS to find its designated target. Their expression in the adult CNS after injury is surprising. Why is it necessary to express guidance molecules at the lesion when regeneration is only limited? Mainly, it was thought to prevent the possibility of an aberrant, erroneous connection with other circuits. Extensive tracing studies demonstrated enormous reorganization of descending motor tracts, such as the CST and rubrospinal tract and the reticulospinal tract (Hill et al., 2001; Raineteau et al., 2002), which was functionally confirmed by electrophysiological and behavioral testing (Bareyre et al., 2004). These observations indicate significant intraspinal reorganization following SCI and suggest a role for lesionally expressed cues in guiding and/or stabilizing newly formed sprouts.

14.5. Perspectives to Stimulate Axon Regeneration

With advances in our understanding of the mechanisms that mediate the inhibitory activity of myelin and other potential inhibitors, a number of strategies may be conceived to block the activity of myelin-associated inhibitors, thus stimulating axon regeneration *in vivo*. At the level of the axonal surface, blocking the activity of individual inhibitors using such neutralizing antibodies as the IN-1 antibody (Bregman et al., 1995; Schnell and Schwab, 1990; Schnell et al., 1994) and anti-myelin antibodies (Huang et al., 1999) has been examined for many years. The observation that these structurally distinct molecules converge on a single receptor suggests a common set of downstream signaling mechanisms mediating myelin inhibition, and offers the promising prospect that affecting a single target may be all that is required to block this inhibition. Such potentially useful reagents include recombinant soluble NgR proteins that have been demonstrated to block the inhibitory activity of CNS myelin at least in *in vitro* assays (Domeniconi et al., 2002; Fournier et al., 2002; Liu et al., 2002), recombinant soluble p75 protein that can block the interaction of NgR1 and p75 and reduce myelin inhibition (Wang et al., 2002a,b; Wong et al., 2002), and overexpression of truncated NgR1 that can bind all ligands but not coreceptor(s) (Domeniconi et al., 2002; Wang et al., 2002b).

In addition to these ligand/receptor interactions, intracellular signaling components are also potential targets for alleviating the inhibitory influences associated with CNS myelin. For example, C3 transferase (a RhoA inhibitor) and Y27632 (a ROCK inhibitor) have been shown to permit a certain degree of axon regeneration when applied to both optic nerve crush and SCI models (Dergham et al., 2002; Fournier et al., 2003; Lehmann et al., 1999; Schwab et al., 2002). Both RhoA and ROCK have also been implicated in the inhibitory signaling pathways of CSPGs (Monnier et al., 2003). Nevertheless, as these signaling molecules are also components of other cellular pathways, the application of these agents in different CNS lesion models may not only reflect their effects on myelin-mediated inhibition. For example, a study suggests an additional role of RhoA activation in operating cell death after CNS injury (Dubreuil et al., 2003). Besides, it needs to be envisaged that the second messenger system Rho regulates also phsysiological functions such as activity-dependent neuronal plasticity, essential for structural refinement of neuronal circuits in the cortex and for learning and memory (Ramakers, 2002). With the identification of other signaling components of myelin inhibitor signaling pathways, it is certain that additional relevant reagents will be considered for similar use in the future.

Studies suggest that pharmacological or genetic blockade of the inhibitory activities may not be enough to allow functionally meaningful axon regeneration. For example, locally introduced PKC inhibitors, which are able to block the inhibition of both the myelin inhibitors and CSPGs, promote the regeneration of ascending dorsal column axons but not descending CST fibers. These results suggest that mature CST fibers may not be able to regenerate even in the minimized inhibitory environment. This could be explained by the decreased intrinsic capacity of the mature cortical neurons. Thus, combinatorial strategies to combine blocking inhibition with promoting growth ability may represent new directions for maximizing the extent of axon regeneration in the adult CNS.

Acknowledgments. This work is supported by grants from the National Institute of Health (NIH), International Spinal Research Trust, and The John Merck Fund.

JMS is a research fellow supported by the German Research Council (DFG, 1164) and the NIH (# P50-DE016191).

References

Arquint, M., Roder, J., Chia, L.S., Down, J., Wilkinson, D., Bayley, H., et al., 1987, Molecular cloning and primary structure of myelin-associated glycoprotein, *Proc. Natl. Acad. Sci. USA* **84:** 600–604.

Bareyre, F.M., Kerschensteiner, M., Misgeld, T., and Sanes, J.R., 2005, Transgenic labeling of the corticospinal tract for monitoring axonal responses to spinal cord injury, *Nat. Med.* **11:** 1355–1360.

Bareyre, F.M., Kerschensteiner, M., Reinetaeu, O., Mettenleiter, T.C., Weinmann, O., and Schwab, M.E., 2004, The injured spinal cord spontaneously forms a new intraspinal circuit in adult rats. *Nat. Neusosci.* **7:** 269–277

Bartsch, U., Bandtlow, C.E., Schnell, L., Bartsch, S., Spillmann, A.A., Rubin, B.P., et al., 1995, Lack of evidence that myelin-associated glycoprotein is a major inhibitor of axonal regeneration in the CNS, *Neuron* **15:** 1375–1381.

Benson, M.D., Romero, M.I., Lush, M.E., Lu, Q.R., Henkemeyer, M., and Parada, L.F., 2005, Ephrin-B3 is a myelin-based inhibitor of neurite outgrowth. *Proc. Natl. Acad. Sci. USA* **102:** 10694–10699.

Bradbury, E.J., Moon, L.D., Popat, R.J., King, V.R., Bennett, G.S., Patel, P.N., et al., 2002, Chondroitinase ABC promotes functional recovery after spinal cord injury, *Nature* **416:** 636–640.

Bregman, B.S., Kunkel-Bagden, E., Schnell, L., Dai, H.N., Gao, D., and Schwab, M.E., 1995, Recovery from spinal cord injury mediated by antibodies to neurite growth inhibitors, *Nature* **378:** 498–501.

Bregman, B.S., McAtee, M., Dai, H.N., and Kuhn, P.L., 1997, Neurotrophic factors increase axonal growth after spinal cord injury and transplantation in the adult rat, *Exp. Neurol.* **148:** 475–494.

Brushart, T.M., 1998, Preferential reinnervation of motor nerves by regenerating motor axons, *J. Neurosci.* **8:** 1026–1031.

Cai, D., Shen, Y., De Bellard, M., Tang, S., and Filbin, M.T., 1999, Prior exposure to neurotrophins blocks inhibition of axonal regeneration by MAG and myelin via a cAMP-dependent mechanism, *Neuron* **22:** 89–101.

Cai, D., Deng, K., Mellado, W., Lee, J., Ratan, R.R., and Filbin, M.T., 2002, Arginase I and polyamines act downstream from cyclic AMP in overcoming inhibition of axonal growth MAG and myelin in vitro, *Neuron* **35:** 711–719.

Camand, E., Morel, M.P., Faissner, A., Sotelo, C., and Dusart, I., 2004, Long-term changes in the molecular composition of the glial scar and progressive increase of serotoninergic fibre sprouting after hemisection of the mouse spinal cord, *Eur. J. Neurosci.* **20:** 1161–1176.

Carbonetto, S., Evans, D., and Cochard, P., 1987, Nerve fiber growth in culture on tissue substrata from central and peripheral nervous systems, *J. Neurosci.* **7:** 610–620.

Caroni, P., and Schwab, M.E., 1988a, Antibody against myelin-associated inhibitor of neurite growth neutralizes nonpermissive substrate properties of CNS white matter, *Neuron* **1:** 85–96.

Caroni, P., and Schwab, M.E., 1988b, Two membrane protein fractions from rat central myelin with inhibitory properties for neurite growth and fibroblast spreading, *J. Cell Biol.* **106:** 1281–1288.

Chan, C.C., Khodarahmi, K., Liu, J., Sutherland, D., Oschipok, L.W., Steeves, J.D., et al., 2005, Dose-dependent beneficial and detrimental effects of ROCK inhibitor Y27632 on axonal sprouting and functional recovery after rat spinal cord injury, *Exp. Neurol.* **196:** 352–364.

Chen, M.S., Huber, A.B., van der Haar, M.E., Frank, M., Schnell, L., Spillmann, A.A., et al., 2000, Nogo-A is a myelin-associated neurite outgrowth inhibitor and an antigen for monoclonal antibody IN-1, *Nature* **403:** 434–439.

Crutcher, K.A., 1989, Tissue sections from the mature rat brain and spinal cord as substrates for neurite outgrowth in vitro: Extensive growth on gray matter but little growth on white matter, *Exp. Neurol.* **104:** 39–54.

David, S., Bouchard, C., and Tsatas, O., and Giftochristos, N., 1990, Macrophages can modify the nonpermissive nature of the adult mammalian central nervous system, *Neuron* **5:** 463–469.

Davies, S.J., Fitch, M.T., Memberg, S.P., Hall, A.K., Raisman, G., and Silver, J., 1997, Regeneration of adult axons in white matter tracts of the central nervous system, *Nature* **390:** 680–683.

Davies, S.J., Goucher, D.R., Doller, C., and Silver, J., 1999, Robust regeneration of adult sensory axons in degenerating white matter of the adult rat spinal cord, *J. Neurosci.* **19:** 5810–5822.

de La Houssaye, B.A., Mikule, K., Nikolic, D., and Pfenninger, K.H., 1999, Thrombin-induced growth cone collapse: Involvement of phospholipase A(2) and eicosanoid generation, *J. Neurosci.* **19:** 10843–10855.

De Winter, F., Oudega, M., Lankhorst, A.J., Hamers, F.P., Blits, B., Ruitenberg, M.J., et al., 2002, Injury-induced class 3 semaphorin expression in the rat spinal cord, *Exp. Neurol.* **175:** 61–75.

Dergham, P., Ellezam, B., Essagian, C., Avedissian, H., Lubell, W.D., and McKerracher, L., 2002, Rho signaling pathway targeted to promote spinal cord repair, *J. Neurosci.* **22:** 6570–6577.

Dickson, B.J., 2002, Molecular mechanisms of axon guidance, *Science* **298:** 1959–1964.

Domeniconi, M., Cao, Z., Spencer, T., Sivasankaran, R., Wang, K., Nikulina, E., et al., 2002, Myelin-associated glycoprotein interacts with the Nogo66 receptor to inhibit neurite outgrowth, *Neuron* **35:** 283–290.

Domeniconi, M., Zampieri, N., Spencer, T., Hilaire, M., Mellado, W., Chao, M.V., et al., 2005, MAG induces regulated intramembrane proteolysis of the p75 neurotrophin receptor to inhibit neurite outgrowth, *Neuron* **46:** 849–855.

Dubreuil, C.I., Winton, M.J., and McKerracher, L., 2003, Rho activation patterns after spinal cord injury and the role of activated Rho in apoptosis in the central nervous system, *J. Cell Biol.* **162:** 233–243.

Fawcett, J.W., and Asher, R.A., 1999, The glial scar and central nervous system repair, *Brain Res. Bull.* **49:** 377–391.

Fournier, A.E., GrandPre, T., and Strittmatter, S.M., 2001, Identification of a receptor mediating Nogo-66 inhibition of axonal regeneration, *Nature* **409:** 341–346.

Fournier, A.E., Gould, G.C., Liu, B.P., and Strittmatter, S.M., 2002, Truncated soluble Nogo receptor binds Nogo-66 and blocks inhibition of axon growth by myelin, *J. Neurosci.* **22:** 8876–8883.

Fu, S.Y., and Gordon, T., 1997. The cellular and molecular basis of peripheral nerve regeneration, *Mol. Neurobiol.* **14:** 67–116.

Funk, C.D., Chen, X.S., Johnson, E.N., and Zhao, L., 2002, Lipoxygenase genes and their targeted disruption, *Prostaglandins Other Lipid Mediat.* **6869:** 303–312.

Ghoumari, A.M., Wehrle, R., De Zeeuw, C.I., Sotelo, C., and Dusart, I., 2002, Inhibition of protein kinase C prevents Purkinje cell death but does not affect axonal regeneration. *J. Neurosci.* **22:** 3531–3542.

Goldberg, J.L., Klassen, M.P., Hua, Y., and Barres, B.A., 2002, Amacrine-signaled loss of intrinsic axon growth ability by retinal ganglion cells, *Science* **296:** 1860–1864.

Goldshmit, Y., Galea, M.P., Wise, G., Bartlett, P.F., and Turnley, A.M., 2004, Axonal regeneration and lack of astrocytic gliosis in EphA4-deficient mice, *J. Neurosci.* **24:** 10064–10073.

Graef, I.A., Wang, F., Charron, F., Chen, L., Neilson, J., Tessier-Lavigne, et al., 2003, Neurotrophins and netrins require calcineurin/NFAT signaling to stimulate outgrowth of embryonic axons, *Cell* **113:** 657–670.

GrandPre, T., Nakamura, F., Vartanian, T., and Strittmatter, S.M., 2000, Identification of the Nogo inhibitor of axon regeneration as a Reticulon protein, *Nature* **403:** 439–444.

Grimpe, B., and Silver, J., 2002, The extracellular matrix in axon regeneration, *Prog. Brain Res.* **137:** 333–349.

Guan, K.L., and Rao, Y., 2003, Signalling mechanisms mediating neuronal responses to guidance cues, *Nat. Rev. Neurosci.* **4:** 941–956.

Hannila, S.S., and Kawaja, M.D., 1999, Nerve growth factor-induced growth of sympathetic axons into the optic tract of mature mice is enhanced by an absence of p75NTR expression, *J. Neurobiol.* **39:** 51–66.

Hansson, A., Serhan, C.N., Haeggstrom, J., Ingelman-Sundberg, M., and Samuelsson, B., 1986, Activation of protein kinase C by lipoxin A and other eicosanoids. Intracellular action of oxygenation products of arachidonic acid, *Biochem. Biophys. Res. Commun.* **134:** 1215–1222.

Hempstead, B.L., 2002, The many faces of p75NTR, *Curr. Opin. Neurobiol.* **12:** 260–267.

Hill, C.E., Beattie, M.S., and Bresnahan, J.C., 2001, Degeneration and sprouting of identified descending supraspinal axons after contusive spinal cord injury in the rat, *Exp. Neurol.* **171:** 153–169.

Huang, D.W., McKerracher, L., Braun, P.E., and David, S., 1999, A therapeutic vaccine approach to stimulate axon regeneration in the adult mammalian spinal cord, *Neuron* **24:** 639–647.

Hunt, D., Mason, M.R., Campbell, G., Coffin, R., and Anderson, P.N., 2002a, Nogo receptor mRNA expression in intact and regenerating CNS neurons, *Mol. Cell Neurosci.* **20:** 537–552.

Hunt, D., Coffin, R.S., and Anderson, P.N., 2002b, The Nogo receptor, its ligands and axonal regeneration in the spinal cord; a review, *J. Neurocytol.* **31:** 93–120.

Irizarry-Ramirez, M., Willson, C.A., Cruz-Orengo, L., Figueroa, J., Velazquez, I., Jones, H., et al., 2005, Upregulation of EphA3 receptor after spinal cord injury, *J. Neurotrauma* **22:** 929–935.

Jalink, K., and Moolenaar, W.H., 1992, Thrombin receptor activation causes rapid neural cell rounding and neurite retraction independent of classic second messengers, *J. Cell Biol.* **118:** 411–419.

Jalink, K., van Corven, E.J., Hengeveld, T., Morii, N., Narumiya, S., and Moolenaar, W.H., 1994, Inhibition of lysophosphatidate- and thrombin-induced neurite retraction and neuronal cell rounding by ADP ribosylation of the small GTP-binding protein Rho, *J. Cell Biol.* **126:** 801–810.

Kater, S.B., and Mills, L.R., 1991, Regulation of growth cone behavior by calcium, *J. Neurosci.* **11:** 891–899.

Keirstead, H.S., Hasan, S.J., Muir, G.D., and Steeves, J.D., 1992, Suppression of the onset of myelination extends the permissive period for the functional repair of embryonic spinal cord, *Proc. Natl. Acad. Sci. USA* **89:** 11664–11668.

Kerschensteiner, M., Schwab, M.E., Lichtman, J.W., and Misgeld, T., 2005, In vivo imaging of axonal degeneration and regeneration in the injured spinal cord, *Nat. Med.* **11:** 572–577.

Koprivica, V., Cho, K.S., Park, J.B., Yiu, G., Atwal, J., Gore, B., et al., 2005, EGFR activation mediates inhibition of axon regeneration by myelin and chondroitin sulfate proteoglycans, *Science* **310:** 106–110.

Kury, P., Abankwa, D., Kruse, F., Greiner-Petter, R., and Muller, H.W., 2004, Gene expression profiling reveals multiple novel intrinsic and extrinsic factors associated with axonal regeneration failure, *Eur. J. Neurosci.* **19:** 32–42.

Laskowski, M.B., and Sanes, J.R., 1988, Topographically selective reinnervation of adult mammalian skeletal muscles, *J. Neurosci.* **8:** 3094–3099.

Lehmann, M., Fournier, A., Selles-Navarro, I., Dergham, P., Sebok, A., Leclerc, N., et al., 1999, Inactivation of Rho signaling pathway promotes CNS axon regeneration, *J. Neurosci.* **19:** 7537–7547.

Li, M., Shibata, A., Li, C., Braun, P.E., McKerracher, L., Roder, J., et al., 1996, Myelin-associated glycoprotein inhibits neurite/axon growth and causes growth cone collapse, *J. Neurosci. Res.* **46:** 404–414.

Liu, B.P., Fournier, A., GrandPre, T., and Strittmatter, S.M., 2002, Myelin-associated glycoprotein as a functional ligand for the Nogo-66 receptor, *Science* **297:** 1190–1193.

Mandolesi, G., Madeddu, F., Bozzi, Y., Maffei, L., Ratto, G.M., 2004, Acute physiological response of mammalian central neurons to axotomy: Ionic regulation and electrical activity, *FASEB J.* **18:** 1934–1936.

Manitt, C., and Kennedy, T.E., 2002, Where the rubber meets the road: Netrin expression and function in developing and adult nervous systems, *Prog. Brain Res.* **137:** 425–442.

Matsunaga, E., Tauszig-Delamasure, S., Monnier, P.P., Mueller, B.K., Strittmatter, S.M., Mehlen, P., et al., 2004, RGM and its receptor neogenin regulate neuronal survival, *Nat. Cell Biol.* **6:** 749–755.

McKeon, R.J., Schreiber, R.C., Rudge, J.S., and Silver, J., 1991, Reduction of neurite outgrowth in a model of glial scarring following CNS injury is correlated with the expression of inhibitory molecules on reactive astrocytes, *J. Neurosci.* **11:** 3398–3411.

McKerracher, L., David, S., Jackson, D.L., Kottis, V., Dunn, R.J., and Braun, P.E., 1994, Identification of myelin-associated glycoprotein as a major myelin-derived inhibitor of neurite growth, *Neuron* **13:** 805–811.

Mi, S., Lee, X., Shao, Z., Thill, G., Ji, B., Relton, J., et al., 2004, LINGO-1 is a component of the Nogo-66 receptor/p75 signaling complex, *Nat. Neurosci.* **7:** 221–228.

Mikule, K., Gatlin, J.C., de la Houssaye, B.A., and Pfenninger, K.H., 2002, Growth cone collapse induced by semaphorin 3A requires 12/15-lipoxygenase, *J. Neurosci.* **22:** 4932–4941.

Mikule, K., Sunpaweravong, S., Gatlin, J.C., and Pfenninger, K.H., 2003, Eicosanoid activation of protein kinase C epsilon: Involvement in growth cone repellent signaling, *J. Biol. Chem.* **278:** 21168–21177.

Miranda, J.D., White, L.A., Marcillo, A.E., Willson, C.A., Jagid, J., and Whittemore, S.R., 1999, Induction of Eph B3 after spinal cord injury, *Exp. Neurol.* **156:** 218–222.

Mohajeri, M.H., Bartsch, U., van der Putten, H., Sansig, G., Mucke, L., and Schachner, M., 1996, Neurite outgrowth on non-permissive substrates in vitro is enhanced by ectopic expression of the neural adhesion molecule L1 by mouse astrocytes, *Eur. J. Neurosci.* **8:** 1085–1097.

Moolenaar, W.H., 1995, Lysophosphatidic acid signaling, *Curr. Opin. Cell. Biol.* **7:** 203–210.
Monnier, P.P., Sierra, A., Macchi, P., Deitinghoff, L., Andersen, J.S., Mann, M., et al., 2002, RGM is a repulsive guidance molecule for retinal axons, *Nature* **419:** 392–395.
Monnier, P.P., Sierra, A., Schwab, J.M., Henke-Fahle, S., and Mueller, B.K., 2003, The Rho/ROCK pathway mediates neurite growth-inhibitory activity associated with the chondroitin sulfate proteoglycans of the CNS glial scar, *Mol. Cell Neurosci.* **22:** 319–330.
Moon, L.D., Asher, R.A., Rhodes, K.E., and Fawcett, J.W., 2001, Regeneration of CNS axons back to their target following treatment of adult rat brain with chondroitinase ABC, *Nat. Neurosci.* **4:** 465–466.
Moreau-Fauvarque, C., Kumanogoh, A., Camand, E., Jaillard, C., Barbin, G., Boquet, I., et al., 2003. The transmembrane semaphorin Sema4D/CD100, an inhibitor of axonal growth, is expressed on oligodendrocytes and upregulated after CNS lesion, *J. Neurosci.* **23:** 9229–9239.
Morgenstern, D.A., Asher, R.A., and Fawcett, J.W, 2002, Chondroitin sulphate proteoglycans in the CNS injury response, *Prog. Brain Res.* **137:** 313–332.
Mukhopadhyay, G., Doherty, P., Walsh, F.S., Crocker, P.R., and Filbin, M.T., 1994, A novel role for myelin-associated glycoprotein as an inhibitor of axonal regeneration, *Neuron* **13:** 757–767.
Neumann, H., Schweigreiter, R., Yamashita, T., Rosenkranz, K., Wekerle, H., and Barde, Y.A., 2002, Tumor necrosis factor inhibits neurite outgrowth and branching of hippocampal neurons by a rho-dependent mechanism, *J. Neurosci.* **22:** 854–862.
Niederost, B.P., Zimmermann, D.R., Schwab, M.E., and Bandtlow, C.E., 1999, Bovine CNS myelin contains neurite growth-inhibitory activity associated with chondroitin sulfate proteoglycans, *J. Neurosci.* **19:** 8979–8989.
Niederost, B.P., Oertle, T., Fritsche, J., McKinney, R.A., and Bandtlow, C.E., 2002, Nogo-A and myelin-associated glycoprotein mediate neurite growth inhibition by antagonistic regulation of RhoA and Rac1, *J. Neurosci.* **22:** 10368–10376.
Prinjha, R., Moore, S.E., Vinson, M., Blake, S., Morrow, R., Christie, G., et al., 2000, Inhibitor of neurite outgrowth in humans, *Nature* **403:** 383–384.
Qiu, J., Cai, D., Dai, H., McAtee, M., Hoffman, P.N., Bregman, B.S., et al., 2002, Spinal axon regeneration induced by elevation of cyclic AMP, *Neuron* **34:** 895–903.
Raineteau, O., Fouad, K., Bareyre, F.M., and Schwab, M.E., 2002, Reorganization of descending motor tracts in the rat spinal cord, *Eur. J. Neurosci.* **16:** 1761–1771.
Rajagopalan, S., Deitinghoff, L., Davis, D., Conrad, S., Skutella, T., Chedotal, A., Mueller, B.K., and Strittmatter, S.M., 2004, Neogenin mediates the action of repulsive guidance molecule, *Nat. Cell Biol.* **6:** 756–762.
Ramakers, G.J., 2002, Rho proteins, mental retardation and the cellular basis of cognition, *Trends Neurosci.* **25:** 191–199.
Ramon y Cajal, S., 1928, *Degeneration and Regeneration of the Nervous System*, R.M. May, ed. and tr., Oxford University Press, London.
Roux, P.P., and Barker, P.A., 2002, Neurotrophin signaling through the p75 neurotrophin receptor, *Prog. Neurobiol.* **67:** 203–233.
Sagot, Y., Swerts, J.P., Cochard, P., 1991, Changes in permissivity for neuronal attachment and neurite outgrowth of spinal cord grey and white matters during development: A study with the 'cryoculture' bioassay, *Brain Res.* **543:** 25–35.
Salzer, J.L., Holmes, W.P., and Colman, D.R., 1987, The amino acid sequences of the myelin-associated glycoproteins: Homology to the immunoglobulin gene superfamily, *J. Cell Biol.* **104:** 957–965.

Savio, T., and Schwab, M.E., 1989, Rat CNS white matter, but not gray matter, is nonpermissive for neuronal cell adhesion and fiber outgrowth, *J. Neurosci.* **9:** 1126–1133.
Schafer, M., Fruttiger, M., Montag, D., Schachner, M., and Martini, R, 1996, Disruption of the gene for the myelin-associated glycoprotein improves axonal regrowth along myelin in C57BL/Wlds mice, *Neuron* **16:** 1107–1113.
Schnell, L., and Schwab, M.E., 1990, Axonal regeneration in the rat spinal cord produced by an antibody against myelin-associated neurite growth inhibitors, *Nature* **343:** 269–272.
Schnell, L., Schneider, R., Kolbeck, R., Barde, Y.A., and Schwab, M.E, 1994, Neurotrophin-3 enhances sprouting of corticospinal tract during development and after adult spinal cord lesion, *Nature* **367:** 170–173.
Schwab, J.M., Hirsch, S., Brechtel, K., Stiefel, A., Leppert, C.A., Schluesener, H.J., et al., 2002, The Rho-GTPase inhibitor C2IN-C3 induces functional neuronal recovery in a rat model of severe spinal cord injury, *Soc. Neurosci.* [Abstract 204.7].
Schwab, J.M., Monnier, P.P., Schluesener, H.J., Conrad, S., Beschorner, R., Chen, L., et al., 2005a, Central nervous system injury-induced repulsive guidance molecule expression in the adult human brain, *Arch. Neurol.* **62:** 1561–1568.
Schwab, J.M., Conrad, S., Monnier, P.P., Julien, S., Mueller, B.K., and Schluesener, H.J., 2005b. Spinal cord injury-induced lesional expression of the repulsive guidance molecule, RGM, *Eur. J. Neurosci.* **21:** 1569–1576.
Schwab, J.M., Failli, V., and Chedotal, A., 2005c, Injury-related dynamic myelin/oligodendrocyte axon-outgrowth inhibition in the central nervous system, *Lancet* **365:** 2055–2057.
Schwab, M.E., and Caroni, P., 1988, Oligodendrocytes and CNS myelin are nonpermissive substrates for neurite growth and fibroblast spreading in vitro, *J. Neurosci.* **8:** 2381–2393.
Schwab, M.E., and Thoenen, H., 1985, Dissociated neurons regenerate into sciatic but not optic nerve explants in culture irrespective of neurotrophic factors, *J. Neurosci.* **5:** 2415–2423.
Shearman, M.S., Naor, Z., Sekiguchi, K., Kishimoto, A., and Nishizuka, Y., 1989, Selective activation of the gamma-subspecies of protein kinase C from bovine cerebellum by arachidonic acid and its lipoxygenase metabolites, *FEBS Lett.* **243:** 177–182.
Silver, J., and Miller, J.H., 2004, Regeneration beyond the glial scar, *Nat. Rev. Neurosci.* **5:** 146–156.
Sivasankaran, R., Pei, J., Wang, K.C., Zhang, Y.P., Schields, C.B., Xu, X.M., and He, Z., 2004, PKC mediates inhibitory effects of myelin and chondroitin sulfate proteoglycans on axonal regeneration. *Nat. Neurosci.* **7:** 261–268.
Soileau, L.C., Silberstein, L., Blau, H.M., and Thompson, W.J., 1987, Reinnervation of muscle fiber types in the newborn rat soleus, *J. Neurosci.* **7:** 4176–4194.
Song, H.J., Ming, G.L., and Poo, M.M., 1997, cAMP-induced switching in turning direction of nerve growth cones, *Nature* **388:** 275–279.
Song, X.Y., Zhong, J.H., Wang, X, and Zhou, X.F., 2004, Suppression of p75NTR does not promote regeneration of injured spinal cord in mice, *J. Neurosci.* **24:** 542–546.
Sotelo, C., and Palay, S.L., 1971, Altered axons and axon terminals in the lateral vestibular nucleus of the rat. Possible example of axonal remodeling, *Lab. Invest.* **25:** 653–671.
Spillmann, A.A., Bandtlow, C.E., Lottspeich, F., Keller, F., and Schwab, M.E., 1998, Identification and characterization of a bovine neurite growth inhibitor, bNI-220, *J. Biol. Chem.* **273:** 19283–19293.
Stahl, B., Muller, B., von Boxberg, Y., Cox, E.C., and Bonhoeffer, F., 1990, Biochemical characterization of a putative axonal guidance molecule of the chick visual system, *Neuron* **5:** 735–743.

Stichel, C.C., and Muller, H.W., 1998, The CNS lesion scar: New vistas on an old regeneration barrier, *Cell Tissue Res.* **294:** 1–9.

Tang, S., Woodhall, R.W., Shen, Y.J., deBellard, M.E., Saffell, J.L., Doherty, P., et al., 1997, Soluble myelin-associated glycoprotein (MAG) found in vivo inhibits axonal regeneration, *Mol. Cell Neurosci.* **9:** 333–346.

Thallmair, M., Metz, G.A., Z'Graggen, W.J., Raineteau, O., Kartje, G.L., Schwab, M.E., 1998, Neurite growth inhibitors restrict plasticity and functional recovery following corticospinal tract lesions, *Nat. Neurosci.* **1:** 124–131.

Tom, V.J., Steinmetz, M.P., Miller, J.H., Doller, C.M., and Silver, J., 2004, Studies on the development and behavior of the dystrophic growth cone, the hallmark of regeneration failure, in an in vitro model of the glial scar and after spinal cord injury, *J. Neurosci.* **24:** 6531–6539.

Vanek, P., Thallmair, M., Schwab, M.E., and Kapfhammer, J.P., 1998, Increased lesion-induced sprouting of corticospinal fibres in the myelin-free rat spinal cord, *Eur. J. Neurosci.* **10:** 45–56.

Walsh, G.S., Krol, K.M., Crutcher, K.A., and Kawaja, M.D., 1999, Enhanced neurotrophin-induced axon growth in myelinated portions of the CNS in mice lacking the p75 neurotrophin receptor, *J. Neurosci.* **19:** 4155–4168.

Wang, K.C., Koprivica, V., Kim, J.A., Sivasankaran, R., Guo, Y., Neve, R.L., et al., 2002a, Oligodendrocyte-myelin glycoprotein is a Nogo receptor ligand that inhibits neurite outgrowth, *Nature* **417:** 941–944.

Wang, K.C., Kim, J.A., Sivasankaran, R., Segal, R., and He, Z., 2002b, P75 interacts with the Nogo receptor as a co-receptor for Nogo, MAG and OMgp, *Nature* **420:** 74–78.

Wang, X., Chun, S.J., Treloar, H., Vartanian, T., Greer, C.A., and Strittmatter, S.M., 2002c, Localization of Nogo-A and Nogo-66 receptor proteins at sites of axon-myelin and synaptic contact, *J. Neurosci.* **22:** 5505–5515.

Wehrle, R., Camand, E., Chedotal, A., Sotelo, C., and Dusart, I., 2005, Expression of netrin-1, slit-1 and slit-3 but not of slit-2 after cerebellar and spinal cord lesions, *Eur. J. Neurosci.* **22:** 2134–2144.

Willson, C.A., Irizarry-Ramirez, M., Gaskins, H.E., Cruz-Orengo, L., Figueroa, J.D., Whittemore, S.R., et al., 2002, Upregulation of EphA receptor expression in the injured adult rat spinal cord, *Cell Transplant* **11:** 229–239.

Willson, C.A., Miranda, J.D., Foster, R.D., Onifer, S.M., and Whittemore, S.R., 2003, Transection of the adult rat spinal cord upregulates EphB3 receptor and ligand expression, *Cell Transplant* **12:** 279–290.

Wizenmann, A., Thies, E., Klostermann, S., Bonhoeffer, F., and Bahr, M., 1993, Appearance of target-specific guidance information for regenerating axons after CNS lesions, *Neuron* **11:** 975–983.

Wong, S.T., Henley, J.R., Kanning, K.C., Huang, K.H., Bothwell, M., and Poo, M.M., 2002, A p75(NTR) and Nogo receptor complex mediates repulsive signaling by myelin-associated glycoprotein, *Nat. Neurosci.* **5:** 1302–1308.

Yamashita, T., Higuchi, H., and Tohyama, M., 2002, The p75 receptor transduces the signal from myelin-associated glycoprotein to Rho, *J. Cell Biol.* **157:** 565–570.

Index

C

D

E

MIX
Papier aus verantwortungsvollen Quellen
Paper from responsible sources
FSC® C105338
FSC
www.fsc.org

If you have any concerns about our products,
you can contact us on
ProductSafety@springernature.com

In case Publisher is established outside the EU,
the EU authorized representative is:
Springer Nature Customer Service Center GmbH
Europaplatz 3, 69115 Heidelberg, Germany

Printed by Libri Plureos GmbH
in Hamburg, Germany